SUPPLEMENTAL IRRIGATION
IN THE NEAR EAST AND NORTH AFRICA

Supplemental Irrigation in the Near East and North Africa

Proceedings of a Workshop on Regional Consultation on Supplemental Irrigation.
ICARDA and FAO, Rabat, Morocco, 7-9 December, 1987

Edited by

E.R. PERRIER and A.B. SALKINI

The International Center for Agricultural Research in the Dry Areas (ICARDA), Aleppo, Syria

Consultant editor

C.F. WARD

The Institute of Irrigation Studies, University of Southampton, England

Springer Science+Business Media, B.V.

Library of Congress Cataloging-in-Publication Data

Workshop on Regional Consultation on Supplementary Irrigation (1987 :
Rabat, Morocco)
  Supplemental irrigation in the Near East and North Africa :
proceedings of a Workshop on Regional Consultation on Supplementary
Irrigation, ICARDA and FAO, Rabat, Morocco, 7-9 December, 1987 /
[edited] by Eugene R. Perrier and Abdul Bari Salkini.
       p.   cm.
  Includes index.
  ISBN 978-0-7923-1006-8   ISBN 978-94-011-3766-9 (eBook)
  DOI 10.1007/978-94-011-3766-9
  1. Supplemental irrigation--Congresses. 2. Supplemental
irrigation--Middle East--Congresses. 3. Supplemental irrigation-
-Africa, North--Congresses.   I. Perrier, Eugene R. II. Salkini,
Abdul Bari, 1942-   . III. International Center for Agricultural
Research in the Dry Areas. IV. Food and Agriculture Organization of
the United Nations. V. Title.
S619.S96W67 1987
333.91'3'0956--dc20   90-48008

ISBN 978-0-7923-1006-8

Printed on acid-free paper

All Rights Reserved
©1991 Springer Science+Business Media Dordrecht
Originally published by Kluwer Academic Publishers in 1991

No part of the material protected by this copyright notice may be reproduced or utilized in any form
or by any means, electronic or mechanical, including photocopying, recording or by any information
storage and retrieval system, without written permission from the copyright owners.

Table of Contents

Prologue	xvii
Foreword	xix
List of National Scientists Contributing Papers	xxi

Chapter 1
INTRODUCTION	1
GOALS AND OBJECTIVES	3
IRRIGATION TECHNOLOGY	3
Definitions of Irrigation Technologies	3
Selection of an Irrigation Method	5
Techniques For Supplemental Irrigation	5
Drainage of Supplemental Irrigation Lands	10
Concepts of Irrigation Technology	10
BOOK ORGANIZATION	12

Chapter 2
SYSTEMS APPROACH TO SUPPLEMENTAL IRRIGATION	15
Potential for Agricultural Modernization	16
Logic of Inquiry	18
A CONCEPTUAL FRAMEWORK FOR IRRIGATION TECHNOLOGY	24
Activity Paradigm	24
Computer-Aided Planning and Design	28
ADDENDUM 2A: RESEARCH PLOT DESIGNS	29

Chapter 3
WATER BALANCE CALCULATIONS	39
Precipitation	40
Effective Rainfall	41
Evaporation	42
Evapotranspiration	44
Supplemental Irrigation Methods	45
Water Balance	45

PROBABILITY ANALYSIS	58
Analytical Method	59
Graphical Method	65
ADDENDUM 3A: PENMAN'S EQUATION	67
ADDENDUM 3B: BASIC WATER BALANCE CALCULATION	70
ADDENDUM 3C: BASIC PROBABILITY CALCULATIONS	74

Chapter 4

REGIONAL APPLICATION OF WATER BALANCE METHODS	79
General Application of Water Balance	83
Water Requirements	86
Phenology of Cereals	88
Method for Determining Growth Stages of Cereals	88

Chapter 5

SOIL WATER RELATIONSHIPS	97
Soil Texture	97
Soil Structure	100
Bulk Density	100
Soil Porosity	104

Chapter 6

MOVEMENT OF WATER IN SOILS	107
Saturated Flow	109
Unsaturated Flow	113
Capillarity	113
Soil Moisture Characteristic Curve	114
Units of Potential	115

Chapter 7

DARCY EQUATION	117
Field Capacity	118
Wilting Point	120
Available Water	120

Chapter 8

SOIL WATER MEASUREMENT	123
Gravimetric Method	123
Tensiometer Method	124
Neutron Probe Method	126
Gypsum Blocks	129
Other Techniques	130

Chapter 9

INFILTRATION	133

Infiltration on Crusting Soils	134
Techniques to Improve Infiltration	135
Factors Affecting Infiltration	138
Approaches to Infiltration	138
Green and Ampt Equation	142
Units of Infiltration	144

Chapter 10
FIELD MEASUREMENT OF INFILTRATION — 145

Single Ring Infiltrometers	145
Double Ring Infiltrometers	146
Ponding Basins	151
Furrow Method of Infiltration Measurement	151
Blocked Furrow Infiltrometer	152
Infiltration Calculations	153

Chapter 11
GROUNDWATER SUPPLY — 155
GROUNDWATER OCCURRENCE — 155

Locating Groundwater	155
Groundwater Terminology	156
Aquifer Terminology	158
Age and Origins of Groundwater	162

WELLS — 164

Cone of Depression	166
Methods of Well Construction	167
Life Expectancy of Wells and Pumps	171

GROUNDWATER POLLUTION — 171
GROUNDWATER ABSTRACTION AND DELIVERY — 173

Pumping Using Suction	173
Pressure and Head	174
Flow and Lift	175

Chapter 12
WATER QUALITY, IRRIGATION MEASUREMENT AND EFFICIENCY — 177

Quality of Water for Supplemental Irrigation	177
Determination of Water Quantity for Irrigation	178
Irrigation Efficiency	181
Micro-Irrigation	187

Chapter 13
LAND LEVELING AND SIMPLISTIC SURVEYING — 191

Equipment	191
Survey of the Land Surface	193
Measurement of Horizontal Distances	198

Chapter 14
ECONOMICS OF SUPPLEMENTAL IRRIGATION 203
ECONOMIC APPRAISAL FOR SUPPLEMENTAL IRRIGATION PROJECTS 204
  Purpose 204
  Identification of Relevant Benefits and Costs 205
  Financial Analysis 206
  Economic Analysis 210
  Sensitivity and Risk Analysis 216
FARM LEVEL ECONOMIC PROCEDURES 217
  Farm Level 217
  Small Farm Budgeting 218
  Cash Flow Budget 221
  Partial Budgets 222
  Discounting and Depreciation 224
  Wells and Pumping 225

Chapter 15
EVALUATION OF SUPPLEMENTAL IRRIGATION 229
  Cyclical Process for Data Acquisition 230
  Application of the Working Hypothesis 230
  Resolution Method for Evaluation 232
  Strategies of Involvement 232
SURVEY METHODS, GENERAL INFORMATION AND FORMS 233
  Survey Questionnaire for Baseline Data 234
  Technical Data Collection 250
EVALUATION OF SPECIFIC METHODS OF IRRIGATION 261
  Furrow Irrigation 262
  Graded Border Strip Irrigation 264
  Hand Move or Side-Roller Sprinkler Irrigation 266
  Drip/Trickle Irrigation 273

Chapter 16
INTRODUCTION TO TECHNOLOGY TRANSFER 279
  What is the Future of Supplemental Irrigation? 279
DECISION FIELD, AN "IDEAL MODEL" FOR TECHNOLOGY TRANSFER 282
  Points of Intervention for Technology Transfer 282
  Linkage to Strategies of Involvement 287

Chapter 17
VERIFICATION OF SUPPLEMENTAL IRRIGATION OF SPRING WHEAT 293
  Characteristics of Plant Growth for Spring Wheat 294
DESIGN FOR SUPPLEMENTAL IRRIGATION RESEARCH, TEL HADYA, SYRIA 294
  Weather Data for Tel Hadya, 1985-89 295
  Results of Supplemental Irrigation Studies 298
ECONOMICS OF SUPPLEMENTAL IRRIGATION OF SPRING WHEAT 309

DISCUSSION AND CONCLUSIONS	311
REFERENCES	312

Chapter 18 by Larbi Baghdali

IRRIGATION OF CEREALS IN ALGERIA	315
CHARACTERISTICS OF FIELD CONDITIONS	315
Agro-Climatic Zones	315
Water Resources	317
MAIN CHARACTERISTICS OF ALGERIAN AGRICULTURE	318
Land Potential	318
Development of Agricultural Production	319
THE ROLE OF IRRIGATION IN AGRICULTURE	320
Structure of Irrigation Planning	320
Irrigation Methods	321
Supplemental Irrigation	321
DEVELOPMENT PROSPECTS FOR SUPPLEMENTAL IRRIGATION	323
Areas for Intensive Cultivation	324
Areas for Extensive Cultivation	325
CONCLUSIONS	325

Chapter 19 by V.C. Krentos

SUPPLEMENTAL IRRIGATION SYSTEMS IN CYPRUS	327
THE NATIONAL PERSPECTIVE	327
Geographical, Morphological, and Geological features	327
Climate	328
Land Resources and Land Use	332
Water Resources and Use	335
Structure and Development of Agriculture	336
Institutional Framework	340
THE DISTRICT PERSPECTIVE: CHARACTERIZATION OF 3 DISTRICTS	343
Climatic Aspects	344
Farming Practices	345
Cropping Patterns	346
Mechanization	347
Fertilizer Use	347
Plant Protection	348
Yield Relationships	348
Livestock and Forage Production	350
Cost of Production	352
COMPARISON OF SUPPLEMENTAL IRRIGATION RESOURCES AND FARMING PRACTICES IN 3 DISTRICTS	352
Demographic Information	353
Land Area and Land Use	353
Soils	353
Infrastructure and Rural Support Services	354

Water Resources and Management	354
Irrigation Practices	356
Water Harvesting	356
Cropping Practices and Crop Production	357
Livestock Production	359
Socio-Economic Aspects	360
FUTURE POTENTIAL FOR SUPPLEMENTAL IRRIGATION	361
SUMMARY AND CONCLUSIONS	362
ACKNOWLEDGEMENTS	363
REFERENCES	364

Chapter 20 by Hamid Siadat

POTENTIAL OF SUPPLEMENTAL IRRIGATION IN IRAN	367
GEOGRAPHICAL CHARACTERISTICS	367
Physiography	367
Climate	367
Land and Water Resources	369
STATE-OF-THE-ART OF WATER MANAGEMENT IN IRAN	369
Supplemental Irrigation	370
Water Harvesting	370
National Goals and Objectives	371
AGRICULTURAL ZONES	372
THE DISTRICT PERSPECTIVE	372
Miandorood District	373
Agh-Ghola District (south)	375
Bala-Darband (Sarab-Nilofar) District	377
Mahidasht District	379
POTENTIAL FOR SUPPLEMENTAL IRRIGATION IN IRAN	381
Potential Areas for Implementation	381
Potential Crops	383
Potential Yields	383
PROBLEMS AND CONSTRAINTS	385
Economic Factors	385
Research and Training	386
Fragmentation of Land Ownership	386
Mechanization Factors	386
Agro-Ecological Constraints	387
SUMMARY	387
ACKNOWLEDGEMENTS	387
REFERENCES	388

Chapter 21 by Shifa'a A. Mahmood

SUPPLEMENTAL IRRIGATION SYSTEMS OF IRAQ	389
IRRIGATION SYSTEMS AND OBJECTIVES	389
COLLECTION AND ASSESSMENT OF DATA	390

PRESENT LAND AND WATER USE	392
Rainfed Farming	392
Land Ownership and Farming Practices	393
Irrigation	393
Crops	393
IRRIGATION METHODS	394
CONCLUSIONS	395
REFERENCES	396

Chapter 22 by A.A. Jaradat

THE FARMING SYSTEMS IN JORDAN: RAINFED, WATER HARVESTING, AND SUPPLEMENTAL IRRIGATION	399
GEOGRAPHICAL CHARACTERISTICS	399
Climate	399
Land Resources	401
THE AGRICULTURE CONTEXT	401
Agriculture Zoning	401
The Agrarian Sector	402
Production Environment of the Dryland Regions	402
Supplemental Irrigation	403
Goals and Objectives of Agriculture	403
RAINFED FARMING PRACTICES	404
Sowing Operations	404
Fertilizer Use	405
Tillage	406
Weed Control	407
SUPPLEMENTAL IRRIGATION AND WATER HARVESTING PRACTICES	408
Characterization of 5 Districts	408
WATER RESOURCES	415
Surface Water	416
Groundwater	417
PROBLEMS AND CONSTRAINTS LIMITING SUPPLEMENTAL IRRIGATION	417
GOVERNMENTAL POLICIES IN THE DRYLAND REGIONS	418
FUTURE PLANS	420
SUMMARY AND CONCLUSIONS	420
REFERENCES	421

Chapter 23 by Saad Ahmed Al Ghariani

SUPPLEMENTAL IRRIGATION AND WATER HARVESTING SYSTEMS IN LIBYA	425
GEOGRAPHICAL CHARACTERISTICS	425
AGRO-ECOLOGICAL CLASSIFICATION	429
Jabal Al-Akhdar and Surrounding Plains	429
The Jafara Plain and Surrounding Western Mountains	433
EVALUATION OF COLLECTED INFORMATION	440
The Jafara Plain	440

The Jabal Al-Akhdar Zone	442
POTENTIAL FOR DEVELOPMENT	444
Development of Jafara Interior	444
Comprehensive Water Harvesting Systems	445
Crop Productivity and Water-Use Efficiency	446
SUMMARY AND CONCLUSIONS	446
REFERENCES	447

Chapter 24 by Ambri Abdel Ilah

SUPPLEMENTAL IRRIGATION SYSTEMS IN MOROCCO	449
GEOGRAPHICAL CHARACTERISTICS	449
Climate	449
Land Use	450
Water Resources	450
EXISTING AREAS WITH SUPPLEMENTAL IRRIGATION	450
Plains and Valleys: Middle and High Atlas Mountains	451
Atlantic Plains and Plateaus of the Eastern Areas	452
Had Kourt District, Kenitra Province	452
RESEARCH FINDINGS	452
Oulad Gnaou Experimental Station	452
Settat Experimental Station	453
SUPPLEMENTAL IRRIGATION 3 DISTRICTS	454
Berrechid District, Settat Province	454
Ain Taoujdate District	456
Taourirt District	458
POTENTIAL FOR DEVELOPMENT	459
CONCLUSIONS	460
REFERENCES	461

Chapter 25 by M. Rafiq

SUPPLEMENTAL IRRIGATION IN PAKISTAN	463
GEOGRAPHICAL CHARACTERISTICS	463
Climate	463
Geology and Hydrogeology	463
Land Resources	464
Water Resources of the Indus Plain	465
ENVIRONMENT AND CROPPING SYSTEMS	465
Existing Supplemental Irrigation	465
Potential Zones for Supplemental Irrigation	466
Existing Rainfed Farming Systems	471
Supplemental Irrigation Farming Systems	481
CHARACTERIZATION OF 3 LOCATIONS	485
Demography	485
Soils	485
Infrastructure and Rural Support Services	486

Water Resources for Supplemental Irrigation	488
Crop Production	489
Livestock Production	490
Socio-Economic Factors	490
POTENTIAL FOR SUPPLEMENTAL IRRIGATION	492
Water Resources	492
Component Needs	493
Problems and Constraints	494
Actions Needed	494
SUMMARY AND CONCLUSIONS	495
REFERENCES	496

Chapter 26 by George Soumi

SUPPLEMENTAL IRRIGATION SYSTEMS OF THE SYRIAN ARAB REPUBLIC (SAR)	497
Agro-Climatic Classification	497
WATER RESOURCES AND THEIR USE	497
Rainfall	497
Aquifers	499
Water Use in Agriculture	499
LAND USE AND CROPPING PATTERNS	500
Land Use	500
Crop Rotations	500
FOOD SECURITY	502
Domestic Demand	503
Production	504
Projected Food Deficit, 1990	504
SUPPLEMENTAL IRRIGATION	504
Productivity of Existing Supplemental Irrigation	504
Potential for Further Development	505
MAAR/ICARDA Research	506
TECHNICAL AND ECONOMIC FEASIBILITY OF SUPPLEMENTAL IRRIGATION OF WHEAT	508
CONCLUSIONS	510

Chapter 27 by A. Bouzaidi

CEREAL CROPPING AND SUPPLEMENTAL IRRIGATION IN TUNISIA	513
CHARACTERISTICS OF RAINFED CEREAL CULTIVATION AND THE DEVELOPMENT PROGRAM	513
Importance of Cereal Cultivation to the National Economy	513
Areas of Cereal Cultivation	515
Farming Systems	515
Crop Rotation	516
Major Constraints	517
Development Program for the Cereal Sector	517
SUPPLEMENTAL IRRIGATION OF CEREALS	518

Major Findings of Experimental Results	518
Large Scale Application of Supplemental Irrigation	521
National Project for Supplemental Irrigation of Cereals	522
Results of the 1986-87 Season	523
CONCLUSIONS	526
REFERENCES	526

Chapter 28 by Necati Gulbahar

SUPPLEMENTAL IRRIGATION IN TURKEY	529
GEOGRAPHICAL CHARACTERISTICS	529
Climate	529
Land Resources	530
Water Resources	532
OVERVIEW OF SUPPLEMENTAL IRRIGATION PRACTICES	534
National Goals and Objectives	535
EVALUATION OF EXISTING FARMING SYSTEMS	535
Weather Station Data	535
Current Farming and Cropping Practices	537
Mechanization	543
Fertilizer Use	545
Weed and Pest Control	546
Yield Relationships	547
Livestock and Forage	548
Cost of Production	551
CHARACTERIZATION OF EXISTING SUPPLEMENTAL IRRIGATION SYSTEMS	552
Karapinar District, Konya Province	553
Central District, Eskisehir Province	554
Ceyhan District, Adana Province	555
Cubuk District, Ankara Province	555
POTENTIAL AREAS FOR SUPPLEMENTAL IRRIGATION DEVELOPMENT	556
CONCLUSIONS	556
REFERENCES	557
ADDENDUM 28A: WATER AND LAND RESOURCE POTENTIAL	558

Chapter 29 by Abdulrahman M. Bamatraf

SUPPLEMENTAL IRRIGATION IN YEMEN ARAB REPUBLIC (YAR)	561
GEOGRAPHICAL CHARACTERISTICS	561
Agro-Ecological Zones	562
Climate	564
Geology	566
Soil Resources	566
Water Resources	567
ASSESSMENT OF PRESENT AGRICULTURAL ACTIVITIES	571
Farming Systems	571
Cultural Practices	573

Mechanization	576
Fertilizer Use	577
Weed and Pest Control	578
Livestock	579
Socio-Economic Aspects	580
EVALUATION OF SELECTED SUPPLEMENTAL IRRIGATION SYSTEMS	582
Tropical Lowland	583
Sub-Tropical Upland	586
Temperate Highlands	591
POTENTIAL FOR SUPPLEMENTAL IRRIGATION	595
SUMMARY AND CONCLUSIONS	596
ACKNOWLEDGEMENTS	598
REFERENCES	598

Chapter 30

CONCLUSIONS	599
RECOMMENDATIONS	600
FOOD FOR THOUGHT	601

APPENDIX

SOURCE MATERIALS FOR CHAPTERS 1-16	603
INDEX	609

Prologue

"Once upon an unfortunate time, there was a hairy thing called man. Along with him was a hairier thing called animal. Man had a larger brain which made him think he was superior to animal.

Some men thought they were superior to men. They became leader men. Leader men said 'We have no need to work, we will kill animals to eat.' So they did.

Man increased, animals decreased. Eventually leader men said 'There are not enough animals left to eat. We must grow our own food.' So man grew food.

Now, the only animals man had not destroyed were tiny ones, like rabbits and mice, and these little animals were caught eating some of man's crops. 'These animals are a menace. They must die.'

In China they killed all the sparrows. In Australia they killed all the rabbits. Everywhere man killed all wild life. Soon there were none, and all the birds were poisoned. Leader men said 'At last! We are free of pests.'

Man's numbers increased. The world became crowded with men. They all had to sleep standing up. One day a leader man saw a new creature eating his crops. This creature's name was starving people.

'This creature is a menace!' said leader man . . . ."

(Milligan, 1963)

While on an explorer's mission to study water harvesting methods in Australia's Barossa Valley, I met a young farmer who had been recounting all of his assets. We were sauntering along in a well managed vineyard with the morning sun brightening for an intense day of heat. I continued with an aimless conversation while kicking dry soil clods with the toe of my shoe and admiring the impact.

"Life's like the weather – what comes, comes," I expanded feeling a bit philosophical.
"It's all in God's power," he added stonily.
"It rains. You can't stop it," I continued.

Now his eyes widened and he straightened upright and said, "But when it's in *my domain*, it's in *my power*. I alone can manage it, control it, bottle it, store it or irrigate with it." Raising his arms he continued, "At that point of arrival, all the rainfall on my farm is mine. It's my atmosphere, it's my soil all the way to France."

The conversation ended.

Foreword

This book is the product of an ICARDA project to define supplemental irrigation in the Near East and North Africa. In cooperation with the Food and Agriculture Organization of the United Nations (FAO) a meeting was held in Rabat, Morocco, on 7-9 December 1987, entitled "Regional Consultation on Supplemental Irrigation"; specialists from 11 different countries were brought together to discuss priorities for supplemental irrigation within their specific regions. The participants were asked to focus on developing an information base using both primary data, results of surveys administered to district level agricultural personnel, and secondary data sources with a particular interest in the application of state-of-the-art knowledge and technology to the problems of supplemental irrigation.

The authors have willingly and thankfully responded to the suggestions and criticisms of Ms Kate Ward, Institute of Irrigation Studies, Department of Civil Engineering, University of Southampton, U.K., who accepted the soporific position of Review Editor and performed miracles. Chapter 2 and parts of chapters 15 and 16 are a partial rendering of a forthcoming book on systems analysis by Janice R. Perrier. The authors recognize the inclusion of this material which outlines the basic philosophical perspective of supplemental irrigation as utilized in the book. The assistance of Mr. Maurice Saade, Agricultural Economist is greatly appreciated for the understanding of Chapter 14. The section on the phenology of cereals near the end of chapter 4 was written by Mr. Pierre Hayek, FRMP, ICARDA and the authors wish to express their thanks for permitting the section to be used in this book.

We would like to express our appreciation to Drs. P. J. M. Cooper, Program Leader of the Farm Resource Management Program, FRMP, ICARDA, and A. Arar, Senior Regional Officer, FAO, Rome, Italy. We thank the ICARDA staff and others who contributed to discussions and editorial comments. In addition, we thank the management of ICARDA, the Ford Foundation, and IDRC for making available the funds to obtain the manuscripts, perform the survey, hold the meeting, and to publish this book, as well as the national scientists for their excellent papers that this book contains.

E. R. Perrier
A. B. Salkini
ICARDA, FRMP
Water Management Project

List of National Scientists Contributing Papers

Dr. Larbi Baghdali
Deputy Director
Ministry of Hydraulics,
Environment, and Forestry
B. P. 17 – Apt. 18
Cite Garidi Kouba, Algeria

Dr. V. C. Krentos
Consultant (Ex-Director)
Agricultural Research Institute
Nicosia, Cyprus

Dr. H. Siadat
Deputy Director
Soil and Water Research Institute
North Karegar Ave.
Tehran, Iran

Dr. Shifa'a A. Mahmood Said Al-Kazzaz
c/o FAO Representative
P. O. Box 10085
Baghdad, Iraq

Dr. Abdullah A. Jaradat
College of Agriculture
Department of Biology
Jordan University of Science and Technology
Irbid, Jordan

Dr. Saad Ahmed Al-Ghariani
Assistant Professor of Irrigation
and Drainage
Agricultural Research Center
P. O. Box 2480
Tripoli, Libya

Mr. Ambri Abdel Ilah
Researcher
National Agricultural Research Institute
P. O. Box 415
Rabat, Morocco

Dr. Mohammad Rafiq
Director
Soil Survey
74 Aurangzeb Block
New Garden Town
Lahore, Pakistan

Dr. George Soumi
Director
Directorate of Irrigation and Water Use
Ministry of Agriculture and
Agarian Reform
Douma (Damascus), Syria

Mr. Abdul Aziez Bouzeidi
Director, P.I. du CRGR
Agricultural Engineering Research Center
Ministry of Agriculture
P. O. Box 10
Ariana, 2080 Tunisia

Mr. Necati Gulbahar
Agricultural Engineer
Ministry of Agriculture, Forestry, and Rural Affairs
Research Planning Coordination Council
Ankara, Turkey

Dr. Abdulrahman M. Bamatraf
Water Management Agronomist
Head, Soil and Water Research
Agricultural Research Authority
Ministry of Agriculture and Fisheries
Taiz, Yemen Arab Republic

Chapter 1

Introduction

Nearly four-fifths of the world's surface is covered by vast oceans. The sun supplies the energy to operate this enormous evaporative surface to produce fresh water that evaporates into the atmosphere. Eventually this is returned as rain, 75% of which falls directly into the sea. Except for insignificant amounts obtained through desalination, the sea has never been tapped as a source for fresh water. Another resource which has not been adequately developed is water lost by excessive evaporation in arid and semi-arid regions. The collection of rainfall at sea and the suppression of evaporation are, in principle, economic problems which might attract greater interest were there cheaper technologies practical for their exploitation. Even now, ideas abound for the utilization of super tankers to capture rainfall when the oil glut runs out, as well as for transportation from the polar caps of icebergs which contain about three-quarters of the world's fresh water supply.

Irrigation technology using the techniques of supplemental irrigation and water harvesting can alleviate climatic risk factors in arid and semi-arid regions by increasing choices for soil and crop management which can stabilize crop water requirements and therefore yields. Supplemental irrigation is being developed in the Near East and North Africa; however, the level and extent of development have not been previously documented. A limited quantity of water in extended areas coincides with extensive use of marginal natural resources which must be related to possible economic benefits and costs. Within this text, supplemental irrigation is examined in detail from both technical and socio-economic perspectives to identify potential areas of improvement and development.

The importance of management in the sense of manipulating the hydrologic cycle manifests itself in the disparity between the extent of irrigated land and its contribution to production. Only 16% of all the world's cultivated land is irrigated (Figure 1.1) but this 16% supplies more than 40% of the world's crop production (Figure 1.2). Clearly, water is the limiting factor. Particularly in semi-arid and arid environments, it *is* the independent variable: farm resources are dependent on water:

E.R. Perrier and A.B. Salkini (eds), Supplemental Irrigation in the Near East and North Africa,
1-14.
© 1991 *ICARDA.*

farm resources = F(Water)

soil, plant, fertilizer, agronomic practices, etc. = F(Water)

Water takes priority since without it, *nothing* will create an economic setting for agricultural enterprise. Sustainable production is possible only if water resources are naturally renewable. An alliance between the management of water and land would operate towards the conservation of these vital resources and their use without waste in order to sustain as high a level of production as is possible for any particular environment.

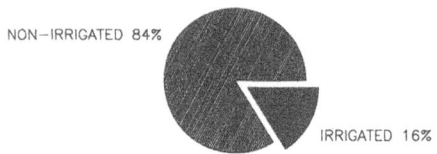

Figure 1.1. Graphical comparison of the world's irrigated and non-irrigated cultivated land.

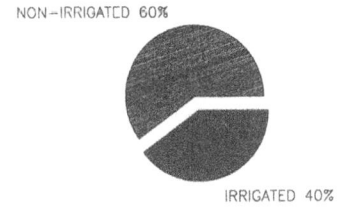

Figure 1.2. Graphical comparison of the world's irrigated and non-irrigated crop production.

Water is the primary resource of any farming system and determines the sustainability of agricultural production for economic development in the Near East and North Africa, an area which has a high human growth rate and a low amount of poorly distributed rainfall. The constraints to existing water supplies must be identified from an agricultural perspective, then prioritized, and action taken to alleviate future problems and chaotic losses for the farmer. There are many examples of destitute agriculture within the region, but no realistic evaluation has been completed towards a solution or an active attack through better management of water resources to prevent agricultural degradation.

"Compared to developed countries, ... underdeveloped countries of the world have a higher population, about twice the population growth rate, a much lower economic growth rate per capita, and a much higher need for an increase in food production." The engineering problems involved in resource conservation can be prioritized into 5 phases: water resource development, irrigation and drainage, water conservation, flood control, and erosion control. Maybe in developed countries, such as North America, Australia, and Europe where governmental economic impacts are accessible, they may reorder these 5

Introduction

phases. However, at international research organizations such as ICARDA, the focus of activities remains on the plight of the small farmer whose means of production are at, or nearly at, the subsistence level.

Goals and Objectives

The *mission* of increasing agricultural productivity to sustain food requirements of rapidly increasing populations could be facilitated through management of water, land and human resources. Poignant to note, however, is the constraint of water supplies for agriculture which results from competition with urban and industrial needs. Realistically, the cost of developing irrigation technology is often prohibitive in the less endowed countries where more food is painfully needed.

The *goal* is to plan and design for the modernization of agriculture through the implementation of irrigation technologies such as supplemental irrigation and water harvesting. New large projects using conventional irrigation (monuments to development) and mobilization of water resources are being relegated to a lower priority in favor of rehabilitating existing facilities and structures. Scarcity of water resources dictates refinement of irrigation practices to maximize productivity. These are welcome trends in the short-term to preserve initial capital investment. However in the long-term irrigation technology offers a positive potential for accelerating the modernization of agriculture and sustaining the ensuing productivity with comprehensive water use planning, expansion of extension services, and training of farmers in alternative irrigation techniques.

The *objectives* of irrigation technology are to establish representative ideas of the potential for modernization of agriculture with environmental stability. The assessment of each country's water resources focuses on land-use patterns and water quality and supply which are reviewed in a regional setting for implementation of advanced irrigation technology. Computer and information systems offer a ready means for manipulation and processing of data and information on a national basis within a regional framework, to establish a data pool which is accessible for the coordination and implementation of irrigation technology.

Irrigation Technology

Definitions of Irrigation Technologies

Irrigation is historically an integral component of civilization and mankind's development. Civilizations have been dependent on irrigation and when the system failed for one reason or another, these civilizations decayed and disintegrated and only the physical or stone remnants of their societal organization remains. If people were so dependent on irrigation, why has it not become an unalterable or permanent feature of successful civilizations? Historically, ancient civilizations based on irrigation have declined. Documentation of why is not forthcoming; however, conclusions usually point to mismanagement and occasionally to mankind's purposeful destruction.

Modern irrigation technology combines the basic theories for water, soil, and plant with climate and incorporates the philosophy of time and space to integrate the domains of water (source, supply, allocation, and recharge) to the farming system. Irrigation encompasses the technical disciplines of conventional irrigation, supplemental irrigation and water harvesting.

Conventional irrigation delivers the entire plant-water needs because rainfall cannot be relied upon during all or a large part of the growing season. Conventional irrigation is employed where water is not a scarce resource and a ready and ample supply is on hand. Practice of this type of irrigation requires a lower level of training and experience. Often these systems are burdened with overextended field development, requiring intensive management for maintenance. In the Near East and North Africa, these systems encompass the Nile Valley of Egypt, Sudan, and Ethiopia, the Indus River of Pakistan and the Euphrates River of Iraq, Syria, and Turkey. Irrigation efficiency, water requirement/water applied $\times$ 100, ranges from 40 to 60% within these systems; however locally efficiencies can be higher.

Supplemental irrigation is defined as a technique used where a crop can be grown by natural rainfall alone but additional water by irrigation stabilizes and improves yield. Whether to irrigate or not is decided purely on the estimated profitability of doing so. This underscores the importance of scheduling supplemental irrigation by minimum, *not maximum* crop water requirements. Water is a scarce resource and rainfall is the principle supply which must be developed and managed. This type of irrigation is more sophisticated than conventional irrigation and requires more experience and a higher level of training to operate. Under these irrigation systems, adding water when it is required gives an efficiency range of 60 to 75%. Irrespective of seasonal rainfall, supplemental irrigation can provide conditions suitable for using higher technology inputs, such as high yielding varieties, fertilizer and herbicides as well as more intensive cropping.

Water harvesting is an irrigation management technique for growing crops in arid and semi-arid areas where rainfall is inadequate for rainfed production and irrigation water is lacking. Rainfall is collected from a modified or treated area to maximize runoff for a specific site such as a cultivated field, cistern, stored in dams or soils, or used for aquifer recharge. Water harvesting ensures that a greater percentage of rainfall is of beneficial use to a water efficient agricultural system. This type of irrigation is labor intensive and requires the most advanced training and experience to be effective. Water harvesting ensures that a greater percentage of rainfall is put to beneficial use to a water efficient agricultural system. Using water harvesting techniques, a rainfall of a few millimeters collected on a catchment basin can be equivalent to several hundred millimeters of rainfall when supplied to a restricted cultivated field.

Introduction

Selection of an Irrigation Method

Selection of a supplemental irrigation method is based upon technical and economical feasibility as well as traditional values of agriculture. Normally, investment in an irrigation system is to increase cropping efficiency and applies to high valued summer crops as well as cereal crops. When considering the economics of a supplemental irrigation system, all costs, benefits, and subsidies should be included in the evaluation. Annual costs per hectare compared to annual returns per hectare are the best economic measure of an irrigation system. The expected return from any supplemental irrigation system should reflect the savings resulting from the following elements:
1. increased yield and quality of product;
2. less land out of production;
3. reduction in land preparation, tillage, and harvesting costs;
4. saving in labor, operations, repair and maintenance; and,
5. conservation of water and power costs.

Techniques for Supplemental Irrigation

Surface irrigation

If water is inexpensive and soil types and topography are adequate, surface irrigation methods are the least technical and least expensive to install but they are labor intensive.

For efficient irrigation by a surface method slopes should be uniform with no high or low spots. For this, some land grading may be required, the extent of which depends on depth of soil, natural topography, and the infiltration rate at the soil surface. Furrows entail the construction of small parallel ridges as part of tillage operations for row crops. Each furrow acts like a miniature channel terrace. Furrows may be on a level grade to increase the residence time for water to infiltrate into the soil. If irrigation water contains sediment or if leaching for salinity control is important, then surface methods of irrigation may be preferred.

Ideally a surface irrigation system (Figure 1.3) should apply an equal depth of water along the run or furrow; however, in practice and without precision land grading equipment, this may be impossible to achieve. Sandy soils with low water storage capacities and high percolation rates can require frequent light irrigations which are difficult with surface methods. Surface irrigation may not be economical for clay soils with low infiltration rates or if the amount or flow of water for irrigation is low.

In the Near East and North Africa, farmers use a combination of the border strip and furrow method where the border strip is made perpendicular to slopes of more than 3.0%. No land leveling is performed and the width of the border strip is dependent on slope, e.g. narrow strips (about 2 m in width with ridges

Figure 1.3. Border irrigation in Northern Syria.

40-50 cm in height) are constructed on the steeper slopes. Within these sloping border strips, 2-4 furrows with ridges 15-20 cm in height are formed by bed shapers and skilled tractor drivers. This method of bed construction restricts the lateral movement of water down the cross-slope.

Water for irrigation may be pumped to the top of a field for gravitational flow. The water enters the horizontal end of a border strip and circulates down the furrows. The total amount of water absorbed is a function of the soil infiltration rate and the quantity of water applied. When the furrows are filled, the border strip is sealed and the water is diverted to the next border strip. Grain is usually broadcast by hand across both border strips and furrows. The success of this method is dependent on the skill of the tractor driver who uses bed shapers and duck foot cultivators to form both borders and furrows. Although this method may be economically feasible, labor is intensive with irrigation operations shared by cooperating farmers or members of the extended family.

Sprinkler irrigation

A major argument in favor of non-surface irrigation is ease of operation of sprinkler systems. A single person hand-move system is not uncommon and permits private individual operation. Such systems are efficient not only for scheduling and estimating volume requirements of irrigation, but also for the distribution of water.

Introduction

For effective operation of a sprinkler system at each farm site, the farmer must examine the pressure head, ensure the correct type and spacing of the sprinklers, check the wind conditions, and know the nature of the soil and crops. The maximum wind speed for effective sprinkler operation is between 5-6 km/hr.

The main elements of a hand-move sprinkler system are:
1. the source of water (well or borehole, regional pipe line, reservoir, canal, river or stream, or natural lake);
2. the main line (conveyance of water from the source to the field by concrete canal, steel pipe or plastic pipe);
3. the sub-mains (which can follow field boundaries or center-line of field);
4. the laterals (which convey water to risers and sprinklers at regular intervals); and,
5. the sprinkler (rotating, whirling, or "guns").

Every sprinkler system is composed of a pipe network and sprinklers. The rotating sprinkler has 1 or 2 inclined nozzles mounted on a body which rotates by the action of a hammer blade about a vertical axis. In operation, one jet impinges on the blade and thrusts it aside. The blade is restrained and returned by a light spring. The return is terminated by a stop on the body which rotates by impulse through a small angle. Then the cycle is repeated. The water supply should be clean for rotating sprinklers as they can clog and for this reason filters are usually placed somewhere in the system.

Design of sprinkler systems for supplemental irrigation consists of selecting a layout where laterals of equal length are placed uniformly along the supply line. Some features should be incorporated into the system design:
1. move the pipe twice a day during an irrigation, once in the morning and once in the evening;
2. use a pump switch to stop irrigation (could be automatic);
3. use only small diameter laterals (75-100 mm) with quick-connect joints; and,
4. limit sprinkler stands to a minimum of 1 m.

For supplemental irrigation, a sprinkler system should be totally portable. Positioning of main lines could affect the portability and the possibility of system expansion. The design should not place irrigation pipes in the path of mechanized equipment but should incorporate the working pattern of farmers.

Three high-energy types of sprinklers have become widely used: the sprinkler gun, center-pivot, and linear-move systems. These systems have the advantage of being portable and require a minimum amount of equipment over a large area. Gun sprinklers operate at high pressures (5-10 bars) and can cover an area of 0.5-1.0 ha per sprinkler if the system is properly designed. The center-pivot system (Figure 1.4) has an aluminum pipe on wheels which are at regular intervals and which support the pipeline high above the ground; water is distributed using rotating sprinklers or micro-irrigators. The pipeline rotates by hydraulic or electrical power around a fixed end or central pivot at a selected speed. The center-pivot system can be designed to irrigate 2-50 ha. The linear-move system is supported by triangular wheel supports which do not pivot about a center but run linearly; water is supplied from a ditch or flexible plastic pipe. All three systems are adaptable to rough ground and rolling fields with slopes of up to 20%. The major advantage of these systems is their highly efficient design for irrigation. The major disadvantages are the initial capital investment, large pumps at high pressures, large supply pipes, the need for skilled technicians for maintenance, and sometimes problems with high wind velocities.

Figure 1.4. Center-pivot irrigation system.

Drip/trickle irrigation

Drip or trickle irrigation is a technique which delivers frequent, slow application of water to soils through mechanical devices called emitters at selected points along a water line. This method allows more precise control of water in the plant available moisture range than do surface or sprinkler irrigation techniques. In addition, fertilizer and micro-nutrients can be supplied on plant demand for efficient application and increased production.

Drip irrigation (Figure 1.5) uses perforated plastic pipes which are laid along

Introduction

the soil surface (or sometimes underground) at the base of a plant row with water supplied from a field main. For the duration of a growing season, all plastic pipes remain in place with water supplied on demand. The emitters or outlets are designed to release a trickle or drip of water and not a jet. Spacing of emitters is selected to produce a wetted strip along the crop row or a wetted bulb of soil at each plant.

Figure 1.5. Drip/trickle irrigation system.

The main advantage of drip irrigation is excellent control over water application. The soil moisture deficit can be controlled at a low level and soil aeration is maintained. Drip irrigation is highly beneficial to plant growth and improvements in yield and quality have been achieved for a wide range of crops. This technique is particularly advantageous when saline water has to be used for supplemental irrigation. The most successful crops in the Near East and North Africa using drip irrigation have been almonds, grapes, citrus, stone fruit, avocados, walnuts, pistachios, olives, pecans, apples, pears, figs, dates, vegetable crops, nursery plants, berries, tropical fruit, and sugar cane.

Unfortunately, the capital investment for drip irrigation equipment can be higher than that of surface or sprinkler irrigation. The high cost of drip irrigation has limited its use for supplemental irrigation; however, if a movable drip system has been purchased already for a high valued crop then its use for grain crops can be economically feasible. The problem of nozzle blockage at the outlets has not been completely solved and drip irrigation systems require extensive line filtration and continuous maintenance. Economical emitters have been developed which are not dependent on water pressure. Sometimes animals, e.g. rodents, dogs, present a problem by damaging plastic pipe.

Drainage of Supplemental Irrigation Lands

Under conventional irrigation, water is usually applied to land over and above the needs of evapotranspiration and drainage. For supplemental irrigation when only 1 or 2 irrigations are applied, drainage is of less importance. The application of excess water is a consequence of inefficient irrigation or is applied intentionally to leach harmful salts below the root zone. The purpose of drainage is to remove unwanted water, to maintain soil structure and aeration, and to assure access to the field for cultivation and harvesting. Saline irrigation water or brackish groundwater can exacerbate the drainage situation.

Drainage requirements can be assessed for different categories of land use and applied to typical soil profiles. Ideally, a drainage survey should be performed at each site to estimate percolation, leaching requirements, and type of drainage system. The drainage system design will specify which method will be needed:
1. tile or tubes;
2. mole drain;
3. open ditch;
4. subsurface (wells);
5. pump-out; or,
6. no drainage method.

Concepts of Irrigation Technology

Technology exists for managing water and land as renewable resources. It is possible to develop more efficient utilization of rainfall and groundwater for land-use diversification. The challenge is to create an economic water supply of an optimal quality and quantity with a capacity for delivery on demand. National yield averages could be improved by 1-2 t/ha of staple, rainfed cereal crops by fertilization, varietal improvement, and water use efficiency techniques. Modern irrigation technology has the potential for a forward leap in production of 4-6 t/ha. At specific locations there are excellent opportunities to utilize and maintain surface and groundwater supplies, install large isolated irrigation systems using center-pivot and side-roller equipment, and to develop long-term water harvesting systems.

Figure 1.6 shows the salient characteristics of rainfed areas which have unstable yields and low cropping intensities. Together with the demographic explosion and demands for a rise in standard of living, these factors contribute to degradation of natural resources which perpetuates low yields, inadequate food supplies, and poor public health. Supplemental irrigation affects social

Introduction

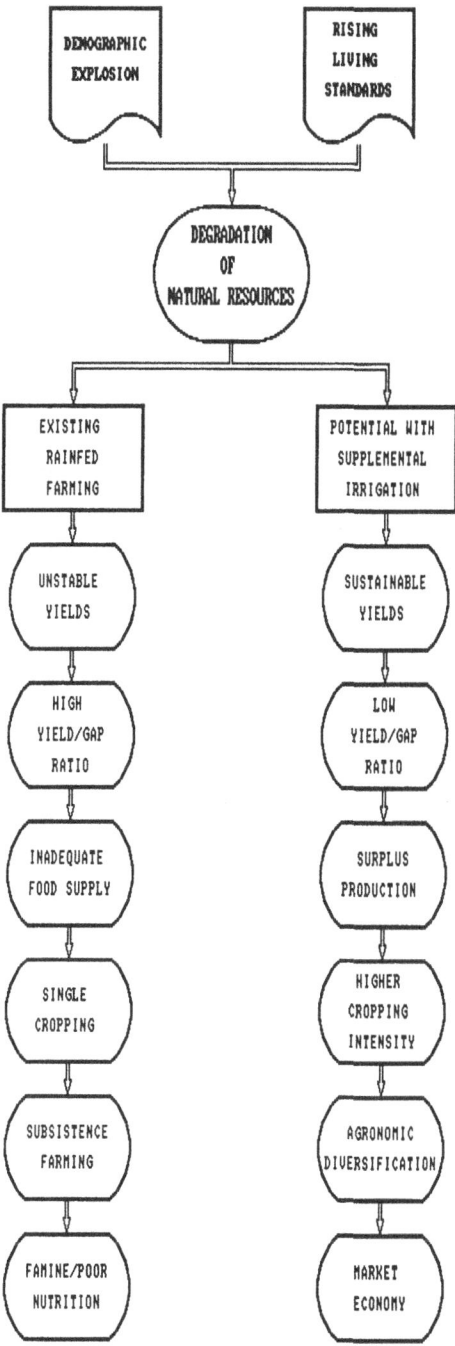

Figure 1.6. Impact of supplemental irrigation on agricultural productivity.

organization and brings about stratification if there is a shortage in water supply, or if conditions exist that, in effect, restrict access to water. The introduction of supplemental irrigation into traditional rainfed farming can alleviate the farmer's dilemma of risk through prospects of stable crop production, increased yields and water use efficiency, surplus food supply, a cash crop program, and a market economy.

Farmer acceptance of supplemental irrigation is an important factor in the success of technology transfer. Farming with supplemental irrigation is more sophisticated and always requires more physical effort than rainfed farming under comparable conditions. Farming practices supported by supplemental irrigation increase the food supply but do not involve the patterns of organization and social control that characterize large-scale irrigation enterprises. If the design of the system presents the farmer with too big a burden and too little profit, the system will likely fail. In areas where supplemental irrigation is not fully understood or accepted because of various socio-economic factors, system design becomes extremely critical. The design must accommodate the local labor supply and equipment which has a minimum maintenance requirement and maximum effectiveness. The selected supplemental irrigation system must support a positive economic alternative to existing conditions.

Risk and uncertainty, inherent in semi-arid and arid environments, weave a precarious physiological balance in support of farmers in their social and ecological contexts. Since farmers in these environments have little, if any, risk-bearing capacity, it becomes crucial for them to choose a crop and management system that can make the best use of rainfall and alternative water supplies. The choice for increasing agricultural production for these farmers who are at, or nearly at, the subsistence level requires synchronization of all farm resources. At an individual farmer level, pursuing simple rainfed farming practices could lead to a less than optimal allocation of scarce water resources. Using rainfed farming practices to achieve the technically most efficient level of production could be at variance with farmers' economic objectives. Alternatively, the maximum output may not be a rational choice if a surplus is produced using supplemental irrigation without an adequate infrastructure to market and distribute excess production.

Book Organization

Supplemental irrigation in the Near East and North Africa was the subject of an FAO/ICARDA seminar. Topics focused on existing practices which would offer ways and means of improving rainfed agriculture with the addition of supplemental irrigation when water supplies from water harvesting, surface water, or groundwater sources exist or have been recognized. A limited quantity of water in extended areas coincides with extensive use of marginal natural resources and the scope of the investigations was related to possible economic benefits of developing the technology. Supplemental irrigation was to be examined in detail from both technical and socio-economic perspectives to identify potential areas of improvement and development.

This book is written in the context of current irrigation technology. The systems approach incorporates various societal levels of each country: the farmer, the extension worker, the scientist-engineer, and the policymaker. The text is intended for use by teachers and lecturers to develop informative discussions of pertinent material relating to supplemental irrigation technology preparatory to field application. The subject material of irrigation, an integral component of a total farming system, is presented with a view to assist small farmers to improve, elevate and enrich their realm of subsistence living.

The text is in 2 parts: chapters 2 through 15 have been developed from field research, classroom experience, and literature reviews for a state-of-the-art perspective of supplemental irrigation farming. These chapters are written as a potential series of 2-hour lectures. Chapter 2 is a philosophical perspective for technology transfer of supplemental irrigation using the logic of inquiry to ensure that the precision of scientific discovery is integrated into planning, development, and production. Chapters 3 through 13 are technical explanations of methodologies for using climatic characterization in conjunction with water, soil, and plant data, to provide a quantitative framework for scheduling irrigation in the semi-arid and arid environments. Chapter 14 concerns the economics of production using irrigation technology with emphasis on supplemental irrigation. Chapter 15 discusses evaluational methods with examples for monitoring and assessing the impact of irrigation technology.

Selected national scientists prepared papers for presentation and inclusion in this book. They were asked to review the literature, examine secondary data sources, and administer questionnaires to agricultural extension personnel in uniform agricultural ecological zones to develop a state-of-the-art summary of supplemental irrigation for their individual countries. A structured questionnaire was provided for collecting baseline data in each country to determine parameters for the improvement of rainfed farming and integration of supplemental irrigation and water harvesting into local farming systems. The scope of these surveys were related to possible economic benefits as well as conservation of natural resources. The categories of information were grouped under the general focus of the strategies of involvement and included pertinent areas of concern to agriculture in the 12 countries of the region.

In the second part of the text, chapters 16 through 30 consider research at ICARDA and each country's investment in supplemental irrigation and the interactions of government institutions on the immediate project environment. Chapter 16 delineates the "ideal model" of the decision field for research and development of irrigation technology. Chapter 17 presents detailed analysis of the supplemental irrigation program at ICARDA, Syria. Chapters 18 through 29 provide baseline data for 12 countries of the ICARDA mandate region. These chapters point out the need to measure the technical, economic, social, and environmental effects of supplemental irrigation to assess regional planning and future land capability for implementation of irrigation technology. The systems approach emphasizes that actions outside the technical scope of supplemental irrigation projects can be critical for sustainable crop production. Assessment can be achieved by integrating an evaluation component into each plan or design and these results can be compared analytically on a system wide basis.

Chapter 30, summary and recommendations, highlights pertinent findings and constraints for future development of supplemental irrigation throughout the region. The chapter suggests future implications for the systematic development of irrigation technology towards evaluating the modernization of agriculture for securing food self-sufficiency in these countries and increased food security for the region.

Appendix I gives the list of literature sources used for chapters 1 through 16. These chapters were written without citations included to permit ease of reading and for lecture preparation.

Chapter 2

Systems Approach to Supplemental Irrigation

Humans have lived on this planet from 10 million to 10 thousand years ago in a hunter-gatherer social economy. Between 8,000 and 10,000 years ago a rapid transformation happened to farming practices with the addition of irrigated agriculture; this occurred mainly in the great valleys of the Yellow River of China, the Indus River of Pakistan, the Ganges River of India, the Irrawaddy River of Burma, the Chao Phraya River of Thailand, and the Rivers Euphrates and Tigris of ancient Mesopotamia. Yet the science of water resource management started merely 200 years ago and is still in the embryonic stage of development.

In view of the enormous complexity and dynamic nature of climate, the study of water resource management incorporates many dimensions of information associated with activities of farming systems that use irrigation technology. A priority exists for a conceptualized framework to organize information pertaining to water resource management that permits comprehensive data analysis.

The systems approach is a means of managing knowledge and provides a format for visualizing system components and their environment as an integrated unity. The approach allows recognition of the position of each element of irrigation technology relative to each system component and the larger environment. The most important concept of systems theory is boundary definition, that is determining where each component ends and the next begins while distinguishing system linkages. Supplemental irrigation should be identified as an integral component of the farming system recognizing of the catenation (linkage) within the network of the social system (local, regional, national, and international). The common denominator of these various linkages is the idea of interaction within the social system: supplemental irrigation does not stand alone. Interaction becomes a continuing debate between various attitudes of mind.

The multidisciplinary perspective of the systems approach is central to upgrading traditional farming systems with modernized technologies of conventional irrigation, supplemental irrigation, and water harvesting. Knowledge depends on the perspective from which it is observed and is not a reflection of real things but rather the result of interaction between "the

E.R. Perrier and A.B. Salkini (eds), Supplemental Irrigation in the Near East and North Africa,
15-38.
© 1991 *ICARDA.*

knowers exchanging the known". Problems are interconnected and overlapping. By whatever means data and information are collected, the analytical results must be viewed as tentative. For technology transfer, inquiry is continuous, cyclical, and repetitive: systems thinking of irrigation technology postulates a new frame of mind – a change in the basic categories of thought.

Transformation from traditional to modern agriculture implies a weakening and elimination of ingrained attitudes. Social stability needs to be maintained but change has to occur if agricultural productivity is to be optimized. A dilemma of modernization is the balancing of public efficiency and individual equity; i.e. gaining maximal economic returns on a national scale or distributing benefits to the underprivileged in the population. Rigid adherence to either is not tenable. Ultimate success will depend upon an organized set of rational goals and objectives for social change laced with fairness and justice.

Even though goals and objectives of exploiting water resources may appear contradictory at different system levels of organization and operation, the eventual motivation is to economically increase productivity of basic food crops, concurrently safeguarding resources. This ensures the effective and efficient use of water to improve crop yields and stabilize production. In establishing the technology of irrigation for sustainable agriculture, conflicts may exist between conception of goal attainment, perception of objective fulfillment, and recognition of human essentials.

Development activities are implemented with quantified objectives designating measures of effectiveness and efficiency to ascertain when an effort has been successful and what percentage of the objective has been fulfilled. These activities are measured to ensure farmer acceptance of irrigation technology. Specialists in agriculture are currently being asked to compress the trial and error process of agricultural development into an accelerated, one-generation endeavor incorporating a complexity of organization and an escalation of agricultural productivity that took nearly two centuries to achieve in Europe and North America. The issue is not whether the development of water resources is possible, but how and when it will be accomplished. Systems theory with monitoring and evaluation can be employed to appraise the process of implementing irrigation technology. This appraisal is based upon the logic of inquiry with the support of computer technology for data classification and information categorization. Coordination of efforts can become a reality for policy, research, extension, and production using systems analysis predicated on scientific observation with logic of inquiry applying evaluation.

Potential for Agricultural Modernization

Traditional agriculture is an extractable process where all resources – human, water, and land – are taken and applied to immediate use. Modern agriculture uses planned technology and emphasizes management practices of conservation and renewability of resources. Modernization precipitates growth of an

Systems Approach to Supplemental Irrigation 17

infrastructure concomitant with rural development, urbanization, and industry. Complete economic food security relies on the efficacious use of human and water resources which are both important to the environment.

An argument can be made that modernization is not improvement intrinsically but increased opportunity for choice among alternatives. Irrigation technology offers greater control of individual destiny; farm improvement coincides with higher incomes to reduce risk through achievement of sustainable productivity. The exigency for agricultural modernization is to quantify farmers' objectives based upon the delineation of a paradigm, an ideal model, and a method founded on theory and technique. Standards and recommendations established from research findings and verification trials with production records can coalesce in an ideal model of irrigation technology with the capacity to produce "enough food as a human right".

Informational categories are arranged in a matrix (a rectangular array of components) for detecting similarities, differences, and trends among system components. These categories bolster system planning and classification, program development, and resource procurement and distribution for irrigation technology. Logging and cataloging with correlation of information is multi-dimensional. The arrangement of the components into a matrix increases the probability that dependency, independency, and/or interdependency of variables will not be overlooked or spurious results accepted. Moreover, this method limits imprecise decisions which could be detrimental to optimal application of irrigation technology. Information is organized to supply specific facts on demand (data retrieval); interim information capable of manipulation (descriptive profile); comparison of data to provide indicators of system need (correlated variables); and, comprehensive analyses of processed data (analytical results).

The classification of system endeavors for data collection for planning and development, can be divided into 4 main groupings: effort, process, efficiency, and effectiveness.

1. The *effort* is the amount of decisional action executed from any component or element of that component. The data prove the decisional capability at points of intervention that can contribute to goal attainment.
2. The *process* is the activity output either at the operational or functional level. These activities contribute most towards the fulfillment of the stated objectives of the system.
3. The *efficiency* measure is a budgetary benefit-cost: the relative costs of precision in data analyses for obtaining "the best available data" in contrast to expert opinion. These data for efficiency correlated with effort and process therefore narrow the selection of strategies of involvement for irrigation technology.
4. The *effectiveness* measure of a response to agricultural productivity, whether positive or negative, can be identified with the systems approach and logic of inquiry. The effectiveness of irrigation technology to increase agricultural productivity depends upon the combined achievements of system effort,

process, and efficiency. These values are equated to standards of performance and the corollary is then compared to quantified objectives and goals.

Logic of Inquiry

To change traditional farming systems requires a blueprint for goal establishing, objective setting, decision making, policy formulating, activity planning, organizing, motivating, innovating, controlling, and regulating elements of irrigation technology to reduce instability in production and scarcity in food supply. The logic of inquiry supports a classification scheme which can accurately place data items in their proper categories for innovative research, calculated development, and sustainable production. Conceptually, there resides in any social system following from the logic of inquiry, information to predict the future, describe the past, and explain the present. There are no experts in inquiry; everyone is being educated. The obligation is to tie-in the logic of scientific discovery to the 4 levels of information users: farmer, extension workers, scientist, and policymaker. These users rely on the same statistical development but at different communication levels.

Inquiry becomes the design process for an "ideal model." Activity is an integral part of the system itself and implies both analysis and understanding which leads to new knowledge and, therefore, continuing modernization of irrigation technology. Design is "goal seeking" behavior which distinguishes between different patterns of activity and seeks to introduce alternatives to others in such a manner as to serve the goals of the total group.

The procedure emphasizes continual assessment and adjustment and seeks verification of actions and activities, not solutions to problems, i.e. the process focuses on the integration of information not on problem diagnosis. Information feeds into a multi-level data bank and is classified according to the precision needed by the level of management. The logic of inquiry proceeds beyond "objective, quantified" and "subjective, qualitative" observation and relies upon statistical inference to construct an adequate hypothesis. Hypotheses are then tested following the 8 points presented in the flow chart in Figure 2.1. The first 4 activities in the flow chart are guided by deductive reasoning to test variability over space; whereas, the last 4 activities use inductive reasoning to test variability over time.

An hypothesis is a statement about a future event, or an event the outcome of which is unknown at the time of the prediction, and is set forth so that it can be rejected. For example, an irrigation modernization project partitions farms into 4 technically efficient segments of conventional irrigation, supplemental irrigation, water harvesting, and rainfed farming. The goal is to increase and diversify agricultural farm income within the existing market economy applying these varying technologies. The hypothesis is that agricultural productivity will increase using irrigation technology and that the existing market distribution system can manage this increased production.

Systems Approach to Supplemental Irrigation

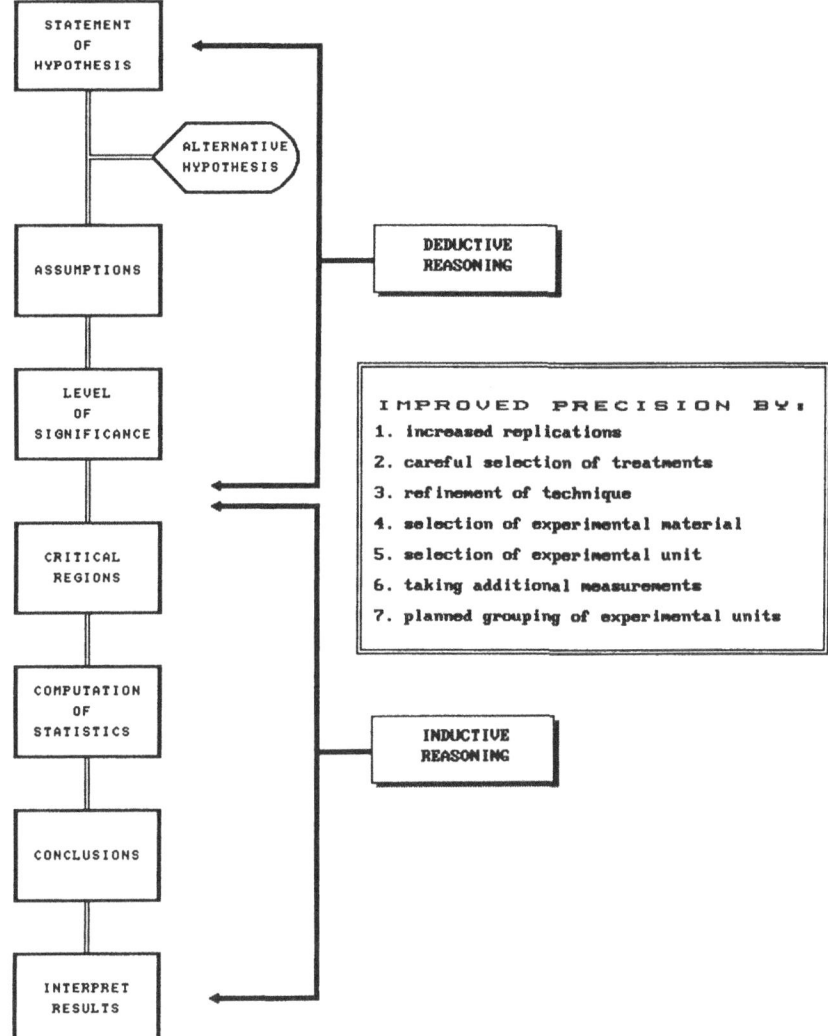

Figure 2.1. Flow chart displaying procedural steps for logic of inquiry.

In the development of an hypothesis, the logical progression from deductive to inductive reasoning as outlined in Figure 2.1 requires 4 sequential steps phrased in the "language of doubt":

Deductive reasoning:
1. All possible outcomes of the experiment or observations are anticipated before the test.
2. Agreement is reached prior to the test as to which operations or procedures will be used in determining which outcomes actually occur.

Inductive reasoning:
3. It is decided in advance which outcomes, should they occur, will result in rejection or non-rejection of the hypothesis. As implied above, rejection must be a possible result.
4. The experiment is performed, or the event observed, the outcomes noted, and the decision made whether to reject the hypothesis.

It follows that all decisions must be made during planning and design. If decisions are not formed prior to implementation or construction, it becomes possible to retain an hypothesis by simply changing the rules as a project proceeds. As in the previous example, if the existing market economy cannot support the diverse agricultural production, then government usually subsidizes the farm development by purchasing and storing non-marketable or surplus products; in reality, the hypothesis that the existing market distribution system was adequate should have been rejected. During project formation, planners and designers should have solicited more information to better evaluate the existing market to present a broader range of alternatives.

The main point is that whenever a project design is formulated the resulting theory implies certain consequences: these consequences and not the theory are subject to verification. The project planners are in the logical position of being able to reject the theory, whereas, the theory cannot be accepted without running the risk of making an error. In planning and designing for technology transfer the vocabulary can become confusing because the "language of doubt" is expressed in goals and objectives not in the hypotheses being tested; nevertheless, the precision as presented in Figure 2.1 must be maintained.

The problem of objective evidence rests with the system design and degree of probability of occurrence of patterns among events. To evaluate the basic problem underlines another predicament; it could be that not enough measures were taken to detect and explain error. As in the example, the market distribution system was not measured sufficiently to detect and explain the error resulting from the level of production.

There are two types of error expressed as variability, random and systematic. Variability over time is random variance and is related to the precision of experimental methodology. The precision of an experiment refers to the rigor of scientific design to detect true treatment effects. Methods to increase precision are intended to lower the unaccounted-for variability per segment. As shown in Figure 2.1, precision is improved by:
1. increased replication;
2. careful selection of treatments;
3. refinement of technique;
4. selection of experimental material;
5. selection of experimental unit;
6. taking additional measurements; and,
7. planned grouping of experimental units.

In the example, refinement of technique (3) would have foreseen the error of relying on the existing market economy or taking additional measurements (6) might have evaluated the capacity of the market distribution system to support increased production.

Two key terms, reliability and validity, provide the essential language of measurement. Reliability is the consistency or repeatability of measurements; validity is the degree to which a set of indicators measures the concept that it is intended to measure. Validity is the correlation between measurement X and measurement Y. Reliability is inversely related to the amount of random error in the measurement process. If all observed variance (by assumption, random variance) is contaminated with error (noise) the reliability equals zero.

For example, at a water division point in a delivery system a farmer uses a water measurement weir constructed of loosely cemented rocks to distribute accurate volume flow rates to the fields. If there is no random error involved in the measurement, reliability equals one, i.e. the farmer's neighbor was honest, no rocks were added to receive a larger share of the water, and the weir was accurately distributing the flow. Conversely, if during subsequent irrigations, a farmer's dishonest neighbor added different sized rocks to the weir, the delivered volume would be randomly less than anticipated, therefore, reliability would equal zero.

The tendency is to assume that all measurement error is random. This is a tenable assumption for research founded on strict experimental design; however, this assumption probably is unjustified for data collected from sample surveys, field studies, or structured observations. Measurements from these data typically include non-random elements of systematic error in addition to random error. The difference between reliability and validity is entirely dependent upon systematic error. Thus, systematic error can give reliable but invalid measurements.

Non-random systematic measurement error does not yield easily to statistical analysis. Indeed, it cannot be detected and estimated unless a series of simplifying assumptions are made about the theoretical structure underlying the empirical measurements. If the construction of the farmer's loosely cemented weir, in the above example, permitted turbulent water action to remove cement slowly over time then a non-random error would occur. Estimating the amount of non-random error in the measurement process – which lies at the heart of assessment of validity – depends more on one's theoretical understanding of the particular substantive area (construction of weirs as a measurement device) than it does on statistical formulas.

When reliability is used in connection with the sample mean, it refers to the closeness of the sample mean to the population mean. Since the variance of the sample mean measures the variation among the sample means, a reduction of its magnitude indicates that the sample mean approaches the population mean more closely. The purpose of having a sample mean is to estimate the population mean.

To illustrate selection of a sample size to estimate the population mean, an example is given of a socio-economic study performed on the northwest coast of Egypt. The land area and land use reflects the variability in rainfall amount and distribution so that the agricultural production strips exhibit differences as land form and climate change. Four parameters were determinants for sample selection: agriculture was divided into 3 agro-ecological production strips; administration of this area was apportioned into 4 political districts; the inhabitants were culturally similar; and the farmers shared the same type of farming system with variations.

Survey instruments were administered by an interviewer to cooperative leaders or extension personnel from 35 cooperatives in 4 districts of the governorate. Travel restrictions prevented farmer interviews in district 4; therefore, farmers were selected in a random sample from the first 3 districts only using the population of 33 cooperatives.

From the 33 cooperative societies, 10 were proportionately selected from within the 3 districts. The selected cooperatives for the farmer level study were randomized using a random number table within each of the 3 districts as follows:
- district 1 used cooperative numbers **1, 3, 8, 13, and 17** from the list of cooperatives in the district;
- district 2 used cooperative numbers **1, 6, and 9** from the list of cooperatives in the district; and,
- district 3 used cooperative numbers **2 and 5** from the list of cooperatives in the district.

From each of the 10 cooperatives, 20 farmers were selected for participation to allow for a minimum sample n = 130. At each cooperative a list of farmer members was used for selection. The number of members was a variable at each cooperative. To select the 20 farmers for study, one random number was selected for each participating cooperative for determination of spacing or interval. The starting random numbers were:
- district 1   **2, 5, 6, 10, and 12**;
- district 2   **1, 15, and 18**; and,
- district 3   **3 and 15**.

The interval was calculated in the following manner:

$$\text{interval} = \frac{\text{total number of members in the cooperative}}{20}$$

If 316 farmers were members of a cooperative then,

$$\text{Interval} = \frac{316}{20} = 15.8 = 15.$$

(Note: round-off was done by truncation). If the starting random number for a cooperative = 9, then the first farmer selected would be number 9; the second farmer selected would be number 24 (9 + 15 = 24); the 3rd farmer would be 39 (24 + 15 = 39); then the 4th would be 54 (39 + 15 = 54); etc., until the 19th = 279, and the 20th farmer would be number 294.

Some problems were encountered in using membership lists as population indicators and to complete sample selection required additional farmers

Systems Approach to Supplemental Irrigation

selecting the farmer in the position immediately before or following the randomly selected farmer.

Generally, at least 12 farmers were drawn from each of the 10 cooperatives for interviewing. At each cooperative a procedure was established. First, there was a membership meeting where the leader introduced the interviewer and explained the purpose of the survey. Then, a schedule was arranged for actual interviews with the farmers.

The sample size at the farm level was n = 130 farmers from 10 cooperatives distributed within 3 districts. The structure and composition of the farmers as a group (Figure 2.2) show that:
1. 91% of the farmers (118) hold barley fields whereas 10% of the group surveyed (12) are not involved in barley production.
2. 97% of the farmers (126) own sheep and goats with only four farmers not involved in livestock production. Of these 126 farmers, 70% (88) have an interest in tree production.
3. 75% of all farmers (97) are involved in tree production whereas the remaining 25% (33) have no fruit trees but do own livestock.

The distribution of the farmers selected by this sampling method accurately reflected the population characteristics. Reviewing pertinent literature about the study area allowed identification of the 4 parameters which could affect the randomness of the sample and the reliability of the final results.

Barley production	No. of farmers	Production involvement			
		Livestock		Tree production	
		+	−	+	−
Farmers +	118	115	3	88	30
−	12	11	1	9	3
Total sample	130	126	4	97	33

+ indicates farmer involvement
− indicates no farmer involvement

Figure 2.2. Structure and composition of a sample of farmers involved in production.

Variability over space tests the validity or the extent to which empirical indicators measure what they are intended to measure and encompasses systematic variance whose magnitude is a function of design and is (normally) of immediate interest only to planners and engineers. Systematic variance (deterministic) does not reduce reliability because the source of error is uniform (logically developed) and not random real time variance (stochastic). However, systematic variance does confound the measurement process because of the theoretical concepts and assumptions which determine operational definitions. In the eventuality of technology transfer, adaptability is the degree of variability over space not a consequence of the validity of a design or a generalization of results, i.e. going from the sample means to the population means. The individual farmer is not concerned with the "variability across time in a given location." Only by engineers and scientists during periods of design and analysis is overt concern given to these concepts.

The logic of inquiry becomes a working hypothesis capable of self-examination to validate activities and the consequences or changes which result from system efforts. Because of the need to generalize beyond the limits of one's data, decision makers are required to develop probability statements applying inductive and deductive reasoning. With probability statements and before implementation, decisions are tested on those results which are plausible so that rejection of theory can be made should anticipated consequences not occur. In effect, if theory is correct (validity), then sample results are within a specified range of outcomes (reliability), i.e. if farmers are told which irrigation technology to incorporate, they should obtain the anticipated productivity.

It is emphasized that purely logical or deductive reasoning, not empirical evidence, is used in going from theory to consequence. Assurance that the theory is true cannot be relied upon unless there is no valid alternative theory. In statistics this type of error, or the error of failing to reject an hypothesis when it is false, is called a *type II error* (affirming the consequence). In rejecting theory when the consequence is false, there is risk of making another kind of error, that of rejecting a true hypothesis. This kind of error is a *type I error* or probability statements expressed as theory. The probability of committing the type I error is called the level of significance. For each test of an hypothesis, the magnitude of the type I error is always specified.

Only in textbooks does an irrigation development project move smoothly from hypothesis to conclusion. In real life, research and production is far less orderly and requires thoughtful effort to achieve success. As in a work of art, what appears natural and simple is usually the product of hard labor. The technical application of this section is presented in examples of research projects. These examples are in Addendum 2A and include plot design and field layouts of supplemental irrigation research with statistical development, goals and objectives (hypotheses), and experimental procedures. The socio-economic application for collection and analysis of baseline data is presented in an example survey questionnaire in Addendum 15A.

A Conceptual Framework for Irrigation Technology

Activity Paradigm

The conceptual framework (activity paradigm), shown in Figure 2.3, is a classification of the level of influence exerted by decisions and activities which are schematically represented in the inverted pyramid image of "controlling

Systems Approach to Supplemental Irrigation

authority" (solid line) and its mirror reflection of "conditioning authority" (dashed line). For example, the controlling authority of the design boundaries has the most potential for change, whereas farmers possess the least. Contrarily, farmers possess the most conditioning authority and design boundaries have the least. Resource exploitation serves as the catalyst for change, i.e. the balance between the controlling and conditioning authorities. The relative power of decision making is founded upon access to skills and information associated with any position. How much bifurcation or consolidation exists within a social system together with available resources can expand or diminish the capacity and flexibility to implement modern irrigation technology.

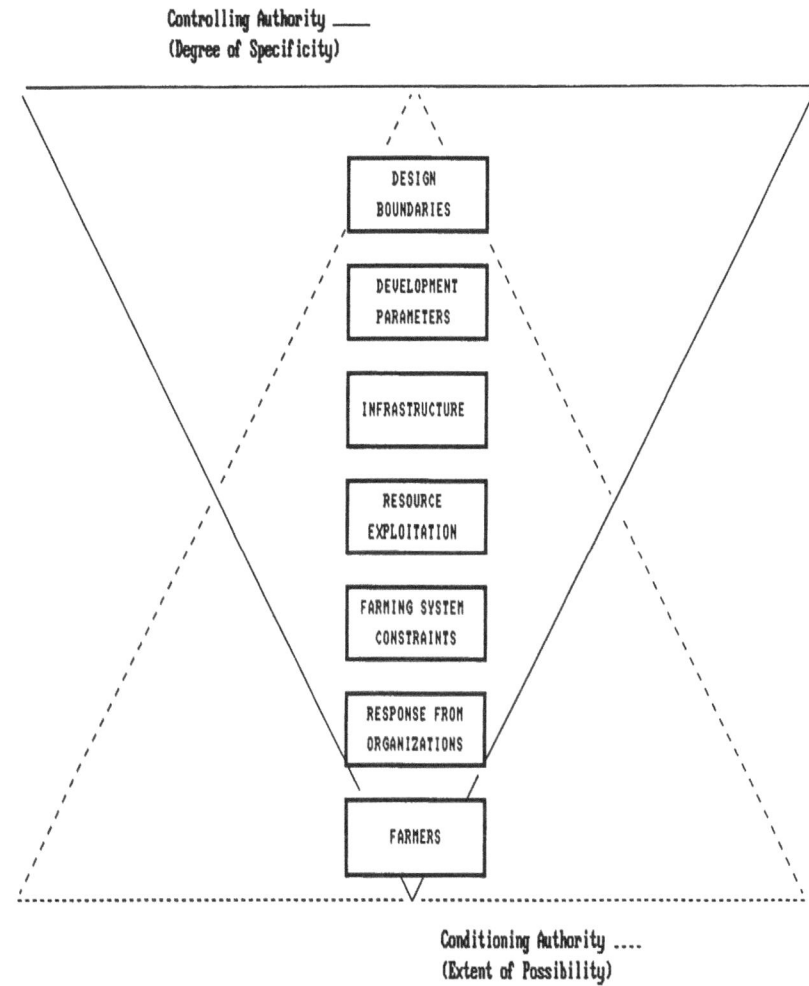

Figure 2.3. A conceptual framework for irrigation technology: a system classification.

The specificity of authority can be defined as the decisive restrictions on the capacity to execute decisions and consummate activities – the downward controlling authority of the system. The degree of specificity is determined by relative position in the triangle. Rule of law is the principal element clarifying jurisdictions for system activity. The extent of possibility of activity can be defined as ability to influence the scope of operation and the scale of achievement – the upward conditioning authority of the system. The extent of possibility is authenticated by the level of attributes or skills possessed by the incumbent, e.g. the farmer, the characteristics of the social institutions, and the complexity of organizational properties.

The design boundaries delineate the structure of authority for decision making. The vitality of the agricultural system emanates from the system's dynamic equilibrium or a balance between the controlling aspects of a country's leaders and the conditioning factors of the population, and more specifically for the agricultural system, the farmer.

The development parameters restrict or establish the vigor for improving and sustaining increased production levels. The articulation of a country's purpose and dedication to planning and development can impede and diminish or guide and accelerate the progress and implementation of irrigation technology.

The dimensions of the infrastructure indicate comparative accessibility to institutional support for achievement. Success of implementation depends on the ease with which a farmer can carry out the everyday functions of the household and concurrently execute the more sophisticated activities of modern irrigated agriculture. The contributing force derives from adequacy of public services of health, education, and welfare; provision of water, electricity, and fuel; convenient allocation of credit and supportive services (veterinarians, mechanics, etc.) for agriculture production; and sufficient distribution of consumer goods to match the modern farmer's life style.

Resource exploitation and availability of natural resources stipulates the planning contingencies of authority and is the catalyst for system balance. Resources are defined by people, not nature. For a constituent of the environment to be classified as a resource, two basic preconditions must be satisfied: first, the knowledge and technical skills must exist to allow its extraction and use; second, there must be a demand for the materials or services produced. Population and water become vital resources for modern agriculture applying irrigation technology only in the context of socio-economic circumstance, public policy goals, and institutional objectives. Human capability and social need, not mere physical presence, create their value.

The farming system constraints articulate curtailments to the enterprise of agriculture. Constraints narrow the capacity of the system and the capability of the farmer to adapt to innovation and to perform the unknown, both facets of change.

The scale of organizational response to modernization of agriculture is a principal indicator of the intention of the system. Discussion is critical for understanding alternative practices and techniques couched in irrigation technology. National organizations are counted on for help in disseminating knowledge and documentation to teach new skills and methods.

The development of irrigation technology focuses on the participation of the farmer. The farmer, in addition to personal talents and attributes, must possess (or be encouraged, through incentives and supports, to acquire) the resources, capital, capability, experience, skills, knowledge, and attitude for concretely demonstrating that modern irrigated agriculture can modify productivity for eventual food security.

Within many large institutions, credence is given to a pyramidal structure of organization with command placed at the peak of the triangle but with reliance on a broadening chain of command and delegation of requisite authority throughout the system. This occurs only if an organization is functioning dynamically, in equilibrium. Reality exhibits that the organization in agricultural production systems of the world is not in balance, almost as if the complimentary authorities of controlling and conditioning were diametrically opposed forces. The vicious cycle in agricultural production of diminishing returns and suppressed benefits is sucking the vitality of production downwards into a whirlpool of unreliable yields and inadequate food supplies.

Leadership appears indistinctly to rise through a fog in a cyclonic vacuum, spiraling ever upwards then outwards, unable to alleviate the food crisis and provide security. Controlling authorities, although recognizing the urgency of the problem, are ignoring the responsibility of delegation and participation which resides in the inverted pyramid of conditioning authority, i.e. with dynamic equilibrium, reflects a clear mirror image of controlling authority. As the spiral gains momentum, authority funnels into infinity and the horizontal plain occupied by administrators, practitioners, scientists, and farmers seeking unilateral communication with topmost authority fails to carry the message that technologies can intercede with constructive change. The organization has a controlling authority but little or no conditioning authority.

In this illustration, agricultural production is caught in the eye of the cyclone struggling to survive but continuously losing ground; resources are imploded and lose the catalytic capacity to bind the opposing forces. The vigor and intensity of the ensuing storms hold sway and the farmer, still bound to rainfed farming and a victim of the environment, is sucked into the vortex. For the farmer, the system's organization for production must resemble a children's top, a cylindrical toy spinning and spinning until it tumbles over, lacking food to fuel the effort of spinning.

An FAO paper prepared for World Food Day in 1986 disputes that world hunger results just from scarcity of food but, contrarily, is a consequence of the way in which food production and distribution is arranged at national and international levels. Furthermore, the text stresses that every individual has a "right to food" and all people have a right to food self-sufficiency. To achieve these noble sentiments, various approaches to agricultural development have been tried and most found wanting. These have ranged from "food self-sufficiency" to fulfill adequate production for local needs, to "food self-reliance" which accepts a level short of self-sufficiency but introduces the concept of encouraging non-food production where justified by a special resource advantage, to the current predominant approach of "food security". According to contemporary thinking, food security associates sufficient production of food with the ability to acquire food.

Food security, to become reality, challenges orthodox application of available resources and alternative technologies. To capture the security of economic and material access to food for a growing population tests the creativity of humankind. The conceptual framework for irrigation technology presents a pathway to potential achievement of surplus food supplies. Moreover, implementation of this technology can provide opportunities for employment, improve the health and welfare of the rural community, offer a fair remuneration to the farmer as a producer, and provide constructive exploitation of resources (especially water), and safeguard strategic reserves and environmental quality.

What the framework does assume is dynamic equilibrium of the system's components (action and activity). Controlling authority should be balanced by conditioning authority; i.e. demands established by agricultural ministries equal the farmers' attributes and the needs of communities. With balance, vertical communication can exist together with the mechanisms to mediate differences and adjust discrepancies or negotiate compromises. However, the data of today's agricultural productivity shows that the system of production is drastically out of balance and future predictions show that the conditions will get worse before they get better. Paralysis of effort is not helping; something however small, has to be achieved but must be preceded by thoughtful, well-documented planning and design.

Computer-Aided Planning and Design

The traditional practice of rainfed farming is the rudiment of inadequate production. Even with integration of technical improvements into rainfed farming, production will not fulfill population demand or, for that matter, fulfill their basic needs. Exposure to the mass media has introduced the world's population to elevated life styles and created expectations of better standards of living. This phenomenon has occurred concurrently with a demographic explosion which has stretched existing resources to the limit and beyond. To counteract the events of accelerated population growth and expectations of a better life, degradation of natural resources has happened at an alarming rate. Continued reliance on rainfed farming with unstable yields has disrupted planning and development efforts until the situation is cataclysmic, forcing concerted action to interrupt the resilience of the vicious cycle of increasing demand and decreasing productivity.

The various techniques of irrigation technology offer alternatives. Efficient but effective techniques are available to make an impact on the crisis of food

Systems Approach to Supplemental Irrigation

shortages and research data confirm the potential of increasing yields and sustaining production levels. The scope of the effort to implement this technology and the scale of development can be coordinated using the conceptual framework and logic of inquiry to monitor the status of the organization and operation of system components. Categories of information with the aid of computer support can be depended upon to assess the effort, process, efficiency, and effectiveness of concerted social change when modernizing traditional agriculture (Figure 2.4).

---

1. Define (accurately and succinctly) the problem of deficient productivity, locally and nationally.
2. Identify the essential questions that need answering to increase production.
3. Compile the methods available for solution of these questions.
4. Delineate the facts, assumptions, and constraints; include the algorithms of solution.
5. Determine the criteria for evaluating selection of alternative solutions.
6. Organize the proposed activities and actions taking advantage of the memory, speed, and precision of computer's hardware and software:
   a. provide graphic support for the planned endeavors that reflect communication for understanding to the varying background and experience of a multi-interest audience of agricultural production.
   b. prepare complete comprehensive designs but with multi-stage units of implementation efforts.
   c. determine required data and analyses as well as system attributes to realize the maximum impact on productivity derived from supplemental irrigation agriculture.
7. Generate creative thinking and judgment for evolving an optimal result for the greatest population with the least cost possible – a totally human endeavor.
8. Use the computer to experiment with alternative combinations of methods, design, and outcome.
9. Assess the selections for fulfillment of objectives (hypotheses): feasible, practical, economic, safe, and reliable; moreover, the legal and moral dimensions of attainment.
10. Proceed with final documents (plot, print, and deliver) for responsible development; also make readily available for future reference and coordination regionally.

Figure 2.4. Ten steps for integration of research methods with elements of planning and design to introduce supplemental irrigation into a modernized agriculture.

Addendum 2A: Research Plot Designs

Line-source Sprinkler System, ICARDA

For the research studies at ICARDA Center, rainbird impact sprinklers No. 30EH with a 3/16 in × 3/32 in (4.7625 mm × 2.38125 mm) nozzle operate best under the prevailing conditions of Tel Hadya, Syria; run at 55-60 psi (3.5-3.87 kg/cm^2), using 3-4 in (7.62-10.16 cm) laterals, on stands up to 1 m in height. The spacing along the lateral should be 30 ft (9.14 m) for soils with medium to low infiltration rates. Closer spacings of 25 ft or 20 ft (7.6 m or 6.1 m) provide better uniformity of water application but high application rates near the lateral are subject to runoff.

Rainbird sprinklers with the E and H combination give a linear pattern of decreasing volume from the lateral outwards to a distance of 50 ft (15.2 m). The designation of model/letter E provides a non-clog vane which increases the sprinkler radius under extremely windy conditions and the designation of H provides a superior hooded bearing which seals out sand and debris to promote longer sprinkler life with uniform application. In general, a plot area, 60 ft (18.3 m) wide, is used on each side of the lateral divided into 6 sections of 10 ft (3.05 m) each. Section 1, nearest the lateral, is wettest with the water application gradually reducing towards section 6 which is essentially dryland cultivation. Division of the area parallel to the line-source into a lesser number of sections results in too large a moisture gradient across the plots; division of the area into a greater number of sections results in moisture overlap problems among the plots.

Plot layout is parallel to the line-source for intensive studies of soil fertility properties; however, for drought screening of many varieties, planting perpendicular to the lateral is used. If localities experience similar weather conditions to those of Tel Hadya, irrigation will have to be done in the evening and early morning hours because of local wind patterns. Wind speeds of 3 km/hr shift the sprinkler pattern by 1 m and for each kilometer greater than 3 km/hr the sprinkler pattern shifts an additional meter. For accurate field plot irrigation when the wind velocity is greater than 3 km/hr the line-source must be shut down. Wind effect can quickly destroy a study if the system is permitted to operate during windy conditions.

At various distances along the irrigation lateral but perpendicular to it, neutron access tubes are placed to measure the soil moisture distribution by depth with tin cans placed to catch water for measuring the distribution of sprinkler application. Soil moisture is measured as a function of time to calculate the water balance for each of the 6 sections parallel to the lateral. The tin can data is correlated with the soil moisture data and is an indication of the irrigation efficiency of the system.

Research Designs for Field Study of Supplemental Irrigation

1. Research managed field study of wheat, 1988-89, line-source

Objectives

To improve supplemental irrigation techniques; to determine wheat varieties that respond to supplemental irrigation and levels of fertility; to estimate consumptive use; and, to determine irrigation scheduling requirements under local conditions to ensure the effective and efficient water use for increased yields and improved crop quality.

Treatments

I. Supplemental irrigation:
   1. Rainfed (no irrigation), I_0;
   2. Irrigate to replenish one-fifth (20%) of water balance requirement, I_1;
   3. Irrigate to replenish two-fifths (40%) of water balance requirement, I_2;
   4. Irrigate to replenish three-fifths (60%) of water balance requirement, I_3;
   5. Irrigate to replenish four-fifths (80%) of water balance requirement, I_4; and,
   6. Irrigate to replenish total (100%) water balance requirement, I_5.

II. Wheat varieties Cham I (Durum), V_1, and Cham IV (Bread), V_2, will be drilled at 125 kg/ha.

III. Four levels of Nitrogen (Urea): N_0 = none; N_1 = 50 kg-N/ha; N_2 = 100 kg-N/ha and, N_3 = 150 kg-N/ha. Nitrogen will be applied at the rate of 30 kg/ha at planting time and the remaining quantities of Nitrogen (20, 70 and 120 kg/ha) will be applied before tillering, near the end of February. Phosphorus (P_2O_5) will be broadcast at a rate of 100 kg/ha at sowing time.

Experimental design

A line-source sprinkler system is used. An 18.8 m wide land area on each side of the lateral is used and further divided into 6 research plots, 3.05 m wide (no parallel alleys) and 20 m long (1.0 m perpendicular alleys). Plot number 5 (nearest the lateral) is wettest with water application gradually reducing towards plot 0 which is the rainfed or dryland treatment. Water movement from surface runoff across the soil surface or diagonal to the plot area **must** be avoided: this is critical.

Plot layout is parallel to the line-source for the supplemental irrigation treatments; however, soil fertility treatments and wheat varieties are placed perpendicular to the lateral. Figure 2A.1 offers a graphic design showing the field layout of a single replication. Wind effect can quickly confound a research study if the irrigation system is permitted to operate during windy conditions, therefore, irrigation operation has to be done in the late evening and early morning hours if local wind patterns occur daily.

In the high fertility plots, F_3, and perpendicular to the irrigation lateral, neutron access tubes are placed to the 1.5 m depth to measure the soil moisture distribution by depth and tin cans are placed at the same locations to measure the distribution of sprinkler application. The soil moisture is usually measured as a function of time (10 day cycle) to calculate the water balance for each of the 6 sections parallel to the lateral. The tin can data correlated with the soil moisture data are used to measure the irrigation efficiency of the system. At each depth, 30 cm and 60 cm, 8 tensiometers are installed for calibration of the lower portion of the soil moisture desorption curve.

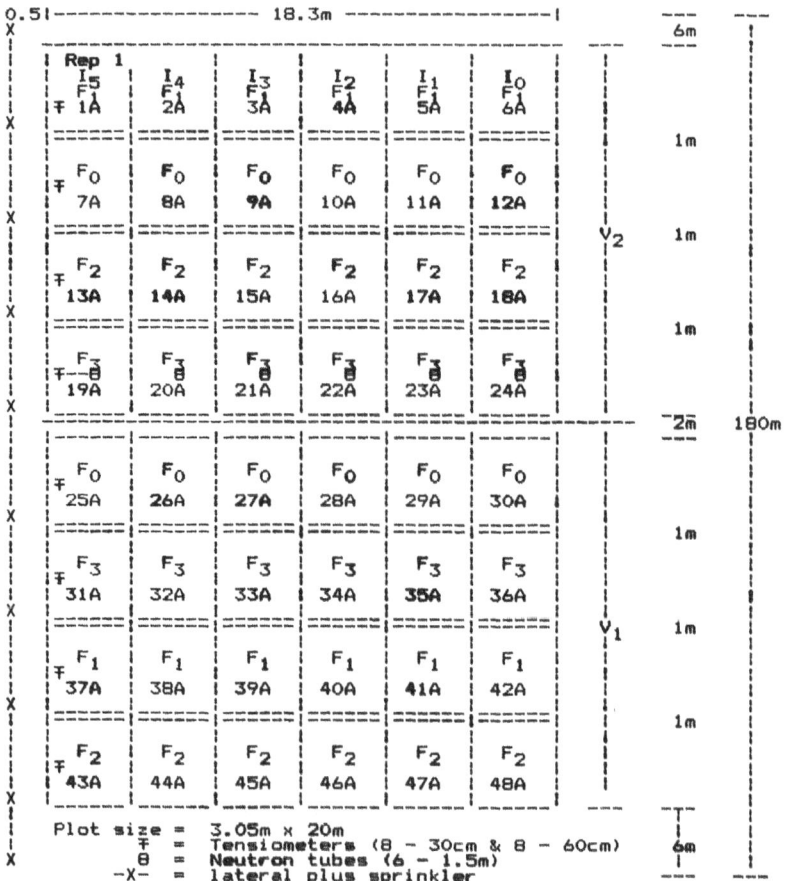

Figure 2A.1. Field plot layout for a line-source study.

Statistical design

The study uses a split-split plot design which is achieved by splitting the subplots of a split-plot design. The main plots for this study are the wheat varieties (V); subplots are the nitrogen fertilizer (N) which are randomized within blocks; and, the sub-subplots of supplemental irrigation (SI) arranged in 4 fixed and replicated blocks.

Analysis of variance (ANOVA)

Source of variance	df
Sub-sub plots	191
Subplots	95

Main plots	23
Replication	3
Wheat variety	1
Main error	19
Nitrogen levels	3
SI × V	3
Subplot error	68
Supp. Irr.	5
V × SI	5
N × SI	15
V × N × SI	15
Sub-subplot error	70

Field plot size and arrangement

A. Four replications and 192 plots will be sown to spring wheat (V_1 = Cham I and V_2 = Cham IV) by the 15th of November.
B. Plot size: 3.05 m × 20 m, requires a research area of 13,536 m² and an irrigated area of 17,136 m². A 5 ha area is required to maintain rotational requirements for multi-year studies.
C. At planting time, all plots will be sown by drilling at seeding rates of 125 kg/ha. No phosphorus (P_2O_5) will be added and nitrogen (Urea) will be applied at a rate of 30 kg/ha on all nitrogen treatments. The remaining quantities of nitrogen of 20, 70, and 120 kg/ha, will be applied before tillering near mid-February.
D. If rainfall is inadequate for germination, an irrigation of 20 mm will be applied using half-circle sprinklers on the supplemental irrigation treatments. *No* irrigation will be applied to the rainfed plots.
E. Irrigation scheduling will follow the water balance method with daily measurements of rainfall and pan evaporation as well as calculations using the Penman technique backed-up by the neutron soil moisture method.

Methodology

A. Use a duckfoot harrow throughout the experimental area and apply pre-emergence weedicide to the field by 1 November.
B. Drill seed and broadcast 30 kg/ha of N *only* on nitrogen treatment plots by 15 November.
C. On 15 November, if rainfall has been inadequate for germination, apply 20 mm of water *only* on the supplemental irrigation treatments.
D. Mark plots to set precise plot dimensions and install line-source laterals to remain until June.
E. Install neutron tubes and catchment cans in the high fertility treatments for

one variety of wheat (24 tubes and cans), paint stakes white, and tag stakes for plot number and treatments. Install tensiometers at the 30 and 60 cm depths in the high fertility treatment of the supplemental irrigation plot, I_s. Place large sign at edge of plot area describing study and cooperators.
F. Add remainder of nitrogen treatment as a top dressing at tillering which occurs during mid-February.

Data to collect

A. Enter all dates of operations into study ledger (daily monitoring of data entry required).
B. Enter daily values of rainfall, evaporation, and max/min temperature into the study ledger and calculate the water balance starting mid-November. Use estimated K_c and root depth from previous years and make corrections from neutron data.
C. Take soil samples on a field basis at intervals of 15 cm to a depth of 1.2 m (or depth of rock layer) for the following laboratory measurements:
   1. determine field capacity, one-bar, and permanent wilting point;
   2. determine Nitrates and Phosphates in total and available forms;
D. On a field basis, take soil bulk density samples at intervals of 15 cm intervals to a depth of 1.2 m.
E. Determine soil moisture percentage by depth on a ten-day cycle and weekly starting in April using the neutron apparatus at 15 cm intervals to a depth of 1.2 m and soil sample for surface moisture content for the supplemental irrigation treatments.
F. Record, service and calibrate tensiometers from 30 and 60 cm depths weekly. To calibrate, take 15 cm soil samples with an Oak Field type sampler starting at 15 cm distance from the tensiometer and with 7.5 cm sampled on top and bottom of the midpoint of the porous cup.
G. Determine infiltration rate (l/sec) using double ring infiltrometers.
H. In December, estimate the percentage of germination and plant population per plot.
I. Measure plant height for each plot monthly starting 1 December and every two weeks starting 1 March until maturity. Record dates of measurement.
J. Check and control for weeds throughout season, especially in spring.
K. Register dates for specific stages of plant growth for each plot: germination; tillering; jointing; booting; heading; flowering; and, harvesting.
L. Select areas within each plot to be harvested with the plot combine taking 2 passes through each plot (44 m²/plot) for grain yield and take subsamples for straw, head length, and number of heads/m².
M. Do the following measurements on the samples from each plot which have been harvested by a combine:
   1. determine total weight of grain;
   2. measure total grain weight/1000 grains (counter);
   3. measure moisture content of grain (moisture meter);

Systems Approach to Supplemental Irrigation 35

    4. take sample fractions of grain for chemical analysis.
N. Do the following measurements on subsamples from each plot:
    1. separate and count total number of heads;
    2. count number of bad and empty heads;
    3. count number of diseased heads and indicate type of disease;
    4. determine total weight of heads and straw;
    5. take sample fractions of straw for moisture content.

2. Split block design, field study of Durum wheat

Objectives

To improve supplemental irrigation practices; to predict level of nitrogen application; to estimate the consumptive use of water; and, to determine irrigation scheduling requirements under local conditions to insure the effective and efficient use of water and fertilizer for increased production of crop yields.

Treatments

I. Supplemental irrigation using the water balance:
    1. Rainfed (no irrigation).
    2. Irrigate to replenish one-third of water balance requirement.
    3. Irrigate to replenish two-thirds of water balance requirement.
    4. Irrigate to replenish total water balance requirement.

II. Levels of Nitrogen: none, 70, 140, 210 kg of N/ha.

Figure 2A.2 is a single replication of a split-block design with four irrigation treatments (I) and four nitrogen treatments (N).

I_3	I_2	I_1	I_4
N_2	N_2	N_2	N_2
N_3	N_3	N_3	N_3
N_0	N_0	N_0	N_0
N_1	N_1	N_1	N_1

Figure 2A.2. Example of split-block research design.

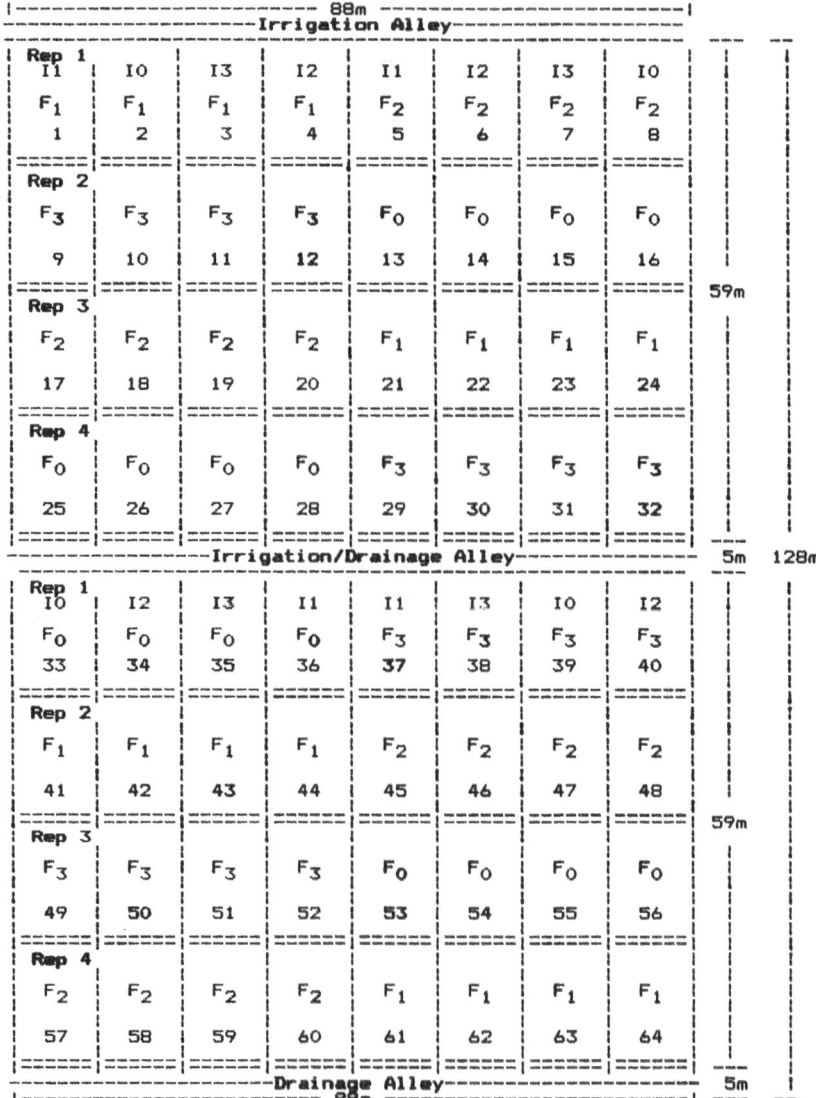

Figure 2A.3. Split block plot layout for supplemental irrigation.

Field plot size and arrangement

A. Four replications and 64 plots are planted to Durum wheat variety, Cham I.
B. Plot size is 10 m × 14 m, with 1 m alleys and 2 m borders requiring an area of 11,264 m² and an irrigated area of 8,960 m².
C. At planting time, all plots will be sown by broadcasting wheat at a seeding rate of 140 kg/ha (if a seed drill is used the rate is decreased to 120 kg/ha).

Phosphorus (P_2O_5) will be banded or broadcast at a rate of 80 kg/ha, and nitrogen (Urea) will be applied at a rate of 40 kg/ha on all nitrogen treatments. The remaining quantities of nitrogen (30, 100, and 170 kg/ha) are applied before tillering (mid-February).

D. Immediately following sowing and fertilizing, if rainfall is inadequate, an irrigation for germination will be applied covering the total plot area. The water balance method will be used to determine the irrigation scheduling (see Chapter 3).

Statistical design

The study is a split-block design (a variation of the split plot design) which examines the effect of the rate of nitrogen fertilizer on yield of Durum wheat for different schedules of supplemental irrigation. In the split-block arrangement, the block of plots with the nitrogen treatments is split so that each treatment of supplemental irrigation occurs in a continuous strip across the block (replication). An independent randomization of the treatments of supplemental irrigation is made for each block of nitrogen plots.

The advantage of the layout is the possibility of greater precision in estimation of the interaction of nitrogen rates on scheduling of supplemental irrigation. Figure 2A.3 shows the field layout for this split-block design. The main plots are 4 nitrogen fertilizer rates arranged in a 4 × 4 Latin square. Subplot treatments are 4 schedules of supplemental irrigation. The supplemental irrigation strips are re-randomized for each column of main plots. Irrigation operations are easier when the plots for the supplemental irrigation treatment form a continuous column. However, this arrangement requires the calculation of a separate error term to test for the main effect of supplemental irrigation levels.

Analysis of variance

Source of variance	df
Subplots	63
Main plots	15
Rows	3
Columns	3
Nitrogen rates	3
Main error	6
Supp. Irr.	3
Error b	9
N rates × Supp. Irr.	9
Error c	27

Methodology

(As for Line-source example)

Data to Collect

(As for Line-source example)

Chapter 3

Water Balance Calculations

Immanuel Kant, the famous German Philosopher, wrote that all scientific observations must be shaped by two dimensions of reason: space and time. The domain of time implies an examination of the process of events within a system just as the domain of space involves examination of the magnitude of design. Some necessary linkage between these dimensions must be assumed by the inquiring scientist to sustain a natural order for agricultural development.

In the Near East and North Africa the demand for water by agriculture, industry, and urbanization is steadily increasing. This necessitates creative development of water supply schemes at all levels of management; i.e. national, provincial, district, and user level. Because weather conditions vary considerably from year to year and also local hydrological conditions are different, no standard procedures for manipulating water supplies are apparent and, in practice, control is done by rough rule-of-thumb guidelines. To induce an increase in agricultural income, farmers will have to invest time and capital and pay the costs of operation and maintenance of supplemental irrigation systems; consequently, the risk of investment must be estimated before an irrigation system is selected and acquired.

The stochastic nature of rainfall in semi-arid and arid regions compels the setting of manageable boundaries to permit crop production to the limit of existing natural resources. The conceptual understanding of water balance methods applied to supplemental irrigation incorporates the domain of time to identify when to irrigate and the domain of space to indicate how much irrigation is required. Relative to space and time, supplemental irrigation is that component of irrigation technology which harnesses the domain of time to restrain the effects of stochasticity for management of agricultural production at deterministic levels. Scheduling of supplemental irrigation encompasses time in union with limited space through deterministic management of natural resources. The required volume of water for supplemental irrigation diminishes space and amplifies time to concentrate natural resources for crop production.

E.R. Perrier and A.B. Salkini (eds), Supplemental Irrigation in the Near East and North Africa,
39-77.
© 1991 *ICARDA.*

Precipitation

The terms, liquid precipitation and rainfall, include drizzle as well as rain. The term, solid precipitation, includes all solid forms, for example, snow, ice pellets, hail, and ice crystals. Precipitation is one of those variables where statistical evaluation rarely exists: spatial distribution is infrequently known. Measurements are made of the vertical depth of water that falls to the soil surface during a known period of time, usually a 24 hr interval between daily precipitation observations. Precipitation is measured with a raingage which is an open-mouthed cylinder with vertical sides. The exposure of a raingage is important for measurement accuracy of precipitation, especially for snowmelt measurements. The ideal exposure would eliminate all turbulence and eddy currents near the gage that tend to carry away the precipitation. The loss of precipitation in this manner increases with wind velocity.

Consider a raingage as shown in Figure 3.1 with an orifice or opening of 20 cm and an area of 314.16 cm². In semi-arid regions this raingage may estimate the rainfall for a 400 km² area. The measured rainfall implies that the raingage represents only 0.000000007856% of the area:

$$(100.0 \times 314.16 \text{ cm}^2 / 4,000,000,000,000.0 \text{ cm}^2 = 7.856 \times 10^{-9}\%).$$

This means that the area measured by the raingage is less than a dot when compared to the total area represented by the raingage. Nonetheless, complete confidence is placed in the magnitude of the recorded rainfall event.

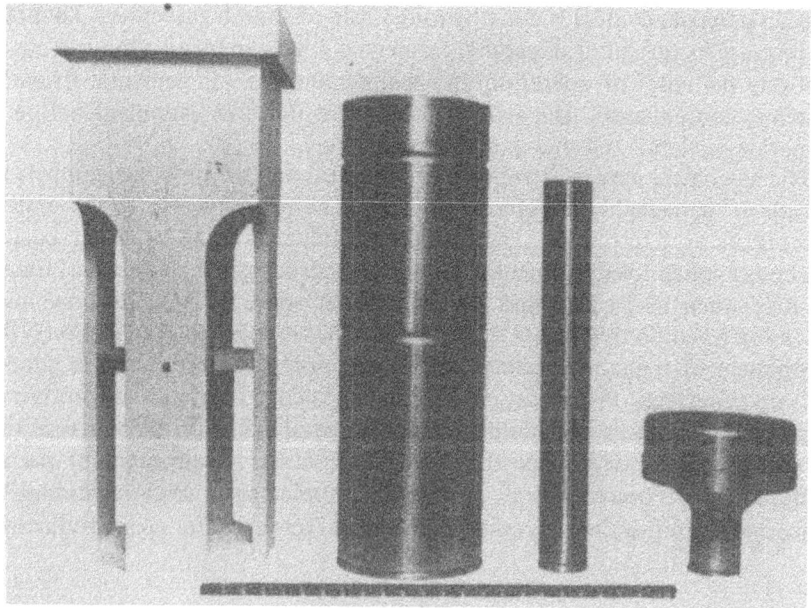

Figure 3.1. Nonrecording 20 cm raingage showing the wooden box used as a support for the overflow can, the overflow can, the measuring tube, the funnel, and the measuring stick.

Water Balance Calculations

When solid precipitation falls, wind shields may be used to help minimize loss in precipitation catch. Wind effects on catch losses are much greater during snowfall than rainfall. For continuous measurement, the tipping bucket raingage is most commonly used for digital recording of rain. This type uses a simple mechanism and requires pulse counting capability in the recording equipment. Still, the measurement of solid precipitation remains a problem for the tipping bucket raingage because of an inaccurate catch. Some systems are heated to reduce the measurement error of solid precipitation.

The exact location of the raingage is an important consideration in precipitation measurement and data evaluation. Of primary importance in processing the data is the tabulation of precipitation at regular intervals. For raingages, this activity should be done daily with time of observation noted. When an observer is available regularly at a site, the times of occurrence of snowfall and hail should be noted so that more accurate use of the data can be made.

Effective Rainfall

Effective rainfall (Table 3.1) is rain which infiltrates into the ground. Usually it is given as an annual value and should include both winter and summer effective rain. Several methods estimate effectiveness of precipitation but the simple method shown below is available from the U.S. Bureau of Reclamation, WMO records, and employs monthly estimates for determining effective rainfall during the growing season only for cultivated fields.

Table 3.1. Effective rainfall estimates for the growing season

Precipitation (mm)		
Monthly	Effective	
	Northern hemisphere	Mediterranean
20	17.9	17.9
40	36.0	36.0
60	52.1	52.1
80	65.6	68.0
100	75.9	83.5
120	82.3	95.2
140	86.0	105.0
> 140	86.2	107.2

In arid and semi-arid regions under conditions of winter fallow, the estimates presented in Table 3.2 are better suited.

Table 3.2. Effective monthly rainfall estimates for winter (fallow) conditions.

Precipitation (mm)	
Monthly	Effective
20	8.0
40	23.3
60	39.6
80	56.0
100	72.0
120	80.2
≥ 140	84.0

Evaporation

Equipment normally used to measure evaporation includes the following items:
a. an evaporation pan (Class A);
b. a fixed point gage and appropriate measuring tube; or a micrometer hook gage and stilling well;
c. a suitable cover to protect the pan from animals and birds; and,
d. a water-storage tank (when necessary) to provide a reserve supply of water for the pan.

The evaporation pan shown in Figure 3.2 is of cylindrical design, 25.4 cm deep and 120.7 cm in diameter (inside dimensions). These dimensions eliminate the need for seams across the bottom of the pan which should be constructed of monel metal but galvanized sheet metal works satisfactorily. The platform support or pallet should be constructed of lumber, 5 cm × 10 cm or heavier. Rot resistant lumber, or lumber treated with creosote or other effective wood preservative should be used.

The evaporation pan is usually located on non-grassed sites which are surrounded by a short crop or bare non-cultivated area to provide standard measurements. However, established weather stations, monitored daily, give the most reliable data. The pan should be centered on the platform. The soil surface should be filled sufficiently to level the support and keep the bottom of the pan above surface water in rainy weather. To anchor in position, earth fill should be placed around the platform support and tamped firmly between the top members to within 1 cm of the top of the support leaving an air space between the bottom of the pan and the earthen fill to simplify inspection of the pan for leaks during use. During inspections the level of the platform should be checked and corrected, if necessary.

During monthly maintenance inspection, the evaporation pan should be checked carefully for leaks since any leaks will render the measurements valueless. The date on which the leak was discovered and the date of the leak's repair or pan replacement must be recorded. Since the heat characteristics of all

Water Balance Calculations

Figure 3.2. Evaporation pan installed on wooden support.

evaporation pans must be identical, they should be painted white or aluminum and should be scrubbed and repainted annually.

The pan must be cleaned as often as necessary to keep it free from sediment, scum, and oil films. An oil film will significantly reduce the rate of evaporation. The pan should be emptied by siphoning or by dipping the water out. Under no circumstances should the pan be lifted and emptied with any significant amount of liquid remaining in the pan. The growth of algae in the pan can be discouraged by the addition of small amounts of copper sulphate to the water. Algae already present must be removed by a thorough cleaning of the pan.

The fixed-point gage has two knife-edged blades attached to the edge of the stilling well. Generally, the stilling well is fixed to the side of the evaporation pan so that any movement does not disturb the gage. The stilling well should be cleaned occasionally to remove any sediment.

To read the fixed-point gage, either add water equivalent to that lost by evaporation or remove the water of precipitation from the pan until the top of the fixed points coincides with the water surface. In either case, a graduated measuring cup should be used to record the amount added or removed. Reflection of the sky in the water helps in determining where the point first breaks through the surface. If only the pan's surface freezes during periods of frost or freezing, then chipping and cracking the ice for removal from the stilling well and placing it in the center of the pan does not disturb the reading. The mass of the ice and not its volume displacement governs the reading at the fixed points in the stilling well.

Piche evaporometers (atmometers) can be used to measure evaporation. They are essentially a test tube filled with water, inverted with a blotter over the end, and installed in a conventional weather shelter with other weather instruments. Like evaporation pans, the Piche is a passive device that loses water in some way related to the absorption of radiant energy, wind strength, and dryness of the ambient air. Evaporometers will break if freezing occurs. Because of the small size of Piche units, rates of evaporation are in excess of the rates of water use by crops and the values must be adjusted for comparability to other evaporation data.

In addition, the potential evaporation can be calculated from Ritchie's reduced form of Penman's Equation (see Addendum 3A):

$$ET_o = \frac{1.28 \times \Delta \times H_o}{\Delta + \varrho};$$

where ET_o is the potential evaporation; Δ, mb/°C, is the slope of the saturation vapor pressure curve at the mean air temperature; H_o is the net solar radiation, cal/cm²/day; and, $\varrho = 0.59$ mb/°C is a psychrometric constant. Then:

$$\Delta = \frac{5304}{(T + 273)^2} \; e^{(21.255 - 5304/T + 273)}$$

where T is the daily temperature in degrees Celcius and:

$$H_o = \frac{(1 - \alpha)(R)}{2.44}$$

where R is the daily solar radiation, cal/cm²/day, and, in this instance, $\alpha = 0.23$ for the solar radiation.

Evapotranspiration

The evapotranspiration process requires energy to convert water from the liquid to the vapor phase. At 100°C the latent heat of vaporization is 540 cal/gm, the energy requirement. At air temperatures up to 40°C:

$$\lambda = 597.3 - 0.564 \times T_a$$

where T_a = mean air temperature (°C).

Irrigation planners rely on estimates of E_t when no direct measurements have been made from lysimeters or soil moisture measurements using gravimetric, neutron probes, or tensiometers. A common method for estimating E_t utilizes weather data for calculating energy converted to E_t.

Supplemental Irrigation Methods

Two important questions for supplemental irrigation must be satisfied to manage the large variation in seasonal water use by plants at the stages of development especially when they are sensitive to soil moisture stress:

1. When should supplemental irrigation be applied to give the highest return? Traditional thinking considers irrigation when plant available water in the root zone is below 50%. When to schedule this additional water can be estimated by calculating the crop water balance and examining the stage of plant growth. Investigation should compare the indigenous farmers' method of adding one or two light supplemental irrigations near the end of each season and verify these local methods with alternatives which have been tested experimentally to determine recommendations for scheduling.
2. How much water should be applied when plant available water is below 50%? This question is best answered by analyzing historical climatic data from the region. These data can be analyzed using the water balance method which computes the number and amount of irrigations required. Historical data from the Near East and North Africa show that, in most cases, 30 mm of water applied when the available water is below 50% would avoid a yield depression in cereals. If, however, the end of season were particularly droughty, then a second application of 30 mm could deliver enough water to avoid a decrease in yield.

For example, at Aleppo, Syria, the number of essential supplemental irrigations can vary from 0-5, depending upon the variability of seasonal rainfall; however, most seasons require only 1-2 irrigations to avoid a depression in yield of wheat. When wheat approaches the maturation stage of growth in the region, rainfall ceases and soil moisture declines; therefore small quantities of supplemental irrigation could be applied.

Water Balance

Agronomists and agricultural engineers concentrate on measurements of soil water availability to schedule irrigation and tend to disregard growth characteristics of plants. Scientists who emphasize economic concepts for scheduling may focus on plant indicators to schedule irrigation at critical growth stages where "limited water or deficit irrigation" become primary techniques for scheduling. The integration of soil water availability and plant indicators with climatic effects into a water balance method allows effective irrigation scheduling which maximizes economic production.

The crop water requirement to achieve optimal production refers to the timing and quantity of water needed to replace moisture used by a crop growing under specific environmental conditions. The water balance method is calculated to determine the crop water requirement under local conditions to ensure the efficient use of water with supplemental irrigation. When applying the water balance method in a predictive mode (before actual measurements have been made) there are three coefficients which must be estimated to predict soil moisture deficits: evaporation pan coefficient, K_p, crop coefficient, K_c, and active root depth, ARD. Various measures of the status of moisture in the soil profile can effectively estimate the water balance without the measurement of climatic parameters. Tensiometers which measure soil moisture suction (tension) in the soil between 0 and 1 bar can be used to estimate water requirements of plants. Neutron probe devices can also be used to measure the volume moisture content in the soil profile; and, in conjunction with the soil bulk density and the soil moisture desorption curve, the same soil moisture desorption value at 1.0 bar can be estimated to determine the crop water requirement and when to irrigate. In general, climatic methods for predicting the water balance are used instead of these measures because of the time required for obtaining and analyzing data from field measurements using soil-moisture samplers, tensiometers, lysimeters and for calibration of equipment such as gypsum blocks, and neutron probe apparatus.

When possible, and for rational allocation of limited water supplies, supplemental irrigation should be scheduled at the moisture sensitive stages of plant growth. For example, for rainfed cereals in the region the 3 most sensitive growth stages for supplemental irrigation are:
1. at sowing time, near mid-November;
2. at tillering, from mid-February to mid-March; and,
3. at heading, from mid-March to mid-April.

Irrigation is discontinued when cereals reach the end of the soft dough stage. Scheduling of supplemental irrigation by water balance calculations should coincide with these sensitive periods to make certain that root zone moisture does not limit growth.

A water balance for scheduling irrigation of a field crop is calculated by estimating or measuring the major input and output components of water movement on a field area. Rainfall and water quantities are expressed by depth of water and it is convenient to express the water balance in similar terms, i.e. in millimeters (mm). The water balance equation can be written as:

$$R + I = ET + RO + S,$$

where R = rainfall on a field (mm);
I = water added by irrigation (mm);
ET = evapotranspiration (mm);
RO = runoff (mm); and,
S = soil-water storage (mm).

Simple calculations estimate the water requirement and time of irrigation for a particular crop. The flow chart in Figure 3.3 describes the data input and calculations needed to determine plant water requirements with time.

To illustrate the computation of the water balance method, 1984-85 climatic data for Aleppo, Syria, are used: inputs are daily rainfall, pan evaporation, and soil and plant growth characteristics. The field selected for illustration, has an

Water Balance Calculations

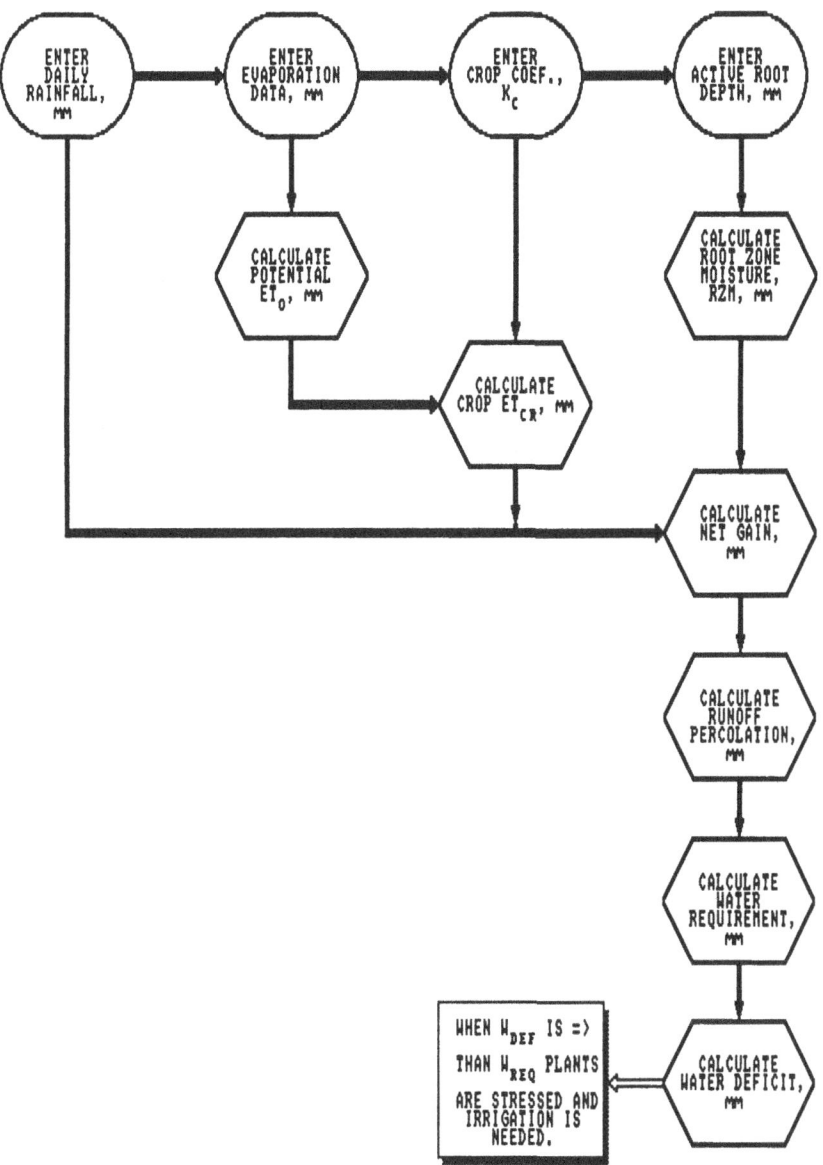

Figure 3.3. Water balance flow chart.

expanding clay soil 1.05 m in depth, a clay content of 70%, a bulk density of 1.01 g/cm³, and an infiltration rate of 8.5 mm/hr. Table 3.3 presents the computations of water balance for 11 days following seedling emergence.

The computations for the variables are ordered as indicated in Table 3.3:
1. **Daily rainfall, rain (mm):** Daily rainfall is measured using standard raingages which are monitored daily at 0800 hr.

Table 3.3. Example worksheet for calculation of supplemental irrigation scheduling and water quantities from rainfall and pan evaporation data starting at seedling emergence, 6 December, 1984, Aleppo, Syria.

Date	6	7	8	9	10	11	12	13	14	15	16
Rain							8.7	4.6			8.8
E_{pan}	1.4	1.0	0.7	1.5	1.4	1.1	1.0	0.6	1.2	0.7	0.2
ET_o	0.98	0.70	0.49	1.05	0.98	0.77	0.70	0.42	0.84	0.49	0.14
K_c	0.35	0.35	0.35	0.35	0.35	0.35	0.35	0.35	0.35	0.35	0.35
ET_{cr}	0.34	0.24	0.17	0.36	0.34	0.26	0.24	0.14	0.29	0.17	0.04
RD	150	150	150	150	150	150	150	150	150	150	150
RZM	28.64	28.64	28.64	28.64	28.64	28.64	28.64	28.64	28.64	28.64	28.64
WB	28.64	28.29	28.04	27.87	27.50	27.16	26.89	28.63	28.63	28.34	28.16
Net Gain	28.29	28.04	27.87	27.50	27.16	26.89	35.35	33.08	28.34	28.16	36.92
Perc/Runoff							6.71	4.45			8.28
W_{req}	14.32	14.32	14.32	14.32	14.32	14.32	14.32	14.32	14.32	14.32	14.32
WD	0.34	0.58	0.75	1.12	1.47	1.73			0.29	0.46	
I_{Appl}											

2. **Evaporation data, E_{pan} (mm):**
3. **Potential evapotranspiration, ET_o (mm):** The potential evapotranspiration, ET_o, is the maximum quantity of water that can be evaporated by a uniform cover of dense short grass when the water supply to the plants is not limited. The pan coefficient, K_p, which is estimated for each location, is multiplied by the pan evaporation data, E_{pan}, to obtain potential evapotranspiration. The pan coefficient is determined by direct measurement of the potential evapotranspiration at the site of the evaporation pan by use of a lysimeter or by calculation using Penman's Equation. K_p is affected by different groundcovers, relative humidity, and wind. For the Aleppo example, the universal pan coefficient of $K_p = 0.7$ was used throughout the growing season for wheat; therefore,

$$ET_o = 0.7 \times E_{pan}.$$

Potential evapotranspiration cannot exceed free water evaporation under the same weather conditions. As the evaporation pan coefficient, K_p, is a constant and infrequently known or only estimated, it is not required if the crop coefficient curve has been developed for the region of concern. The crop coefficient, K_c, is simply multiplied by the constant (if previously verified) and this term is ignored in the new water balance. The actual crop evapotranspiration,

$$Et_{cr} = K_c \times K_p \times E_{pan}.$$

If the crop coefficient curve is being developed for a region, then the pan coefficient is simply ignored and the variance accrues in the crop coefficient curve. This saves one step and greatly simplifies the concepts necessary for technology transfer of the water balance method to technicians and farmers.

Water Balance Calculations

If a Piche evaporometer is used, the formula by Espinar to calculate daily potential evaporation, ET_o, for an irrigated grassed area is:

$$ET_o = \frac{(T_{min} + T_{max} - 36)}{3218} \times D_j \times (D_j - 5) \times E_p^{1/3}$$

where T = daily temperature (°C);
D_j = maximum sunshine duration (hr/day); and,
E_p = daily evaporation (mm/day) with Piche.

To assist in using the Piche equation, data is presented showing the maximum possible hours of sunshine per day on a monthly basis and for different latitudes. The data in Table 3.4 is the same for the southern latitudes but the months must be rearranged.

Table 3.4. Duration of maximum possible hours of sunshine per day on a monthly basis and for different latitudes.

Lat.	Jan	Feb	Mar	Apr	May	Jun	Jul	Aug	Sep	Oct	Nov	Dec
50	8.5	10.1	11.8	13.8	15.4	16.3	15.9	14.5	12.7	10.8	9.1	8.1
48	8.8	10.2	11.8	13.6	15.2	16.0	15.6	14.3	12.6	10.9	9.3	8.3
46	9.1	10.4	11.9	13.5	14.9	15.7	15.4	14.2	12.6	10.9	9.5	8.7
44	9.3	10.5	11.9	13.4	14.7	15.4	15.2	14.0	12.6	11.0	9.7	8.9
42	9.4	10.6	11.9	13.4	14.6	15.2	14.9	13.9	12.6	11.1	9.8	9.1
40	9.6	10.7	11.9	13.3	14.4	15.0	14.7	13.7	12.5	11.2	10.0	9.3
35	10.1	11.0	11.9	13.1	14.0	14.5	14.3	13.5	12.4	11.3	10.3	9.8
30	10.4	11.1	12.0	12.9	13.6	14.0	13.9	13.2	12.4	11.5	10.6	10.2
25	10.7	11.3	12.0	12.7	13.3	13.7	13.5	13.0	12.3	11.6	10.9	10.6
20	11.0	11.5	12.0	12.6	13.1	13.3	13.2	12.8	12.3	11.7	11.2	10.9
15	11.3	11.6	12.0	12.5	12.8	13.0	12.9	12.6	12.2	11.8	11.4	11.2
10	11.6	11.8	12.0	12.3	12.6	12.7	12.6	12.4	12.1	11.8	11.6	11.5
5	11.8	11.9	12.0	12.2	12.3	12.4	12.3	12.3	12.1	12.0	11.9	11.8
0	12.1	12.1	12.1	12.1	12.1	12.1	12.1	12.1	12.1	12.1	12.1	12.1

4. **Crop coefficient, K_c:** The value of K_c, determined for each cultivated area, is the ratio of the actual crop evapotranspiration, ET_{cr}, to the potential evapotranspiration, ET_o, which is related to various stages of plant growth and can include a K_p coefficient. K_c is affected by the method of determining ET_o as well as site specific factors such as crop characteristics, sowing date, plant development, length of growing season, and climate. During the growing season, K_c can be adjusted by taking consecutive soil moisture samples to measure ET_{cr} and back-calculating to adjust estimated K_c values. Figure 3.4 illustrates a crop coefficient for wheat (which was not under soil moisture stress) sown on 6 December and harvested at the end of June.

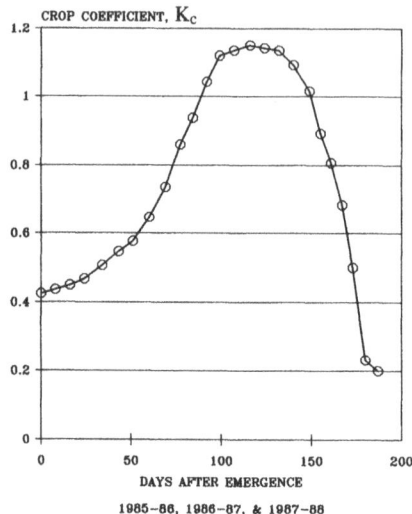

Figure 3.4. Crop coefficient, K_c, for wheat sown on 6 December and harvested at the end of June.

5. **Crop evapotranspiration, ET_{cr} (mm):** The actual amount of water used by the crop, ET_{cr}, can be measured directly or can be calculated using the potential evapotranspiration, ET_o, and a crop coefficient, K_c, where:

$$ET_{cr} = K_c \times ET_o.$$

However, with this equation, at the start of the season when there are no plants (ARD = seed depth, variable 6, and RZM = 0, variable 7) then ET_{cr} must be computed using K_c without a crop, e.g. $K_c = 0.4$ for Aleppo during this early period.

6. **Active root depth, ARD (mm):** The active root depth in millimeters or effective depth of water use as a function of time can be determined by collecting root samples in the soil profile, estimating from plant height measurements and other plant characteristics, or measuring moisture desorption patterns in the soil profile. The active root depth as a function of time for wheat at Aleppo is presented in Figure 3.5.

7. **Root zone moisture at field capacity, RZM (mm):** The percent available water that a soil will hold is estimated by the difference between the percent field capacity and the percent wilting point on a dry weight basis (% Available Water = % Field Capacity – % Wilting Point). For the Aleppo clay soil (70% clay, 15% silt, and 15% sand), the difference between field capacity (44.6%), and wilting point (25.7%), gives the available water equal to 18.9%. When the soil profile is at field capacity, the total available moisture, TA, (mm/m) in a soil of 1.0 m depth is found by multiplying the percent available moisture by the apparent specific gravity (note: soil bulk density, BD, in units of g/cm^3 divided by the density of water which is 1.00

Water Balance Calculations

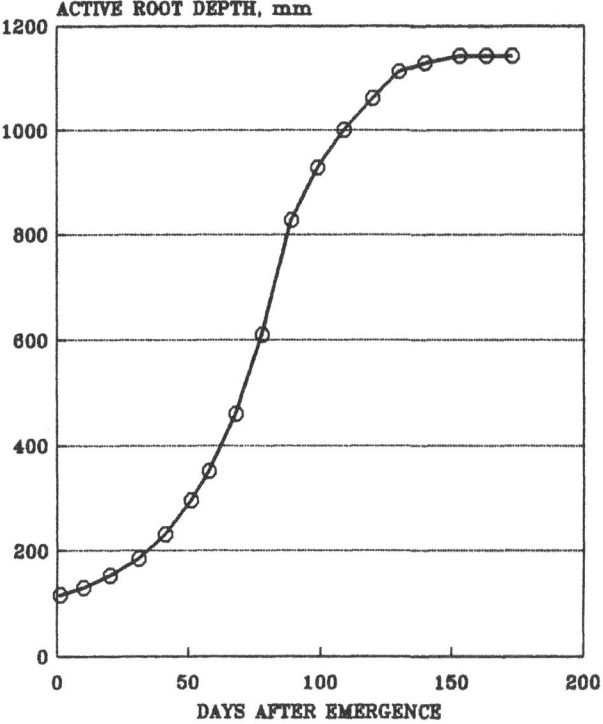

Figure 3.5. Active root depth (mm) as a function of time for spring wheat at Aleppo, Syria.

g/cm³ gives the dimensionless apparent specific gravity for soil). For convenience, TA is written as:

TA = BD × % Available moisture × 1,000 mm/m.

The unit of 1,000 mm/m, is to correct the values to millimeters, and percentage values are divided by 100. For the Aleppo clay soil:

TA = 1.01 × 18.9/100 × 1,000 mm/m = 190.9 mm/m.

If the water balance is to be calculated prior to emergence or if the soil profile is not at field capacity at sowing time, then TA must be determined by direct measurement of soil moisture to the depth of the soil profile or expected root depth. The total available moisture in the root zone, RZM, is then given by TA multiplied by the active root depth, ARD:

RZM mm = ARD mm/1,000 mm/m × TA mm/m.

For wheat at Aleppo, the total available water in the root zone of the soil profile at emergence (root depth equal to seeding depth) which must be available for optimal crop growth is given by:

RZM = 150 mm/1,000 mm/m × 190.9 mm/m = 28.64 mm.

8. **Water balance, WB (mm):** The daily amount of available moisture in the root zone is estimated by the water balance, WB, which can be an indication of plant water stress. At the start of the water balance computations, WB = RZM; but thereafter, WB is equal to the previous daily value for Net Gain (variable 9). Table 3.3 displays the calculations from the time of plant emergence. As calculations continue, Net Gain may exceed RZM if rainfall is high; the difference between the two values, Net Gain − RZM, is surface runoff or deep percolation. Then, WB becomes the previous value of RZM. For the Aleppo example, the first value of WB = 28.64 mm at seedling emergence. Table 3.3 indicates the effect of light rains on the water balance as apparent by computed percolation/runoff values. Although December does not have large ET_{cr} values, the process of calculating the water balance and the potential water deficit in the soil profile can be followed easily. Verification of these calculations can be made by measuring the moisture in the soil profile as a function of time to the estimated depth of active root development. These values can be used to adjust the coefficients used in the water balance calculations of rainfall, pan evaporation, or Piche evaporation data.

9. **Net Gain (mm):** The Net Gain is computed from the daily value of water balance plus rainfall minus ET_{cr}. Net gain is computed as:

   Net Gain = WB + Rain + I_{Appl} − ET_{cr}.

   For Aleppo on 6 December 1984, the Net Gain at emergence was computed as:

   Net Gain = 28.64 − 0.34 = 28.30 mm.

10. **Deep percolation or surface runoff, Perc/Runoff (mm):** The daily amount of water lost to the plant growth system is computed from the difference between the Net Gain and RZM:

    Perc/Runoff = Net Gain − RZM.

    For the Aleppo example on 12 December 1984:

    Perc/Runoff = 35.35 mm − 28.64 mm = 6.71 mm.

11. **Water requirement, W_{req} (mm):** The water requirement is determined from the amount of available water permitting unrestricted evapotranspiration, i.e. the plant is not under soil moisture stress. On most soils, when the moisture in the soil profile has been reduced to at least 50% of the available water (soil moisture suction exceeds 1-2 bars), plants begin to show stress; therefore, RZM is multiplied by 0.5 to estimate the daily value of W_{req} for the season,

    W_{req} = 0.5 × RZM.

    For the Aleppo example, W_{req} = 14.32 mm at seedling emergence.

Water Balance Calculations

12. **Water deficit, W_{Def} (mm):** The amount of water needed to replenish soil moisture used by crop evapotranspiration is the difference between RZM and Net Gain for each day:

    W_{Def} = RZM − Net Gain.

    If Net Gain is greater than RZM, then W_{Def} = 0. When W_{Def} is equal to or greater than W_{req}, the plants are experiencing stress and irrigation is needed, i.e. irrigation should be applied when $W_{Def} = W_{req}$. For the Aleppo example on 14 December 1984:

    WD = 28.63 mm − 28.34 mm = 0.29 mm,

    which is much less than 14.32 mm and no irrigation is required.

13. **Irrigation to be applied, I_{Appl} (mm):** The irrigation applied, I_{Appl}, is the net amount of water applied before correcting for the efficiency of the irrigation system. For the Aleppo example, the soil profile was at field capacity at seedling emergence and no irrigation was needed during the 11 day period (Table 3.3).

Subsequently, to calculate the irrigation application and the infiltration rate, the following techniques can be applied. The **irrigation application**, IA (mm), is calculated by the following:

IA = $(2 - I_{eff}/100) \times I_{Appl}$.

The irrigation efficiency, I_{eff}, is the percentage ratio of the crop evapotranspiration to the irrigation application:

I_{eff} = 100 × ET_{cr} mm/IA mm.

For the furrow method of irrigation to attain a uniform distribution, IA would be adjusted for the water application efficiency, I_{eff}, which is usually 60% to 70% for medium to heavy textured soils. For the Aleppo example, I_{eff} = 70%. The **infiltration rate** (mm/hr) determines the length of time required to wet a soil and the time required in hours to apply a given irrigation. For the Aleppo heavy clay soil, the infiltration rate is 8.5 mm/hr.

The above variables can be computerized so that several years of water balance data can be analyzed to calculate the probability of irrigations required as related to climatic data (see Addendum 3B). Figures 3.6 and 3.7 are illustrations of worksheets for calculating the water balance for a month on a daily basis. For Aleppo with 23 years of rainfall and pan evaporation data, Table 3.5 presents the number of irrigations specified, seasonal rainfall, evapotranspiration, quantity of water for supplemental irrigation, and percolation/runoff. For each of the 23 years, the soil profile is at field capacity at sowing time; emergence date is assumed fixed at 6 December; and, K_c and ARD (Figures 3.4 and 3.5) are the average values for wheat in the area. Percolation/runoff is not partitioned because the internal drainage of the soil or hydraulic conductivity is unknown.

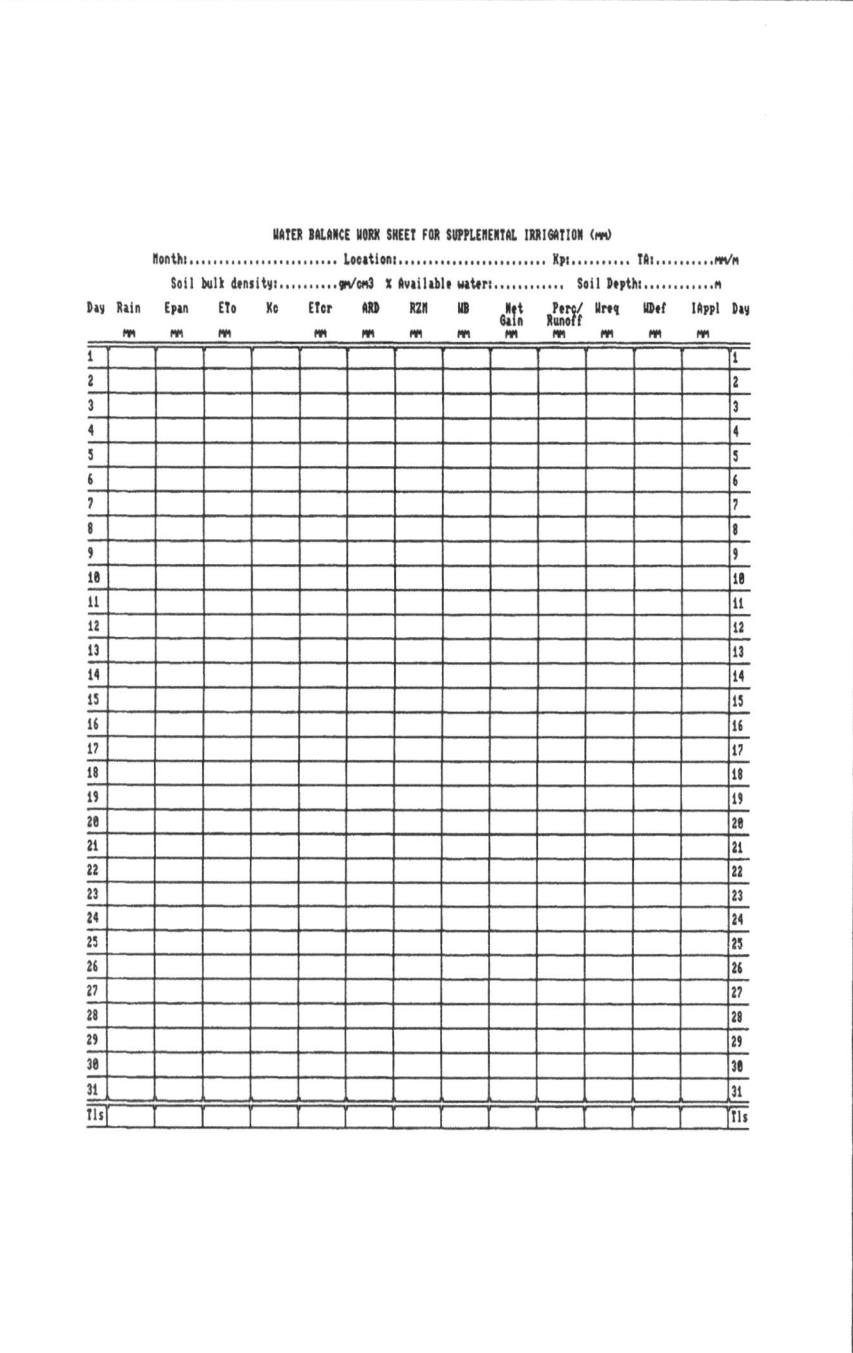

Figure 3.6. Example worksheet for water balance data of class-A evaporation pan.

Water Balance Calculations

WATER BALANCE WORK SHEET FOR SUPPLEMENTAL IRRIGATION (mm)

Month:..................... Location:...................... Kp:.......... TA:..........mm/m
Soil bulk density:..........gm/cm3 % Available water:............ Soil Depth:...........m

Day	Rain mm	Tmax oC	Tmin oC	Day Length hours	EPiche mm	ETo mm	Kc	ETcr mm	ARD mm	R2M mm	WB mm	Net Gain mm	Perc/ Runoff mm	Wreq mm	WDef mm	IAppl mm	Day
1																	1
2																	2
3																	3
4																	4
5																	5
6																	6
7																	7
8																	8
9																	9
10																	10
11																	11
12																	12
13																	13
14																	14
15																	15
16																	16
17																	17
18																	18
19																	19
20																	20
21																	21
22																	22
23																	23
24																	24
25																	25
26																	26
27																	27
28																	28
29																	29
30																	30
31																	31
Tls																	Tls

Figure 3.7. Example worksheet for water balance data of Piche evaporometer.

Table 3.5. Date of supplemental irrigations, seasonal rainfall, evapotranspiration, quantity of water required for irrigation, and total percolation/runoff for 23 years at Aleppo, Syria.

Growing season	Supplemental irrigation no.					Seasonal rainfall (mm)	Evapo-trans (mm)	Water quantity (mm)	Total perc/run (mm)
	1	2	3	4	5				
1963	5/4					373.5	211.5	93.5	128.5
1964	5/1	1/4	2/5			238.4	274.5	215.3	55.0
1965	12/3	3/5				310.5	267.9	180.2	104.6
1966	22/2	11/3	6/4	1/5		136.7	362.3	335.8	14.1
1967	16/4					437.7	196.4	97.2	174.3
1968	20/3	8/4	1/5			324.7	297.6	278.9	150.3
1969	1/3	14/4				399.3	230.2	166.3	211.7
1970	9/2	27/3	16/4	10/5		135.6	332.6	337.2	14.8
1971	6/1	16/2	26/3			234.2	281.1	170.4	42.0
1972	25/2					326.9	223.8	66.3	56.9
1973	28/12	9/2	16/3	15/4	11/5	136.4	341.4	348.7	2.3
1974	21/2	1/5				364.0	233.5	162.3	179.7
1975	16/3	10/4				271.0	283.0	177.5	64.5
1976						366.3	184.2		67.2
1977	24/2	5/4				262.5	254.9	158.8	25.6
1978	26/3	27/4				266.1	260.6	187.8	108.0
1979	6/2	11/3	9/4	9/5		182.2	269.0	322.1	82.7
1980	25/2	24/5				279.1	285.6	167.0	89.9
1981	2/4					322.0	204.9	92.2	67.6
1982	12/3	19/4				267.1	225.0	177.6	72.3
1983	1/2	29/4				245.5	214.7	143.3	38.3
1984	3/1	25/2	4/4	22/5		134.6	240.7	282.0	10.1
1985	10/3	16/4				247.7	253.7	175.4	89.1

The stochasticity (random real time variability) of climate is easily recognizable by observing the variability of dates for supplemental irrigation. The years of 1973 and 1976 suggest that extremes can be close together and, for yield stabilization of food production, management must plan for this variation to avoid chaos. The coefficient of correlation (R) between rainfall and the number of irrigations is 0.83 and the linear regression equation is:

rainfall = 413.8 − 61.4 × number of irrigations;

and, R = 0.72 for the correlation of rainfall with evapotranspiration of the climatic system of Aleppo where the linear regression equation is:

rainfall = 575.0 − 1.17 × ET_{cr}.

These data suggest that some of the stochasticity can be managed as deterministic elements of the farming system.

Water Balance Calculations

Research at the ICARDA Center has revealed that the Cham I variety of wheat, receiving only 1/3 of the water requirement (120 mm versus 360 mm), with the addition of a moderate amount of nitrogen (70 kg/ha), can produce as much as 8 t/ha if irrigations are scheduled by water balance calculations. This increase in yield represents a 400% increase in production over rainfed farming. Studies have shown that scheduling of supplemental irrigation is more important than the quantity of water applied when the quantity is at least 30 mm per irrigation. The total water requirement for crop production can be somewhat lower, 1/3 the volume computed, than is indicated by the calculated value of water deficit, W_{Def}. Although irrigation quantities calculated for the non-stressed plant condition can be reduced by 2/3 the volume computed; scheduling (timing) of irrigation applications should be for a non-stressed plant condition.

An example of water balance calculations is given in Figure 3.8 for the 1984-85 season for wheat at the ICARDA Research Center, Tel Hadya. The zero soil water content indicates the wilting point at the root zone depth for the clay soil (% available water = 18.9, maximum root depth = 1,050 mm, bulk density = 1.01 gm/cm³, total active root zone moisture = 200.5 mm). The soil profile was at field capacity at sowing time. The water requirement curve (dashed line) indicates the soil water in the root zone when 50% of the available water has been used. The curve for 100% replenishment indicates the amount of soil water available to the plant on any given day. Seedling emergence started on 18 December 1984 following the sowing date, 6 December. Rainfall essentially

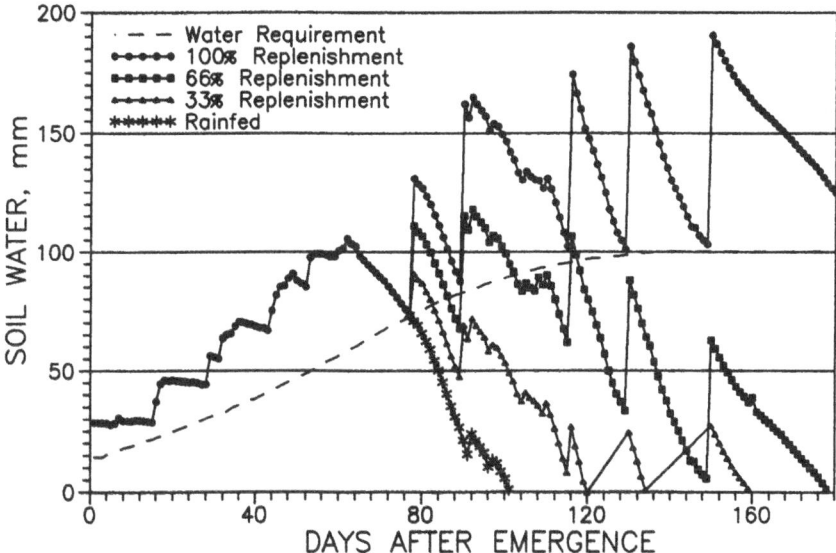

Figure 3.8. Water balance computations for the 1984-85 growing season at Tel Hadya showing the water requirement to maintain the 50% available water content in the root zone and the soil water in the root zone for each treatment.

ceased 65 days after emergence with the accumulative rainfall of 192.2 mm by 20 February 1985 and it was necessary to irrigate 77 days after emergence on 4 March.

Five supplemental irrigations were required to maintain the 100% replenishment treatment at or above the 50% available moisture requirement. The total rainfall during the growth period was only 229 mm; and, had not the soil profile been at field capacity at sowing time, there would have been a serious water shortage for the remainder of the treatments. The water treatments of 66% and 33% replenishment were well below the 50% available moisture level and the 33% replenishment treatment was below the wilting point by 120 days after emergence. However, for this latter treatment (33% replenishment), supplemental irrigation would bring it above that low level for short periods of time. The rainfed treatment did not receive enough rainfall to maintain the water level above the wilting percentage and, in consequence, produced average yields of only 1.0 t/ha. The average yield for the 100%, 66%, and 33% replenishment treatments were 3.2 t/ha.

Probability Analysis

Frequency curves are most commonly used to determine the economic value of supplemental irrigation. Common uses include the determination of rainfall recurrence amounts to design an irrigation system, i.e. size of pumping plant, scale of sprinkler or drip/trickle equipment, storm drainage design, size of farm pond, etc. In almost all locations, there are growing seasons where rainfall or

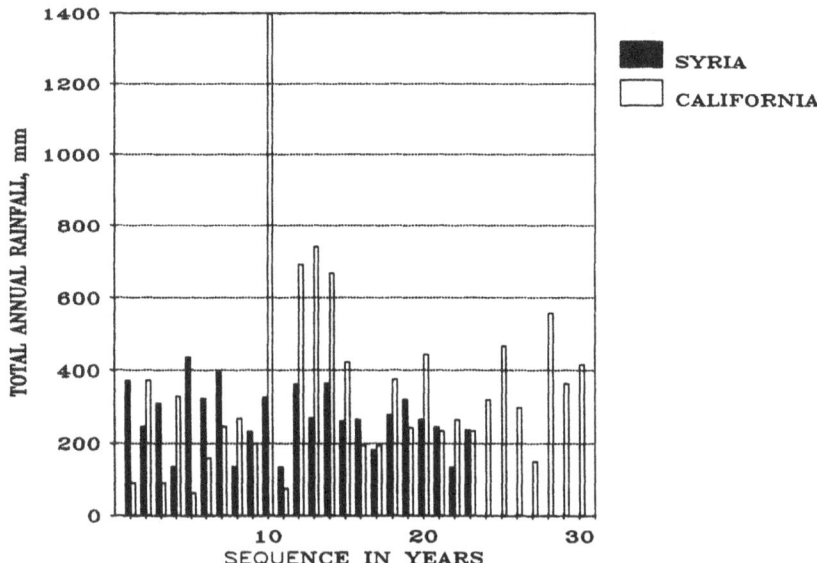

Figure 3.9. Chronological sequence of rainfall for a 23-year period at Aleppo, Syria.

Water Balance Calculations

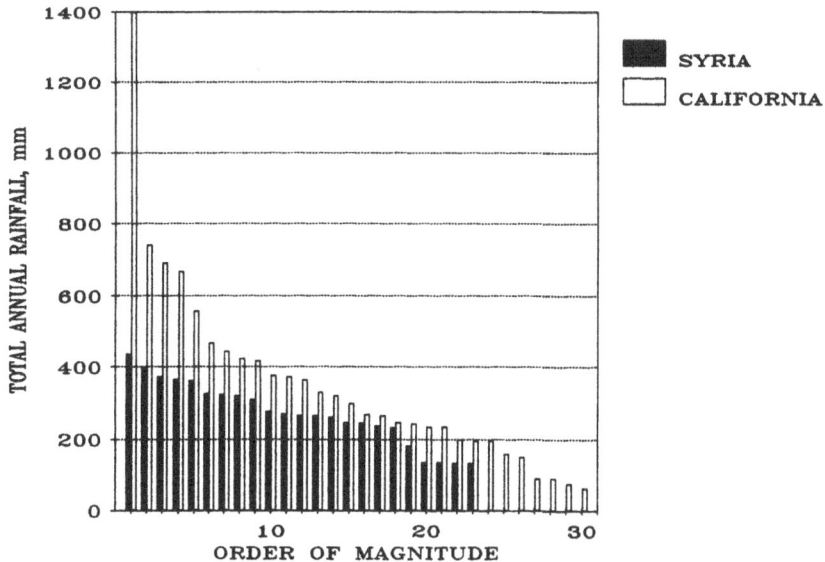

Figure 3.10. Annual array of rainfall in decreasing order of magnitude for a 23-year period at Aleppo, Syria.

storms do not create damage. However, the design criteria to assess seasonal variation of storm damage can be shown by a graphical demonstration of a comparison between a chronological rainfall record as shown in Figure 3.9 and an annual rainfall array as shown in Figure 3.10. Frequency curves evaluate whether to refill the soil profile by an early irrigation before sowing to test the probability of an economic return to counteract drought effects, or, to assess if this early irrigation is a waste of water and energy exacerbating leaching of fertilizer and accelerating soil erosion.

Natural or random rainfall events can be analyzed using probability techniques. One way to define a random event (such as rainfall) is as a variable whose values are associated with some probability of being observed. Also, probability can be defined as the study of random or non-deterministic variables. Therefore, in probability theory, a mathematical model can be defined for the phenomenon of rainfall by assigning probabilities to the events associated with rainfall. The probability of the amount and timing of rainfall to meet agricultural production can be estimated from analysis of daily values of rainfall and evaporation. Uncertainties of rainfall events are difficult to reconcile with crop water requirements but supplemental irrigation reduces this incertitude of risk.

Analytical Method

When applying various statistical methods to estimate rainfall probability, the frequency distribution of the data must be known, e.g. normal, logarithmic, γ, or other skewed distributions. The frequency distribution does not quantify the variability of measurements but distributes the values about their relative magnitude independent of position. The mean of a data set, Y, is usually obtained by taking many samples so that the population mean can be estimated. In irrigation terminology, the reciprocal of the probability is frequently used; this is termed the recurrence interval or return period. For example, recurrence means the probability of a flooding rainfall occurring on the average of 1 year in 20 years (1/20 or 0.05), or a percent probability of $0.05 \times 100 = 5\%$ of being exceeded in any 1 year. It does not mean that every 20 years a rainfall of that magnitude will occur. To calculate the probability of the annual rainfall data (see Table 3.7), general statistics were computed on a monthly basis for each of 5 variables using the 5 steps presented in Figure 3.11.

---

1. Transform the rainfall to their logarithms (base 10),

    $X = \log_{10} R$.

2. Compute the mean of the logarithms,

    $M = (\Sigma X)/n$,

    where n = number of years.

3. Compute the standard deviation of the logarithms,

    $s = [\Sigma (X_i - Y)^2/n - 1]^{1/2}$;

    or, when using hand calculations:

    $s = [\Sigma (X^2) - 1/n(\Sigma X)^2/n - 1]^{1/2}$.

4. Compute the skewness of the logarithms,

    $$g = \frac{n^2(\Sigma X^3 - 3n(\Sigma X)(\Sigma X^2) + 2(\Sigma X)^3}{n(n-1)(n-2) s^3}.$$

5. Compute the curve from the relationship,

    $\log R = M + Ks$,

    where K is selected from Table 3.4.

Figure 3.11. Five steps to compute statistics for calculating probability of data presented in Table 3.7.

The Pearson K value depends upon the probability and skew. Computation uses the log Pearson type III distribution which is equivalent to the γ distribution (Table 3.6). When the skew is zero, this distribution reduces to the log-normal distribution. The data from the distribution can be plotted as continuous relations on logarithmic probability paper but as straight lines with zero skewness. The process and equations have been written in BASIC for Personal Computers and are included in Addendum 3C.

As an example, using the annual rainfall data in Table 3.5 for Aleppo (1963-1985), data are ranked in descending order for plotting as illustrated in Table 3.7. The logarithms, X^2 and X^3, are presented. The rainfall is transformed by

Water Balance Calculations

Table 3.6. Values of K in the Pearson Type III distribution at given exceedance probabilities and specific skewness.

Skew	Exceedance probability					
(g)	0.99	0.90	0.50	0.10	0.02	0.01
3.0	−0.667	−0.660	−0.396	1.180	3.152	4.051
2.5	−0.799	−0.771	−0.360	1.250	3.048	3.845
2.0	−0.990	−0.895	−0.307	1.302	2.912	3.605
1.5	−1.256	−1.018	−0.240	1.333	2.743	3.330
1.2	−1.449	−1.086	−0.195	1.340	2.626	3.149
1.0	−1.588	−1.128	−0.164	1.340	2.542	3.022
0.9	−1.660	−1.147	−0.148	1.339	2.498	2.957
0.8	−1.733	−1.166	−0.132	1.336	2.453	2.891
0.7	−1.806	−1.183	−0.116	1.333	2.407	2.824
0.6	−1.880	−1.200	−0.099	1.328	2.359	2.755
0.5	−1.955	−1.216	−0.083	1.323	2.311	2.686
0.4	−2.029	−1.231	−0.066	1.317	2.261	2.615
0.3	−2.104	−1.245	−0.050	1.309	2.211	2.544
0.2	−2.178	−1.258	−0.033	1.301	2.159	2.472
0.1	−2.252	−1.270	−0.017	1.292	2.107	2.400
0.0	−2.326	−1.282	0.000	1.282	2.054	2.326
−0.1	−2.400	−1.292	0.017	1.270	2.000	2.252
−0.2	−2.472	−1.301	0.033	1.258	1.945	2.178
−0.3	−2.544	−1.309	0.050	1.245	1.890	2.104
−0.4	−2.615	−1.317	0.066	1.231	1.834	2.029
−0.5	−2.686	−1.323	0.083	1.216	1.777	1.955
−0.6	−2.755	−1.328	0.099	1.200	1.720	1.880
−0.7	−2.824	−1.333	0.116	1.183	1.663	1.806
−0.8	−2.891	−1.336	0.132	1.166	1.606	1.733
−0.9	−2.957	−1.339	0.148	1.147	1.549	1.660
−1.0	−3.022	−1.340	0.164	1.128	1.492	1.588
−1.2	−3.149	−1.340	0.195	1.086	1.379	1.449
−1.5	−3.330	−1.333	0.240	1.018	1.217	1.256
−2.0	−3.605	−1.302	0.307	0.895	0.980	0.990
−2.5	−3.845	−1.250	0.360	0.771	0.798	0.799
−3.0	−4.051	−1.180	0.396	0.660	0.666	0.667

taking the common logarithms and the mean, standard deviation, and skewness of the transformed variables are found:

$$\text{Mean (log R)} = 55.4387/23 = 2.4104$$

$$s(\log R) = \left[\frac{134.1636 - 1/23\,(55.4384)^2}{22} \right]^{1/2} = 0.1525861$$

$$g = \frac{23^2 \times 325.9069 - 3 \times 23 \times 55.4384 \times 134.1623 + 2 \times 55.4384^3}{23 \times 22 \times 21 \times 0.1525861^3}$$

$g (\log R) = -0.671$.

Table 3.7. Annual rainfall for Aleppo, 1963-85.

Rainfall	$\log R = X$	X^2	X^3
437.7	2.6412	6.9759	18.4248
399.3	2.6012	6.7662	17.6003
373.5	2.5724	6.6172	17.0222
366.3	2.5638	6.5731	16.8520
364.0	2.5611	6.5592	16.7989
326.9	2.5144	6.3222	15.8966
324.7	2.5115	6.3076	15.8416
322.0	2.5079	6.2896	15.7736
310.5	2.4921	6.2106	15.4773
279.1	2.4458	5.9819	14.6306
271.0	2.4330	5.9195	14.4021
267.1	2.4267	5.8889	14.2905
266.1	2.4250	5.8806	14.2605
262.5	2.4191	5.8520	14.1567
247.7	2.3939	5.7308	13.7189
245.5	2.3901	5.7126	13.6536
238.4	2.3773	5.6516	13.4354
234.2	2.3696	5.6150	13.3053
182.2	2.2606	5.1103	11.5524
136.7	2.1358	4.5616	9.7428
136.4	2.1348	4.5574	9.7291
135.6	2.1323	4.5476	9.6949
134.6	2.1291	4.5331	9.6514
Total	55.4387	134.1636	325.9115

The mean rainfall is determined with a 10% probability of being exceeded using values from Table 3.6. With the exceedance probability of 0.10, K is 1.200:

$$X_{10} = \log R_{10} = 2.4104 + 1.200 (0.1526) = 2.5935$$

and

$$R_{10} = 10^{2.5935} = 392.18 \text{ mm}.$$

The 10-year recurrence rainfall is the standard established for designing supplemental irrigation systems. Storms of higher values of recurrence could demand storage facilities beyond economic feasibility even though eventual damage to a smaller facility is probable. Figure 3.12 illustrates the percent probability of exceedance in relation to the logarithm of annual rainfall taken from Table 3.7. The curve is drawn from the values in Table 3.8 for the log Pearson Type III fit.

Table 3.8. Results of probability analysis, Aleppo example.

Probability	K	Ks	Log R	R (mm)	Recurrence (yr)
0.99	−2.833	−0.442	1.9684	93.0	1.010
0.90	−1.333	−0.208	2.2024	159.4	1.111
0.50	0.118	0.018	2.4284	268.2	2.000
0.10	1.168	0.182	2.5924	391.2	10.000
0.02	1.613	0.252	1.6624	459.6	50.000
0.01	1.743	0.272	2.6823	481.2	100.000

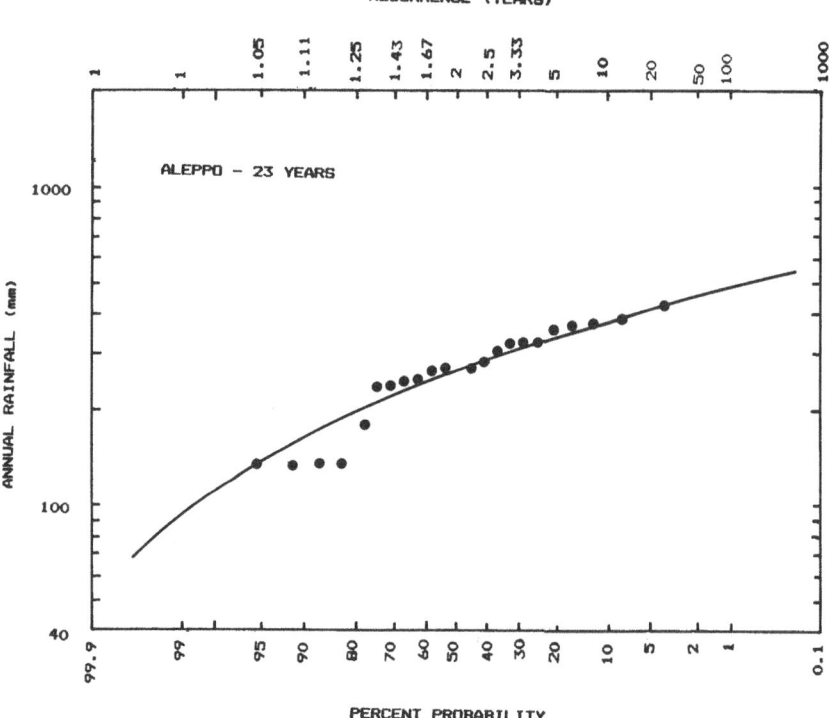

Figure 3.12. Probability curve of the annual rainfall for Aleppo, Syria.

Table 3.9 indicates the relation of the 4 moments along with the probability of recurrence in years for the 23 year data set. These data disclose that, on the average, seasonal rainfall is greater than evapotranspiration by a margin of 13.2 mm which implies that agronomic production for this level of rainfall should not be restricted.

Table 3.9. Relation of seasonal rainfall, evapotranspiration, ET_{cr}, percolation/runoff, irrigation amount, and number of irrigations to the mean, standard deviation, coefficient of skewness, coefficient of kurtosis × coefficient of variation, median, and the maximum and minimum values for 5, 10, 25, and 50 year recurrence.

Variable	Mean	Std. Dev.	Coeff. Skew.	Coeff. Kurtosis	%COV	Median	Recurrence			
							5	10	25	50
Rainfall	272.3	85.3	−0.09	2.25	31.3	267.1	215	169	128	102
							354	394	435	462
ET_{cr}	259.1	52.3	0.59	2.51	20.2	254.9	305	330	356	373
							212	186	163	145
Perc/run	80.4	56.1	0.69	2.70	69.8	67.6	123	147	174	190
							55	6	0	0
Irr. amt.	188.5	90.3	0.20	2.49	47.9	175.4	271	316	363	394
							100	57	20	0
No. of irr.	2.3	1.2	0.40	2.86	50.3	2.0	3.6	4.2	4.9	5.3
							1.2	0.6	0	0

However, data from Table 3.5 show that ET_{cr} = 362.3 mm for 1966 which exceeded the 25 year recurrence value for ET_{cr} = 356 mm and the seasonal rainfall = 136.7 mm which was less than the 10 year recurrence value for the minimum rainfall = 169 mm. Only once, in 1967, was the seasonal rainfall (437.7 mm) greater than the 25 year recurrence value (435 mm). Also, notice should be taken that the high rainfall of 1967 followed a low rainfall in 1966 with a 20 year recurrence. The results of analyses indicate that the data are only slightly skewed and kurtotic which speculates that the mean may be a good estimate of the central tendency. Notwithstanding, the high values for the standard deviation and percent coefficient of variation suggest the trend of a non-normal data set.

The values for runoff establish that water storage for supplemental irrigation is feasible regardless of the method of storage or means of application. With systematic conservation, surplus water from wet years could be made available during dry periods or years of drought. Probability analysis demonstrates that the size of storage facility can be estimated to ensure an adequate supply of water for supplemental irrigation whether it is a check dam, pond, catchment basin, recharge well, or other facility. The average irrigation amount needed on a yearly basis suggests that a storage facility could be constructed to collect the runoff from a water harvesting catchment basin design of 3:1 or 240.2 mm annually. Once in 5 years, according to the probability analysis, an adequate supply of water would not be available for supplemental irrigation; of course, an alternative water source could alleviate this condition.

Water Balance Calculations

Graphical Method

Even though results can be attained entirely by analytical methods described above, every derived frequency relation should be plotted graphically to observe data visually for comparison with the derived curve. The principal advantages of graphical methods are:
1. immediate applicability;
2. rapid visualization; and,
3. comparison of observed data with analytical computations.

However, graphical methods have the disadvantage that any two persons might not achieve identical frequency curves and, occasionally, these differences can

Table 3.10. Rainfall frequency analysis.

Order recorded		Decreasing order		
Growing season	Rainfall (mm)	Order, m	Plotting position	Rainfall (mm)
1928-29	92.5	1	3.2	1,399.6
29-30	365.1	2	6.5	742.4
30-31	91.3	3	9.7	693.7
31-32	331.0	4	12.9	669.4
32-33	65.7	5	16.1	558.6
33-34	160.0	6	19.4	469.2
34-35	244.0	7	22.6	445.4
35-36	266.5	8	25.8	424.1
36-37	201.4	9	29.0	418.7
37-38	1,399.6	10	32.3	376,1
38-39	76.7	11	35.5	373.6
39-40	693.7	12	38.7	365.1
40-41	742.4	13	41.9	331.0
41-42	669.4	14	45.2	321.3
42-43	424.1	15	48.8	298.8
43-44	195.9	16	51.6	269.6
44-45	196.6	17	54.8	266.5
45-46	376.1	18	58.1	247.7
46-47	247.7	19	61.3	244.0
47-48	445.4	20	64.5	235.5
48-49	235.5	21	67.7	235.5
49-50	269.6	22	71.0	201.4
50-51	235.5	23	74.2	196.6
51-52	321.3	24	77.4	195.9
52-53	469.2	25	80.6	160.0
53-54	298.8	26	83.9	150.9
54-55	150.9	27	87.1	92.5
55-56	558.6	28	90.3	91.3
56-57	373.6	29	93.5	76.7
57-58	418.7	30	96.8	65.7

be important. Also, these methods provide no means of evaluating the reliability of the estimates. Nonetheless, the method is commonly applied and is quick for small data sets.

For annual seasonal rainfall, the amounts are ranked in order of magnitude. The largest rainfall is given the order number m = 1, and the smallest, order number N where N is the number of years of record. Plotting positions, P, for the probability axis of the graph are assigned to each rainfall event, the plotting position for the percent chance of an event is determined only by the order number m and N. For example, the largest rainfall in 50 years of record might be assigned the plotting position of 1 in 51 if the formula

$$P = \frac{100 \times m}{(N + 1)}$$

is used to approximate the average. There are several alternative formulas in the literature, but this formula will be used for the example in this chapter to assign plotting positions. Table 3.10 is a tabulation of sample rainfall data, taken from chronological records and ranked in decreasing magnitude, and of corresponding plotting positions.

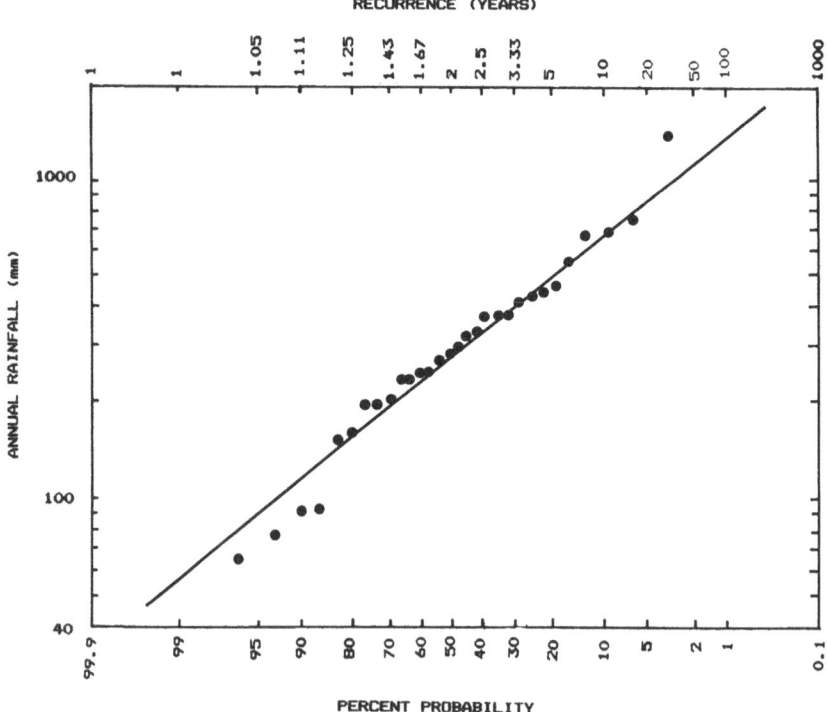

Figure 3.13. Logarithmic probability curve of data in Table 3.10 for rainfall frequency data of 30 year record.

Water Balance Calculations

The plotting graph paper used in rainfall distribution curves is logarithmic probability paper designed to plot data as a straight line. A graphical illustration of the logarithmic probability data is shown in Figure 3.13. The curve is based on rainfall records of 1928-59. The graph shows the percent probability that rainfall will be exceeded; the recurrence time period in years, the logarithmic plot of the plotting numbers P, and a straight line fitted by eye to the points.

Addendum 3A: Penman's Equation

Although many empirical approaches have been used to estimate potential evapotranspiration, only Penman's Equation is physically based. Penman combined the energy balance and vapor transfer equations to determine evaporation from a wet surface. There are many paths to follow in deriving the Penman equation: only one is presented here. The energy balance is given either in water equivalents of mm/day or cal/cm²/day (also Langleys/day or kJoules/m²/day) as:

$$R_n = E_t + G + H + M$$

where R_n = net short wave and long wave radiation at a surface;
E_t = energy used for evapotranspiration (latent heat);
H = energy which heats the air (sensible heat);
G = energy which heats the soil (soil sensible heat); and,
M = negligible miscellaneous energy terms, i.e. photosynthesis and heat stored in plant-air layer.

Rewriting the energy balance for convenience:

$$E_t = R_n - G - H.$$

As R_n and G can be measured with a reasonable degree of accuracy, only the terms E_t and H require immediate examination. In this instance, the Bowen ratio is used:

$$\beta = H/E_t = \gamma (T_s - T_a)/(e_s - e_a);$$

where γ = psychometric constant = $\dfrac{c_P \times P}{\epsilon \times \lambda}$;

c_P = specific heat of air at constant pressure
 = 0.242 cal/gm/°C;
P = atmospheric pressure (mb);
ϵ = ratio of molecular weight of air to water = 0.622;
λ = latent heat of vaporization (cal/gm);
T_s = temperature at the surface (°C);
T_a = temperature in the air (°C);
e_s = saturated vapor pressure at temperature T_s at a free water surface (a lake) (mb); and,
e_a = actual vapor pressure at dewpoint temperature of air at some point above a free water surface (mb).

Substitute $H = \beta \times E_t$ into the energy balance equation and rearrange terms:

$$E_t = (R_n - G)/(1 + \beta),$$

or rewrite,

$$E_t = \frac{R_n - G}{1 + \gamma \times \dfrac{T_s - T_a}{e_s - e_a}}$$

At this juncture, the saturation vapor pressure curve is defined in Figure 3A.1 which shows the constant relationship between the maximum amount of water vapor that air can hold at any given temperature. The slope of the saturation vapor pressure curve, Δ (mb/°C), is given by:

$$\Delta = \frac{e_s - e_o}{T_s - T_a},$$

where e_o = saturation vapor pressure at air temperature, T_a, at a free water surface. The slope of the water vapor curve, Δ, is usually calculated at the mean air temperature, T_o, since the mean temperature at the surface, T_s, is not known.

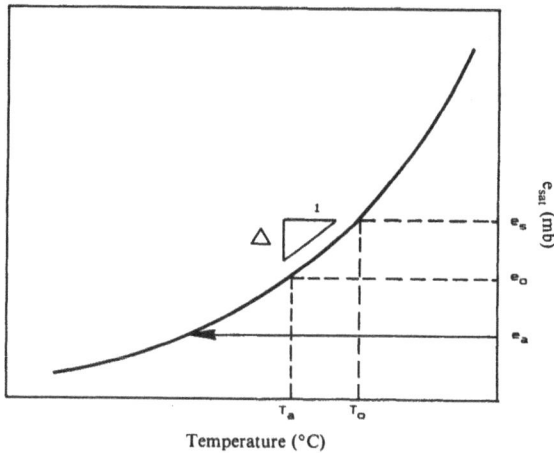

Figure 3A.1. Psychrometric curve showing relationship of saturation vapor pressure to temperature and definition of slope.

Now by manipulation:

$$T_s - T_a = \frac{e_s - e_o}{\Delta}.$$

Substituting, the equation can be written:

$$E_t = \frac{R_n - G}{1 + \dfrac{\gamma}{\Delta} \times \dfrac{e_s - e_o}{e_s - e_a}}.$$

Water Balance Calculations

At this point, it is recognized that as T_s is often unknown then e_s is also unknown. As e_a can be subtracted from each term in $e_s - e_o$, it follows:

$$e_s - e_o = (e_s - e_a) - (e_o - e_a),$$

upon substitution and rearrangement of terms;

$$\frac{(e_s - e_a) - (e_o - e_a)}{e_s - e_a} = 1 - \frac{e_o - e_a}{e_s - e_a}$$

Then substituting into the equation for energy balance:

$$E_t = \frac{R_n - G}{1 + \frac{\gamma}{\Delta} \times \left(1 - \frac{e_o - e_a}{e_s - e_a}\right)}.$$

The Dalton Equation predicts evaporation as a function of vapor pressure in air and is written:

$$E_a = (e_o - e_a) \times f(u),$$

where $f(u)$ defines the windspeed above the surface. Rewriting the Dalton Equation for evaporation from a saturated surface, e_s replaces e_o:

$$E_t = (e_s - e_a) \times f(u).$$

By division of E_a by E_t the terms become:

$$\frac{E_a}{E_t} = \frac{(e_o - e_a)}{(e_s - e_a)}.$$

It is possible to rewrite the equation in the form:

$$E_t = \frac{R_n - G}{1 + \frac{\gamma}{\Delta} \times \left(1 - \frac{E_a}{E_t}\right)}.$$

Solving for E_t by rearranging terms:

$$E_t = \frac{R_n - G}{1 + \frac{\gamma E_t - \gamma E_a}{\Delta E_t}},$$

and

$$E_t = \frac{(R_n - G) \times \Delta E_t}{\Delta E_t + \gamma E_t - \gamma E_a}$$

and

$$\Delta E_t + \gamma E_t - \gamma E_a = \Delta \times (R_n - G);$$

whereupon,

$$E_t = \frac{\Delta \times (R_n - G) + \gamma E_a}{\Delta + \gamma}.$$

Substituting, $f(u_2) = 0.35 \times (0.5 + u_2/160)$ for wind run, km/day, measured at a height of 2 m above the surface gives Penman's Equation for an open-water surface:

$$E_t = \frac{\Delta \times (R_n - G) + \gamma \times (e_o - e_a) \times f(u_2)}{\Delta + \gamma}.$$

Penman, as well as Bowen, ignored the soil heat flux, G, because it is only a small fraction of R_n when soil moisture is not limiting. However, as measurement of G is not limiting it is included in the above form. Monteith adjusted the equation to account for both aerodynamic resistance and stomatal resistance.

Addendum 3B: Basic Water Balance Calculation

```
    SCREEN 0
    COLOR 7, 1, 3
    REM    ****   PROGRAM FOR CALCULATING WATER BALANCE WORK SHEET   ***
    :
    REM    ****          PROGRAM NAME IS --- WBALCOM.BAS ---
    :
    PRINT CHR$(12)
    PRINT
    PRINT "        WATER BALANCE WORK SHEET FOR SUPPLEMENTAL IRRIGAT
    PRINT
    PRINT "              FOR MICROSOFT QuickBASIC, VERSION 4.0"
    PRINT
    PRINT
    PRINT "                         WRITTEN March, 1990"
    PRINT
    PRINT "                                   BY"
    PRINT
    PRINT "                            EUGENE R. PERRIER"
    PRINT : PRINT : PRINT : PRINT
    PRINT "            PRESS '**CONTROL BREAK**' TO QUIT PROGRAM"
    PRINT
    PRINT "            YOU CAN RUN ONLY ONE -- 1 -- FILE AT A TIME"
    PRINT "         YOU MUST RESTART PROGRAM TO RUN ANOTHER FILE":
    PRINT "                  PRINTER ++MUST++ BE TURNED ON": PRINT
    PRINT "            PRESS 'RETURN' OR <CR> TO CONTINUE"
    LINE INPUT R$
    PRINT CHR$(12)
320 CLEAR
    KNT = 1: I = 1: COUNT = 0
    IF CB$ = "Y" THEN Q$ = "N": GOTO 350
330 PRINT : INPUT "IS THIS THE START OF A 'NEW' DATA SET? TYPE Y OR N";
    IF Q$ = "" OR Q$ <> "Y" AND Q$ <> "N" THEN 330
    IF Q$ = "N" THEN KNT = 2: PRINT : PRINT
    PRINT
350 IF CB$ = "Y" THEN X$ = "F": GOTO 355
352 INPUT "IS DATA TO BE INPUT FROM 'K'EYBOARD OR 'F'ILE? TYPE K OR F";
    IF X$ = "" OR X$ <> "K" AND X$ <> "F" THEN 352
    IF X$ = "K" THEN I = 2: GOTO 360
355 IF CB$ = "Y" THEN 370
    PRINT : PRINT "ENTER THE COMPLETE FILE NAME WITH PATH AND DRIVE"
    INPUT H$
360 PRINT
    PRINT "ARE YOU ADDING DATA TO A FILE AND SAVING TO DISK? TYPE Y OR
    INPUT AB$: PRINT
    IF AB$ = "" OR AB$ <> "Y" AND AB$ <> "N" THEN 360
    IF AB$ = "N" THEN 370
    IF X$ = "F" THEN BA$ = H$: GOTO 365
    PRINT "ENTER NAME OF NEW OR OLD FILE WITH PATH AND DRIVE"
    INPUT BA$: PRINT : PRINT
365 OPEN BA$ FOR APPEND AS #3
    LPRINT BA$: LPRINT
370 PRINT : INPUT "ENTER THE PAN COEFFICIENT FOR THIS REGION"; YZ
380 PRINT
390 PRINT "ENTER TOTAL AVAILABLE WATER IN mm/m FOR TOTAL ROOT ZONE";
    PRINT " SOIL PROFILE"
    INPUT TA
    IF Q$ = "N" AND AB$ = "Y" THEN 402
    IF Q$ = "N" THEN 402
    INPUT "ENTER TOTAL WATER IN THE SOIL PROFILE mm/m AT SEASON START";
    IF TB < 0 THEN TB = 0
    TDAB = TA - TB: IF TDAB <= 0 THEN TDAB = 0
```

Water Balance Calculations

```
402 PRINT : PRINT : IF CB$ = "Y" THEN 422
    DIM D1(225), R1(225), E1(225), K1(225), RD1(225), E2(225), E3(225)
    DIM RZM(225)
    DIM WB(225), PR(225), NG(225), WD(225), IAMT(225), IA(225), WR(225)
    DIM RD2(225)
422 IF X$ = "K" THEN 425
    OPEN H$ FOR INPUT AS #1
    I = 2
425 IF Q$ = "Y" AND AB$ = "Y" THEN CLS : GOTO 412
    IF Q$ = "Y" THEN 460
    IF Q$ = "N" AND AB$ = "Y" THEN PRINT CHR$(12): GOTO 412
    GOTO 427
412 IF Q$ = "Y" AND X$ = "K" THEN 460
    PRINT : PRINT : PRINT : PRINT : PRINT : PRINT : PRINT : PRINT
    COLOR 7 + 16, 1
    PRINT "                        JUST A MINUTE PLEASE": PRINT
    PRINT "                        SIT BACK AND RELAX"
    FOR I = 1 TO 10000: NEXT I
    CLOSE #3
    GOSUB 1200
410 OPEN BA$ FOR APPEND AS #3
    IF Q$ = "Y" THEN 470
427 PRINT "ENTER 3 VALUES FOR THE DAY 'BEFORE' THE DATA IS TO START."
    PRINT "ENTER THE PREVIOUS DAYS JULIAN DATE, ROOT ZONE MOISTURE,";
    PRINT " AND NET GAIN": PRINT
    INPUT "PREVIOUS JULIAN DAY THAT DATA IS TO START,"; XZX
    INPUT "PREVIOUS DAYS ROOT ZONE MOISTURE,"; XZZ
430 INPUT "PREVIOUS DAYS NET GAIN VALUE."; DTA
440 PRINT
450 :
460 IA(I) = 0
    IF X$ = "K" THEN 470
    IF X$ = "F" AND AB$ = "Y" THEN GOTO 470
    IF Q$ = "Y" AND AB$ = "Y" THEN GOTO 470
465 IF EOF(1) THEN 1085
    INPUT #1, D1(I), R1(I), E1(I), K1(I), RD1(I), IAMT(I)
    IF Q$ = "Y" THEN 570
    IF D1(I) <= XZX THEN I = I + 1: GOTO 465
    IF KNT = 2 AND DTA > XZZ THEN WB(I - 1) = XZZ
    IF KNT = 2 AND DTA <= XZZ THEN WB(I - 1) = DTA
    KNT = 5
    IF CB$ = "Y" THEN 580
    GOTO 570
470 PRINT "ENTER 3 VALUES: DAY, RAIN, AND EPAN"
480 PRINT : PRINT "*** NOTE *** DAY MUST BE JULIAN DAY": PRINT
490 INPUT "JULIAN DAY ="; D1(I)
500 IF D1(I) < 1 OR D1(I) > 366 THEN PRINT "ERROR IN DAY": GOTO 490
510 INPUT "RAIN mm ="; R1(I)
520 INPUT "EVAP PAN mm ="; E1(I)
    IF KNT = 2 AND DTA > XZZ THEN WB(I - 1) = XZZ
    IF KNT = 2 AND DTA <= XZZ THEN WB(I - 1) = DTA
530 PRINT
540 PRINT "ENTER 2 VALUES: Kc AND ACTIVE ROOT DEPTH, ARD"
550 INPUT "KC ="; K1(I)
560 INPUT "ARD mm ="; RD1(I)
570 IF AB$ = "N" THEN 580
571 PRINT : INPUT "ARE THESE VALUES CORRECT? -- TYPE Y OR N"; AA$
    IF AA$ = "" OR AA$ <> "Y" AND AA$ <> "N" THEN 571
    IF AA$ = "N" AND AB$ = "Y" THEN 427
    IF AA$ = "N" THEN 470
    B1$ = "###.   ": B2$ = "###.#   ": B3$ = "###.#.#  "
    B4$ = "#.###,  ": B5$ = "###.#.,  ": B6$ = "###.#.#  "
        PRINT #3, USING B1$; D1(I);
        PRINT #3, USING B2$; R1(I);
        PRINT #3, USING B3$; E1(I);
        PRINT #3, USING B4$; K1(I);
        PRINT #3, USING B5$; RD1(I);
        PRINT #3, USING B6$; IAMT(I)
572 PRINT : INPUT "DO YOU WISH TO ENTER MORE DATA -- TYPE Y OR N"; BC$
    IF BC$ = "" OR BC$ <> "Y" AND BC$ <> "N" THEN 572
    IF BC$ = "Y" THEN 600
573 IF BC$ = "N" THEN CLOSE #3
580 PRINT : PRINT
600 E2(I) = FIX(YZ * E1(I) * 1000) / 1000
610 E3(I) = FIX(K1(I) * E2(I) * 1000) / 1000
    RD1(I) = FIX(RD1(I)) / 1000
620 RZM(I) = FIX(RD1(I) * TA * 100) / 100
630 IF KNT = 1 THEN WB(I - 1) = RZM(I)
640 KNT = KNT + 3
650 IF IAMT(I) > 0 THEN IA(I) = IA(I) + IAMT(I)
660 NG(I) = FIX((WB(I - 1) + R1(I) + IA(I) - E3(I)) * 100) / 100
670 IF NG(I)>RZM(I) THEN PR(I)=FIX((NG(I)-RZM(I))*100)/100: WD(I) = 0
680 IF NG(I)<=RZM(I) THEN PR(I)=0: WD(I)=FIX((RZM(I)-NG(I))*100)*10 ^ -2
    IF PR(I) >= 0 AND TDAB >= 0 THEN TDAB1 = TDAB - PR(I)
    IF TDAB1 => 0 THEN PR(I) = 0: TDAB = 0
    IF TDAB1 < 0 THEN PR(I) = PR(I) - TDAB: TDAB = 0
    IF TDAB1 > 0 THEN PR(I) = 0: TDAB = TDAB1
    TDAB1 = 0
690 WR(I) = FIX(.5 * RZM(I) * 100) / 100
700 IF WD(I) < WR(I) THEN IA(I) = 0
710 IF WD(I) >= WR(I) THEN IA(I) = WR(I)
```

```
720 :
730 COUNT = COUNT + 1
    PRINT " DAY"; "  RAIN"; "  P-EVAP"; " EVAPP "; "    KC   ";
    PRINT "    EVAPT   "; "   ARD";
    PRINT "        RZM    "
    PRINT D1(I); TAB(6); R1(I); TAB(12); E1(I); TAB(19); E2(I);
    PRINT TAB(28); K1(I);
    PRINT TAB(35); E3(I); TAB(47); RD1(I) * 1000; TAB(57); RZM(I)
    PRINT
    PRINT "   WB     "; "NET GAIN "; "PERC/RUN"; "      WREQ ";
    PRINT "    W DEF  ", "    IRR AP";  "   IMDEF";
    PRINT WB(I - 1); TAB(10); NG(I); TAB(20); PR(I); TAB(30); WR(I);
    PRINT TAB(40); WD(I); TAB(50); IAMT(I); TAB(57); TDAB
    PRINT : LPRINT
    LPRINT " DAY"; "  RAIN"; "  P-EVAP"; " EVAPP "; "    KC   ";
    LPRINT "    EVAPT   "; "   ARD";
    LPRINT "        RZM    "
    LPRINT D1(I); TAB(6); R1(I); TAB(12); E1(I); TAB(19); E2(I);
    LPRINT TAB(28); K1(I);
    LPRINT TAB(35); E3(I); TAB(47); RD1(I) * 1000; TAB(57); RZM(I)
    LPRINT
    LPRINT "   WB     "; "NET GAIN "; "PERC/RUN"; "      WREQ ";
    LPRINT "    W DEF  ", "    IRR AP";  "   IMDEF";
    LPRINT WB(I - 1); TAB(10); NG(I); TAB(20); PR(I); TAB(30); WR(I);
    LPRINT TAB(40); WD(I); TAB(50); IAMT(I); TAB(57); TDAB
    IF WD(I)>=WR(I) THEN PRINT "*IRRIGATION*": LPRINT "*IRRIGATION NEEDE
    LPRINT "-----------------------------------------------": LPRI
820 IF WD(I) > WR(I) THEN IAMT(I + 1) = CINT(WR(I))
    IF NG(I) > RZM(I) THEN WB(I) = RZM(I)
    IF NG(I) <= RZM(I) THEN WB(I) = NG(I)
    PRINT
930 NF = I
    IF BC$ = "N" THEN 1085
    IF X$ = "F" THEN 1050
940 INPUT "DO YOU WANT TO VIEW THE TOTAL INPUT, Y OR N"; A$
950 IF A$ = "" OR A$ = "N" THEN 1000
    PRINT : LPRINT
    PRINT "THIS IS THE INPUT + CALCULATED INPUT DATA"
    LPRINT "THIS IS THE INPUT + CALCULATED INPUT DATA"
960 FOR J = 2 TO NF
970 PRINT USING "###.###";D1(J);R1(J);E1(J);E2(J);K1(J);E3(J);RD1(J);RZ
    LPRINT USING "###.###";D1(J);R1(J);E1(J);E2(J);K1(J);E3(J);RD1(J);RZ
980 NEXT J
990 PRINT : LPRINT
1000 INPUT "DO YOU WANT TO VIEW THE TOTAL OUTPUT, Y OR N"; B$
1010 IF B$ = "" OR B$ = "N" THEN 1060
     PRINT "THIS IS THE OUTOUTPUT DATA"
     LPRINT "THIS IS THE OUTPUT DATA"
1020 FOR J = 2 TO NF
1030 PRINT USING "###.###";WB(J - 1);NG(J);PR(J);WR(J);WD(J);IAMT(J);TD
     LPRINT USING "###.##";WB(J - 1);NG(J);PR(J);WR(J);WD(J);IAMT(J);TDA
1040 NEXT J
1050 :
1060 I = I + 1
     IF X$ = "F" AND COUNT = 30 THEN 1082
     IF X$ = "F" AND COUNT < 30 THEN 1095
1070 INPUT "IF IRRIGATION OCCURRED, ENTER NET WATER APPLIED, mm"; IAMT(I
     IF Q$ = "N" AND AB$ = "Y" THEN 470
1082 COUNT = 0: PRINT : PRINT
     INPUT "DO YOU WISH TO QUIT THE PROGRAM? -- TYPE Y OR N"; Z$
     IF Z$ = "" OR Z$ <> "Y" AND Z$ <> "N" THEN 1082
     IF Z$ = "N" THEN 460
1084 PRINT : PRINT
1085 RESET: END
1090 IF AB$ = "N" THEN 1095
1092 INPUT "DO YOU WANT TO RUN THE DATA? -- TYPE Y OR N"; CB$
     IF CB$ = "" OR CB$ <> "Y" AND CB$ <> "N" THEN 1092
     IF CB$ = "Y" THEN H$ = BA$: GOTO 320
1095 IF X$ = "K" THEN 470
     IF AB$ = "Y" THEN GOTO 470
     IF X$ = "F" AND AB$ = "Y" THEN 470
     IF X$ = "F" THEN 465
1100 END
1200 CLOSE #3: CLS : COLOR 7, 1
     PLAY ON
          ON PLAY(2) GOSUB REFRESH
     PLAY "MB"
          GOSUB REFRESH
     I = 0: KNT = 1: CNT = 1
     OPEN BA$ FOR INPUT AS #3
     DO UNTIL EOF(3)
         I = I + 1
         INPUT #3, D1(I), R1(I), E1(I), K1(I), RD1(I), IAMT(I)
         CNT = CNT + 1
     LOOP
     CLOSE #3: PRINT : CLOSE #3: RESET: PRINT : I = 1
1210 IF D1(I) = 0 THEN I = I - 1: GOTO 1205
     E2(I) = FIX(YZ * E1(I) * 1000) / 1000
     E3(I) = FIX(K1(I) * E2(I) * 1000) / 1000
     RD1(I) = FIX(RD1(I)) / 1000
```

Water Balance Calculations

```
            RZM(I) = FIX(RD1(I) * TA * 100) / 100
            IF KNT = 1 THEN WB(I - 1) = RZM(I)
            KNT = KNT + 1
            IF IAMT(I) > 0 THEN IA(I) = IA(I) + IAMT(I)
            NG(I) = FIX((WB(I - 1) + R1(I) + IA(I) - E3(I)) * 100) / 100
            IF NG(I)>RZM(I) THEN PR(I)=FIX((NG(I)-RZM(I))*100)/100: WD(I)=0
            IF NG(I)<=RZM(I) THEN PR(I)=0: WD(I)=FIX((RZM(I)-NG(I))*100)/100
            IF PR(I) >= 0 AND TDAB >= 0 THEN TDAB1 = TDAB - PR(I)
            IF TDAB1 = 0 THEN PR(I) = 0: TDAB = 0
            IF TDAB1 < 0 THEN PR(I) = PR(I) - TDAB: TDAB = 0
            IF TDAB1 > 0 THEN PR(I) = 0: TDAB = TDAB1
            TDAB1 = 0
            WR(I) = FIX(.5 * RZM(I) * 100) / 100
            IF WD(I) < WR(I) THEN IA(I) = 0
            IF WD(I) >= WR(I) THEN IA(I) = WR(I)
            IF I < CNT - 1 THEN 1220
1205        PRINT "PREVIOUS DAYS VALUES": PRINT
            PRINT " DAY";" RAIN";" P-EVAP";" EVAPP ";"    KC    ";
            PRINT " EVAPT      ";"   ARD";
            PRINT "        RZM      "
            PRINT D1(I); TAB(6); R1(I); TAB(12); E1(I); TAB(19); E2(I);
            PRINT TAB(28); K1(I);
            PRINT TAB(35); E3(I); TAB(47); RD1(I) * 1000; TAB(57); RZM(I)
            PRINT
            PRINT "   WB        ";"NET GAIN ";"PERC/RUN";"       WREQ ";
            PRINT "   W DEF  ";"     IRR AP";"   IMDEF"
            PRINT WB(I - 1); TAB(10); NG(I); TAB(20); PR(I); TAB(30); WR(I)
            PRINT TAB(40); WD(I); TAB(50); IAMT(I); TAB(57); TDAB
            PRINT
            IF WD(I) >= WR(I) THEN PRINT "***IRRIGATION NEEDED***": PRINT
1220        IF NG(I) > RZM(I) THEN WB(I) = RZM(I)
            IF NG(I) <= RZM(I) THEN WB(I) = NG(I)
            I = I + 1: IF I < CNT THEN GOTO 1210
            PLAY STOP
            PRINT "WRITE THE VALUES DOWN FOR PREVIOUS DATE, RZM, AND NET GA
            PRINT "          AND THEN CONTINUE -- PRESS <ENTER> TO CONTINUE"
            LINE INPUT R$
            CLOSE #3
            RETURN 410
REFRESH:
            ' Beethoven's Fifth Symphony:
            Lizzie$ = "o3 L8 E D+ E D+ E o2 B o3 D C L2 o2 A"
            Listen$ = "t180 o2 p2 p8 18 GGG L2 E-"
            Fate$ = "p24 p8 L8 FFF L2 D"
            PLAY Lizzie$ + Listen$ + Fate$

            RETURN
150  '
160  :
170  REM                     PROGRAM IS NAMED    ---PROBAB.BAS---
180  '
190  :
200  PRINT CHR$(26)
210  PRINT "            FIELD PLOT UNIFORMITY STUDY FOR FREQUENCY DISTRIBUT
220  PRINT
230  PRINT "           FOR MICROSOFT BASIC (BASIC-80) VERSION CP/M, REV. 5
240  PRINT "           KAYPRO II COMPUTER WITH 64K MEMORY, CP/M VERSION 2
250  PRINT : PRINT
260  PRINT "                              WRITTEN May, 1983"
270  PRINT : PRINT : PRINT : PRINT : PRINT
280  PRINT "            INSERT DATA DISK IN DRIVE B OR ONE (1)"
     INPUT"PRESS <ENTER> TO CONTINUE";C$
310  LINE INPUT R$
320  CLEAR : PRINT CHR$(26)
330  INPUT "ENTER NAME OR DESCRIPTION OF PROGRAM"; A$
340  PRINT A$: PRINT : PRINT
350  LPRINT A$: LPRINT : LPRINT
360  INPUT "ENTER NUMBER OF HORIZONTAL ROWS ON DATA SET"; K
370  PRINT
380  INPUT "ENTER NUMBER OF VERTICAL COLUMNS ON DATA SET"; H
390  PRINT
400  DIM V(65, 10), P(65), T(65), C(65), P1(65, 10), K1(65, 10)
     DIM P2(8), K2(8), D1$(25), CN(30), D2$(25), D3$(25)
420  PRINT CHR$(26): INPUT "ENTER DATA FILE NAME"; F$
430  '
440  REM     *****        DISK READ SEGMENT        *****
450  '
     B$ = "###.##"
460  OPEN "I", #1, F$
490  '
500  FOR I = 1 TO K
     IF EOF(1) GOTO 560
     INPUT #1, D1$(I), D2$(I), D3$(I)
530  V(I,1) = VAL(D1$(I)):V(I,2)=VAL(D2$(I)):V(I,3)=VAL(D3$(I))
     FOR J = 1 TO H
     PRINT USING B$; V(I,J);
     LPRINT USING B$; V(I,J);
540  NEXT J
     PRINT:LPRINT
550  NEXT I
560  CLOSE #1
     INPUT"PRESS <ENTER> TO CONTINUE";C$
```

```
570 PRINT : PRINT
590 :
600 REM     ***      READ PEARSON TABLE FOR K VALUES     ***
610 :
620 FOR EE = 1 TO 31
630 FOR E1 = 1 TO 6
640 READ P1(EE, E1), K1(EE, E1)
650 NEXT E1
660 NEXT EE
670 FOR I = 1 TO 8
680 READ P2(I), K2(I)
690 NEXT I
700 PRINT : PRINT : PRINT : PRINT
710 PRINT "                           ENTERING COMPUTATIONS"
720 PRINT : PRINT : PRINT : PRINT : PRINT : PRINT
730 :
740 REM     ***     COMPUTING FIRST 4 MOMENTS, SKEWNESS AND KURTOSIS     ***
750 '
760 FOR J = 1 TO H
770 A1 = 0: B1 = 0: C1 = 0: D1 = 0
780 A2 = 0: B2 = 0: C2 = 0: D2 = 0
790 PRINT : PRINT : PRINT "RUNNING VARIABLE NUMBER "; J
800 LPRINT : LPRINT : LPRINT "RUNNING VARIABLE NUMBER "; J
810 LPRINT
820 FOR I = 1 TO K
830 A1 = A1 + V(I, J)
840 B1 = B1 + V(I, J) ^ 2
850 C1 = C1 + V(I, J) ^ 3
```

Addendum 3C: Basic Probability Calculations

```
100 REM ***               PROGRAM FOR FREQUENCY DISTRIBUTIONS    ***
110 REM                    INCLUDES PEARSONS TYPE III
120 REM                              GIVES
130 REM                    PROBABILITY OF EXCEEDANCE ALSO
140 '
860 D1 = D1 + V(I, J) ^ 4
870 IF V(I, J) = 0 THEN V(I, J) = 1: UQ = 5
880 A2 = A2 + LOG(V(I, J))
890 B2 = B2 + (LOG(V(I, J))) ^ 2
900 C2 = C2 + (LOG(V(I, J))) ^ 3
910 D2 = D2 + (LOG(V(I, J))) ^ 4
920 IF UQ = 5 THEN V(I, J) = 0: UQ = 0
930 NEXT I
940 Z = K
950 T1 = A1 / Z
960 T2 = (B1 / Z) - T1 ^ 2
970 T3 = (C1 / Z) - 3 * T1 * B1 / Z + 2 * T1 ^ 3
980 T4 = (D1 / Z) - 4 * T1 * C1 / Z
990 T5 = T4 + 6 * T1 ^ 2 * B1 / Z - 3 * T1 ^ 4
1000 ZZ = SQR(T2)
1010 ZW = ZZ ^ 3
1020 T6 = T3 / ZW
1030 T7 = T5 / (T2 ^ 2)
1040 W1 = A2 / Z
1050 W2 = (B2 / Z) - W1 ^ 2
1060 W3 = (C2 / Z) - 3 * W1 * B2 / Z + 2 * W1 ^ 3
1070 W4 = (D2 / Z) - 4 * W1 * C2 / Z
1080 W5 = W4 + 6 * W1 ^ 2 * B2 / Z - 3 * W1 ^ 4
1090 ZX = SQR(W2)
1100 QX = ZX ^ 3
1110 W6 = W3 / QX
1120 W7 = W5 / (W2 ^ 2)
1130 BZZ = B1 / Z - (A1 / Z) ^ 2
1140 ST = SQR(BZZ)
1150 BXX = B2 / Z - (A2 / Z) ^ 2
1160 SD = SQR(BXX)
1170 COV = 100 * ST / T1
1180 LCOV = 100 * SD / W1
1190 PRINT "MEAN OF V(J,I) = "; T1
1200 LPRINT "MEAN OF V(J,I) = "; T1
1210 PRINT "STANDARD DEVIATION OF V(J,I) = "; ST
1220 LPRINT "STANDARD DEVIATION OF V(J,I) = "; ST
1230 PRINT "SKEWNESS COEFFICIENT = "; T6
1240 LPRINT "SKEWNESS COEFFICIENT = "; T6
1250 PRINT "KURTOSIS COEFFICIENT = "; T7
1260 LPRINT "KURTOSIS COEFFICIENT = "; T7
1270 PRINT "COEFFICIENT OF VARIATION = "; COV
1280 LPRINT "COEFFICIENT OF VARIATION = "; COV
1290 LPRINT : PRINT : LPRINT : PRINT
1300 PRINT : PRINT
1310 '
1320 PRINT "LOGARITHMIC DATA SET"
1330 LPRINT "LOGARITHMIC DATA SET"
1340 LPRINT : PRINT
1350 PRINT "LOG OF MEAN = "; W1
```

Water Balance Calculations

```
1360 LPRINT "LOG OF MEAN = "; W1
1370 PRINT "LOG OF STANDARD DEVIATION = "; SD
1380 LPRINT "LOG OF STANDARD DEVIATION = "; SD
1390 PRINT "LOG OF SKEWNESS COEFFICIENT = "; W6
1400 LPRINT "LOG OF SKEWNESS COEFFICIENT = "; W6
1410 PRINT "LOG OF KURTOSIS COEFFICIENT = "; W7
1420 LPRINT "LOG OF KURTOSIS COEFFICIENT = "; W7
1430 PRINT "LOG OF COEFFICIENT OF VARIATION = "; LCOV
1440 LPRINT "LOG OF COEFFICIENT OF VARIATION = "; LCOV
1450 PRINT CHR$(26): LPRINT "-----------------------------------------"
1460 LPRINT
1470 :
1480 REM    ***    CALCULATING THE PROBABILITY FOR NON-LOGARITHMIC DATA
1490 :
1500 LPRINT : PRINT
1510 PRINT "PROBABILITY FOR NON-LOGARITHMIC VALUES ": PRINT
1520 LPRINT "PROBABILITY FOR NON-LOGARITHMIC VALUES ": LPRINT
1530 FOR N = 1 TO 8
1540 NQ = T1 + K2(N) * ST
1550 PRINT "EXCEEDANCE PROBABILITY ="; K2(N)
1560 PRINT "VARIABLE V(J,I) = "; NQ; "  AT PROBABILITY = "; P2(N)
1570 LPRINT "EXCEEDANCE PROBABILITY ="; K2(N)
1580 LPRINT "VARIABLE V(J,I) = "; NQ; "  AT PROBABILITY = "; P2(N)
1600 PRINT : PRINT
1610 '
1620 :
1630 REM    ***    CALCULATION OF TABULAR K VALUES    ***
1640 :
1650 IF W6 > 3 THEN PRINT "ERROR IN DATA SET": M = 1: GOTO 1980
1660 IF W6 <= 3 AND W6 > 2.5 THEN M = 1: GOTO 1980
1670 IF W6 <= 2.5 AND W6 > 2 THEN M = 2: GOTO 1980
1680 IF W6 <= 2 AND W6 > 1.5 THEN M = 3: GOTO 1980
1690 IF W6 <= 1.5 AND W6 > 1.2 THEN M = 4: GOTO 1980
1700 IF W6 <= 1.2 AND W6 > 1 THEN M = 5: GOTO 1980
1710 IF W6 <= 1 AND W6 > .9 THEN M = 6: GOTO 1980
1720 IF W6 <= .9 AND W6 > .8 THEN M = 7: GOTO 1980
1730 IF W6 <= .8 AND W6 > .7 THEN M = 8: GOTO 1980
1740 IF W6 <= .7 AND W6 > .6 THEN M = 9: GOTO 1980
1750 IF W6 <= .6 AND W6 > .5 THEN M = 10: GOTO 1980
1760 IF W6 <= .5 AND W6 > .4 THEN M = 11: GOTO 1980
1770 IF W6 <= .4 AND W6 > .3 THEN M = 12: GOTO 1980
1780 IF W6 <= .3 AND W6 > .2 THEN M = 13: GOTO 1980
1790 IF W6 <= .2 AND W6 > .1 THEN M = 14: GOTO 1980
1800 IF W6 <= .1 AND W6 > 0 THEN M = 15: GOTO 1980
1810 IF W6 <= 0 AND W6 > -.1 THEN M = 16: GOTO 1980
1820 IF W6 <= -.1 AND W6 > -.2 THEN M = 17: GOTO 1980
1830 IF W6 <= -.2 AND W6 > -.3 THEN M = 18: GOTO 1980
1840 IF W6 <= -.3 AND W6 > -.4 THEN M = 19: GOTO 1980
1850 IF W6 <= -.4 AND W6 > -.5 THEN M = 20: GOTO 1980
1860 IF W6 <= -.5 AND W6 > -.6 THEN M = 21: GOTO 1980
1870 IF W6 <= -.6 AND W6 > -.7 THEN M = 22: GOTO 1980
1880 IF W6 <= -.7 AND W6 > -.8 THEN M = 23: GOTO 1980
1890 IF W6 <= -.8 AND W6 > -.9 THEN M = 24: GOTO 1980
1900 IF W6 <= -.9 AND W6 > -1 THEN M = 25: GOTO 1980
1910 IF W6 <= -1 AND W6 > -1.2 THEN M = 26: GOTO 1980
1920 IF W6 <= -1.2 AND W6 > -1.5 THEN M = 27: GOTO 1980
1930 IF W6 <= -1.5 AND W6 > -2 THEN M = 28: GOTO 1980
1940 IF W6 <= -2 AND W6 > -2.5 THEN M = 29: GOTO 1980
1950 IF W6 <= -2.5 AND W6 > -3 THEN M = 30: GOTO 1980
1960 IF W6 = -3 THEN M = 31: GOTO 1980
1970 IF W6 < -3 THEN PRINT "ERROR IN DATA (NEGATIVE)": M = 31: GOTO 1980
1980 PRINT : PRINT : LPRINT : LPRINT
1990 PRINT "PROBABILITY FOR LOGARITHMIC VALUES": PRINT
2000 LPRINT "PROBABILITY FOR LOGARITHMIC VALUES": LPRINT
2010 PRINT : LPRINT
2020 FOR N = 1 TO 6
2030 LQ = W1 + K1(M, N) * SD
2040 Q1 = EXP(LQ)
2050 PRINT "EXCEEDANCE PROBABILITY VALUE = "; K1(M, N)
2060 LPRINT "EXCEEDANCE PROBABILITY VALUE = "; K1(M, N)
2070 PRINT "VARIABLE V(J,I) = "; Q1; "  AT PROBABILITY = "; P1(M, N)
2080 LPRINT "VARIABLE V(J,I) = "; Q1; "  AT PROBABILITY = "; P1(M, N)
2090 NEXT N
2100 NEXT J
2110 :
2120 REM    ***    SORT ROUTINE    ***
2130 PRINT CHR$(26): LPRINT
2140 PRINT "ENTERING SORT ROUTINE -- DATA NOT LOGARITHMIC"
2150 LPRINT "ENTERING SORT ROUTINE -- DATA NOT LOGARITHMIC"
2160 PRINT "REARRANGEMENT OF DATA INTO ASCENDING SEQUENCE"
2170 LPRINT "REARRANGEMENT OF DATA INTO ASCENDING SEQUENCE"
2180 LPRINT
2190 FOR I = 1 TO H
2200 PRINT : PRINT "RUN NUMBER ="; I
2210 LPRINT : LPRINT "RUN NUMBER ="; I
2220 NS = 1
2230 LI = K - 1
2240 IN = NS
2250 FOR L = NS TO LI
2260 IF V(L, I) <= V(L + 1, I) THEN GOTO 2310
2270 TE = V(L, I)
2280 V(L, I) = V(L + 1, I)
2290 V(L + 1, I) = TE
```

```
2300 IN = L
2310 NEXT L
2320 IF IN = NS THEN GOTO 2380
2330 LI = IN - 1
2340 GOTO 2240
2360 REM    ***    RANK AND PROBABILITY    ***
2370 :
2380 PRINT TAB(2); "RANK"; TAB(16); "ORDER"; TAB(32); "PROBABILITY";
2390 PRINT TAB(46); "RECURRENCE"
2400 LPRINT TAB(2); "RANK"; TAB(16); "ORDER"; TAB(23); "PROBABILITY";
2410 LPRINT TAB(46); "RECURRENCE"
2420 FOR KK = 1 TO K
2430 NK = K - KK + 1
2440 P(NK) = NK / (K + 1)
2450 T(NK) = 1 / P(NK)
2460 PRINT V(KK, I), NK, P(NK), T(NK)
2470 LPRINT V(KK, I), NK, P(NK), T(NK)
2480 NEXT KK
2490 :
2500 REM    ***    DETERMINATION OF THE INTERVAL    ***
2510 :
2520 NN = 5
2530 PRINT CHR$(26)
2540 ZI = (V(K, I) - V(1, I)) / NN
2550 FOR IJ = 1 TO NN
2560 IF IJ > 1 THEN GOTO 2590
2570 C(IJ) = V(1, I) + ZI
2580 GOTO 2600
2590 C(IJ) = C(IJ - 1) + ZI
2600 CN(IJ) = 0
2610 NEXT IJ
2620 PRINT : LPRINT
2630 FOR II = 1 TO NN
2640 PRINT "INTERVAL = "; C(II), II
2650 LPRINT "INTERVAL = "; C(II), II
2660 NEXT II
2670 :
2680 REM    ***    FREQUENCY ROUTINE    ***
2690 :
2700 PRINT CHR$(26): LPRINT : PRINT
2710 PRINT "DETERMINING FREQUENCY"
2720 PRINT
2730 FOR MM = 1 TO NN
2740 FOR AA = 1 TO K
2750 IF V(AA, I) > C(MM) THEN GOTO 2790
2760 IF MM < 2 THEN GOTO 2780
2770 IF V(AA, I) <= C(MM - 1) THEN GOTO 2790
2780 CN(MM) = CN(MM) + 1
2790 NEXT AA
2800 NEXT MM
2810 TW = 0: TL = K - 1
2820 FOR BG = 1 TO NN
2830 TW = TW + CN(BG)
2840 NEXT BG
2850 IF TW > TL THEN GOTO 2870
2860 CN(NN) = CN(NN) + 1
2870 FOR CC = 1 TO NN
2880 PRINT "FREQUENCY COUNTER = "; CN(CC), CC
2890 LPRINT "FREQUENCY COUNTER = "; CN(CC), CC
2900 NEXT CC
2910 NEXT I
2920 LPRINT : LPRINT "PROGRAM IS TERMINATED"
2930 PRINT : PRINT "PROGRAM IS TERMINATED"
2940 PRINT CHR$(26)
2950 PRINT : PRINT : PRINT : PRINT
2960 PRINT "              NORMAL PROGRAM END": END
2970 DATA 0.99,-0.667,0.90,-0.660,0.50,-0.396,0.10,1.180,0.02,3.152,0.01
2980 DATA 4.051
2990 DATA 0.99,-0.799,0.90,-0.771,0.50,-0.360,0.10,1.250,0.02,3.048,0.01
3000 DATA 3.845
3010 DATA 0.99,-0.990,0.90,-0.895,0.50,-0.307,0.10,1.302,0.02,2.912,0.01
3020 DATA 3.605
3030 DATA 0.99,-1.256,0.90,-1.018,0.50,-0.240,0.10,1.333,0.02,2.743,0.01
3040 DATA 3.336
3050 DATA 0.99,-1.449,0.90,-1.086,0.50,-0.195,0.10,1.340,0.02,2.626,0.01
3060 DATA 3.149
3070 DATA 0.99,-1.588,0.90,-1.128,0.50,-0.164,0.10,1.340,0.02,2.542,0.01
3080 DATA 3.022
3090 DATA 0.99,-1.660,0.90,-1.147,0.50,-0.148,0.10,1.339,0.02,2.498,0.01
3100 DATA 2.957
3110 DATA 0.99,-1.733,0.90,-1.166,0.50,-0.132,0.10,1.336,0.02,2.453,0.01
3120 DATA 2.891
3130 DATA 0.99,-1.806,0.90,-1.183,0.50,-0.116,0.10,1.333,0.02,2.407,0.01
3140 DATA 2.824
3150 DATA 0.99,-1.880,0.90,-1.200,0.50,-0.099,0.10,1.328,0.02,2.359,0.01
3160 DATA 2.755
3170 DATA 0.99,-1.955,0.90,-1.216,0.50,-0.083,0.10,1.323,0.02,2.311,0.01
3180 DATA 2.686
3190 DATA 0.99,-2.029,0.90,-1.231,0.50,-0.066,0.10,1.137,0.02,2.261,0.01
3200 DATA 2.615
3210 DATA 0.99,-2.104,0.90,-1.245,0.50,-0.050,0.10,1.309,0.02,2.211,0.01
3220 DATA 2.544
3230 DATA 0.99,-2.178,0.90,-1.258,0.50,-0.033,0.10,1.301,0.02,2.159,0.01
```

Water Balance Calculations

```
3240 DATA 2.472
3250 DATA 0.99,-2.252,0.90,-1.270,0.50,-0.017,0.10,1.292,0.02,2.107,0.01
3260 DATA 2.400
3270 DATA 0.99,-2.326,0.90,-1.282,0.50,0.000,0.10,1.282,0.02,2.054,0.01
3280 DATA 2.326
3290 DATA 0.99,-2.400,0.90,-1.292,0.50,0.017,0.10,1.270,0.02,2.000,0.01
3300 DATA 2.252
3310 DATA 0.99,-2.472,0.90,-1.301,0.50,0.033,0.10,1.258,0.02,1.945,0.01
3320 DATA 2.178
3330 DATA 0.99,-2.544,0.90,-1.309,0.50,0.050,0.10,1.245,0.02,1.890,0.01
3340 DATA 2.104
3350 DATA 0.99,-2.615,0.90,-1.317,0.50,0.066,0.10,1.231,0.02,1.834,0.01
3360 DATA 2.029
3370 DATA 0.99,-2.686,0.90,-1.323,0.50,0.083,0.10,1.216,0.02,1.777,0.01
3380 DATA 1.955
3390 DATA 0.99,-2.755,0.90,-1.328,0.50,0.099,0.10,1.200,0.02,1.720,0.01
3400 DATA 1.880
3410 DATA 0.99,-2.824,0.90,-1.333,0.50,0.116,0.10,1.183,0.02,1.663,0.01
3420 DATA 1.806
3430 DATA 0.99,-2.891,0.90,-1.336,0.50,0.132,0.10,1.166,0.02,1.606,0.01
3440 DATA 1.733
3450 DATA 0.99,-2.957,0.90,-1.339,0.50,0.148,0.10,1.147,0.02,1.549,0.01
3460 DATA 1.660
3470 DATA 0.99,-3.022,0.90,-1.340,0.50,0.164,0.10,1.128,0.02,1.492,0.01
3480 DATA 1.588
3490 DATA 0.99,-3.149,0.90,-1.340,0.50,0.195,0.10,1.086,0.02,1.379,0.01
3500 DATA 1.449
3510 DATA 0.99,-3.330,0.90,-1.333,0.50,0.240,0.10,1.018,0.02,1.217,0.01
3520 DATA 1.256
3530 DATA 0.99,-3.605,0.90,-1.302,0.50,0.307,0.10,0.895,0.02,0.980,0.01
3540 DATA 0.990
3550 DATA 0.90,-3.845,0.90,-1.250,0.50,0.360,0.10,0.771,0.02,0.798,0.01
3560 DATA 0.799
3570 DATA 0.90,-4.051,0.90,-1.180,0.50,0.396,0.10,0.660,0.02,0.666,0.01
3580 DATA 0.667
3590 :
3600 DATA 0.99,-2.326,0.90,-1.282,0.70,-0.524,0.50,0.000,0.30,0.524,0.10
3610 DATA 1.282
3620 DATA 0.025,1.960,0.01,2.326
```

Chapter 4

Regional Application of Water Balance Methods

The subject material of irrigation, an integral component of a total farming system, is organized with introspection for farmers to improve, elevate, and enrich their realm of subsistence living. To achieve these objectives, this chapter emphasizes 4 categories of data management, each with increasing precision of measurement and sophistication of technique: rainfall plus evaporation; water balance calculations; systems simulation; and, probability analysis. These 4 categories can be used to transfer the technology of supplemental irrigation to existing farming systems.

Rainfall plus evaporation

Daily measurements of rainfall and open water evaporation are available in most regions where supplemental irrigation is practiced or can be developed. As a first approach, these two measurements acknowledge the farmer's dilemma in answering the questions of when and how much to irrigate. By accumulating rainfall and evaporation on a 10 day interval, the farmer can observe the differences and see that supplemental irrigation can replace the deficit moisture in the soil when evaporation is greater than rainfall. Operators of drip irrigation equipment commonly practice this method to replenish the exact amount of water lost on a decade basis.

For example, Tables 4.1 and 4.2 show computations for spring wheat during a growing season for Aleppo, Syria using rainfall and class-A pan evaporation data. In these instances, wheat was sown on 15 November and emergence occurred about 10-12 days later. The average depth of soil was 110 cm with wheat harvested on 5-7 June. The standard practice is not to irrigate until at least 30 mm of moisture has been lost because the efficiency of water distribution becomes much lower at application amounts of less than 30 mm. For the 1983-84 season, irrigation does not start until after decade 110 (4 March) and continues until the wheat grains are in the hard dough stage, after decade 190 (23 May), totaling 5 irrigations of 207.9 mm. However, for the 1984-85 season, irrigation started after decade 170 (3 May) and ended on decade 190

E.R. Perrier and A.B. Salkini (eds), Supplemental Irrigation in the Near East and North Africa, 79-95.
© 1991 *ICARDA.*

(23 May) for a total of 3 irrigations of 147.6 mm. The method appears to give increased amounts of water late in the season, albeit low amounts, but does not give adequate attention to the early growing season when wheat is in the vegetative and active growth stages.

An assumption is made that all rainfall and irrigation infiltrates uniformly into the soil and is stored there for plant demand, i.e. no surface runoff or deep percolation occurs for an irrigation efficiency of 100%. This method does not allow adjustment for the depth of active roots or for periods of high water use during the vegetative and heading stages of plant development. Whether or not the plants suffer from water stress conditions is not known. Although the method lacks finesse and sophistication, it is a step in the right direction and the farmers become alert to climatic phenomena and the gains and losses that are affecting the crops and soils.

Table 4.1. Calculation for irrigation using rainfall and pan evaporation decade data for Aleppo, Syria, 1983-84.

Decade	Decade date	Rain (mm)	Accum. rain (mm)	Evap. (mm)	Accum. evap. (mm)	Rain-evap. (mm)	Irrig. amt. (mm)
10	24/11	28.8		10.3			
20	4/12	1.5	30.3	12.9	23.2	+ 7.1	
30	14/12	9.3	39.6	6.4	29.6	+10.0	
40	24/12	0.3	39.9	13.6	43.2	− 3.3	
50	3/1	0.0	39.9	14.3	57.5	−17.6	
60	13/1	9.1	49.0	8.8	66.3	−17.3	
70	23/1	22.0	71.0	5.2	71.5	− 0.5	
80	2/2	15.3	86.3	12.5	84.0	+ 2.3	
90	12/2	14.6	100.9	16.4	100.4	+ 0.5	
100	22/2	0.0	100.9	18.2	118.6	−17.7	
110	4/3	0.0	100.9	24.0	142.6	−41.7	41.7
120	14/3	15.7	15.7	25.1	25.1	− 9.4	
130	24/3	21.8	37.5	19.3	44.4	− 6.9	
140	3/4	0.0	37.5	33.7	78.1	−40.6	40.6
150	13/4	7.2	7.2	31.8	31.8	−24.6	
160	23/4	17.7	24.9	26.8	58.6	−33.7	33.7
170	3/5	1.6	1.6	41.6	41.6	−40.0	40.0
180	13/5	0.0	0.0	51.9	51.9	−51.9	51.9
190	23/5	0.0	0.0	69.4	69.4	−69.4	
200	2/6	0.0		73.6			
Totals		164.9		515.8			207.9

Table 4.2. Calculation for irrigation using rainfall and pan evaporation decade data for Aleppo, Syria, 1984-85.

Decade	Decade date	Rain (mm)	Accum. rain (mm)	Evap. (mm)	Accum. evap. (mm)	Rain-evap. (mm)	Irrig. amt. (mm)
10	24/11	22.4		6.0			
20	4/12	19.0	41.4	13.3	19.3	+ 22.1	
30	14/12	13.3	54.7	10.6	29.9	+ 24.8	
40	24/12	13.7	68.4	9.0	38.9	+ 29.5	
50	3/1	42.0	110.4	6.2	45.1	+ 65.3	
60	13/1	14.9	125.3	6.6	51.7	+ 73.6	
70	23/1	16.8	142.1	7.4	59.1	+ 83.0	
80	2/2	57.1	199.2	8.3	67.4	+131.8	
90	12/2	21.9	221.1	9.2	76.6	+144.5	
100	22/2	13.8	234.9	17.7	94.3	+140.6	
110	4/3	0.0	234.9	14.9	109.2	+125.7	
120	14/3	0.0	234.9	30.7	139.9	+ 95.0	
130	24/3	24.2	259.1	24.2	164.1	+ 95.0	
140	3/4	8.0	267.1	22.4	186.5	+ 80.6	
150	13/4	3.1	270.2	44.9	231.4	+ 38.8	
160	23/4	12.2	282.4	41.1	272.8	+ 9.6	
170	3/5	0.0	282.4	48.2	321.0	− 38.6	38.6
180	13/5	1.6	1.6	55.8	55.8	− 54.2	54.2
190	23/5	5.1	5.1	59.9	59.9	− 54.8	54.8
200	2/6	0.0		65.9			
Totals		289.1		502.3			147.6

Water balance calculations

When daily rainfall and evaporation measurements are available, then a more refined and sophisticated technique is feasible. The water balance using the calculations described in chapter 3 includes pan evaporation, E_{pan}, the growth characteristics of wheat using a crop coefficient, K_c, and root/soil characteristics of the active root depth, ARD. In addition, when the soil profile is filled, runoff and deep percolation occur. For the previous example of Aleppo, Syria, the water balance calculations for spring wheat in a 1.1 m depth of soil showed that runoff/percolation = 28 mm for the 1983-84 season and 121 mm for 1984-85. For both seasons, the field required 3 irrigations as shown in Table 4.3 with 250 mm of water added during the 1983-84 season and 290 mm added in 1984-85. In this instance, irrigation was scheduled when 50% of the available water had been withdrawn from the soil profile. Under this management system, plants would not undergo stress conditions throughout the growth period.

Table 4.3. Irrigation dates and amounts.

1983-84		1984-85	
Date	Irrigation amt. (mm)	Date	Irrigation amt. (mm)
21 Feb	56	1 Mar	73
8 Mar	84	4 Apr	107
26 Apr	110	4 May	110
Totals	250		290

Distribution of rainfall has a significant effect on the amount of water applied. For instance, there was only 200.2 mm of rainfall during the 1983-84 season, whereas, there was 335.8 mm rainfall in 1984-85. Nevertheless, more irrigation was required in the 1984-85 season. The difference in rainfall of 135.6 mm is noted in the percolation/runoff values but the irrigation demand was higher by 40 mm. The potential error from the rainfall – evaporation method during the 1983-84 season was not too large but did not accurately predict the timing of irrigation; however, for the 1984-85 season, the error would have been significant with an irrigation difference of 142.4 mm.

Systems simulation

Computer technology and information systems offer a ready means of achieving success in the manipulation and processing of data and information from experimental development research. Crop simulation models can evaluate supplemental irrigation to increase grain yields. The Crop Estimation through Resource and Environment Synthesis (CERES-N) wheat production model incorporates readily available weather, soil, agronomic, fertility, and genetic information. This model simulates wheat growth and grain production on a daily basis as affected by rainfall, temperature, and radiation at specific sites and incorporates water balance calculations to determine when and how much to irrigate. Using the model presents the operator with an opportunity to evaluate various wheat cultivars for long term climatic and nitrogen fertilizer effects on production. The output can then be printed on an annual basis and retained for further evaluation and probability analysis. It offers an opportunity for agricultural planners, administrators, and technicians to understand the growth process of wheat as well as establishing predictions of potential wheat production.

Probability analysis

Probability analysis is required when the data of rainfall, evapotranspiration, yield, or determination of irrigation requirement are difficult to interpret. Output from water balance calculations or the CERES model can be further evaluated to establish when and how much to irrigate. The possibility exists of analyzing climatic variables for probability events to support activities of planning, designing, and implementing the optimal supplemental irrigation system at a specific site. For example, consideration can be given to whether an irrigation before seeding always supports increased crop production, and to how much water should be applied before seeding.

General Application of the Water Balance

The management strategy for supplemental irrigation is dominated by the cyclic withdrawal by plants and infiltration, redistribution, and deep percolation of water from within the active root zone. The water balance uses measurements of rainfall and atmometer (Piche) or pan evaporation. The relationships for the various crop, soil and pan coefficients are included to estimate the actual crop evapotranspiration. Calculations can be performed daily, weekly, at 10 day intervals, or on a monthly cycle to provide information of the soil moisture status in the active root zone. The data required, which are typical estimates for the Mediterranean Region, are listed in Table 4.4.

Table 4.4. Potential evaporation (ET_o), crop coefficient (K_c), active root depth (ARD) and monthly total evapotranspiration (TET_o) for wheat in the Mediterranean Region.

	Nov	Dec	Jan	Feb	Mar	Apr	May	Jun
ET_o (mm)	1.95	1.09	1.01	1.56	2.42	5.90	7.33	9.75
K_c (mm)	0.40	0.42	0.48	0.65	0.95	1.18	0.97	0.26
ARD (mm)	12.0	14.0	20.0	50.0	92.0	105.0	120.0	120.0
TET_o (mm)	58.5	33.8	31.3	43.7	75.0	177.0	227.2	292.5

The potential evaporation, ET_o, is determined either by Piche or pan evaporation measurements. Crop coefficients, K_c, and active root depths, ARD, for farmer use can be determined by national research centers and obtained from extension agencies. Monthly total evapotranspiration, TET_o, is the total of daily records for each month from the date of sowing to the date of senescence or harvest. TET_o can be interpreted as the monthly amount of water required to maintain plant growth. Values are variable for specific locations and are presented here only for clarification.

In general, the amount of water that can be held by the soils of a region are given as:

- heavy (clay) .... 200 mm/m
- medium (loam)............ 150 mm/m
- light (sandy) .... 100 mm/m

The water balance calculation is based upon the amount of water used in the active root depth of the crop. When 50% of the available water has been extracted, it is time to irrigate before plants undergo stress. For example, during April, the ARD is nearly 1.05 m (105 cm) and the amount of water stored before the critical limit is:
- heavy (clay) .... 200 mm/m × 1.05 m × 50/100 = 105 mm
- medium (loam)............ 150 mm/m × 1.05 m × 50/100 = 79 mm
- light (sandy) .... 100 mm/m × 1.05 m × 50/100 = 53 mm

By measuring rainfall and calculating total potential evapotranspiration, the approximate amount of water in the active root zone can be determined.

For example, if wheat is sown on 15 November, 29.25 mm of water will be used by the crop in clay soil (200 mm/m). The initial available moisture for plant growth in the soil profile can be measured by soil sampling or can be estimated. Often it is assumed to be zero because little or no rainfall is received before sowing. Following the example shown in Table 4.5, 88 mm of rainfall occurred before sowing which is enough moisture for sowing and germination if the soil is at field capacity in the active root zone. A work sheet is indispensable for keeping track of the values starting on 15 November.

Table 4.5. Calculation of water balance from monthly values of rainfall and evapotranspiration.

Month	Rain (mm)	ET_o (mm)	K_c	ET_{cr} (mm)	ARD (cm)	RZM (mm)	WB (mm)	NG (mm)	R/P (mm)	W_{req} (mm)	WD (mm)	Irrig. (mm)
Sep	14											
Oct	39											
1-14 Nov.	35				sowing of wheat on 15 November							
Nov	15	29.3	0.40	11.7	12	24	24	27	3	123		
Dec	96	33.8	0.42	14.2	14	28	27	109	81	14		
Jan	73	31.3	0.48	15.0	20	40	109	167	127	20		
Feb	71	43.7	0.65	28.4	50	100	167	210	110	50		
Mar	72	75.0	0.95	71.2	92	184	210	211	27	92		
Apr	34	177.0	1.18	208.9	105	210	211	36	–	105	174	180
May	15	227.2	0.97	220.4	120	240	36	11	–	120	229	200
Jun	–	97.5	0.26	25.4	120	240	11	186	–	120	54	–
Totals	464	376.0										380

During the season until April, monthly rainfall is more than adequate to replenish the soil profile and the summation of the potential evaporation is less than that stored in the soil at the critical limit of 50% available. During April and May, the opposite is true; rainfall is less than evapotranspiration. Wheat is usually in the hard dough stage by the end of May or beginning of June and no further irrigation is required, but the amount of water lost from the soil is the volume of water required for a non-stressed plant condition.

As a further example, a farmer who is production oriented can buy a raingage and some device to measure evaporation, say an evaporation pan. The farmer will need to keep some records to use this equipment. Normally, farmers do not begin with the complexity of the daily water balance calculations as outlined in Chapter 3 but they can start with computations of rainfall and ET_o using the universal pan coefficient of 0.7 for correction.

As noted in Table 4.5, 15 mm of additional rainfall occurred in the latter part of November during the germination period. Even though ET_o was 29.3 mm, this would be enough water to avoid plant stress because of earlier rains which added to the soil moisture storage in the root zone. Keeping track of the daily rainfall and the ET_o, by the end of March, the total rainfall since sowing wheat would be 327 mm and ET_o would be 213 mm.

The farmer can understand that moisture has infiltrated or runoff occurred because more rainfall was available than was lost to evapotranspiration. During April, the picture begins to change and measurements of rainfall (34 mm) and ET_o (177 mm) affect the moisture in the soil profile. Adding the values on a weekly or 10 day interval would show that irrigation was required. For the example, a deficit of 143 mm (177 − 34 = 143) was observed for the month and irrigation was required for replenishment of soil moisture. Using the same technique for the month of May, the deficit was 212 mm, requiring an additional irrigation.

Because irrigation efficiency is only 60-80% for most farmer operated systems, measurements from the raingage and evaporation pan are usually adequate. These calculations point out when irrigation would be required. However, there would be a tendency to over-irrigate in the early season and under-irrigate when the seasonal demand is high. This can be seen in the example given in Chapter 3 when values of K_c and ARD are also included in the calculations. Many drip/trickle irrigation systems rely on values of rainfall and evaporation without resorting to the higher efficiency of calculating the total water balance. Table 4.6 presents the estimated amount of irrigation required by month as a function of soil type when an irrigation is required, i.e. at least 50% of the available water has been used by evapotranspiration.

Table 4.6. Amount of water (mm) to be irrigated each time the plant water requirement is needed for a given month.

Soil type	Dec	Jan	Feb	Mar	Apr	May	Jun
Heavy	30	30	40	60	90	100	40
Medium	30	30	30	40	60	90	40
Light	20	20	25	30	40	75	30

Water Requirements

The water supply for supplemental irrigation systems includes the water requirements for crop and livestock production as well as for domestic use. For agronomic applications of supplemental irrigation, water will be needed during the growing season and the supply should be adequate to support water requirements of the selected crop. Water balance calculations forecast the water requirements and fortify system design for projecting production levels according to the magnitude and distribution of expected rainfall. Selected crops of the Near East and North Africa are presented in Table 4.7; the growth periods and evapotranspiration values are guidelines for predicting water requirements for supplemental irrigation.

Table 4.7. Seasonal evapotranspiration and growth periods for selected crops.

Crop	Growth period	Seasonal evapotranspiration (mm)
Wheat	Nov – May	300– 600
Barley	Dec – Apr	200– 450
Faba beans	Jan – May	300– 600
Cotton	Apr – Nov	550–1150
Sugar beet	Oct – Jul	450–1100
Maize	Mar – Jun	400– 750
Potatoes	Feb – Jun	350– 650

This information also acts as a guide to water storage requirements. Table 4.8 presents average annual volumes of irrigation water for plant growth in specific climatic zones. The water could be available either in an aquifer, river, reservoir, or an embankment type farm pond.

Table 4.8. Required capacities of irrigation farm ponds assuming the pond is filled annually.

Climate zone	Annual rainfall (mm)	Volume of water for irrigation (m^3/ha)		
		Vegetable crops	Field crops	Perennial crops
Humid	1000–1500	3.000	4,500	5,300
Moist	800–1000	4,500	6,000	7,600
Semi-arid	300– 800	9,000	12,250	18,300
Arid	Under 200	Small farm ponds are not reliable		

Plants respond positively when soil-water is available during a sensitive growth stage. Table 4.9 shows the best potential of limited water supplies for selected crops with water application scheduled at the moisture sensitive stages of plant growth. To use this information, the soil profile is assumed to be at field capacity at sowing time.

Regional Application of Water Balance Methods

Table 4.9. Moisture sensitive growth stages for selected crops.

Crop	Moisture sensitive period			
	Shooting	Rooting	Heading/flowering/earing	Grain/fruit formation
Wheat	————		————	
Barley	————		————	
Lentils			————	
Broad beans			————	
Maize				————
Sorghum		– – – – – – – –		
Millet			– – – – – – – – –	
Groundnuts				– – – – – – – – – –
Tomatoes				————
Cotton				————
Sugar beet	....			
Potatoes	....			

———— Clearly defined sensitive phase
– – – – – – – – Plant insensitive but responds
.... No clear indication

Estimated water requirements for household use and livestock in the Near East and North Africa are shown in Table 4.10. In general, the water requirement per farm unit is divided into 4 classes with relative percentages for:
– domestic purposes 10%
– farm and animals 5%
– irrigation 80%
– waste 5%

Waste is from water conveyance system losses, e.g. open ditches, pipe joints, general leaks, and defective equipment. Seepage and evaporative losses of water from storage must also be included as part of the water requirement during planning.

Table 4.10. Daily water requirements for domestic and livestock use in the Near East and North Africa.

Use	Water requirements (l/day)
Domestic:	
Per person (includes cooking, drinking, and washing)	10–60
Animals	
Beef cattle	35
Dairy cattle	45
Mature sheep	10
Goats	8
Horses	30–45
Chickens/100 head	8–15

Phenology of Cereals

In the Near East and North Africa, cereals are usually sown during November and occasionally during October and December. After germination, if air and soil are dry, the exposed cereal plants suffer little and can readily survive winter temperatures which do not fall below $-10°C$. For the most satisfactory growth and development of grain, a cool, moist growing season is best followed by a bright, dry and warm ripening period of 6-8 weeks with mean temperatures of 18-19°C.

With supplemental irrigation, germination of cereals in a moist soil takes about 5-7 days. Cereal plants are extremely resistant to drought during the first stages of growth. Crops that germinate during mid-November with adequate soil moisture enter the winter with good vegetative cover and resistance to low winter temperatures. During February, rapid vegetative growth and oncoming drought (increasing vapor pressure deficits) can accelerate the rate of water use. When anthesis (flowering) is complete, the rate of water use rapidly declines near the beginning of May, coinciding with diminishing rainfall.

Grain head (spikelet) development depends upon 4 main factors to insure a stable yield:
1. early germination results in early heading; however, late germination may not delay heading which is dependent on phototropism and other characteristics of cultivars;
2. lack of water during the boot stage (vegetative development) hastens maturation; however, deficiency of water early in the vegetative growth period affects the time of heading;
3. nitrogen deficiency causes early heading; and,
4. low winter temperatures and phosphate deficiency may delay heading.

Method for Determining Growth Stages of Cereals

In the past, comparisons between research findings have been confused by imprecise reference to growth stages of cereal plants at the time of treatment or assessment. The confusion was reflected in vague recommendations for the timing of pesticide applications. A notation system (decimal code) was designed to consider the needs of all disciplines concerned with cereal production and has achieved recognition by research workers, extension advisors, and farmers. The notation system provides descriptions of the earlier stages of cereal growth and is applicable to all small grain cereal species growing in a broad range of climatic zones.

Principal growth stages

The decimal code in Table 4.11 is based on 10 principal growth stages with approximate dates of occurrence for spring wheat at Tel Hadya, Syria:

- 00 – Seeding (15/11)
- 0 – Germination (17/11)
- 07 – Emergence (28/11)
- 1 – Seedling growth (10/12)
- 2 – Tillering (20/1)
- 3 – Stem elongation (1/2)
- 4 – Booting (25/2)
- 5 – Inflorescence emergence (5/3)
- 6 – Anthesis (25/3)
- 7 – Milk development (5/4)
- 8 – Dough development (25/4)
- 9 – Ripening (10/5)
- Harvesting (5/6)

Such broad descriptions are adequate for many purposes. Greater detail is often necessary and is provided by further sub-divisions into secondary growth stages.

Table 4.11. Descriptions of the principal and secondary growth stages of the decimal code for cereals.

```
0  Germination                          5  Inflorescence (ear/panicle) emergence
   00  Dry seed                            50  ---
   01  Start of imbibition                 51  First spikelet of inflorescence
       (water absorption)                      just visible
   02  ---                                 52  ---
   03  Imbibition complete                 53  1/4 of inflorescence emerged
   04  ---                                 54  ---
   05  Radicle (root) emerged              55  1/2 of inflorescence emerged
       from caryopsis (seed)               56  ---
   06  ---                                 57  1/4 of inflorescence emerged
   07  Coleoptile (shoot) emerged          58  ---
       from caryopsis                      59  Emergence of inflorescence
   08  ---                                     completed
   09  Leaf just at coleoptile tip

1  Seedling growth                      6  Anthesis (flowering)
   10  First leaf through coleoptile       60  ---
   11  First leaf unfolded                 61  Beginning of anthesis
   12  2 leaves unfolded                   62  ---
   13  3 leaves unfolded                   63  ---
   14  4 leaves unfolded                   64  ---
   15  5 leaves unfolded                   65  Anthesis halfway
   16  6 leaves unfolded                   66  ---
   17  7 leaves unfolded                   67  ---
   18  8 leaves unfolded                   68  ---
   19  9 or more leaves unfolded           69  Anthesis complete

2  Tillering                            7  Milk development
   20  Main shoot only                     70  ---
   21  Main shoot and 1 tiller             71  Caryopsis (kernel) water ripe
   22  Main shoot and 2 tillers            72  ---
   23  Main shoot and 3 tillers            73  Early milk
   24  Main shoot and 4 tillers            74  ---
   25  Main shoot and 5 tillers            75  Medium milk
   26  Main shoot and 6 tillers            76  ---
   27  Main shoot and 7 tillers            77  Late milk
   28  Main shoot and 8 tillers            78  ---
   29  Main shoot and 9 or more tillers    79  ---

3  Stem elongation                      8  Dough development
   30  Pseudostem (leaf sheath)            80  ---
       erection                            81  ---
   31  First node detectable               82  ---
   32  2nd node detectable                 83  Early dough
   33  3rd node detectable                 84  ---
   34  4th node detectable                 85  Soft dough
   35  5th node detectable                 86  ---
   36  6th node detectable                 87  Hard dough
   37  Flag leaf just visible              88  ---
   38  ---                                 89  ---
   39  Flag leaf ligule just visible

4  Booting                              9  Ripening
   40  ---                                 90  ---
   41  Flag leaf sheath extending          91  Caryopsis hard
   42  ---                                     (difficult to divide)
   43  Boots just visibly swollen          92  Caryopsis hard
   44  ---                                     (not dented by thumbnail)
   45  Boots swollen                       93  Caryopsis loosening in daytime
   46  ---                                 94  Overripe, straw dead & collapsing
   47  Flag leaf sheath opening            95  Seed dormant
   48  ---                                 96  Viable seed giving
   49  Awns swollen                            50% germination
                                           97  Seed not dormant
                                           98  Secondary dormancy induced
                                           99
```

Secondary growth stages

Each of the principal stages has provision for a sub-division into 10 secondary stages. These are a second series of digits extending the scale from 00 to 99. Positions are left blank when sub-divisions have no description. Figures 4.1, 4.2, and 4.3 in schematic detail illustrate a range of these values from seed (germination) to anthesis complete (flowering).

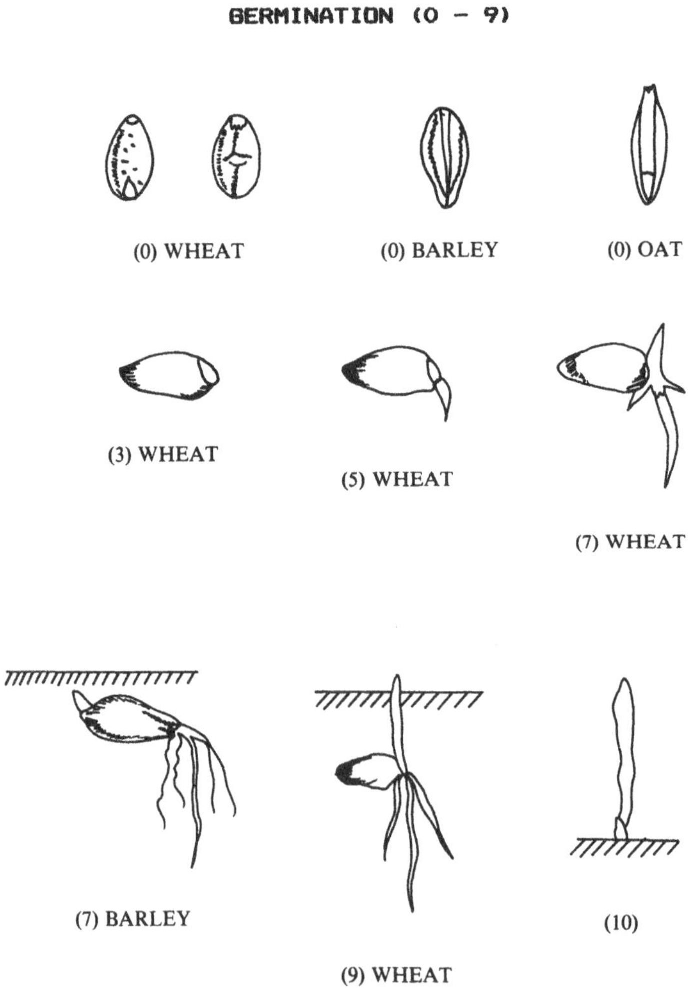

GERMINATION (0 – 9)

(0) WHEAT   (0) BARLEY   (0) OAT

(3) WHEAT   (5) WHEAT   (7) WHEAT

(7) BARLEY   (9) WHEAT   (10)

Figure 4.1. Illustration of decimal code germination (0.00-0.09) and first leaf through coleoptile (1.10) for wheat, barley, and oats.

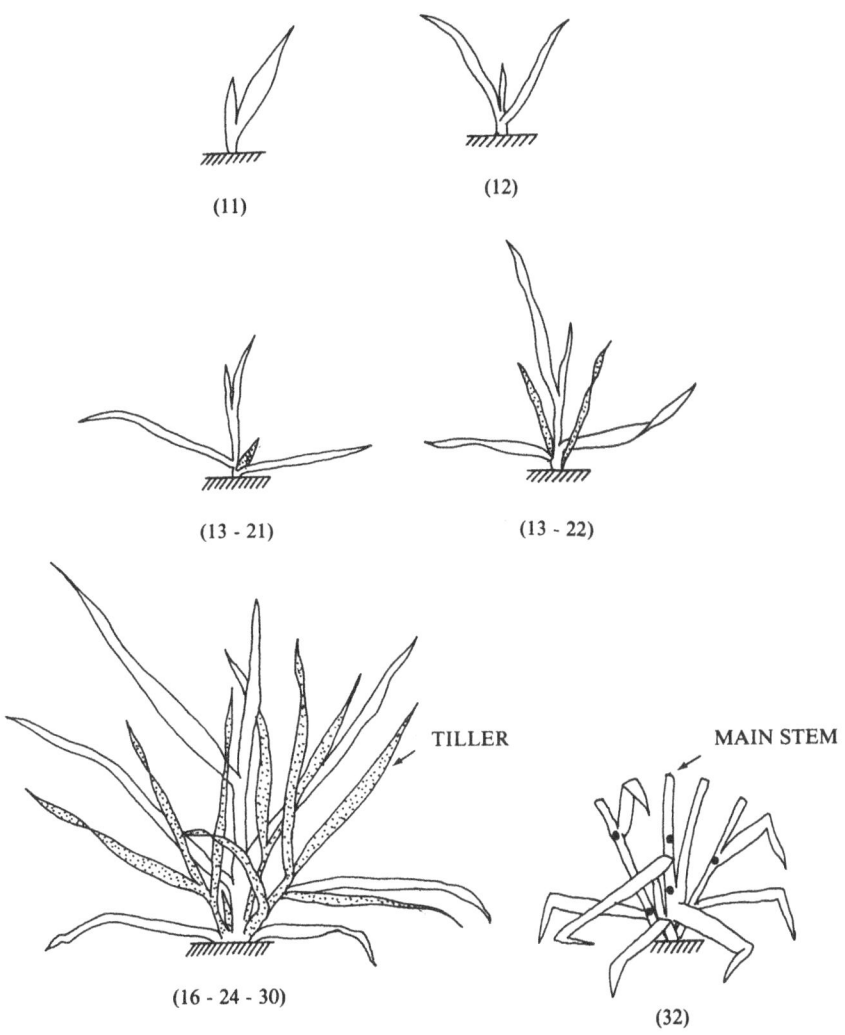

Figure 4.2. Illustration of decimal code for seedling growth to stem elongation (1.11-3.32) for cereals.

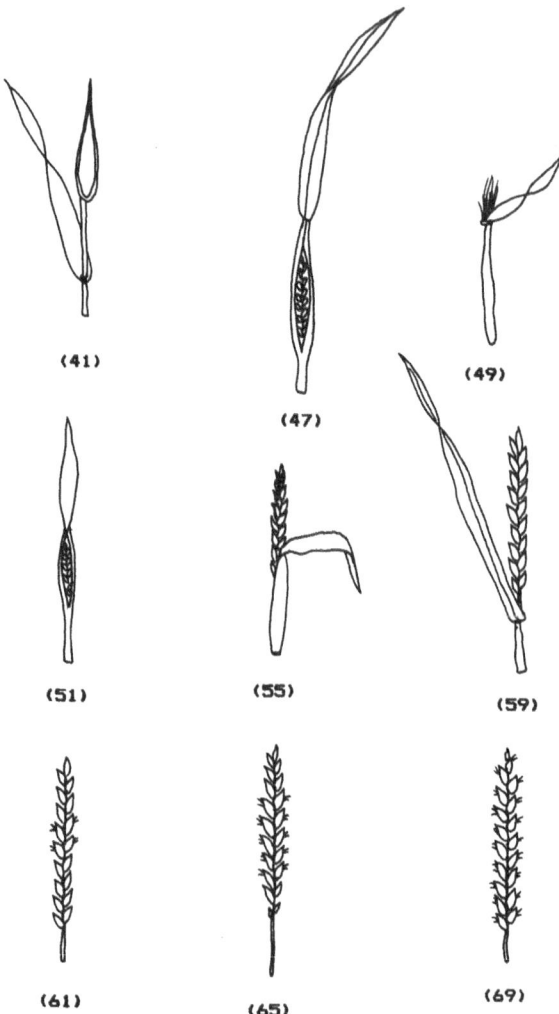

Figure 4.3. Illustration of decimal code for booting to anthesis complete (4.41-6.69) for wheat.

The decimal code differs from previous keys by describing individual plants rather than classifying crop growth stages. Figure 4.4 shows a general key for a cereal plant with two types of coding, decimal code and a notation system showing stages of growth which correspond to the decimal code for use by farmers without botanical training. Figure 4.5 illustrates the parts of an unfolded leaf at decimal code 1 with the stem and node code 3. Descriptions from different parts of the code can be used concurrently. For example, a plant might have 6 leaves unfolded (16), a main shoot and 4 tillers (24), and the first node detectable (31), where only an indication of the plant's development is needed. In this case, the highest decimal code (31) would be an adequate description.

Regional Application of Water Balance Methods 93

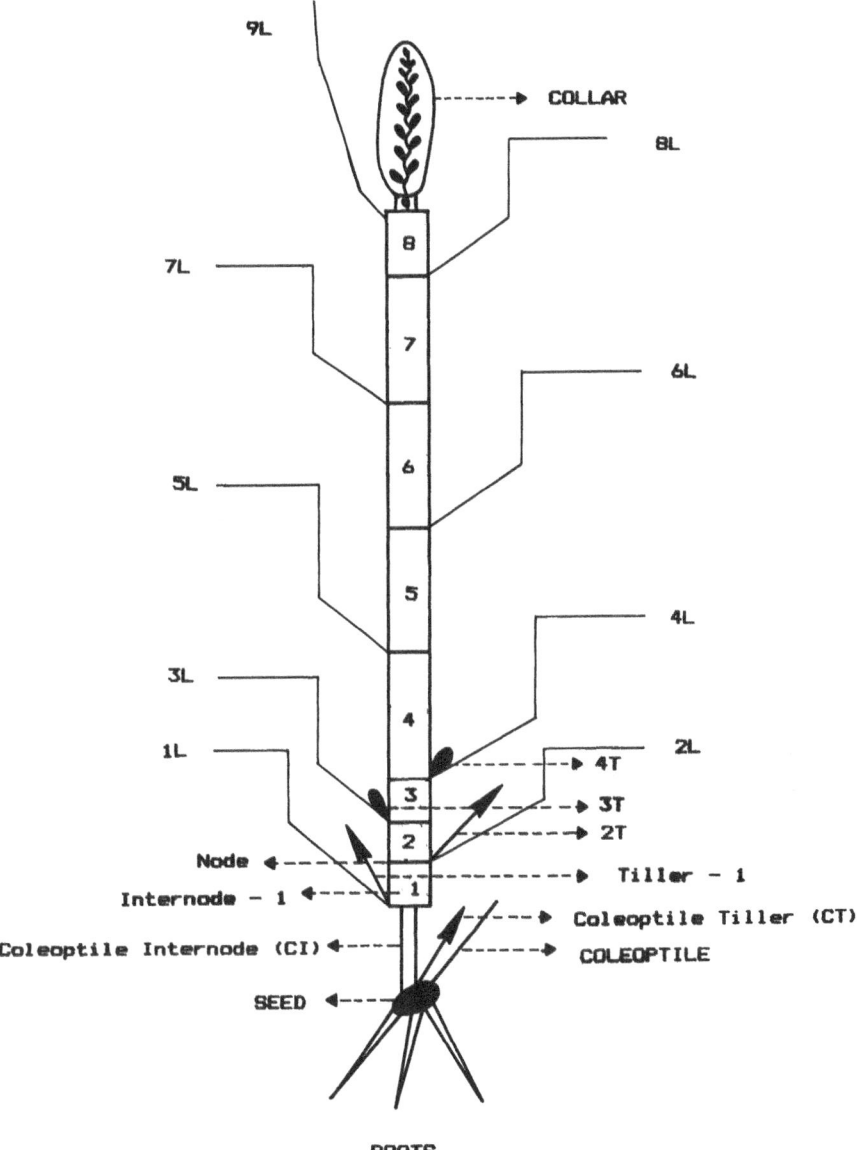

Figure 4.4. Diagram of cereal plant showing decimal code and common equivalents.

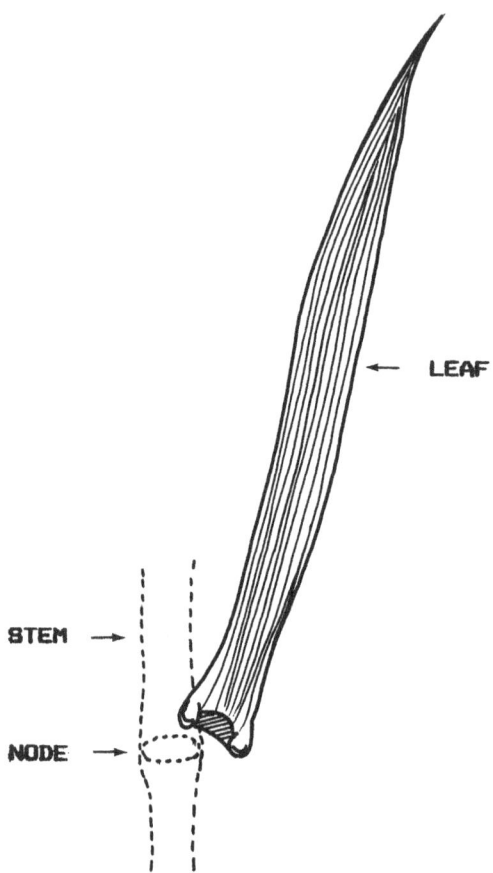

Figure 4.5. Unfolded leaf with stem and node.

The notation system deviates during the inflorescence emergence and anthesis (decimal code 5 and 6) from the principle of describing individual plants. In some instances, for a uniformly developed crop, odd numbers have been assigned to growth stages reached simultaneously by all ears and even numbers to irregularly developed crops when 50% of the ears are at the stage described by the decimal code. Even for irregularly developed crops, describing single ears is still simpler at these growth stages in the same manner as descriptions of earlier stages of single plants which use only odd number positions of the secondary decimal code.

When specifying the stage of development of a crop or plant population, each plant or a random sample of plants is described using the decimal code and the central tendencies (mode and range) of these observations is recorded. For example, the range of a sample of plants from an irregularly developed crop may vary from 5 leaves unfolded (1.15) with a main shoot and 2 tillers (2.22) to 7 leaves unfolded (1.17). a main shoot with 5 tillers (2.25), and the first node detectable (3.31). The most commonly occurring stage (mode) can be represented by a plant with 6 leaves unfolded (1.16), a main shoot with 3 tillers (2.23), and an erect pseudostem (3.30). The decimal code chosen to represent the growth stages of the crop would be 3.30.

The descriptive phases of the crop can be used to indicate when treatments should be applied. The significant feature of a typical plant can be specified and the crop treated when most plants conform to that description. If timing of treatment is critical for crop protection, treatment recommendations should confirm that application of fertilizer, herbicide, pesticide, irrigation, etc., at other stages of development can cause damage to the plants.

Chapter 5

Soil Water Relationships

An understanding of the interacting relationships between soil and water is imperative for irrigators who strive for the most efficient use of water for their crops and soils. The complex interaction between the components of water, soil, and plant narrows technical options from which to choose alternative management regimes. What occurs in the soil profile can be explored in the context of soil water being a medium or solution to transport essential elements to plant roots. Figure 5.1 depicts soil as a medium for nutrients, water, air, micro-organisms, and plant roots.

The soil profile is usually thought of as a natural body which is porous to water and air and provides an environment for plant roots. Soils can be conceptualized as 1-2 m in depth with fine granular pores, weathered, with the surface interspersed with organic residue from dead and decaying plant life. Soil properties vary significantly with space and time and, therefore, require extensive sampling and analysis for a thorough evaluation.

Soil Texture

Soil texture is a function of grain size:

soil texture = F(grain size).

Grain size is determined using sieves: this process is termed mechanical analysis. The size fraction of the larger particles is confirmed by drying the soil, pulverizing the large aggregates, and sifting this material through varying sizes of sieves. The weight of soil particles caught on each sieve is plotted into a particle size distribution curve (Figure 5.2). The mixture or arrangement of soil particle sizes affects the water holding capacity and the relationship between hydraulic conductivity and soil moisture content. The percentage of sand, silt, and clay in soil determines the texture class presented in Figure 5.3.

E.R. Perrier and A.B. Salkini (eds), *Supplemental Irrigation in the Near East and North Africa*, 97-105.
© 1991 *ICARDA*.

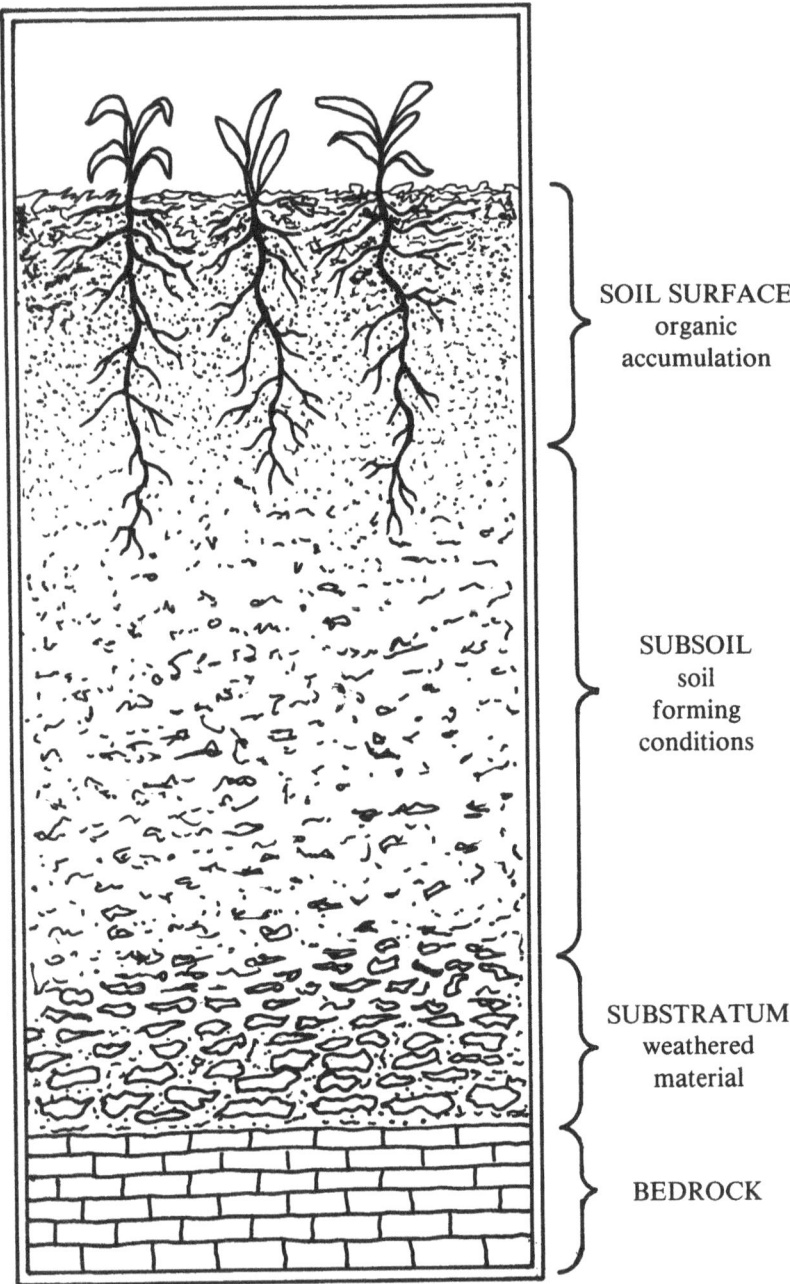

Figure 5.1. Diagram of a soil profile showing the relative positions of the soil horizons.

Soil Water Relationships

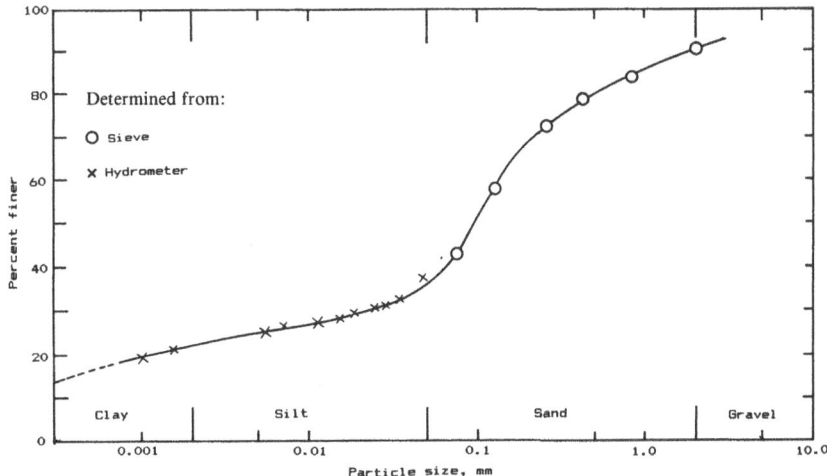

Figure 5.2. Particle size distribution curve for well-graded gravel (GW) and sand (SW).

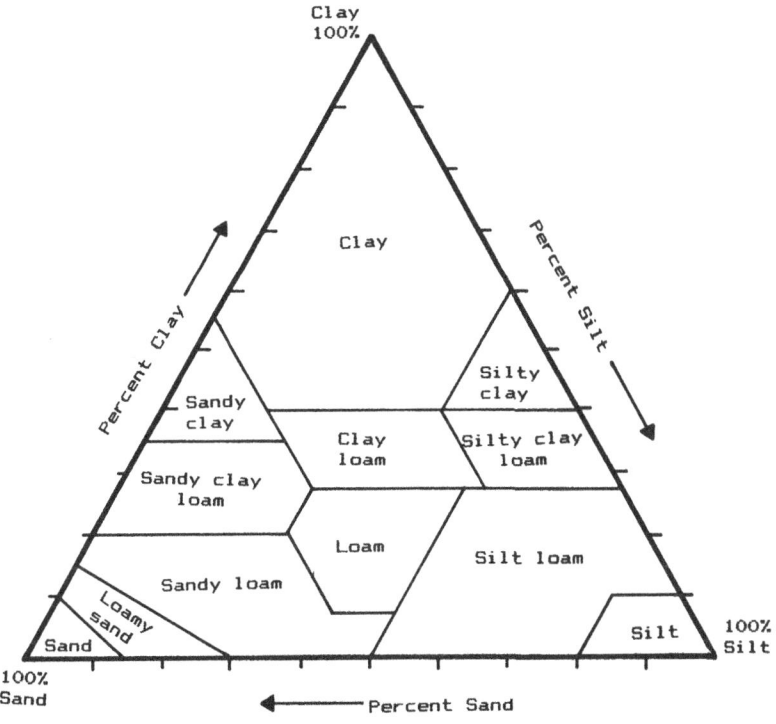

Figure 5.3. Diagram for textural name of a soil as determined from a mechanical analysis.

Soil Structure

Groups of fine soil particles can adhere to form coarse aggregates. Specific groups of these aggregates determine the structure. Large or macro-sized aggregates in the soil structure are called peds. Structure is a field term descriptive of the grouping and arrangement of aggregates. Soil particles which form these aggregates are bound by surface tension of water, weak cementing agents, and organic matter. Plants are affected by structure because the size of the soil pores influences root development, e.g. soil strength, water-holding capacity, aeration, and drainage characteristics.

Figure 5.4 shows an Aridisol with a structural development within the soil profile. Cracks between aggregates and macro particles disorder water movement (initial infiltration) and aeration. For example, a clay soil with structure (such as the red Vertisols at ICARDA's research center) can have a saturated hydraulic conductivity more than one hundred times greater than a soil of the same clay content but with little or no structure.

Soil structure is affected by tillage, compaction, organic matter, formation history, and climate. Structure is preserved and improved by careful cultivation and irrigation but can be rapidly destroyed by mismanagement.

Bulk Density

Bulk density of mineral soils is a measure of the total soil space which is occupied by solids and air. The bulk density of the soil, ϱ_b (g/cm³) is the average weight, W (g), of dry soil solids divided by the unit volume of soil, V (cm³), defined as:

$$\varrho_b = \frac{W}{V} = \frac{\text{weight of soil}}{\text{volume of soil}}.$$

The bulk volume includes the actual volume occupied by both the solid particles or soil grains and the pore space. Bulk densities vary with the actual particle density of the solids and the packing of the particles and void spaces. They are dependent on the degree of compaction, soil moisture content, clay type, aggregation, aeration, etc. Common values of bulk density range from 1.00 to 1.80 g/cm³ for clay through sandy soils.

Bulk density measurements can be used to calculate soil porosity, η, which approximates the maximum percentage of water that the soil will hold (saturation percentage):

$$\eta = 100 - \frac{\varrho_b \times 100}{\varrho_s}.$$

The soil particle density, ϱ_s (g/cm³), is normally assumed to be 2.65 g/cm³, but can range from less than 2.40 g/cm³ in organic soils to more than 2.75 g/cm³ in heavy mineral soils. Exact values of soil particle density are determined by laboratory analysis.

Soil Water Relationships

Figure 5.4. Soil profile with cracks and aggregates: (a) soil pit and (b) panorama of soil pit area.

Field method for determining bulk density

Soil bulk density is determined by collecting a sample of a known volume of soil (volume = 3.14 d² h/4 where d = diameter and h = height), oven drying the sample, and determining the weight of the dry sample. The most difficult problem is accurately collecting the soil sample as it exists in the cultivated state without disturbing the walls or losing soil from the hole.

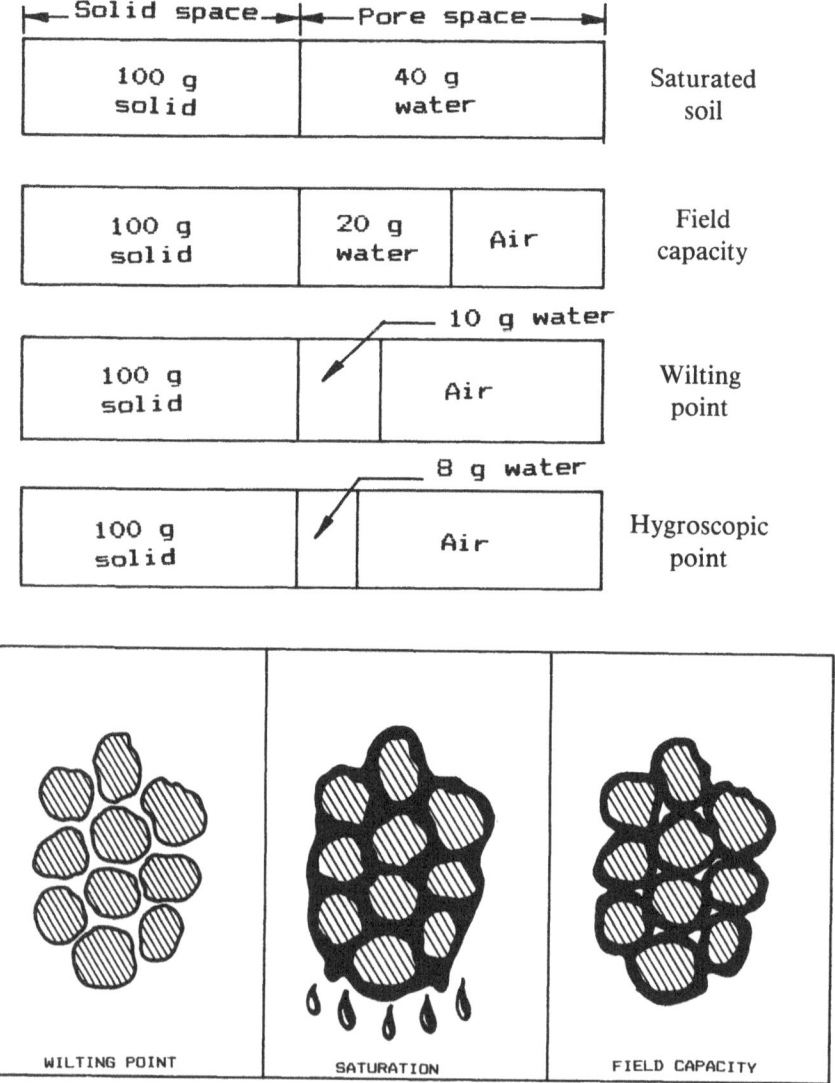

Figure 5.5 Diagram showing an elemental volume of soil which delineates the soil particles and voids as well as air and water.

Soil Water Relationships

Equipment and materials needed for the field method are:
1. at least 10 kg of standard sized sand: washed sand particles which have passed a number 20 or 60 sieve are usually recommended;
2. 1,000 ml graduated cylinder filled to the mark with clean sand;
3. a flat metal field plate 30 cm on the square and 6-8 mm thick with a 12 cm hole in the center and a large sheet of polyethylene plastic (1 m square) with an 18 cm diameter hole at the center to assist soil sampling;
4. a 30 cm ruler;
5. a large spoon, a teaspoon, and a trowel;
6. several large soil sample cans;
7. a drying oven set at 105°C, to be used 24 hr; and,
8. a gram-balance with a weighing sensitivity of 0.1 g.

The procedure for collecting the soil sample and determining bulk density requires 4 steps.
1. Prepare the soil surface by removing vegetation and loose soil; smooth the surface; lay the plastic cover over the selected surface area; place the field plate over the hole in the plastic. Usually, only surface samples are required for irrigation; no sub-surface samples are taken.
2. Through the hole in the plate, carefully excavate a hole 12 cm in diameter and about 5 cm in depth. Place the excavated soil sample in the large soil cans and determine the oven-dry weight. Use any utensil to excavate the hole which gets the job done conveniently with the least disturbance to the natural condition of the soil. If the soil is loose, the large sheet of clean polyethylene plastic will catch that part of the sample which would otherwise be lost.
3. To determine the volume of the previously excavated hole, refill the hole with sand to the bottom of the field plate. Pour the sand from the graduated cylinder.
4. Calculate the soil bulk density knowing the oven-dry weight of the excavated soil and the volume of sand used to refill the excavated hole.

Figures 5.5 and 5.6 display the relation of solids, voids, and water within an elemental volume of soil from a profile.

Some additional terms, definitions, and comments for a better understanding of soil bulk density are:

soil bulk density = mass of dry soil/bulk volume
("bulk" refers to the total volume)
particle density = mass dry soil/volume of soil particles
= 2.65 g/cm³ for typical agricultural soil
(real specific gravity)
soil moisture content = mass of water/mass of dry soil (weight basis)
soil moisture content = volume of water/bulk volume of soil
(volume basis)

Soil moisture content calculated on a volume basis is a convenient form for depth of water calculations and, typically, ranges from 0-60%. Soil moisture content on a volume basis is equivalent to a depth of water per unit depth of soil, i.e. 30% or 0.3 = 0.3 m (30 mm) of water per meter depth of soil.

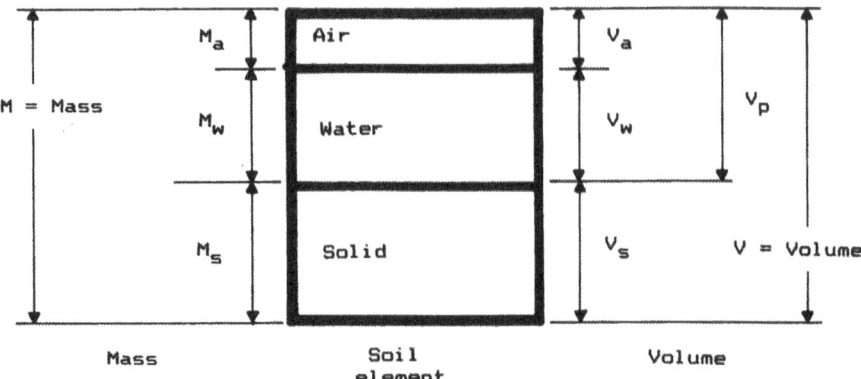

Figure 5.6 Diagram showing a typical arrangement of an elemental volume of soil for the solids, water and air.

Soil Porosity

The ratio of the volume of voids (non-solids) in a fixed total volume of soil (including water and air) to the volume of solids, is termed the porosity (pore space) and when multiplied by 100 is expressed as a percentage. There is a wide variation in the shape and size of soil particles and, for irrigated soils, the porosity ranges between 35-55%. Soil porosity (η) = volume of pores/bulk volume of soil. For common agricultural soils:

$$\eta = 1.00 - 1.33/2.65 = 0.50$$

or, when multiplied by 100 to determine percentage, 50% of the bulk soil.

Example for bulk density and porosity

Figure 5.7 presents an example to explain how to determine bulk density and soil porosity.

Soil Water Relationships

Determine bulk density and porosity for the top 6 cm of a soil.

The measured values are:

Sample volume = 678.6 cm³
Total wet mass (weight) = 1111.5 gm
Total oven dried mass = 855 gm

To calculate bulk density, ϱ_b:

$$\varrho_b = \frac{M_s}{V} \quad 855/678.6 = 1.26 \text{ gm/cm}^3 \text{ (common value)}$$

To calculate soil moisture content (dry weight basis), θ_m:

$$\theta_m = \frac{M_w}{M_s} \quad (1111.5-855)/855 = 256.5/855 = 0.30 \text{ or } 30.0\%.$$

To calculate the soil moisture content (volume basis), θ_v:

$$\theta_v = \frac{V_w}{V} \quad (256.5/1)/678.6 = 0.378 \text{ or } 37.8\%.$$

Another technique to calculate soil moisture content (volume basis), θ_v:

$$\Theta_v = \Theta_m \times \frac{\varrho_b}{\varrho_w} = 30.0(1.26/1) = 37.8\%.$$

Now, to calculate the porosity, η:

$$\eta = 1 - \frac{\varrho_b}{\varrho_s}$$

assume $\varrho_s = 2.65$ (common value for agricultural soil)

$$\eta = 1 - (1.26/2.65) = 0.475 = 47.5\%$$
(common value for agricultural soil).

Agricultural soils have a porosity of about 50% of which the soil pore volume is air or water. A low porosity can seriously affect the productive capability of soils because of its influence upon the water-holding capacity and transport of air, water, and movement of roots through the soil.

The total soil moisture in the profile is given by:

$$37.8/100 \times 1000 \text{ mm/m} = 378 \text{ mm/m}.$$

To calculate the amount of air in the bulk sample:

$$\% \text{ air} = \frac{V - V_s - V_w}{V}.$$

If $\varrho_s = 2.65$ g/cm³, then $V_s = 855/2.65 = 322.6$ cm³ and it follows:

$$\% \text{ air} = \frac{678.6 - 322.6 - 256.5/1.0}{855} \times 100 = 99.5/855 \times 100 = 11.6\%.$$

Figure 5.7. Example calculation for determining bulk density and soil porosity.

Chapter 6

Movement of Water in Soils

Investigation of water movement within soil profiles is essential to a better understanding of the process of irrigation. Soils must be capable of cyclic rewetting when roots need water, otherwise there will be a reduction in plant growth and a decline in production. The total water potential may be defined in terms of energy levels of soil water:

$$\psi_T = \psi_M + \psi_g + \psi_p + \psi_O + .. + .. + ..$$

where M is the matric potential, g is the gravitational potential (elevation), p is the pressure potential, and O is the osmotic potential. Additional terms are theoretically possible denoted here on the right-hand side of the equation by the use of plus and dot.

Figure 6.1 shows pressure, matric, gravitational, and osmotic potentials are the forces which create moisture flow when a difference in potential occurs within a soil medium. The potential difference between two points in a given soil profile regulates the velocity and volume flow between these points; the potentials are specified by the moisture content or degree of saturation. When soil water potentials are at equilibrium within the soil profile (no moisture flow), ψ_h is constant. Therefore, following Figure 6.1, as ψ_z (elevation) increases, ψ_p decreases, e.g. at 20 m above the water table ψ_p = –200 cm = –2 bars; whereas, ψ_p = 0 at the water table or free water surface.

The matric (or matrix) potential is that portion of the water potential attributed to the attraction of the soil matrix (i.e. particle distribution, aggregation, and peds) to water and includes the capillary pressure. As soil water exhibits either a saturated or an unsaturated condition, but not both simultaneously, the matric potential specifies only the unsaturated soil which has no pressure potential. The term, soil matric potential, is used because pressures are limited to –1.0 bar; however, moisture tension forces can exceed –15.0 bars. The matric potential of water above the water table is always negative but becomes zero when the soil is saturated at or below the water table. The matric potential is measured directly by a tensiometer. The measurement units are work per unit volume, which is the determinate quantity for pressure. In Table 6.1, pressure is expressed in "bars" (1 bar = 0.9869 atmospheres =

E.R. Perrier and A.B. Salkini (eds), Supplemental Irrigation in the Near East and North Africa,
107-115.
© 1991 *ICARDA.*

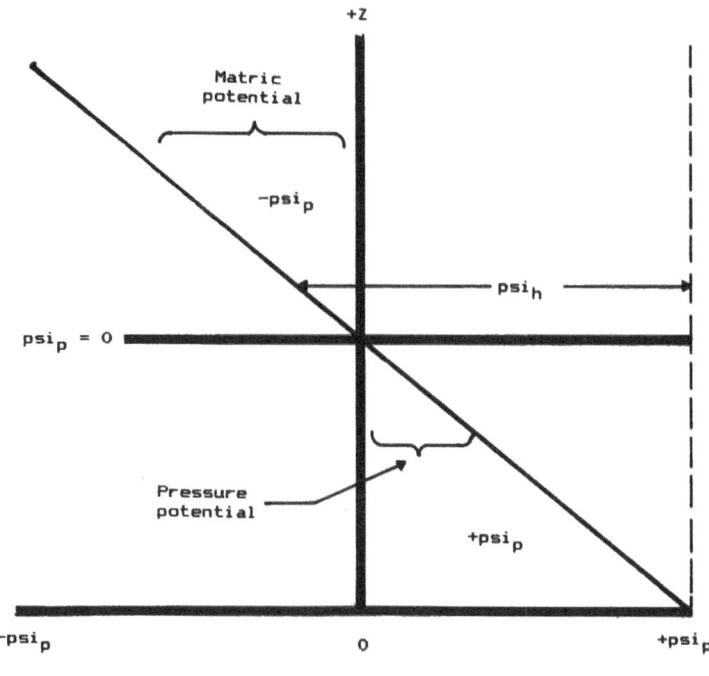

Figure 6.1. Diagram of soil water potential as a function of position, Z.

14.5 pounds/inch² (psi)) but can be expressed on the pF scale where pF = $\log_{10}h$ and h is the tension head expressed as the height of a water column in centimeters. Problems of the pF scale are that it has no zero (saturation) and it is a logarithmic scale.

The gravitational potential is energy associated with the vertical location of water or the elevation of water compared to an arbitrary reference level. If the reference level is below the point of measurement, the gravitational potential is

Table 6.1. Relationship between soil water tension (suction), centimeters of water, pF, and relative humidity.

Soil water suction		pF	Relative humidity
(bars)	(cm of water)		(%)
0.01	10.2	1.01	100.0
0.1	102.0	2.01	99.993
0.5	510.0	2.71	99.964
1.0	1020	3.01	99.927
5.0	5100	3.71	99.637
15.0	15,300	4.18	98.9
30.0	30,600	4.49	98.0

positive; if the reference level is above the point of measurement, the gravitational potential is negative. For example, if water is above the reference level, the force is downward and is pushing; if water is below the reference level, the force is still downward but is pulling.

Pressure potential is important when measuring hydraulic conductivity, depth to water table, and deep percolation. The sum of the pressure potential and the gravitational potential (elevation, z) is the piezometric head, h = ψ_p + z and can be measured in the field with a piezometer. The hydraulic potential is stated:

$$\psi_h = \psi_g + \psi_p$$

and is used to measure free water movement within a soil profile.

The osmotic potential, sometimes called solute potential, expresses the level of solutes in water and is a major consideration for saline and sodic soils. In these soils, leaching of excess salts may be necessary. For supplemental irrigation, the presence of salts would be minimal with 1-2 irrigations per year; however, if the salt content of irrigation water is high, then some monitoring of salinity is required. The osmotic potential is zero for pure water and becomes increasingly negative as concentration of solutes rises. Osmotic potential can be estimated from $\psi_o = 0.36 \times EC$, where EC is the measured electrical conductivity in mmho/cm of a saturated extract from a sample of the soil profile.

Saturated Flow

Soil water moves in response to differentials in soil water potentials. Uptake of water by plant roots can cause this difference in potential which forces soil water to move towards the root. Also, during irrigation, soil water content near saturation has a different potential than the water in the soil profile which is in a drier state. Consider a soil segment of a cross sectional area A (cm²), with length L (cm), saturated with water. If there is a piezometric head difference (pressure potential gradient or difference) across the soil section Δh (cm), then the volume discharge of water Q (cm³/hr), is given by Darcy's Law (discussed in Chapter 7) as:

$$Q = KA \frac{\Delta h}{L}.$$

K (cm/hr) is called the hydraulic conductivity and is the rate of flow through a unit cross section which is normal to the flow under a unit gradient (unit pressure potential difference).

The hydraulic conductivity depends upon soil properties of texture, porosity, shape of pores, and properties of the soil water solution. The value can vary in time and from point to point in the field. A swelling or shrinking soil will alter the hydraulic conductivity. Hydraulic conductivity is concerned with flows at, or near, saturation (0 bars of matric potential or pF = 1.0) such as the wetting front of water moving through the soil profile during irrigation.

Field piezometer method

The piezometer method is used to measure the hydraulic conductivity of a given depth in a soil profile. A hole is bored to the desired depth below the water table with a smaller diameter than the steel piezometer tube of 2 to 3 cm to be installed. The tube is sharpened at one end and a small plug or rivet is placed in the bottom before the tube is driven into the hole. When the tube has reached the bottom of the hole, a rod is placed into the center of the tube and the plug or rivet is driven 5 cm beyond the base of the tube. The cavity below the tube is where the hydraulic conductivity is measured and this is a function of the rate of entry of water into the cavity from the surrounding soil.

A plastic tube is pushed into the piezometer hole to the plug or rivet. Then a finger is put on the top of the tube which is quickly removed containing the water (under suction) from the 5 cm deep hole. This procedure permits the soil pores of the cavity walls to be flushed out. Flushing is repeated until the rate of rise in the piezometer is the same as a previous extraction. The plastic tube is used to determine the depth of water in the piezometer by sealing the outer end then lowering it slowly into the piezometer while moving the plastic tube up and down to hear a popping sound when the air-tight tube comes in contact with the water. The elevation of the water level is marked and noted. The water levels and times of observations are recorded and the piezometer formula is used to calculate the hydraulic conductivity:

$$K = 3769 \times \frac{r^2}{C} \times \frac{dh}{dt} \times \frac{1}{A + R - 2B},$$

where K = hydraulic conductivity (cm/hr);
  r = outside diameter of plug or rivet (cm);
  C = determined from Figure 6.2 (cm);
  dh = rise of water level at timed intervals (cm);
  dt = change in time required for increments in h (min);
  A = depth to water level at start of test (cm);
  R = depth to water level at end of test (cm); and,
  B = depth to static water level (water table depth) (cm).

Here the quadratic relation for the C function (Figure 6.2) was given by:

$$C = 13.3922 + 7.7511 \times r - 0.272562 \times r^2$$

Movement of Water in Soils

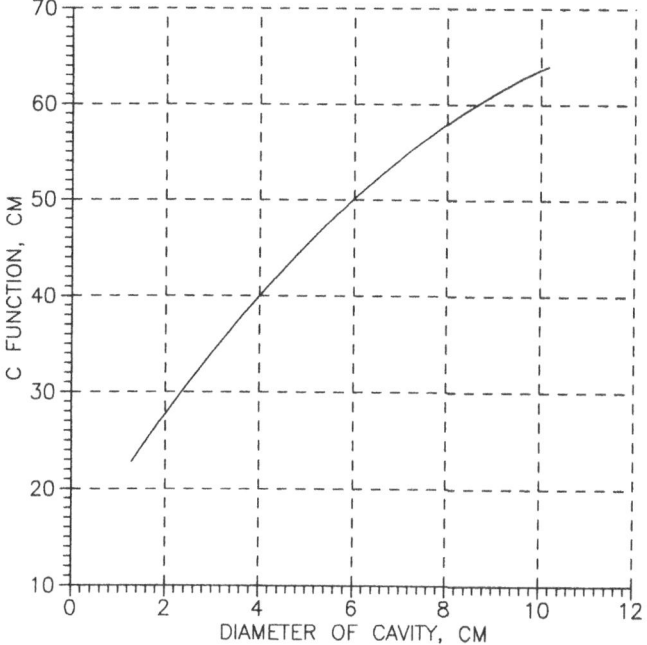

Figure 6.2. C function.

Example for determining hydraulic conductivity

The following data are proposed as typical values for measurements of hydraulic conductivity and used in conjunction with Figures 6.3 and 6.4:

r = 1.3 cm,
C = 31.75 cm,
dh = 43.28 cm,
dt = 20 min,
A = 123.44 cm,
R = 80.16 cm,
B = 60.96 cm,

where,

$$K = 3769 \times \frac{1.3^2}{31.75} \times \frac{43.28}{20} \times \frac{1}{123.44 + 80.16 - 121.92} \text{ cm/hr}$$

K = 5.32 cm/hr.

Here the cubic equation for the residual drawdown, Rd, (Figure 6.4) versus time in minutes, T, was given by:

$$Rd = 61.5711 - 1.87719 \, T - 0.028305 \, T^2 - 0.00103353 \, T^3$$

Field hydraulic conductivity test: piezometer method

Piezometer Number _____

Location _____ Date _____

Piez. Depth _____ cm   Auger Dia. _____ cm   Piez. Dia. _____ cm

Cavity Length _____ cm   Times flushed _____

time	Elapsed time (min)	dt (min)	Distance to water surface from reference point			dh A − R (cm)	Residual drawdown R − B (cm)
			Before pumping B (cm)	After pumping A (cm)	During recharge R (cm)		
11:00			60.96				
:01	0			123.44			62.48
:06	5				112.78		51.82
:11	10				102.11		41.15
:16	15				91.44		30.48
:21	20	20			80.16	43.28	18.59
:26	25				71.63		10.67
:31	30				67.97		7.01
:36	35				65.53		4.57
:41	40				64.01		3.05

Figure 6.3. Data sheet for determining field hydraulic conductivity using the piezometer method.

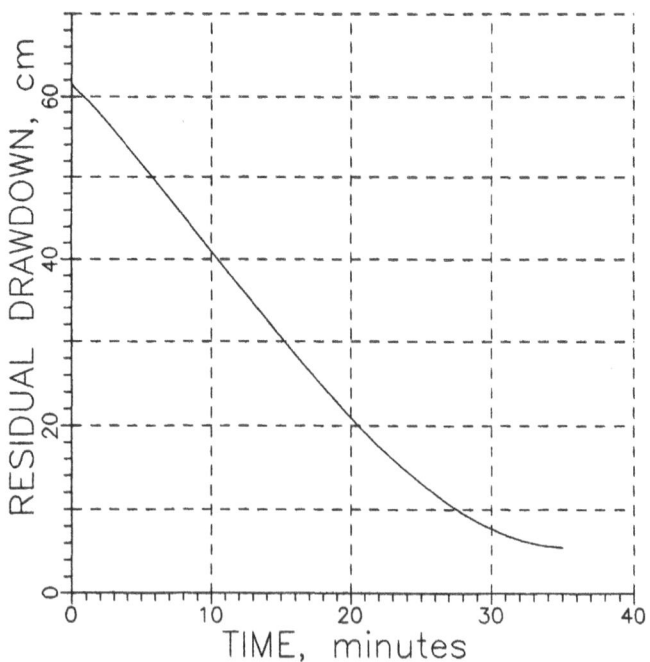

Figure 6.4. Residual drawdown for a piezometer as a function of time.

Unsaturated Flow

An equation similar to that used for saturated flow can be used for unsaturated flow. However, the hydraulic gradient is regarded as the suction gradient and the hydraulic conductivity, K, is no longer a constant but decreases with decreasing water content. When the soil pores are saturated all the pores conduct water, but when the soil becomes unsaturated, the larger pores are emptied first with the remaining water held more tightly in the smaller pores causing the water to move at a slower rate through the soil. During desorption, the effective cross sectional area for conveyance of water becomes smaller and the unsaturated conductivity decreases. In unsaturated flow, water moves through smaller pores which impart a much higher resistance to flow.

Field methods for unsaturated conductivity

On a plot of bare soil, using a combination of the neutron probe for measuring the volume moisture content and tensiometers for measuring soil water suction, the unsaturated conductivity can be determined over the range of −0.1 to −0.3 bars. Following an initial irrigation, the soil surface is covered with a plastic tarpaulin or sheet to prevent evaporation. Continuous measurements are recorded over time with both instruments. These data can be used with Darcy's Law to obtain the unsaturated conductivity over this narrow range. For lower values of unsaturated conductivity, the Guelph Permeameter can be used.

Capillarity

Capillarity is the phenomenon of a rise of water above a free water surface, such as a water table, caused by the forces of surface tension. The term surface tension is used loosely to identify the apparent stress in the surface layer of water. This surface layer behaves like a stretched membrane and can give rise to a pressure difference across a curved water surface (i.e. an air-water interface). Water above the free water surface is at a negative pressure or under tension. Figure 6.5 shows a capillary tube and the direction of forces acting on the water in the tube. The water in the glass capillary forms an acute contact angle with the glass, concave with the air. Therefore, the pressure in the water at the point of contact is less than the pressure in the air. To offset this pressure difference, the water in the tube rises from the free water surface until the pressures at the point of contact are at equilibrium. Surface tension decreases with increasing temperature which decreases the density of water, and therefore the forces of cohesion. Although the capillary tube is a rather simplistic example of what occurs within a soil profile, the complexities of the total potential acting on a soil solution can be partitioned into understandable components which when put together define the whole.

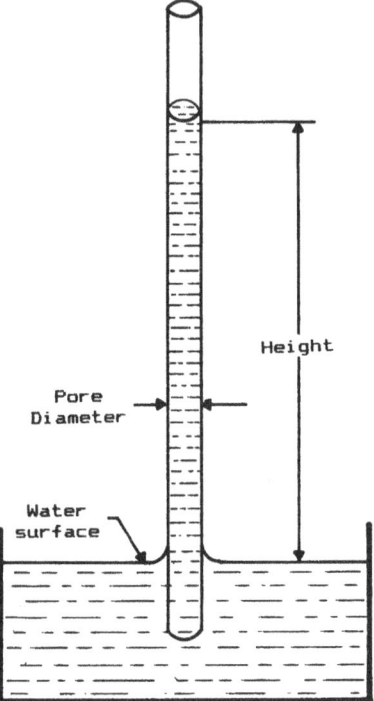

Figure 6.5. Diagram of capillary tube, outlining forces in balance.

Soil Moisture Characteristic Curve

Soils have a unique relationship between the soil pressure potential (ψ_p) and the soil moisture content (% saturation). This curve, as shown in Figure 6.6, is called the "characteristic" or desorption curve of the soil. If a suction is applied to soil water in a saturated soil ($\psi_p = 0$), then the pressure potential starts to rise. As suction increases, water is drawn out of the larger pores ($\psi_p = -0.3$ bars or near field capacity) and as suction further increases, water is retained in only the smallest pores and the pressure potential can be as low as -15 bars (near the wilting point for most plants). These values of suction are plotted and connected by a curve whose shape is "characteristic" of the water holding capacity of a specific type of soil as shown in Figure 6.6.

Movement of Water in Soils

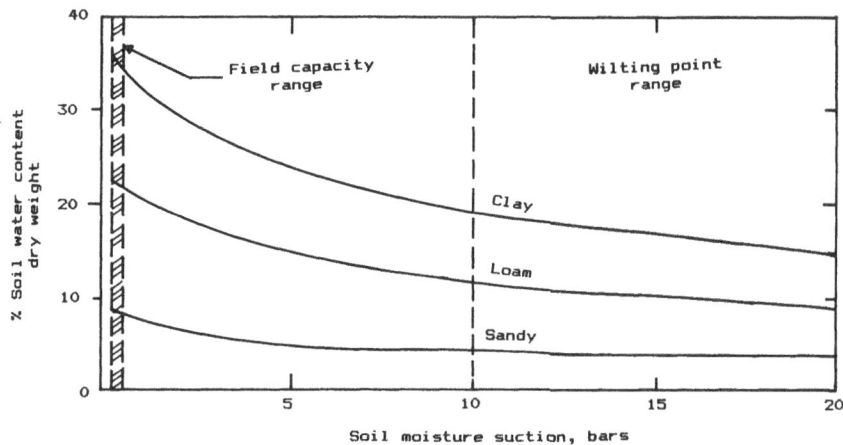

Figure 6.6. General relationship for the soil water characteristic curves.

Units of Potential

1 bar = 100 kPa (kilo-Pascals)
      = 0.99 atm (atmospheres)
      = 14.5 psi (pounds per square inch)
      = 102 cm head (centimeters of head)
      = 10.2 meters head

1 psi = 2.31 feet head
      = 7.03 cm head
      = 6.9 kPa

1 kPa = 1 kJ/m³ (kilogram Joule per cubic meter)
      = 1 J/kg (Joules per kilogram)

1 meter head = 9.8 kPa
             = 9.8 kJ/m³
             = 9.8 J/kg

Chapter 7

Darcy Equation

The terms, laminar flow and purely viscous flow, are used synonymously to mean soil water movement which flows in laminas or layers, as opposed to turbulent flow in which the velocity components have random turbulent fluctuations distorting their mean values. Laminar flow conditions exist in soils because of the low flow velocities restricted by the small pores. In fact, for agricultural soil water conditions, turbulent flow velocities are virtually non-existent. Darcy's Law, describing the laminar flow conditions within the soil profile can be stated as:

$$Q = -K i A$$

where Q = discharge rate of volume per unit time (cm³/sec),
 A = cross sectional area of flow (cm²),
 K = hydraulic conductivity (cm/sec),

 i = hydraulic gradient (cm/cm) = $\dfrac{d\psi_h}{dL}$.

The hydraulic gradient is the driving force acting on the water. The proportionality constant, K, is the property of the conducting soil to transmit water, i.e. the hydraulic conductivity. For vertical, one-dimensional flow, Darcy's Law can be rewritten as:

$$q = -K i = -K \frac{d\psi_h}{dz},$$

where $q = Q/A$ = velocity (cm/sec).

It should be noted that the hydraulic conductivity is fairly constant for saturated soil. However, as soil moisture decreases, there is less cross-sectional area for the flow of water through the soil pores (the passageways become "thinner"), large pores are drained which forces the flow through small pores only: then, the hydraulic conductivity decreases. The result is a relationship between K_θ and θ_M, soil moisture content, for unsaturated soils. Figure 7.1 shows

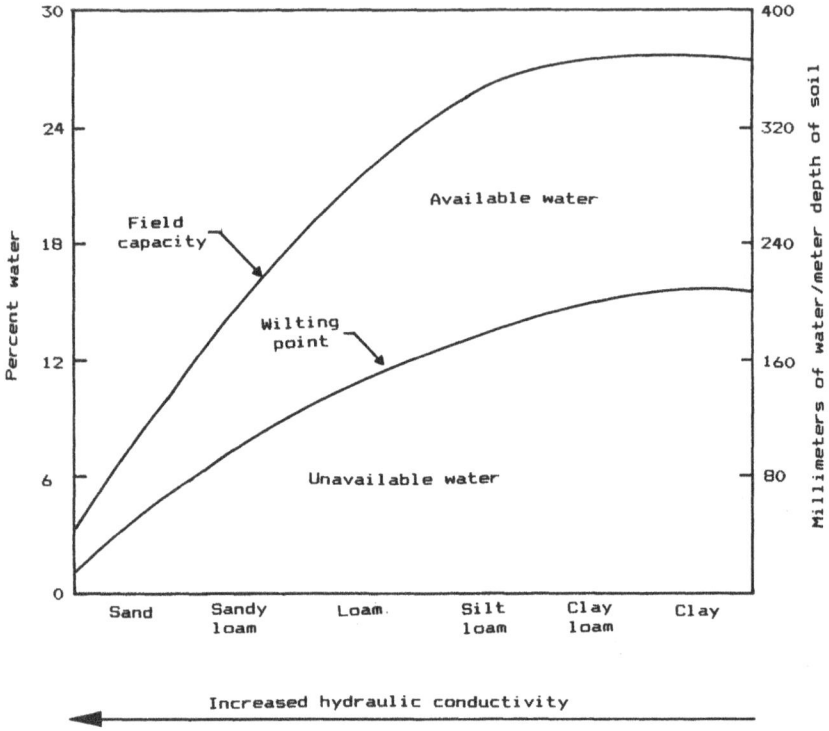

Figure 7.1. General relationship between soil moisture characteristics, soil texture, and hydraulic conductivity.

the general relationship for hydraulic conductivity and unsaturated conductivity for various types of soil texture.

The K_θ versus θ or K_θ versus ψ_M curve is determined by applying a specific amount of suction or negative ψ_M to the soil and then monitoring the flow rate of water through the soil under a specified gradient. Often, as shown in Figure 7.2, when θ is halved, K_θ is decreased by a factor of 10. As θ is quartered, K_θ may be decreased by a factor of 1000. The K_θ versus θ curve can also be determined using diffusivity relationships coupled with a point source drip/trickle irrigation field method. Some of the uses for K_θ versus θ are stated as:
1. to predict the movement of unsaturated water in soils;
2. to estimate infiltration rate from surface and sprinkler systems; and,
3. to estimate values of soil moisture such as wilting point and field capacity.

Field Capacity

When a soil is irrigated, water percolates downward through the soil profile by the force of gravity. If the soil continues to drain, it becomes unsaturated and

Darcy Equation

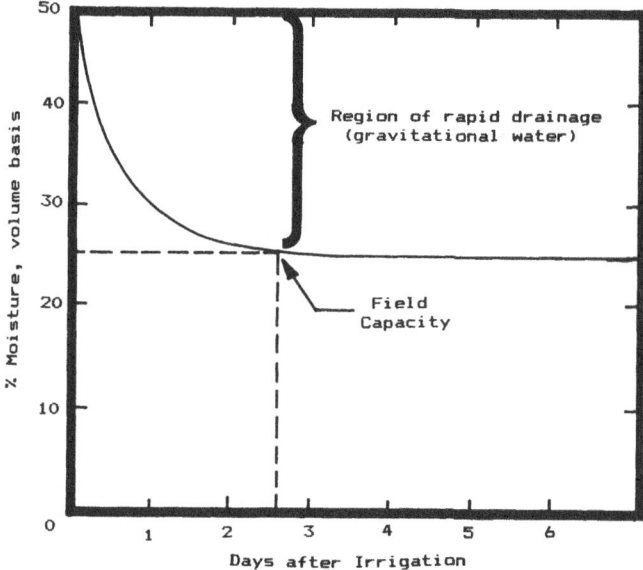

Figure 7.2. Relationship between hydraulic conductivity, soil moisture (volume basis), and soil water potential.

K_θ decreases (Figure 7.3). Using Darcy's Law in the vertical direction, when the soil moisture gradient in the z direction equals a constant, then as θ decreases, K_θ decreases.

At some time, during the drainage process, K_θ becomes small and for all practical purposes, θ (drainage) can be ignored. Then the change in moisture content with time is small and the soil continues to hold the remaining soil moisture against the force of gravity. This juncture is often termed field capacity. Field capacity should be measured in the field; but, *in lieu* of field measurements, soil water content at a suction of –0.33 bars (–0.1 bar for sandy

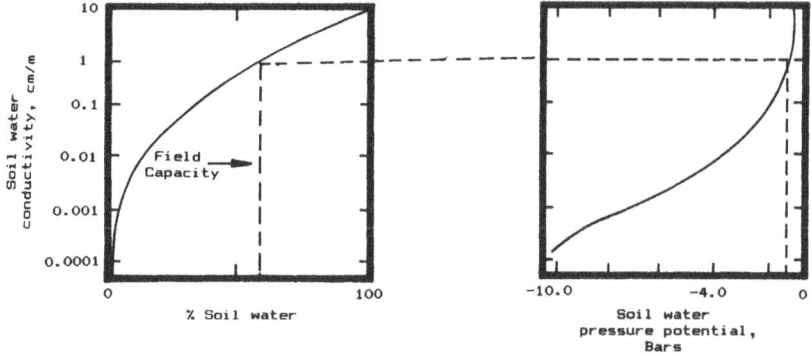

Figure 7.3. Relationship of soil moisture content to days after irrigation to determine the field capacity of a soil.

soils) from the soil moisture characteristic curve are used as estimates. In a very deep soil, the soil could dry down to low values of soil moisture although this could take hundreds of years.

Field capacity can be visualized on curves of K_θ as a function of soil moisture content (Figure 7.2). K_θ decreases rapidly with decreasing soil moisture, thereby restricting additional drainage in significant quantities. When soil moisture is plotted as a function of time after irrigation, it should attain a steady level after 2-3 days of drainage and this value of soil moisture is the field capacity. The term, field capacity, is a general term with the value of soil moisture not rigidly fixed, as normally, some drainage occurs after field capacity has been determined but this additional drainage is small. Water in excess of field capacity is termed gravitational water and drains freely through the soil profile following irrigation.

One factor affecting field capacity is soil texture: clay soils retain more water for a longer period of time than do sandy soils. Also, type of clay can have an effect, for example, montmorillonitic soils slowly adsorb water but retain greater amounts of water for a longer period. If there is a crop actively growing on the soil, then the rate and pattern of upward extraction of water could affect the downward flow gradients and flow directions which modify the redistribution processes.

Wilting Point

The wilting point (loss of turgor in the plant) is the lower limit of soil water availability. As soil continues to dry (moisture decreases), the pressure potential and hydraulic conductivity decrease. When the pressure potential decreases, the soil increasingly holds water tighter making it more difficult for roots to extract water. Since hydraulic conductivity decreases in a drying soil, greater forces (gradients) must be exerted by the plant to obtain moisture from a given body of soil. When the plant is unable to obtain sufficient moisture to maintain the evaporative demand, it cannot sustain life over periods of daylight and may wilt.

The average value of the wilting point is given as -15 bars. Some crops, such as soybeans, can show temporary wilt during daylight hours as the evaporative demand of the plants is higher than the capacity of the soil to supply water to the plant. This can occur when $E_t > KiA$ (Darcy's Law) even at high levels of soil moisture. The plants may recover their turgor later in the day or overnight and continue to produce high yields if soil water does not become more limiting.

Available Water

Available water is a useful concept to specify time and amount of water to be applied for irrigation. Because of the dynamics of the soil-plant-atmosphere system, the concept is not totally accurate when using fixed points of field capacity and wilting point. Nevertheless, it permits a more efficient irrigation scheduling regime than if irrigation were decided based upon experience or with plant characteristics alone. Available water is also termed crop extractable moisture or water holding capacity. Using the concepts of field capacity, FC, and the wilting point, WP, the total amount of water available to plants is termed available water, AW, where:

$$AW = FC - WP.$$

The available water concept is of great importance in irrigation, water balance calculations, and plant yield modeling. Its value defines the maximum amount of time that a soil can provide moisture to a crop without yield depression and the volume of water required to recharge the soil profile.

Chapter 8

Soil Water Measurement

Gravimetric Method

The measurement of soil water by the gravimetric method is difficult and subject to errors. The method depends on sampling, transporting, storing, and repeated weighings which all entail inherent error. Other techniques such as the neutron probe have come into vogue because the gravimetric method is so laborious and time consuming. Water content determination by the gravimetric method means collecting and weighing a moist soil sample, removing the water by oven drying, and weighing the dry sample.

To collect a soil sample use a steel soil tube (King tube sampler) which comes with a 5 kg hammer. These tubes can vary in length but the 1.5 m length is the most useful. Some suggestions for the operation of the tube are as follows.
1. Clean and polish the tube regularly to be certain it is rust free. Cover with a light film of oil if the tube is not used regularly.
2. Select representative sites from the field to be sampled. Areas within the field which appear different from the average should be sampled separately.
3. Vertically align the soil tube and strike the tube downwards into the soil surface so that the pointed edge of the hammer moves inside the tube in an up and down motion. Never swing the hammer as a driving device.
4. After the tube and soil sample have been removed from the soil, invert the tube with the top of the tube over a moisture can (with a tight fitting lid) and force the sample loose by pushing it with the finger. Close the can immediately after collecting the soil sample to prevent loss of moisture. Place cans filled with moist soil in an area not exposed to direct sunlight.
5. Weigh and record the weight of the moist sample. Dry the sample in an oven set at 105°C. Cool and reweigh the sample after an overnight or 24 hr drying time.
6. Water content percentage is determined by:

$$\theta_w = \frac{\text{Wet Wt} - \text{Dry Wt}}{\text{Dry Wt}} \times 100.$$

Do not include the can weight when calculating the percent moisture content. If the sample volume or bulk density information is available, calculate the moisture content on a volume basis using:

$$\theta_v = \theta_m \times \varrho_b/\varrho_w$$

The sun drying method can be used because ovens generally are not available at all sites. Plastic sheets or plastic bags in which the samples are stored are exposed to the sun after the moist soil sample is weighed. The tare weight of plastic bags can be determined by weighing 100 bags and then calculating the average weight per bag. For this method you must spread the soil sample out on the plastic sheet or in the bags and break any clods present to provide maximum exposure of soil to the sun. Examples of exposure times are given in Table 8.1. The tabulated times are for the number of hours of bright sunshine. Drying cannot be done during cloudy or partly cloudy weather. Overnight drying is not recommended because wind or storms can ruin samples quickly.

Table 8.1. Exposure times for moist soil samples to dry where the climate is warm and dry and the sun intensity is high.

Time of year	Plastic sheet	Plastic bag
Hot season	3 hr	5 hr
Cool season	4 hr	7 hr

The accuracy of the sun drying method depends on the care with which the sample is handled, temperature, humidity, and intensity of the sun. The use of the oven is the preferred method and provides a true analytical analysis of soil moisture content.

Tensiometer Method

A tensiometer (Figure 8.1) is an instrument which measures directly the soil moisture suction, ψ. It consists of a porous ceramic cup which is placed in contact with the soil, filled with water, and attached to a vacuum gauge. There is a free transfer of water and solutes between the soil and the cup, but the cup excludes the soil and air. If the soil is not saturated with water, the gauge pressure indicated by the tensiometer will be measured relative to standard or atmospheric pressure; a suction or tension, i.e. a pressure less than atmospheric. The negative pressure is directly proportional to the suction of the soil water. As the soil water is depleted by plants and evaporation, the tensiometer measurement decreases. When the soil is fully saturated, the tensiometer reading is 0.

The range of soil water suction measured by tensiometers is from 0 to −0.85 bar; they are most effective in the high soil moisture range. When the soil water suction approaches −1.0 bar gauge pressure, the air entry value of the cup has

Soil Water Measurement

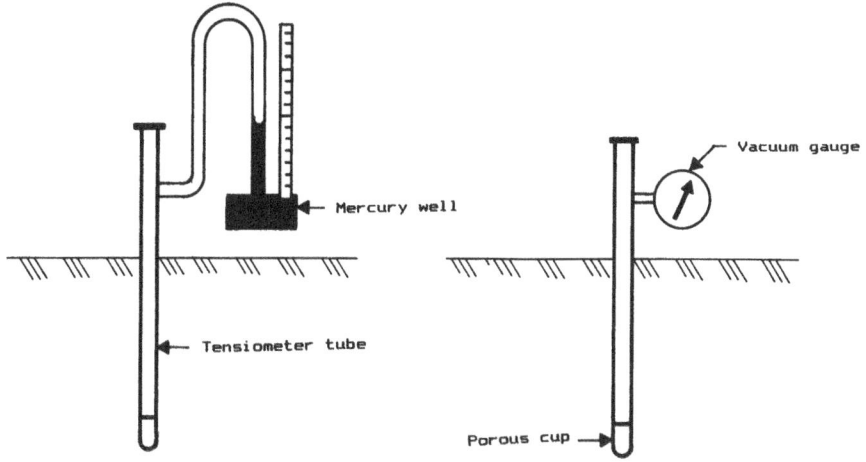

Figure 8.1. Soil moisture tensiometers.

been reached and air comes out of solution to form bubbles within the tensiometer making reliable measurements no longer possible. The tensiometer has a continuous column of water from the porous cup to the gauge and any air in the system is affected by temperature, expanding during the day and contracting during cooler nights. Nonetheless, tensiometers have been found useful for scheduling irrigation. For example, when the tensiometer is at −0.85 bar, soil water content is near 50% of available water for plant growth which is an indicator for irrigation scheduling.

If a tensiometer is installed at about the 8 m depth, the gravitational component would cause the maximum pressure depression to be at −0.8 bar and no additional suction could be measured. If it is installed at the 4 m depth, it could be adjusted to measure between 0 and about −0.4 bar. Consequently, the practical limit for depth of tensiometers is in the 1-2 m range.

To install a tensiometer in a soil, auger a hole of a diameter slightly smaller than the cup to an appropriate depth and push the cup firmly into the hole, taking care that the cup makes direct contact with the surrounding soil. In some soils a little water is placed in the hole to assist in the cup contact; however, in expanding soils, this technique causes puddling which changes the water transmission properties around the cup causing erroneous measurements. Some technicians pour a small pinch of powder-dry soil into the hole which may extend the time for returning to equilibrium and reliable readings. Care must be taken to remove all air from the tensiometer system before reliable readings can be taken. Water previously boiled to remove the air or previously de-aired by a vacuum pump can be used to fill or refill the tensiometer.

Some water must move through the cup to bring the solution in the tensiometer to equilibrium with the soil solution. This transfer of water causes delays in achieving equilibrium between the soil and the tensiometer, so that a short time period is required after refilling before reliable measurements can be made. Tensiometers are sensitive to temperature but tensiometers made from plastic reduce the thermal influence by reducing heat conductivity so that the cup will be at the ambient temperature of the soil.

Difficulties are experienced when these instruments are used in swelling clays; as the soil dries, the ceramic cup can break contact with the soil. Also, the ceramic cup should not be handled because oil from the skin can block water transmission through the porous cup. Usually operational difficulties are noticed as gauge readings remain constant but the soil continues to dry.

Neutron Probe Method

The neutron probe method (Figure 8.2) is undoubtedly one of the best to monitor the soil water content in the soil profile. It is rapid and the investigator obtains immediate results. In addition, repeated measurements at the same location and depths give results on a volume basis for a representative volume of soil. The major drawbacks are the expense of purchase, maintenance and repair, and parts are difficult to find in the Near East and North Africa, and the surface soil to the 15 cm depth must be sampled by the gravimetric method.

The neutron probe system works in the following way. It contains a radioactive source, a detector tube, and a preamplifier. The probe is lowered into a previously installed access tube inserted vertically into the soil. The

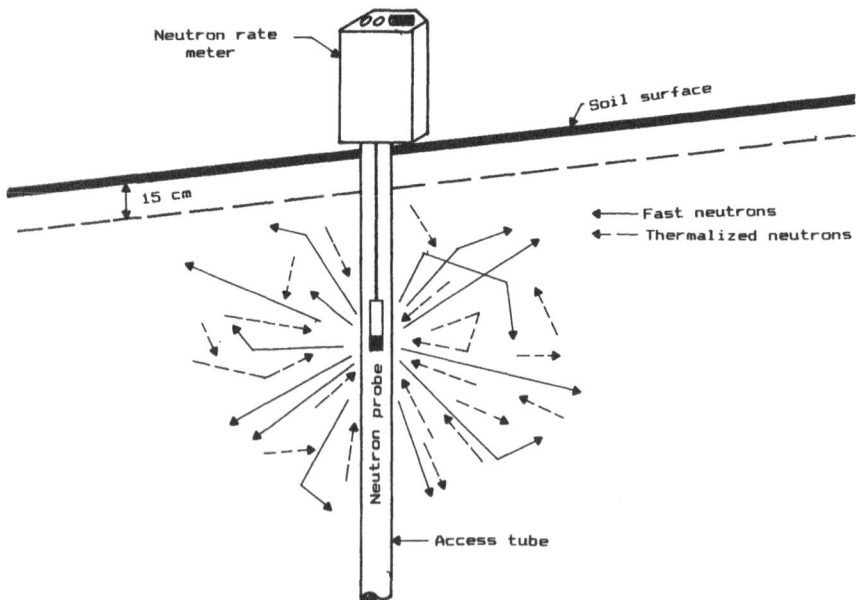

Figure 8.2. Diagram of the neutron probe method including scaler or ratemeter, and effective volume of measurement.

neutron probe source emits a continuous radially directed cloud of high energy fast neutrons. When a neutron collides with a nucleus which has nearly the same mass as the neutron, it loses a large part of its velocity and slows down. Nuclei which have nearly the same mass as neutrons are the hydrogen nuclei of water. As the fast neutrons collide with hydrogen atoms in the soil, they slow down, or are thermalized. Some of the slow or thermalized neutrons will randomly make their way back to the neutron probe detector unit where they are counted. A scaler or ratemeter measures the number of slow neutrons detected by the probe which has been calibrated with the soil moisture content on a volume basis.

The more hydrogen atoms in the soil, the more collisions and the greater the number of slow neutrons created, thereby increasing the scaler detector reading from the probe. Since most hydrogen in the soil is in the water molecule, the detector count can be used as a direct indicator of soil moisture content. There is some hydrogen present in organic matter in the soil and perhaps in the access tube if polyvinyl chloride (pvc) is used. Also some trace elements may slow neutrons down and even capture slow neutrons. Each probe may need to be calibrated for specific soils.

Detection sphere

Typically the sphere of detection in a moist soil is only about 8 cm in radius, whereas it increases in size to about 30 cm in a dry soil due to fewer hydrogen atoms present to slow the velocity of fast neutrons and detect the number of slow neutrons. Measurements should not be closer than 15 cm to the soil surface because of this sphere of influence. The size of the sphere of detection depends on the strength of the radiation source. To reduce the radiation hazard, the most commonly used source is Americium-beryllium of a strength of 100 mc (millicuries). Some manufacturers do use stronger sources to increase the detection sphere and thereby gain a theoretical sales advantage.

Access tube placement

Since the radius of the detection sphere can be relatively small, it is important to place the access tube in a representative location, e.g. in the plant row, next to actively growing plants, and deep enough to allow readings below the root zone. Tubes can be used repeatedly and even for several seasons if they do not interfere with field operations. Steel or aluminum access tubes are installed by drilling or augering an access hole vertically with the diameter of the drill or auger slightly smaller than the access tube in order to avoid air pockets around the tube. The access tube is closed at the bottom with a metal or plastic plug (rubber stopper in an emergency) to prevent water from entering. At least 25 cm of the tubing is left above the soil surface. The access tube is forced into the hole so that the soil fits the tube like a tight glove. The top of the tube is closed with an easily removable rubber stopper or plastic cover to prevent entrance of trash and water. If the tubes are not clearly visible, their position is marked with a tall stake ensuring that this does not interfere with the measurement process. When plastic (pvc) tubes are used they are difficult to install with a tight fit, the plastic contains hydrogen atoms, and if the probe has a small source, the sphere of influence will be reduced.

Calculation of soil moisture with a neutron probe

Neutron sources age and the emission from the neutron probe reduces with time. In addition, the response of the electronic components as well as the detector unit in the probe change with temperature and age. Therefore, the neutron readings should be standardized by calculating a relative ratio of neutron counts to a standard shield or water container count. The standard count is simply a reading by the meter with the neutron source (probe) placed in the shield or container. If this is done in the field, the probe and shield should be suspended above the ground at least 70 cm (or sitting on the travel case or on a pickup tailgate). Readings are usually taken for 60 seconds at 15 cm intervals, starting at 30 cm from the soil surface. The number of fast neutrons that have been slowed and returned is random and varies a small amount with each reading.

Usually there is a linear relationship between moisture and count ratio, R, (reading in access tube/standard shield count) and the following equation can be used to calculate θ_v:

$$\theta_v = a + b \times R$$

where a and b are empirical coefficients (from regression). At ICARDA another form of the linear relationship has been used to calculate the percentage of θ_v:

$$\theta_v = 100 \times (b \times \text{field reading/water count}) - a$$

where a = intercept, and b = slope.

Calibration of the neutron probe

If a calibration curve is not available, gravimetric soil moisture samples are collected from several depths and from 2 holes about 30 cm from the access tube. In addition, the soil bulk density should be measured by depth for the immediate area. The soil water content is calculated on a volume basis for each access tube and each depth. The volumetric water content is plotted as the dependent variable (y-axis) versus the neutron count ratio as the independent variable (x-axis). A large number of gravimetric and bulk density samples

Soil Water Measurement

should be obtained for varying depths and levels of moisture content to provide sufficient information to derive the linear relationship. The neutron probe must be rechecked or recalibrated every 2-3 years using the gravimetric method. Also, some soils contain different minerals as well as having structural differences and this could necessitate a separate calibration curve.

Basic rules and guidelines for neutron measurement

1. Do not trample or disturb the area around the access tube during measurement periods or during initial installation.
2. Follow the same footsteps for each visit or use a ladder suspended above the hole to protect the plants. If the evapotranspiration of trampled plants decreases, the neutron reading will increase relative to less trampled areas, and estimated measurements will not correspond with the rest of the field.
3. Always use a standard size, thickness and type of access tube to which the neutron probe has been calibrated.
4. There is a slight radiation exposure danger, therefore, always wear a radiation exposure sensitive badge. Have it evaluated every 2-3 months.
5. Always place the end of the neutron apparatus away from the body when walking and place the apparatus sideways in a car when driving.

Gypsum Blocks

The electrical resistance of a porous material is a function of its water content. As the moisture content of a gypsum block changes so does the electrical resistance between the electrodes. This is measured by a resistance meter, preferably an AC meter to reduce the effects of polarization. The electrodes are surrounded by a porous material which is in contact with the soil. Water is transferred between the soil and the porous material until equilibrium occurs between the water suction in the two systems. At equilibrium, the electrical resistance of the gypsum block is directly proportional to the soil water suction; nonetheless, the system usually requires calibration.

Today, most gypsum blocks are constructed with concentric stainless steel screen electrodes (Figure 8.3) which exhibit very little electrical flow outside the configuration of the block. Other configurations of electrodes and embedding material can be sensitive to temperature and soluble salts in the soil.

The effective range of gypsum blocks is from -0.5 to -15.0 bars of soil water suction. Gypsum blocks exhibit hysteresis and for reliable measurements they are read after being saturated and only the desorption part of the hysteresis loop is used. Gypsum blocks are soluble in the soil solution and deteriorate in 1-3 seasons of use. The units are generally installed by augering a hole to the desired depth, placing a unit in the hole, backfilling with powder-dry soil, and tamping the soil in place.

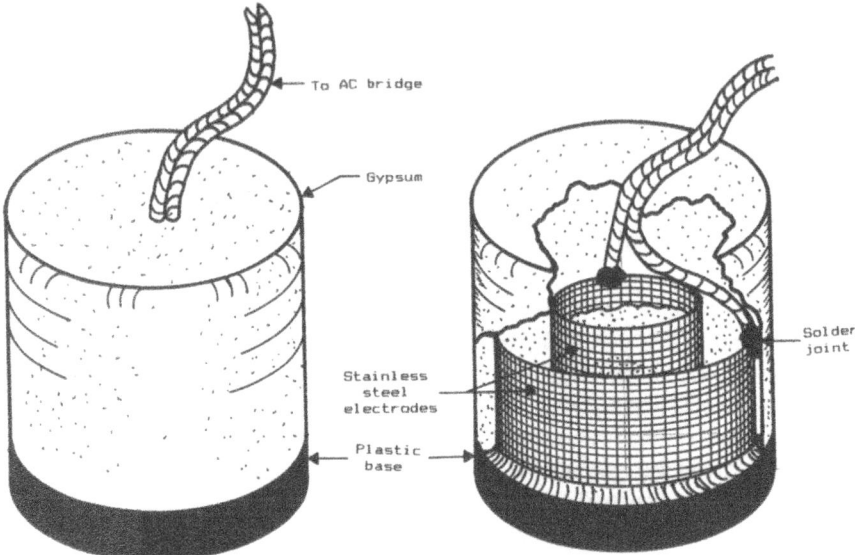

Figure 8.3. Electrical resistance unit of a gypsum block showing the concentric electrode arrangement.

Other Techniques

Three other techniques are presented none of which are entirely satisfactory for field use.

Thermocouple psychrometer

Peltier-effect thermocouple psychrometers measure the total water potential. In this technique, water is condensed on a wet junction (thermocouple) by passing a current through the thermocouple which cools the water below the dew point (the Peltier-effect). The wet-bulb temperature is measured after the current has ceased flowing and the junction has reached a steady temperature. Because only a small quantity of water condenses on the thermocouple junction, the temperature remains steady for only a brief period after wetting. The soil temperature (ambient) is measured to facilitate the calculation of the total water potential (vapor pressure).

Time domain reflectometry

This method measures the propagation velocity of a voltage step to determine the dielectric constant of soil. A voltage pulse is dispatched down a transmission line consisting of 2 parallel wave guides (wire probes) of known length separated by 5 cm. The high frequency waves (10 mHz – 1 GHz) traveling in the soil are reflected from the ends of the wave guides and return along the original path. The analyzer measures the travel time (velocity) of these waves in the soil to determine the average water content. The dielectric constant of soil is dependent on the water content.

Freezing point depression

This method measures the total potential of the soil water, i.e. the osmotic potential plus the matric and pressure potentials. The soil samples are placed in a constant temperature bath with precise control at $-4°C$ plus or minus $0.01°C$. When the soil sample reaches $-4°C$, a needle-thermistor is inserted into the sample to initiate freezing and the sample's temperature measured. During freezing the temperature drops 1-2 degrees below the temperature of the constant temperature bath. This freezing point depression is related to the total potential of the soil water solution.

Chapter 9

Infiltration

One of the basic concepts of irrigation is the downward entry of water into soil: infiltration. For irrigation, the most efficient application depends on the infiltration capacity of the soil. If the process is not fully considered, runoff or poor distribution of water in the soil profile can result with the possibility of water loss through deep percolation. Knowledge of the infiltration characteristics of a soil is basic information required for designing efficient irrigation systems.

Infiltration is normally defined as the vertical, one dimensional flow of water into soil. Infiltration is also expressed as rainfall minus interception, evaporation, and surface runoff – that part of rainfall that enters the soil. The term "intake" may be used to include the effect of two-dimensional infiltration as occurs under drip/trickle or furrow irrigation (Figure 9.1). These two terms, infiltration and intake, are frequently used interchangeably. Infiltration into

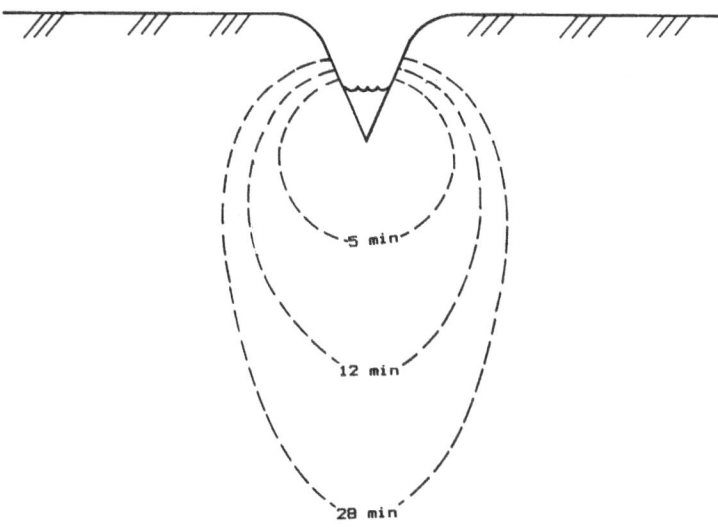

Figure 9.1. Water ponded in a furrow infiltrates into an initially dry soil.

E.R. Perrier and A.B. Salkini (eds), *Supplemental Irrigation in the Near East and North Africa*, 133-144.
© 1991 *ICARDA*.

agricultural soils usually decreases with time. The infiltration rate will affect the advance and recession times during irrigation as well as such important factors as deep percolation and tailwater runoff in furrow and border methods, advance and ponded times in level borders or basins, and maximum allowable application rate for sprinkler or drip/trickle irrigation methods.

Infiltration on Crusting Soils

In the Near East and North Africa, crusting at the surface interferes with infiltration into calcareous soils. Aspects of surface flow and infiltration are important. A simple general concept of hydraulic relationships between variables is helpful for an appreciation of the suitability of different methods of water application. The objective of any water application is to uniformly replenish the root zone moisture with enough deep percolation for the effective leaching of harmful salts. Manageable hydraulic factors of a cultivated field are:
1. water supply, rainfall intensity, irrigated field size;
2. slope of cultivated field;
3. surface roughness;
4. shape of field "channel";
5. resistance to infiltration at the soil surface;
6. vertical permeability;
7. horizontal permeability; and,
8. drainage rate.

Depth of surface flow is affected by hydraulic factors 1-4, and the infiltration rate is influenced by soil factors 5-8. Generally, water is received directly from rainfall and indirectly from supplemental irrigation which together supply enough water to require management of the cultivated field.

Effect of slope on the infiltration rate is small and of minor importance on slopes less than 2%. The scientific literature is filled with data showing bare ground with infiltration rates of 1 mm/hr whereas for the same soil types, pastures can have rates of 20 mm/hr. The old adage of plant to grass is best but at least the cultivated soil should be planted.

Conceptualization of an "ideal" cultivated field presents an invariant slope, an unlimited length, a uniform soil, a constant water supply for supplemental irrigation, and a steady infiltration rate (depth/hour). At the time of rainfall or irrigation, soil is assumed to be of a uniformly low moisture content. When water application begins at a rate not to cause surface sealing, water simultaneously moves down the slope as an advancing wave and infiltrates into the soil to move vertically and laterally as a wetting front. As the wave progresses down the field (furrows or borders), its magnitude is reduced by the water lost to the soil until infiltration has accounted for the total volume of the wave; then the advance terminates at the end of the cultivated field. This is the "model" for which to search, a model that can be mathematically evaluated;

Infiltration

but, the soils will not cooperate with this pictured "ideal steady-state" of surface flow. Rainfall and supplemental irrigation which produce runoff probably seal the powdery soil surface of a cultivated field. With the surface sealed, runoff occurs or water ponds and awaits evaporation before disappearing. This water is lost to crop production.

To promote infiltration, soil surface management techniques should be implemented, for example, various types of cultivation, tillage, and crop residue practices. Without a doubt, mulch in the bottom of the furrow will help; then, water will rise in the furrow ridge by capillarity until the soil surface is wet. The immediate overall effect of raindrop, sprinkler, and surface irrigation compaction has to be considered to avoid undue crusting at the soil surface.

The meager rains of the region do not penetrate the semi-arid and arid soils deeply enough to cause appreciable percolation. Lack of infiltration and percolation through these soils, gives rise to calcium salts remaining at the soil surface and the vicious cycle continues with inadequate applications of water. These soils impair plant growth and remain unproductive. The most effective method for the removal of salt from soil is by means of water which passes through the root zone of the soil; but, if the amount of salt carried away is not enough because of runoff and low permeability, then infiltration will remain low. Any management technique that can maintain a larger pore size at the soil surface and within the soil profile results in a greater infiltration rate. In general, management of the size of the passageways and infiltration into the soil is dependent upon:
1. the size of the particles that make up the soil;
2. the degree of aggregation between the individual particles; and,
3. the arrangement of the particles and aggregates.

The importance of maintaining permanent channels, particularly at the soil surface, is critical. For calcareous soils, 2 important components are involved:
1. the severe breakdown of soil surface structure due in part to the beating action of raindrops, and in part, to a sorting action on the dispersed soil particles by the water flowing over the surface; and,
2. after these fine particles are fitted around larger ones, they form a relatively impervious seal giving the soil a slick appearance and with plenty of calcium to form strong bonds (not unlike casting plaster).

These surface-sealing effects can largely be eliminated when the soil surface is protected by crop cover, mulch, and/or permeable mechanical protection.

Techniques to Improve Infiltration

Techniques to improve infiltration and crop growth on semi-arid and arid soils are based on a strategy of needing to test every soil type and location. A brief review follows of some of the more common types of soil surface reclamation for increasing infiltration.

Deep plowing

Soil samples should be taken of the profile in centimeter depths down to at least 30-40 cm. These samples should be tested for electrical conductivity, total salts, pH, calcium and magnesium carbonates, metallic cations such as Ca, Mg, K, and Na, as well as organic matter.

If the calcareous soil improves with depth (less disruptive soluble salts), then deep plowing may be warranted to bury the high calcium salts of the surface. The assumption of the technique is that infiltration will be increased and leaching of the soluble calcium salts will start as a "natural" percolation process. Once the process is started then it is much easier to continue with leaching and increased plant growth will contribute to better infiltration.

Chisels or subsoilers

Chisels down to the 30 cm depth can help with water entry into soil and are usually better than deep plowing because they are less destructive to the soil structure. However, subsoiling should be performed when the soil is not so dry as to powder but when it is moist enough to preserve the micro structure. In the same manner, vertical mulching with a subsoiler is extremely beneficial to initiate water entry and to continue infiltration throughout the season; notwithstanding, all mulching is bothersome at planting time and if the mulch is buried then its benefits are lost.

Surface tillage

Usually the cheapest method of obtaining satisfactory intake rates for calcareous soils is to minimize the surface tillage, thus, avoiding creation of a powdery surface. Leaving the soil surface rough and cloddy will materially increase the rate of water entry through the soil. Avoiding tillage when soil is either too wet or too dry is essential. Cultivating soil to a dry powder is especially harmful for infiltration. Another technique could be to plant on furrows or run through the field area with a ducksfoot cultivator set about 40 cm width, broadcast seed by hand and then run through the ridges with the ducksfoot cultivator to bury the seed and make furrows of about 15 cm.

Removal of surface crust

If infiltration and soil profile measurements have been made and the results show that the crusting problem is a surface phenomenon, then the "bulldozer" solution of scraping off a thin layer of surface soil may be an alternative.

Although this may appear to solve everyone's surface crusting problem, it has been used for research plots and not always successfully.

Neutralizing the effects of calcium

The rate of water infiltration into the soil depends on soil texture, structure, degree of dispersion, and also on surface residue. For soils in the semi-arid and arid region, caution must be exercised in the use of gypsum to replace the exchangeable sodium with calcium. A chemical analysis is needed before application to find out if sodium is a part of the problem. Adding chemical amendments such as flowers of sulfur to reduce dispersion of soil particles is one method of increasing the intake rate. This usually undergoes vigorous microbial oxidation in the soil when wetted and is very effective for developing acidity. Sometimes ferrous sulfate or aluminum sulfate is used to create a rather immobile gypsum but aluminum may be toxic to some plants and it should be applied with caution. These salts by hydrolysis develop sulfuric acid and can lower pH drastically.

No definite recommendation is made as to the amounts of ferrous sulfate or flowers of sulfur which should be applied since the buffering of soils and their original pH are so variable. Applying flowers of sulfur at a rate of 10 kg/100 m^2 for calcareous soils with pH's of around 8 or 8.2 would be adequate for a medium textured soil. The dosage should be varied according to texture but the pH should not be reduced below 6 for general crop production. In any case, the sulfur should be well mixed with the surface soil with a ducksfoot cultivator making numerous trips back and forth across the field.

Soil additives

The physical characteristics of soil and the infiltration rate can be changed by adding chemicals. There are generally 2 types. The first consists of materials that add to the permanency of the soil aggregate formation, thereby improving soil structure and contributing considerably to increases in the rates of infiltration and percolation. The second additive is essentially a wetting agent that does not change the soil but instead changes the angle of contact of water with the soil surface and thereby the rate at which water moves through soil. The second type can affect water movement at depths greater than the zone of application.

Vegetation and tolerant crops

Surface sealing can be greatly reduced by vegetation. The protective vegetative cover can be grasses and forages or other close-growing vegetation grown with the addition of mulches. The value of hardy, deep-rooted crops like rye, barley, and sorghum invariably will increase infiltration significantly and should not be neglected. Getting plant roots to penetrate soil will not only materially improve soil water permeability but the roots leave cavities when they decay. In addition, plants cause alternate wetting and drying which is essential for building soil structure which occurs in a relatively short period of time.

Manures and plant residues

The value of manures and plant residues, including green manures, cannot be overemphasized. The availability of these materials is the difficulty; however, stubble residue and stubble mulching can significantly affect infiltration.

Factors Affecting Infiltration

1. Soil texture and structure (including tillage, plow pan, clay layers, soil cracking).
2. Soil surface (compaction, wheel tracks, crusting) because of history of wetting and energy from rainfall or sprinkler drops.
3. Soil moisture content and hydraulic gradients within the soil profile at start of irrigation.
4. Time since start of irrigation.
5. Time of season (irrigation, tillage history).
6. Entrapped air.
7. Soil and water temperature (viscosity).
8. Salt content in water and soil (as EC increases, infiltration may increase; except for sodium conditions).
9. Irrigation method (sprinkler versus surface; border versus furrow).
10. Crops and surface mulches.

Approaches to Infiltration

As shown in Figure 9.2, the infiltration rate, I, is the time rate of change at which water is absorbed by the soil. The cumulative depth of infiltration, z, is the total accumulated depth of water infiltrated in a given period of time, t. It follows that the cumulative depth of infiltration is the integration (area under the curve) of the infiltration rate; and conversely, the infiltration rate is the derivative (slope of the line) of the cumulative depth of infiltration:

$$z_t = \int_0^t I \, dt,$$

and,

$$I_t = \frac{dz}{dt}.$$

Infiltration

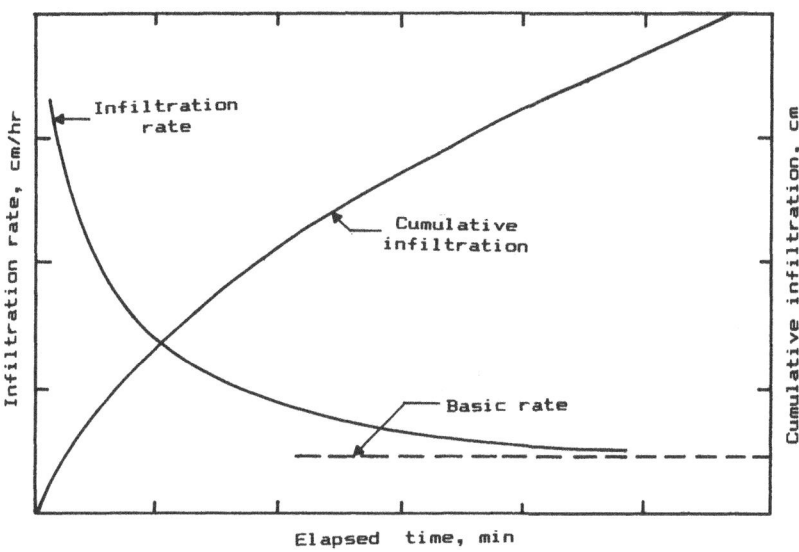

Figure 9.2. Infiltration rate versus time.

The rate at which water infiltrates into a soil decreases with time until a somewhat steady base infiltration rate is reached. In Figure 9.2, the infiltration rate asymptotically approaches the base infiltration rate. The decrease in the infiltration rate is caused by decreasing matric potential as the soil becomes wet. Also, the constant base infiltration rate is caused by the constant gravitational potential and other components of the total potential or driving force.

A typical soil moisture profile during infiltration is shown in Figure 9.3 which defines the various components of the wetting front during infiltration. There is a very thin zone of saturation at the soil surface followed by a rather large

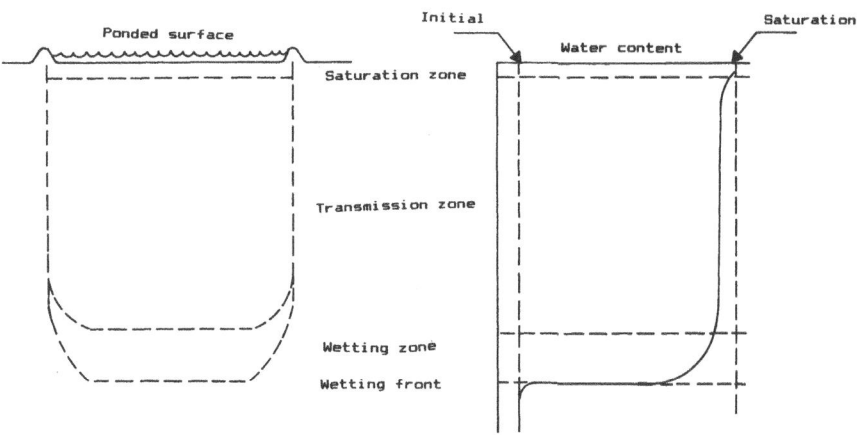

Figure 9.3. Infiltration moisture profiles as a function of depth.

nearly saturated transmission zone which lengthens downward during the process of infiltration. The somewhat drier wetting zone (transition zone) is just behind the wetting front where the gradients of soil water potentials are acting along with the constant gravitational potential.

To gain a better comprehension of the concept of infiltration it becomes necessary to understand why the infiltration rate decreases with time. The Darcy Equation (see Chapter 7) can be used to explain this phenomenon. Since, at the surface, the infiltration rate equals the gross apparent velocity, q, of the wetting front, then the infiltration rate,

$$I = q = -Ki,$$

where K is the hydraulic conductivity and i is the hydraulic gradient. Following the description provided in Figure 9.4, the Darcy Equation can be written for infiltration from a ponded surface to a wetting front at a distance L below the ground surface.

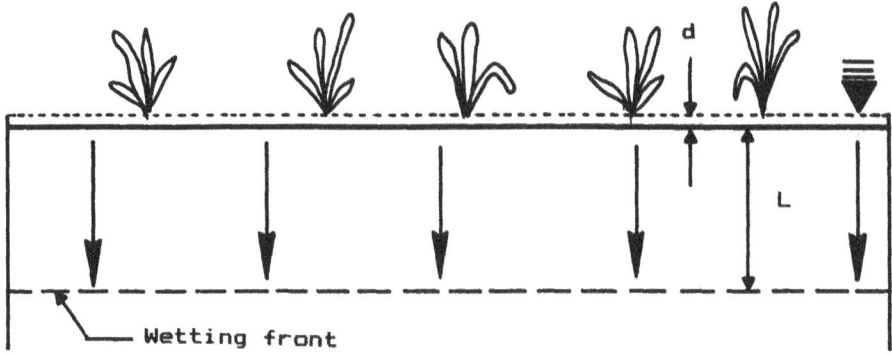

Figure 9.4. Ponded infiltration for applying Darcy's Law.

Assume that the hydraulic conductivity behind the wetting front is equal to or approaches the saturated hydraulic conductivity, K_s, (the matric potential is equal to 0) and that the 0 elevation reference datum is at the depth of the wetting front. Then it follows:

$$I = -K_s \frac{dh}{dL} = -K_s \frac{L + d - \psi_i}{L},$$

where d is the depth of the ponded surface and equal to the pressure potential, ψ_i, which is the negative matric potential at initial θ_i as shown from a characteristic curve. If d is small compared to L, which is the usual condition, and if I is considered positive in the downward direction into the soil, then the infiltration rate can be expressed as:

$$I = K_s (1.0 - \frac{\psi_i}{L}).$$

Infiltration

Since the pressure potential, ψ_i, in the above equation is negative for unsaturated soil, then the hydraulic gradient term will always be greater than or equal to 1. Therefore, as L increases with time of infiltration, the overall gradient (ψ_i/L) decreases and the infiltration rate, I, decreases. Or, restating, as the infiltration rate decreases with time; the hydraulic gradient decreases with time. Then it follows that as ψ_i/L decreases and L becomes larger with time, the difference between the hydraulic gradient and 1.0 becomes closer and closer to 1.0 and I approaches K_s.

As $(1.0 - \psi_i/L) \rightarrow 1.0$,

then,

$I \rightarrow K_s$.

At this point, the infiltration rate becomes the base infiltration rate, the flat portion of the curve, and infiltration progresses downward under a unit gradient (1/1) by the gravitational potential only.

Infiltration rates depend on so many circumstances that few valid generalizations can be made. Notwithstanding, Table 9.1 has been prepared to gain a conceptual understanding of the order of magnitude of basic infiltration

Table 9.1. Agricultural soil characteristics*.

Textural class USDA	BIR (mm/hr)	Porosity (vol/vol)	FC (vol/vol)	WP (vol/vol)	K_s (mm/hr)
1. CoS	12.70	0.351	0.174	0.107	303.53
2. CoSL	11.43	0.376	0.218	0.131	180.09
3. S	10.16	0.389	0.199	0.066	168.15
4. FS	9.91	0.371	0.172	0.050	137.16
5. LS	9.65	0.430	0.161	0.060	70.61
6. LFS	8.64	0.401	0.129	0.075	25.40
7. LVFS	8.13	0.421	0.176	0.090	23.11
8. SL	7.62	0.442	0.256	0.133	17.02
9. FSL	6.35	0.458	0.223	0.092	13.97
10. VFSL	6.35	0.511	0.301	0.184	8.38
11. L	5.08	0.521	0.377	0.221	5.33
12. SIL	4.32	0.535	0.421	0.222	2.79
13. SCL	2.79	0.453	0.319	0.200	2.13
14. CL	2.29	0.582	0.452	0.325	1.65
15. SICL	1.78	0.588	0.504	0.355	1.04
16. SC	1.52	0.572	0.456	0.378	1.65
17. SIC	0.51	0.592	0.501	0.378	0.84
18. C	0.25	0.680	0.607	0.492	0.56

* USDA Soil Classification System:
  Co = coarse, C = clay, SI = silt, S = sand, L = loam, F = fine, and V = very
  BIR = Basic Infiltration Rate
  K_s = Saturated Hydraulic Conductivity
  FC and WP are the field capacity and the wilting point whose difference is the available water capacity, AWC

rates, hydraulic conductivity, porosity, field capacity, and wilting point. The danger of using these values directly can be shown by considering the red vertisol at the ICARDA research center which is a clay soil (70% clay, 15% silt, and 15% sand), with a bulk density of 1.01 g/cm^3, a basic infiltration rate of 10-12 mm/hr, porosity of about 0.5, field capacity of 0.446, wilting point of 0.257 and a hydraulic conductivity of 4.75 mm/hr.

Green and Ampt Equation

A simplified approach was developed on a theoretical basis by Green and Ampt in 1911 to predict infiltration from a ponded surface using the physical model shown in Figure 9.4. The assumptions of the Green and Ampt Equation are that a distinct and precisely definable wetting front exists and the matric potential at this wetting front remains effectively constant regardless of time and position, i.e. there are no large amounts of entrapped air and no significant surface sealing. They started with the Darcy Equation and developed it into a form now known as the Green and Ampt Equation using the definition of effective porosity:

η_e = the percentage of soil which is not saturated at a specific soil moisture content, θ_i, and is available to be filled or saturated with water during the infiltration process.

Here, η_e is the difference between initial and final moisture on a percent volume basis, θ_{vi}, or:

$$\eta_e = (\eta - \theta_{vi})$$

$$= 100 \times \frac{\text{total vol. of pores} - \text{vol. of water in pores}}{\text{total bulk volume of soil}}.$$

For this equation, if a soil is saturated, then $\eta_e = 0$ and if a soil is dry, then $\eta_e = \eta$ (percent porosity). The cumulative infiltration depth, z, was given before as:

z = total infiltrated depth of water with time during the infiltration process,

$$z = \int_0^t I \, dt.$$

Since a uniformly wetted zone is assumed to extend all the way to the wetting front, then the cumulative infiltration, z, should be equal to the product of the wetting front depth, L, and the effective porosity, η_e, which asserts that:

$$z = \frac{L \, \eta_e}{100},$$

or by rearranging terms,

$$L = \frac{100\,z}{\eta_e},$$

where the value 100 is to correct the porosity percentage. The previously defined Darcy Equation for infiltration can be given:

$$I = K_s \left(1 - \frac{\psi_i}{L}\right)$$

which can be rewritten,

$$I = K_s \left(1 - \frac{\psi_i}{z} \times \frac{\eta_e}{100}\right).$$

The above equation is one of the forms of the Green and Ampt Equation. The term ψ_i is often called the "soil suction" parameter and is equivalent to the average matric potential at the wetting front. If necessary, ψ_i can be evaluated or estimated from:

$$\psi_i = \frac{\int_{\theta_i}^{\theta_s} K(\theta) \times \psi(\theta) \times d\theta}{K_s \times (\theta_s - \theta_i)}.$$

This gives an "average" value of soil suction or matric potential in the wetting zone between soil nearly saturated (θ_s) to the zone below the wetting front at its antecedent value (θ_i).

The Green and Ampt Equation is a theoretically based equation which predicts that with time the total depth of infiltration, L, will increase and the infiltration rate, I, will decrease. This equation is not widely applied because of the difficulty in estimating ψ_i and K_s. However, ψ_i can be determined from the characteristic curve and with this initial estimate of θ_i, η_e can be calculated from θ_i, and K_s can be determined from a saturated hydraulic conductivity test.

The K_s in the Green and Ampt Equation is less than the value of hydraulic conductivity for a completely saturated soil because a few large pores contain entrapped air during the infiltration process. Since the size of these large pores contribute significantly to the magnitude of the hydraulic conductivity, K_s is usually reduced in value until the air dissolves in these pores and is replaced by water but this process requires a long period of time. The lower value for K_s helps to explain the effects of surface sealing.

The major benefit of the Green and Ampt Equation is the prediction in the change of rate and shape of the infiltration curve depending on the degree of wetness of the soil at the beginning of infiltration (antecedent moisture content). This can be helpful if the soil moisture is subject to change from one infiltration event to the next as shown in Figure 9.5.

One advantage of the Green and Ampt Equation over other infiltration equations is the ease of use of the soil moisture characteristic curve. In addition, K_s and the antecedent moisture content, θ_v, can be obtained from field measurements. The Equation can be used to estimate the magnitude and shape of the infiltration curve before infiltration occurs. The coefficients are

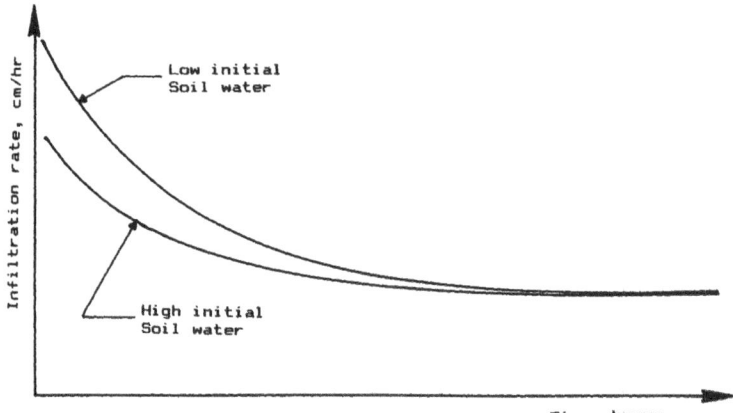

Figure 9.5. Infiltration rate versus time for two levels of initial soil moisture.

physically based; therefore, the equation works over a wide range of θ_i which helps in understanding the physical processes of infiltration.

One disadvantage of the Green and Ampt Equation is the difficulties of calibrating parameters using only observed field infiltration measurements, i.e. 1 equation and 3 unknowns. Another disadvantage is that the equation is difficult to apply to a layered soil. Also, it is difficult to generalize the equation beyond 1-dimensional values to include the vertical and horizontal variability in soil hydraulic parameters.

Units of Infiltration

$$\text{length/time} = \text{cm/hr, mm/day, m}^3/\text{m}^2/\text{hr, l/ha/day.}$$

In furrows, infiltration may be:

$$\text{volume/furrow/time} = \text{l/hr/furrow,}$$

or, using equivalent uniform depth over entire surface,

$$\text{volume/(length} \times \text{furrow spacing)/time.}$$

Chapter 10

Field Measurement of Infiltration

Techniques for the measurement of the infiltration rate rely upon either single or double ring infiltrometers. However, in many soils and even with careful management, the results may not be satisfactory. Other attempts to measure infiltration by increasing the wetted perimeter, e.g. blocked furrow or basin, can present specific measurement problems. Techniques which measure rainfall, evaporation, plant interception, and runoff are probably more accurate for overall evaluation of infiltration occurring in a cultivated field. Notwithstanding, in semi-arid and arid zones, these data are difficult to obtain and, therefore, only the techniques measuring the depth of ponded water are in common usage.

Single Ring Infiltrometers

The ring is a metal cylinder which is driven into the soil. The ring (about 30 cm in diameter and 60 cm in length) is inserted in the soil to the 30 cm depth as shown in Figure 10.1. Water is then ponded on top of the soil surface in the infiltrometer.

The soil surface must be protected with a light burlap cloth to reduce surface sealing or surface destruction. The decrease in depth of water with time can be recorded with a point gauge or ruler taped to the inside of the infiltrometer. Only in soils of high infiltration rates is the effect of falling water level significantly different when compared to the magnitude of the matric potential below the wetting front. In practice, the effect of water depth can be ignored for most soils but it is a good procedure to standardize the water depth and use a Marriotte bottle for control (a method described later).

The advantages of a single-ring infiltrometer are that installation and measurements are rapid with only 1 ring required, providing a measure of 1-dimensional infiltration within the ring. A disadvantage of this device is the required pounding of the ring into the soil (a difficult task) to a depth that can fracture a dry soil. Cracks, or spaces around the inside of the ring, allow water to seep into these spaces which increases the value of the estimated infiltration

E.R. Perrier and A.B. Salkini (eds), Supplemental Irrigation in the Near East and North Africa,
145-153.
© 1991 *ICARDA.*

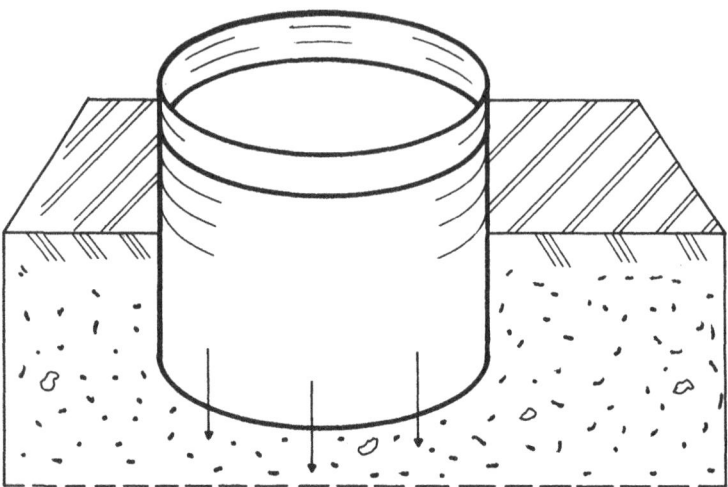

Figure 10.1. A single ring infiltrometer.

rate of the soil. Accurate infiltration can only progress to the depth of the ring insertion; below this point, the infiltrating water can spread out which may invalidate the assumption of 1-dimensional vertical infiltration. The results are difficult to use for prediction of infiltration along furrows because the single-ring infiltrometer is a point measurement device which does not consider spatial variability.

Double Ring Infiltrometers

The double ring infiltrometer shown in Figure 10.2 avoids the requirement of deep insertion into the soil. The outer ring provides a buffer of infiltrating water which forces infiltration below the inner ring to remain vertical and, in theory, 1-dimensional.

The double ring infiltrometer requires that water levels within both rings be kept identical to prevent any cross gradients in soil below the rings which would affect the vertical infiltration. Because of this requirement, Marriotte siphons shown in Figure 10.3 should be used to maintain constant water levels within both rings. Infiltration with time then becomes the volume of water displaced from a Marriotte container divided by the surface area of the inner ring. Except for the vertical infiltration effect, the double ring infiltrometer has the same disadvantages as the single ring.

A Marriott siphon is a closed container with the only access to atmospheric pressure being through an adjustable tube which extends below the water surface. The air tube creates a free water surface at atmospheric pressure at the bottom of the tube regardless of the actual depth of water in the container. The outflow from the reservoir will maintain the desired depth by placing the water

Field Measurement of Infiltration 147

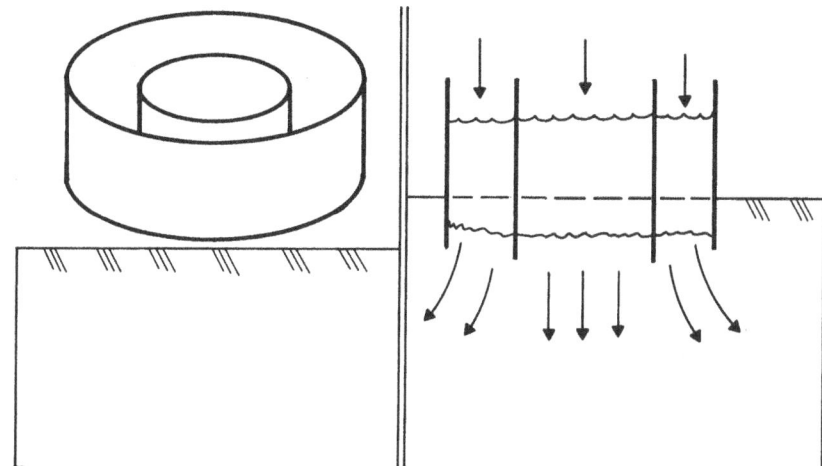

Figure 10.2. A double ring infiltrometer.

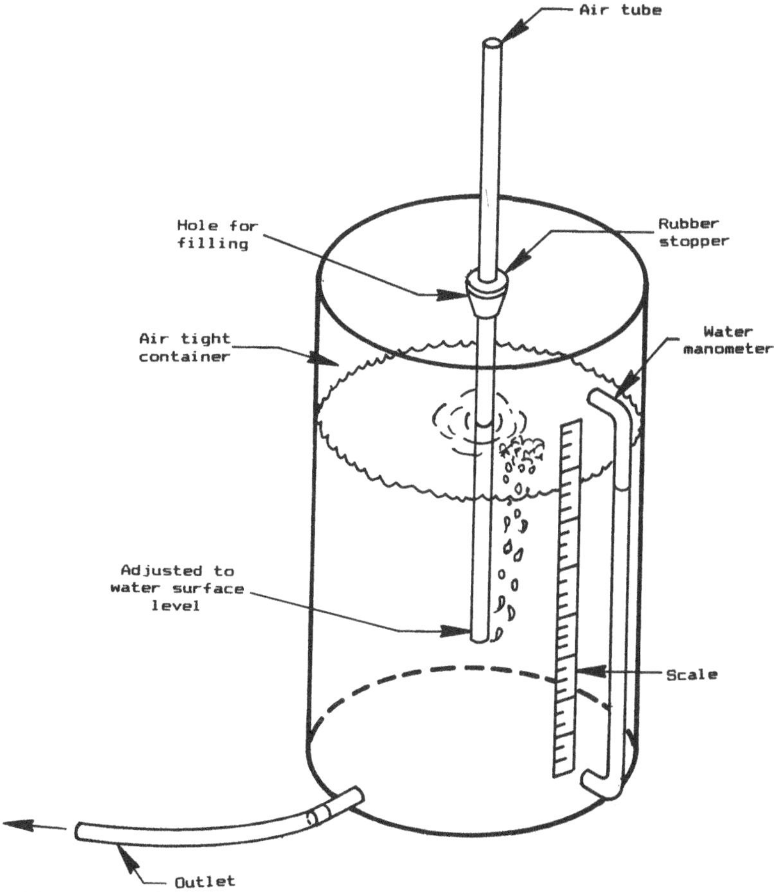

Figure 10.3. An operating Marriotte siphon.

outlet hose in the bottom of the infiltrometer and positioning the lower end of the air tube at the desired elevation of the water surface. This apparatus accounts for the volume of water added; therefore, the volume infiltrated is measured since the volume stored in the infiltrometer is maintained constant.

Equipment for a double ring infiltrometer

1. An infiltrometer is a metal cylinder, 30 cm in diameter, d, and 30 cm in length, h, and made of smooth steel 2 mm thick (14 gauge) with a metal band for reinforcement around the upper edge, Volume = $3.14 \, d^2 h/4$.

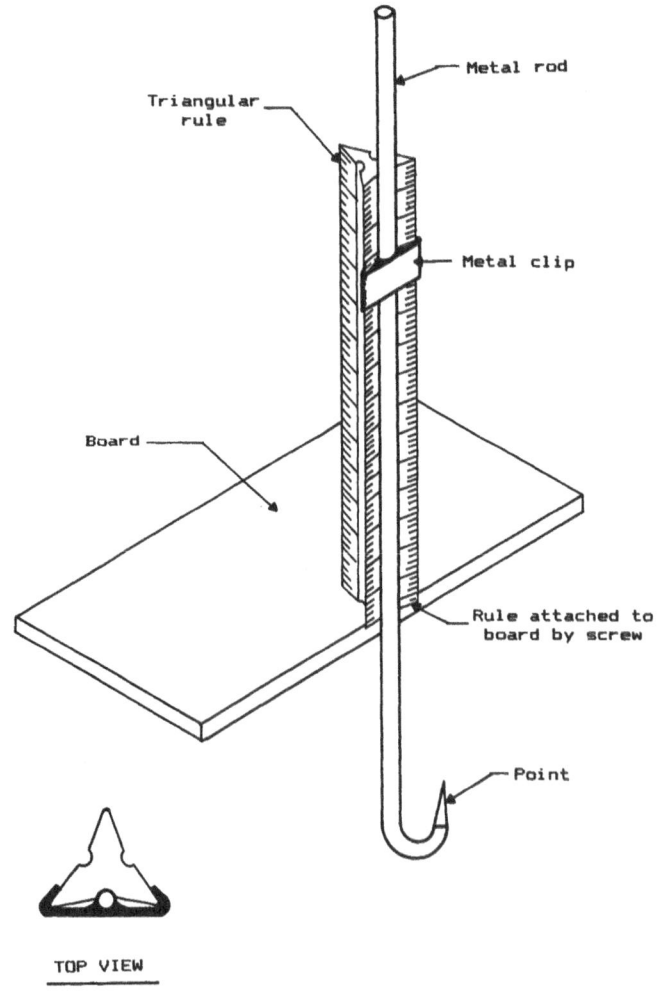

Figure 10.4. Hook gauge for measuring ponded water levels in ring or cylinder infiltrometers.

Field Measurement of Infiltration

2. A buffer cylinder, 60 cm in diameter and 20 cm in length which has been similarly constructed to the infiltrometer.
3. A 15 cm plastic ruler; or, if available, more accurate readings can be made by constructing a hook gauge as shown in Figure 10.4. In soils of moderate to high infiltration rates, the plastic ruler is better and a plastic wash bottle should be used to clean the debris and bubbles from the ruler.
4. A Marriotte siphon or other regulated supply reservoir (if available) and a sheet of plastic or plastic bag. The Marriotte siphon can be constructed from a metal tank as shown or a large glass jar (Demijohn Jar) can be used with glass tubes for the air entry and water outlet. The air tube, in both cases, is adjusted to the desired water level in the ring, furrow, or basin.
5. A metal plate or heavy timber to cover the rings and a 5 kg sledge hammer (mall);
6. A supply of water (about 100 l), a 1,000 ml graduated cylinder, and a bucket.
7. A watch with sweep second hand and data sheets.

Procedure for a double ring infiltrometer

1. Select a representative location for the infiltrometer and examine carefully for any unusual surface disturbances such as cracks, sticks, or stones.
2. Drive the infiltrometer cylinder to a depth of 15 cm. The cylinder should be installed as carefully and as vertically as possible. If it has been driven in the soil so that first 1 side then the other goes down, remove the infiltrometer and re-install it in a nearby location making certain that proper alignment is maintained.
3. Place the buffer cylinder around the infiltrometer and drive it into the soil to a depth of 10 cm. If a buffer cylinder is not available, use a soil dike around the outside of the infiltrometer.
4. Lay a plastic sheet or bag on the soil surface within the infiltrometer and fill the plastic bag with at least a 10 cm depth of water.
5. Fill the buffer area with water to a 10 cm depth and maintain that depth throughout the measurement period.
6. Place the ruler inside the edge of the cylinder and scotch tape it to the infiltrometer (or attach the hook gauge to the surface of the infiltrometer).
7. To start the system, quickly but smoothly remove the plastic bag from inside the infiltrometer to release the water. Read and record depth of water on the ruler (or hook gauge), and record time.
8. When the water level in the infiltrometer has declined to the 5 cm depth, read and record depth of water on the ruler and record the time. Refill the infiltrometer rapidly but smoothly from a graduated cylinder to the 10 cm depth, read and record the depth of water on the ruler and record the time and volume of water added. The less disturbance to the soil surface by the inflowing water, the more reliable the data.

9. Make additional readings on the ruler at periodic intervals and record the readings and the times.
10. A continuous plot of the data should inform you whether or not the infiltrometer is operating properly. The measurements should be continued for at least 8 hr; at which time and if the water level is sufficiently constant, refill to the 10 cm depth and leave overnight. Take a final reading in the morning.
11. Enter results on a data sheet (Figure 10.5), then plot data and determine the infiltration rate.

INFILTROMETER DATA SHEET

Date _____ Location _____

Soil condition _____

Infiltrometer dia. (cm) _____ Buffer dia. (cm) _____

Furrow length (cm) _____ ; Furrow spacing (cm) _____

Area irrigated (cm^2) _____

Reservoir cross-sectional area (cm^2) _____

Wetted perimeter (cm) _____ , _____ , _____ , _____ ; Average (cm) _____

Wetted area (cm^2) _____

Time (hr: min)	Elapsed time (min)	Water level reading (cm)	Volume added (cm^3)	Equivalent depth added (cm)	Cumulative depth infiltrated (cm)	Comments

Figure 10.5. Field data sheet for measurements of infiltration.

Field Measurement of Infiltration 151

Ponding Basins

This method has distinct advantages over the ring infiltrometers but is more difficult to control and measure. Ponding basins are effective for measuring infiltration on level basins with a rapid advance of water. This method tends to integrate infiltration spatially over the basin.

Ridges of at least 30 cm in height are constructed on the square around a 1 m² area. A stiff ruler is pressed into the surface near an edge of the ridge. A square 30 cm piece of cloth or burlap is placed near the edge of the ridge as the location at which water will be added. Water should be added smoothly but carefully to the 10 cm depth in the basin and refilled when the water has lowered to the 5 cm depth.

Furrow Method of Infiltration Measurement

Small weirs or flumes are placed along a furrow as shown in Figure 10.6. This technique may represent real conditions because when the water moves down the furrow and infiltrates into the soil surface it rearranges the surface particles. This erosion and dispersion tends to seal the soil surface and reduce the infiltration rate. Depending on the soil type, selection of the distance between the flumes for accuracy of measurement is the major limitation of the method. Also, the small flumes must be kept in a level position and a stilling basin of about 1 m in length has to be maintained before water enters the flume.

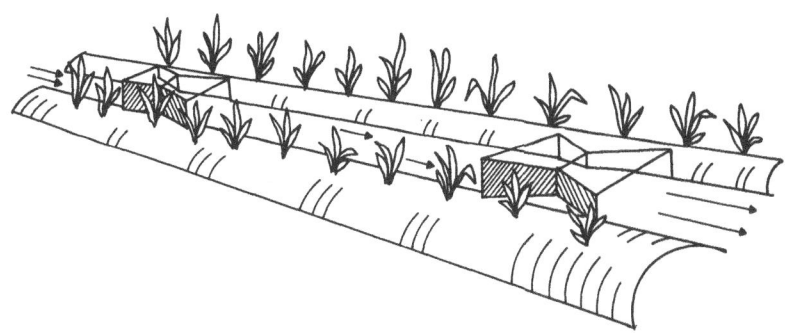

Figure 10.6. Infiltration rate determination using two flumes in a furrow.

With this method, discharge from both devices is recorded with time. The infiltration rate is calculated by dividing volume differences in the discharge from both devices by the length and spacing between furrows. Flumes must be far enough apart, at least 100 m (dependent on soil type), to minimize errors, ± 10%, in flow measurement. Flow measurement error can easily mask measurement of infiltration because long distances between flumes are needed.

As an example, if upstream $q_1 = 40$ l/min and downstream $q_2 = 30$ l/min then the difference $\Delta q = 10$ l/min. If the minimum allowable reading error in measurement for each flume is ± 10%, then determined values of Δq could range from 3-17 l/min or an error in measurement of ± 70%.

Blocked Furrow Infiltrometer

The blocked furrow method as shown in Figure 10.7 blocks-off a furrow and measures the infiltration of water into the soil profile from a short segment of a furrow (about 1 m). The water is ponded in this section by metal barriers driven into the soil across the furrow. In this case, infiltration is 2-dimensional because it includes both vertical and horizontal movement of water into the soil profile. Adjacent furrows can be filled with water to act as buffers and prevent excess lateral movement and simulate actual irrigation.

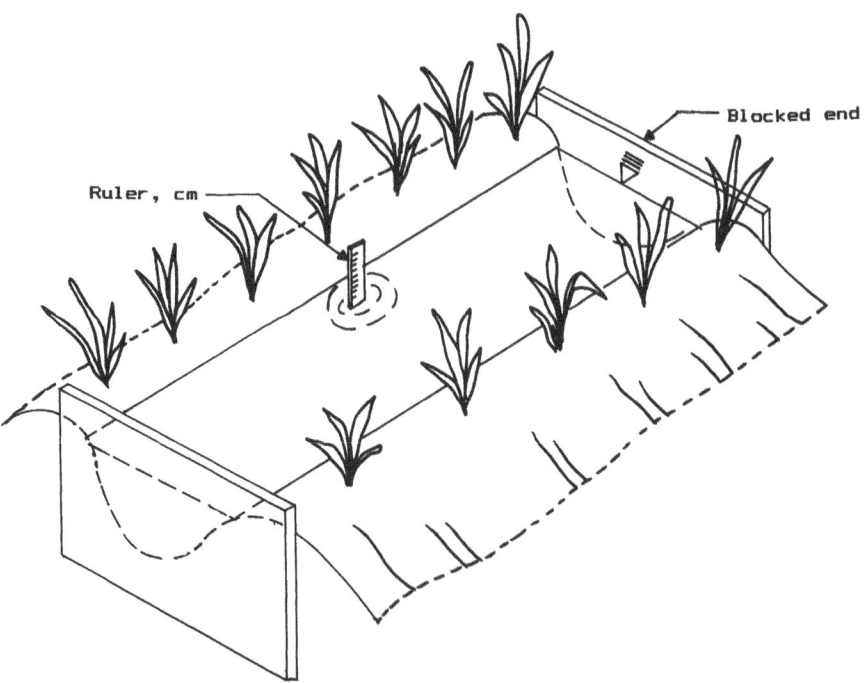

Figure 10.7. Blocked furrow method of measuring infiltration with a ruler in a water reservoir.

With this method, depth of water should be maintained at a constant level because the wetted perimeter may vary as water depth changes. Marriotte siphons (see Figure 10.3) should be used to maintain a constant water depth. Infiltrated depth or volume of water added over time is measured and recorded. The volume infiltrated into the blocked furrow is converted to depth by dividing

Field Measurement of Infiltration

this measured volume by the furrow length and row spacing. To ensure that the infiltrometer is performing accurately, the data should be plotted from the start of water intake.

Infiltration Calculations

The infiltration data should be plotted in the form of a cumulative infiltrated depth, z, as a function of time, t, from the start of water intake. This plot will indicate errors associated with the data and replication is important for all methods of determining infiltration. The infiltration rate is the derivative of the cumulative infiltration curve as shown in Figure 9.2 and can be determined at any time from the slope of the z versus t curve.

Some investigators believe that infiltration characteristics are easier to describe and use in the form of mathematical relationships. The 2 equations of major importance are the infiltration equation:

$$z = Kt^a,$$

and the more general equation called the modified Kostiakov-Lewis Equation:

$$z = Kt^a + I_b t,$$

where: K, a = empirically determined coefficients;
I_b = basic infiltration rate; and,
t = time of water intake.

These equations can be evaluated by graphically using log-log paper, or analytically with regression techniques. The value of the coefficient a in the linearized portion of the infiltration equation is usually close to 1/2 which permits the plotting of I versus the square root of t and the slope near the end of the infiltration measurement determining K.

Graphically, a power curve of the infiltration equation form should plot as a straight line on log-log paper, with the coefficient a as the slope of the line and K as the value at the y-intercept at t = 1. In the Kostiakov-Lewis Equation the y-intercept would be equal to $z - I_b t$. The value for I_b is determined independently from the infiltration rate at the end of measurement as previously discussed. The value of I_b should be adjusted by trial and error until a graph of $z - I_b t$ versus t will plot as a straight line on log-log paper. A plot which curves upwards at a longer time indicates that the I_b value is too small and should be increased. If the line curves during the initial periods of infiltration, a constant needs to be added or subtracted (concave upwards) to all z values. The need for such corrections would indicate that some of the experimental conditions are imprecise.

Chapter 11

Groundwater Supply

The study of water is called hydrology. Hydrologists study its properties and examine how it is distributed and circulated throughout the world. The total volume of water on this planet is finite and constant. The water that has always been here will probably always be here; it may move around in various ways and change form, but the sum of all Earth's water never varies. Inflow into the Earth's hydrologic system arrives as precipitation, in the form of rainfall or snowmelt. Precipitation is delivered to streams both on the land surface as runoff, and by subsurface flow following infiltration into the soil. A component of the subsurface hydrologic system is groundwater. Groundwater is the major source of water supply for supplemental irrigation in the Near East and North Africa. Other sources of supply are used such as reservoirs, check dams, streams, and rivers; but they will not be included in this chapter.

Groundwater Occurrence

Locating Groundwater

Groundwater maps are prepared which describe the existence of varying amounts of water at any given time, where it flows, and what quality is underground; but, geohydrologists admit that mapping of groundwater is an imprecise science. Nevertheless, scientists and engineers use groundwater maps with the same reliance as they do rainfall data for planning and designing groundwater exploitation for irrigation technology.

For practical application, basic criteria are lacking for selecting a specific site for a well in the region. Farmers check with the local extension service and neighbors or may use the water witch or divining rod (dowsing); and, in spite of all this invaluable information, they prefer digging the well in a convenient location on the farm. When the drill rig arrives, the driller puts the well in the most convenient place for drilling operations. These behaviors reflect that there is no foolproof or scientific way to guarantee groundwater at a specific site. Every search is a fascinating calculated risk.

E.R. Perrier and A.B. Salkini (eds), Supplemental Irrigation in the Near East and North Africa, 155-175.
© *1991 ICARDA.*

Notwithstanding, groundwater does leave a few clues about its presence. In the Near East and North Africa, limestone caverns and decomposed calcareous sands and gravels are common locations for groundwater. Land contours and rock structures at the surface can suggest subterranean water flow and aquifers. Wet areas, marshes, and erosion which does not conform to localized surface runoff are all indicators of possible locations for groundwater. Depth and flow information from a neighbor's well may help a farmer's decision on a well site but some of the best information comes from local well diggers. Occasionally, if you know how to read and interpret the evidence, flow records on streams and springs can be beneficial and water-loving vegetation can be a suggestion as to the depth and location of groundwater. During storms, lightning tends to strike the tallest trees in a stand; however, sometimes it strikes shorter trees whose roots have penetrated the groundwater which is a good electrical ground.

Groundwater Terminology

The Earth stores its moisture in water-bearing formations, within loose gravels and soil, and in rock crevices, fissures, and cracks. Underground water can move slowly sometimes as little as a few centimeters a year; but it can move rapidly upward because of capillary action, temperature, and pressure. Porous materials, sand and gravel, accept rainfall easily but deeper strata of limestone and sandstone also accept rainfall. Generally, clay, shale, marl, marble, and granite are impermeable and restrict or inhibit flow. In most parts of the region, water is found in clearly definable layers of sands and gravels within their geologic sequence.

Water that penetrates the soil moistens the earth and can be taken up by plant roots and returned to the atmosphere through the process called transpiration. Water that drains past plant roots (deep percolation) may reach a lower level where all spaces between soil particles are filled with water. This area is termed the zone of saturation, and the liquid here is known as groundwater. The uppermost surface of groundwater when measured in wells is called the water table. The surface of a water table usually mirrors the land surface but can run contrary to the surface. Figure 11.1 illustrates groundwater location showing the dry surface soil as a raised relief.

In the upper portion of the zone of saturation is a thin layer known as the capillary fringe. In this thin layer the water is held by pressure forces and is not available to flow. Trees and some deep rooted plants may take advantage of the capillary fringe and use this water source. Seasonal changes cause differences in the depth to the water table and the thickness of the capillary fringe.

Before consideration of the well, something should be known of the nature, occurrence, and behavior of the groundwater to be extracted. Unlike mechanical phenomena such as the automobile or airplane, groundwater is hidden to the eye; therefore there is no everyday experience with it and so people are blind to the magnitude of quantities involved in flow and storage. Because

Groundwater Supply 157

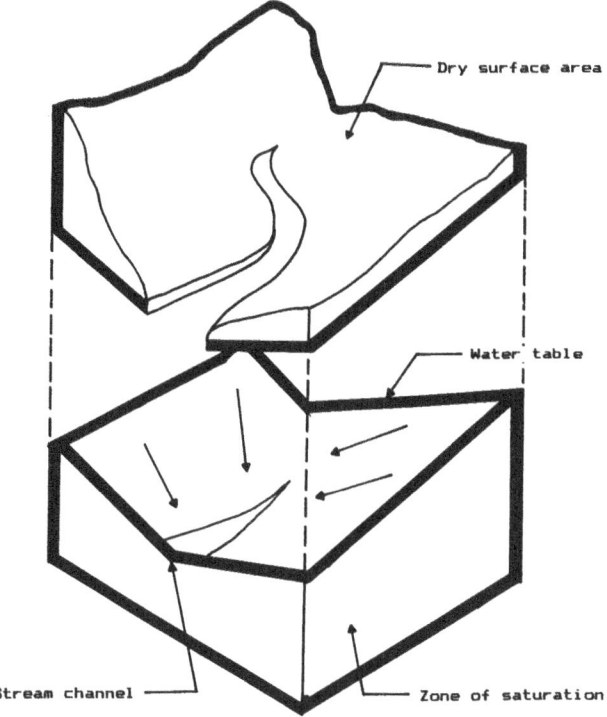

Figure 11.1. Location of groundwater.

it is there but cannot be seen, physical formulas must be relied on to provide estimates of the extent of groundwater and these formulas have a limited range of validity.

Groundwater formations called aquifers lie under a large part of the Near East and North Africa. Whether only a few meters below the soil surface or a kilometer deep, an aquifer is any formation that holds and transmits groundwater under ordinary hydraulic gradients. The amount of the stored groundwater that can be extracted depends on the material of which the aquifer is composed.

Groundwater supplies can be estimated by studying aquifers, including their geology, the characteristics of flow through them, and the climate over their intake or recharge areas. Natural recharge to an aquifer can occur by:
1. direct recharge from precipitation;
2. indirect recharge from infiltrating surface water, from wadi runoff, and from the accumulation of surface water in gypsum-saline marshes; and,
3. subsurface inflow from other areas or other aquifers.

Groundwater measurements are approximate because conditions underground are irregular, tests are costly, and some basic quantities such as evapotranspiration are only estimates. A pumping test can serve 2 general objectives. (1) A pumping test may be performed to determine the hydraulic characteristics of aquifers or water-bearing layers. Such a test is often called an aquifer test because it is the aquifer and not the pump or well, which is tested. Properly planned and carefully done aquifer tests may provide basic information for the solution of many regional, as well as local, groundwater flow problems. (2) A pumping test may provide information about the yield and drawdown of the well. These data can be used for selecting the type of pump best suited for the site and for estimating the cost of pumping.

Groundwater movement is closely linked to gravity, geophysical pressure, and the pressure of the atmosphere. As air pressures change regularly, these pressure changes affect groundwater, particularly as it relates to the height of water in a well. Also, groundwater temperature varies 5-18°C but fluctuates only a few degrees over an entire year. The chemical character of groundwater depends entirely on the material with which it has been in contact. Groundwater at great depths, can be hot, salty or brackish.

Aquifer Terminology

Within the Near East and North Africa, aquifers vary from place to place but are primarily recharged from natural precipitation. Occasionally, pumping for irrigation can be restricted by the month of May and can cease completely by July. The rains of autumn recharge the aquifer; therefore by December, farmers can irrigate once again. A limitation for aquifers in the region is that they have intercalated clay beds. These are thin blue clay layers (impermeable) within the limestone and soil layers. To understand the effectiveness of aquifer recharge, the question of whether the intercalated beds are impervious or leaky should be known.

Recharging of groundwater occurs when precipitation permeates an aquifer at a rate equal to or greater than the discharge flow. Normally, recharge occurs at higher elevations and discharge at lower elevations (Figure 11.2). The overall slope of an aquifer creates the hydrostatic pressure, especially towards the lower end when water pushes against impermeable rock.

Aquifers are of several types. Sometimes, aquifers are overlain by an impermeable layer of soil or rock, in which case they contain water under pressure. The level at which water rises in an unpumped well is called the water table, i.e. the water surface is open to the atmosphere (free) when the surrounding soil and rock strata are saturated. Occasionally low yielding wells have been dug into perched aquifers. Since the quantity of water held by a perched aquifer depends on the area of the impervious layer, wells sunk into perched aquifers are likely to exhaust their capacity and dry up if too much pumping is done. Extensive digging through the impervious layer in search of water at a deeper layer may lose the perched aquifer completely by percolation or drainage through these newly dug wells to a deeper, less accessible aquifer.

Other aquifers are confined between impermeable layers: confined aquifers

Groundwater Supply

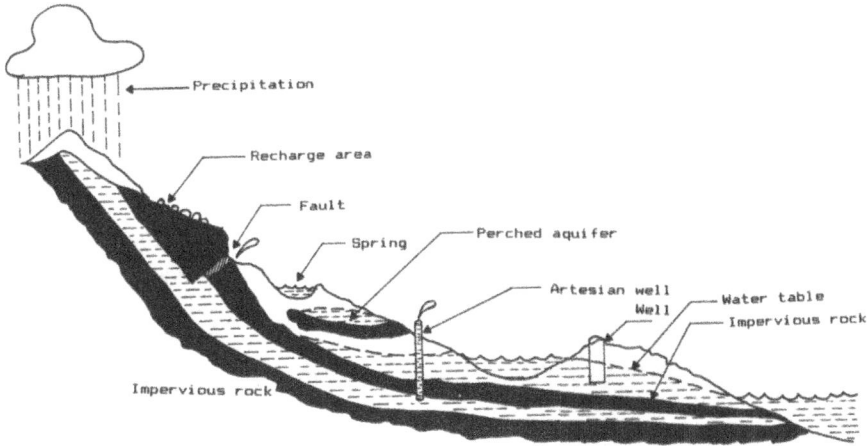

Figure 11.2. Diagram showing groundwater occurrence and terminology.

may be described as artesian or sub-artesian. In wells dug into a confined stratum which is recharged in an outcropping area of the same stratum but at a higher elevation and at some distance away, the water table rises to or above the surface (artesian) or part of the way to the surface (sub-artesian). The term, artesian, comes from the conquering Romans who discovered a large volume of water flowing from an open well near Artesium (now Artois) an ancient province of Northern France. Therefore, the term artesian has nothing to do with depth but everything to do with pressure. The natural pressure of water in an artesian aquifer usually remains constant throughout the year; however, extensive pumping from a confined aquifer will reduce this pressure and the flow.

An aquastat (or aquiclude) is an impervious geologic formation that contains little or no extractable water. To have an aquifer, there must be an aquastat to provide a bottom or impervious layer. Lateral barriers may be the impervious bedrock sides of a buried valley, a fault, or simply lateral changes in the layers of sandwiched materials (lithology) of the aquifer. Water always takes the path of least resistance; but, it may be impossible for water to seek its own level because of the obstruction of impervious barriers and aquastats. These make a risky business of locating a site for a well using the water witch, i.e. the Y-branch of a pomegranate bush or willow tree).

The porosity of an aquifer is a measure of its ability to store water; its permeability is its property of permitting the through-flow or passage of water. Aquifers usually have a greater permeability in the horizontal than in the vertical direction. Some clays are able to store enormous volumes of water and yet these clays are unproductive as aquifers because they do not yield their water, i.e. they have a low permeability and may be aquastats or aquicludes. Small pores, such as in clays, do not contribute to the effective porosity because in small pores the retention forces are greater than the weight of the water and flow is inhibited. In a good aquifer the pores are both large enough to hold groundwater and sufficiently interconnected to allow it to flow freely.

To understand the performance of an aquifer, information is needed on the subsurface hydrogeological conditions. To study aquifers in a specific region, the lithological cross-section or driller's log should be known. The lithosphere is the solid part of the Earth and lithology is the study of rocks or rock formations within that solid part. A lithological cross-section is a profile description of soil and rock materials found when digging a well. The cross-section identifies the depth of stratification of the various layers encountered in the borings to delineate the character, thickness, and boundaries of the various clay and limestone beds overlying and underlying the aquifer. In this way the average grain size, degree of sorting, and the clay/sand content can be estimated to predict the porosity and permeability of the aquifer material. Figure 11.3 shows a driller's log or lithological cross-section for the Tel Dhaman and Mare'a subdistricts in the Aleppo Province, Syria.

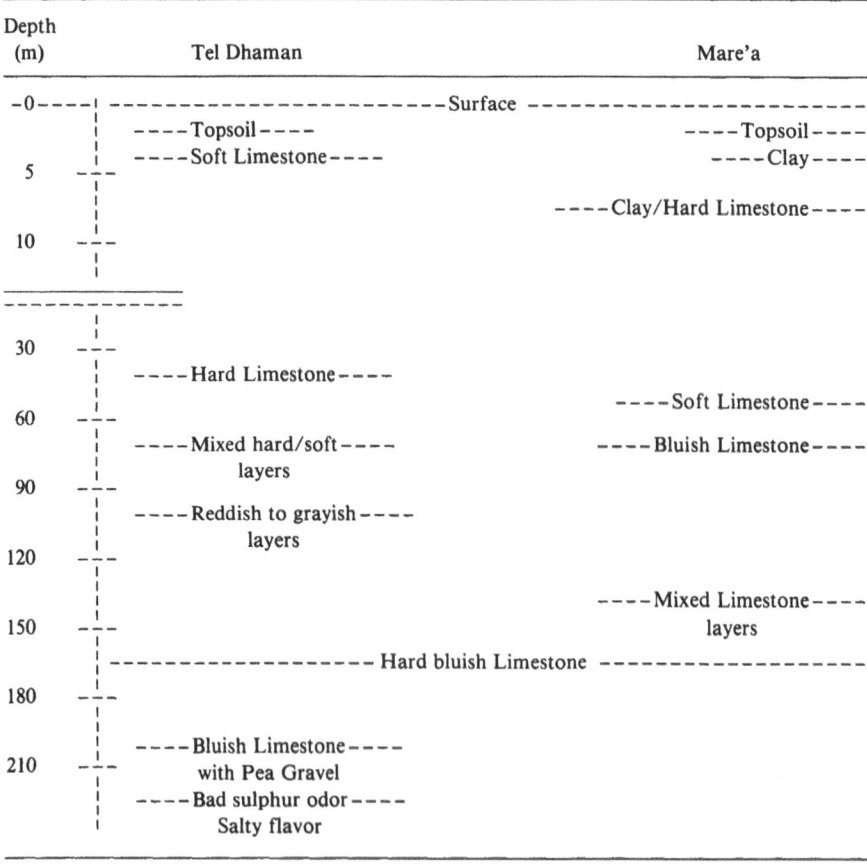

Figure 11.3. Lithological cross-sections from 2 areas of the Aleppo Province, Syria.

Groundwater Supply

The depth to the principal aquifer has not been established for the 2 study sites. Below the 180 m depth, some wells in the Tel Dhaman region have produced sulfurous waters of high salinity. There is a greater depth of clay in the first 30 m of material in the Mare'a region than in the Tel Dhaman region. The cross-sections of these clay and rock materials indicate that recharge by direct precipitation would be restricted by their low permeability and flow would be expected through fissures and cracks in the vertical planes of rock materials.

Figure 11.4 shows 4 types of aquifer generally found in the region. These aquifers (Table 11.1) are determined by the character of the covering layer when the base material is impervious. An unconfined aquifer is a permeable bed overlying an aquastat or impervious layer of hard or crystalline limestone whose upper boundary is formed by a free water table under atmospheric pressure. In fine-grained unconfined aquifers, gravity drainage of the soil pores is not instantaneous and water is released only some time after a lowering of the water table in a well; this phenomenon is called delayed yield.

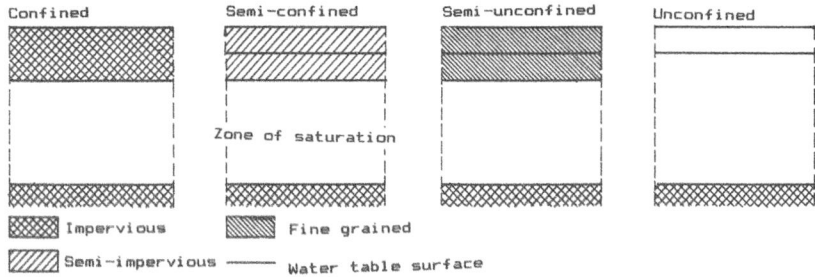

Figure 11.4. Four types of aquifer found in the region.

Table 11.1. Types of aquifer.

Covering layer	Aquifer type
Impervious	Confined (artesian)
Semi-pervious, (horizontal flow is neglected)	Semi-confined
Less pervious aquifer, (horizontal flow not negligible)	Semi-unconfined
Same as main part of aquifer (delayed yield)	Unconfined (open)

An artesian or confined aquifer is a completely saturated aquifer whose upper and lower boundaries are aquastats (see Figure 11.2). In confined aquifers the pressure of the water is usually higher than that of the atmosphere and the water level in wells is above the top of the aquifer. A confined aquifer implies no recharge from the soil surface. Completely impervious layers rarely exist in nature and hence confined aquifers are not common.

A semi-confined or leaky aquifer is a completely saturated aquifer which is bounded above by a semi-pervious layer and below by an aquastat. Since the permeability of the covering layer is usually low, horizontal flow in this layer is negligible. However, if the permeability of the covering layer in a semi-confined aquifer is so great as to make horizontal flow not negligible then such an aquifer, intermediate between the traditional semi-confined aquifer and the unconfined aquifer, is called a semi-unconfined aquifer.

Some aquifers extend over many hundreds of square kilometers with little variation in characteristics. Once such an area has been identified, the performance of a well within its boundaries can be predicted with reasonable accuracy and a well construction program can be planned with confidence. Other aquifers can vary in character over short distances making the potential yield of a well difficult to predict. The structure of aquifers is always changing and geologic stresses change with Earth's movement as well as weight shifts. Cracks open and close and water flows to new locations. Earthquakes, blasting, lightning, strong vibrations, or bulldozing have been said to alter the course of groundwater.

Most water stored underground is recoverable by pumping from boreholes or wells. Recharge wells have been used to raise the level of the water table which has the advantage of storage without evaporation. Aquifers can be recharged from surface water using sedimentation ponds with sand filters or from water collected by water harvesting techniques. The main disadvantages of recharge wells are the costs, continued maintenance, and a high failure rate.

Age and Origins of Groundwater

The age of groundwater in some areas of the Near East and North Africa is estimated to be older than 20,000 yr (fossil water). Such water can be mined but the supply will fail when the reserves are exhausted. In most cases, fossil water should be treated as a strategic reserve, a non-renewable resource. Groundwater recharge occurs in the wadis, but is generally not enough to decrease the salinity of groundwater to the quality of fresh water. Methods using isotopes in groundwater analysis can delineate regional flow systems, show where recharge is currently occurring, determine the recharge processes during earlier periods, and clarify the consequences of over-extraction of groundwater resources.

Direction and rate of flow in aquifers. To measure the direction and rate of flow in a borehole or well, radioactive isotopes are released at the level of the aquifer. For the flow direction, the radioactive isotope solution is released within a cylinder of iron gauze, carried by a rigid tube from the surface. Compass directions are labeled on the gauze which, after a predetermined period, is

Groundwater Supply

removed from the borehole and returned to the laboratory for measurement of the distribution of radioactivity. For measurement of the flow rate, a known quantity of radioactive solution is released at the depth to be investigated and a counter, immersed in the solution, measures the fall in concentration as it is diluted by the flow of groundwater. The flow rate of groundwater is slow, sometimes not more than a few meters per year, and generally, not exceeding 3 m/day.

Age of groundwater: To determine the age of groundwater, C^{14} analysis is performed. During the process of groundwater recharge, infiltrating rainwater on its way through the top soil towards the aquifer first dissolves carbon dioxide (with a C^{14} concentration of 100% modern) and then dissolves carbonate. The carbonate is assumed to be of marine origin and free of radiocarbon (0% modern). The carbonate-bicarbonate equilibrium established during the process of groundwater recharge can be described by:

$$CO_2 + H_2O + CaCO_3 \rightleftharpoons CA(HCO_3)_2 + CO_2.$$

The first C^{14} content of the dissolved carbon dioxide in newly recharged groundwater theoretically ranges from 55-65% modern.

When groundwater has entered the saturated zone of the aquifer, the C^{14} content decreases as the result of 3 processes.

1. The most important is radioactive decay. Age of groundwater is calculated if this is a unique process.
2. Hydraulic conditions can cause continuous mixing of water of differing ages from adjacent aquifers even when separated by aquicludes. In such cases, the C^{14} content reflects regional hydraulic parameters and not the age of groundwater.
3. When groundwater becomes mineralized, the hydrochemical processes may decrease the C^{14} content. Such processes are the oxidation of sulfides and the concurrent dissolution and precipitation of carbonates. Salt dissolution alone does not change the C^{14} age.

Tritium analysis of groundwater is performed to determine the extent of recharge from current infiltration. Natural tritium (H^3) concentrations of the hydrosphere are given in tritium ratios (1 TR = one tritium atom per 10^{18} hydrogen atoms). The natural tritium level of rainwater before the atom bomb test series was between a few TR and just above the detection level of common measuring devices. When production of tritium by the nuclear tests became dominant during the first years of the 1960s, the H^3 level increased by a factor of 1,000. This tritium now acts as a tracer in the hydrosphere. Therefore, tritium in any water sample is a definite indicator that rainwater has percolated into the aquifer during the last 2-3 decades. In some areas of the region, the absence of tritium has confirmed that the groundwater is fossil water which has not been newly recharged.

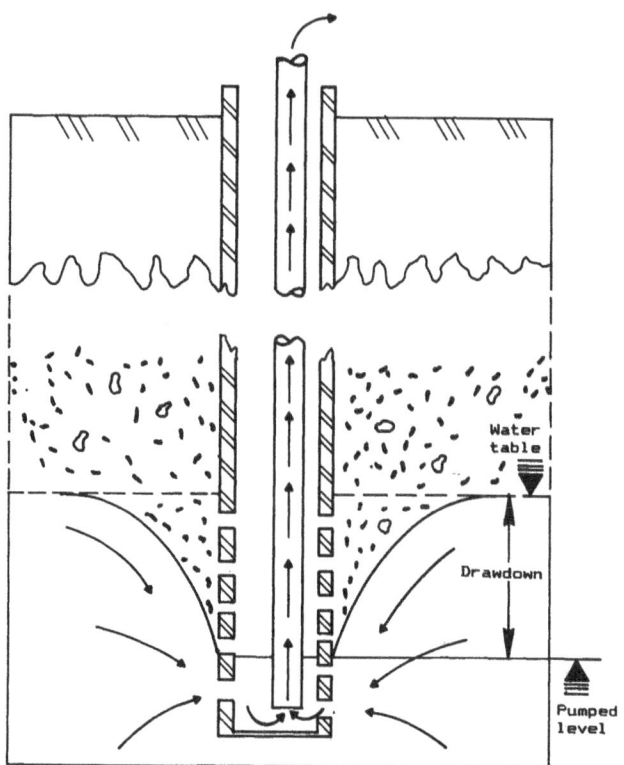

Figure 11.5. Elements of a typical well.

Wells

A well is a hole or pit sunk into the Earth for the purpose of extracting or removing groundwater (Figure 11.5). For the philosopher, a well penetrates the past as it is driven through eons of geologic time. Wells are driven to obtain water not to find water. There are various types of wells. Shallow wells can be dug by hand or with a backhoe and may be only 2-3 m deep. Other wells can be bored, jetted, driven, or drilled for hundreds of meters as illustrated in Figure 11.6. Bored, drilled, and dug wells vary in depth to 400 m and use casings of 5-300 cm in diameter; these 3 methods are suitable for most aquifers. The best place to drill a well is where the water table or aquifer is a known distance below the surface.

As wells normally penetrate a water table or aquifer from above, water must be pumped to the surface unless the well is artesian. Proper management prescribes that water should be pumped no faster than the aquifer can be replenished. If the aquifer is pumped too fast, excessive turbulence can occur and dirty or sandy water will be drawn, the water level will be lowered requiring more energy for pumping, and the safe yield of the water source could be

Groundwater Supply

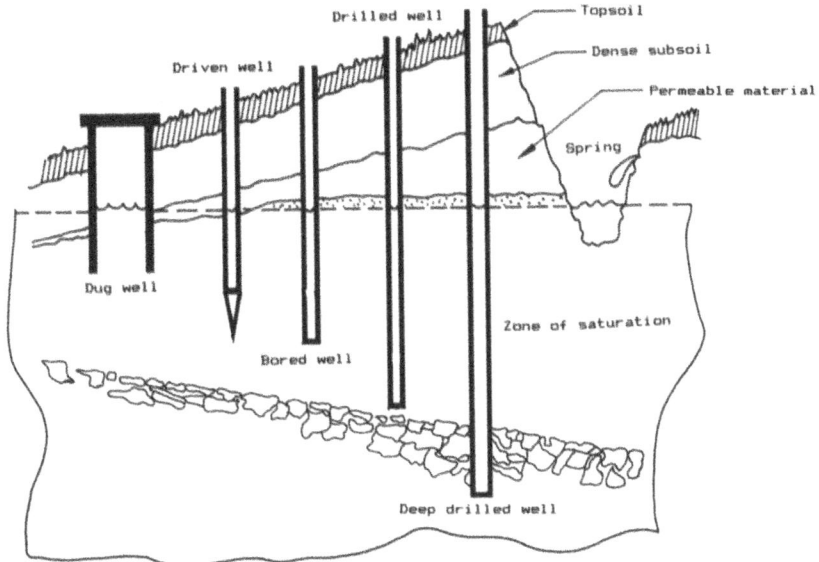

Figure 11.6. Soil and geologic strata with various well types.

pushed beyond reasonable limits. Safe yield is the withdrawal of water without producing an undesired result, e.g. deterioration and depletion of groundwater supplies, depletion of streamflow by induced infiltration, or land subsidence.

A well is a route to a supply of water from the ground but it can also be considered a storage facility. Contemporary thinking suggests that the deeper the well, the greater its capacity as a reservoir. Stored water can be used as needed and the aquifer will be depleted but will recover later as the large porous aquifer fills with water from the surrounding strata. Wells without enough yield to supply water during peak demand times may require holding tanks, cisterns, or ponds.

Several methods for protecting wells from surface contamination are available for drilled, bored, and dug wells. Grout seal is particularly important where the wells pass through potentially movable soils. Sidewall sluffing, collapsing soils, dry sliding sand, wet running sand, perched water table drainage, turbulent water flow, and pollution are a few of the ways in which wells become unusable. Sometimes the problems can be repaired by re-digging and casing or lining the troublesome area of the well with pipe or cement, but usually it is more economical to abandon the well and dig a new one in another location. Sand also limits the expected life of pump impellers and diffusers by excessive wear on the moving parts. Exterior and built-in lightning arrestors should be used at every installation and dielectric bushings can be added to retard sodium and calcium crusting. Bored, drilled, and dug wells which have been correctly cased and sealed are shown in Figure 11.7.

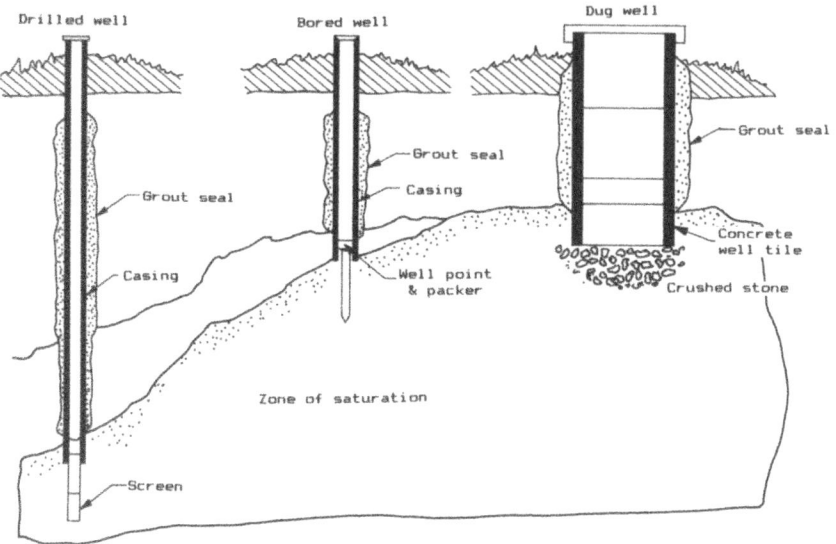

Figure 11.7. Cased and sealed wells.

Cone of Depression

When water is pumped from a well, the drawdown of the water table in the immediate vicinity of the submersed pump is affected at once. When a significant amount of water is removed from the water table, an unsaturated, funnel-shaped area called the cone of depression develops between the static water level and the pumping level as shown in Figure 11.8. The cone will begin to resaturate the moment pumping stops. The shape and width of the inverted cone, and how rapidly it refills, depends on the hydraulic conductivity of the material at the intake area of the well. Shallow, wide cones form in aquifers of permeable sand and gravel; but, when water moves more slowly through less permeable strata, steeper and narrower cones will develop.

Drawdown and the cone of depression have a direct bearing on the well's recovery rate, and on pumping costs. If the pump must work overtime to pull water from the surrounding stratum, then more energy must be used.

A well's radius of influence is the horizontal distance from the borehole to the top of the cone of depression. Although it cannot be observed, the radius of influence expands with more and more pumping. When several wells have been drilled too close together, one well's radius of influence can overlap that of another and so yield unsatisfactory amounts of water; the wells tend to pump each other dry. When too many wells are drawing water from the same aquifer, the permanent water table is lowered, and wells begin to go dry. Drilling deeper wells is only a stop-gap measure. In coastal regions, where salt water is nearby, salty water is drawn into dwindling supplies of fresh groundwater.

Groundwater Supply

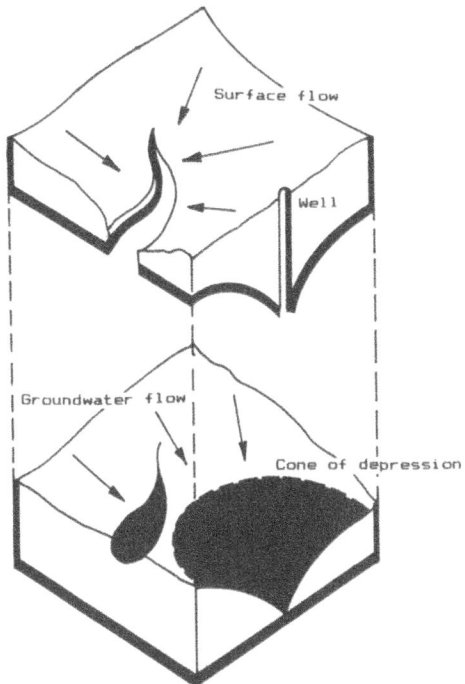

Figure 11.8. Cone of depression resulting from pumping a well.

Methods of Well Construction

Dug wells

Dug wells are generally thought to be undependable as well as vulnerable to pollution. They often fail during periods of extended drought when water is most needed. They are generally shallow, rarely more than 20 m in depth, but can be up to 120 m in depth in some areas of the world. Also, dug wells vary in diameter from 30 cm to more than 10 m to reduce deterioration. Two primary methods of digging wells are used: hand digging well and boring.

To take water from an aquifer, a hole is dug into the zone of saturation and is then lined to prevent collapse of its sides. Either the side lining or the bottom must be porous for water intake. Water from the aquifer will flow through the intake until the level of the water surface within the well coincides with the surrounding water table: the natural flow stops. When water is extracted by pump or bucket, the water level will drop and water will flow from the aquifer to the well. If the withdrawal is steady, inflow equals outflow, then the water level will reach a new constant level known as the drawdown.

Two steps can be taken as shown in Figure 11.9 to increase the yield capacity of a hand dug well:

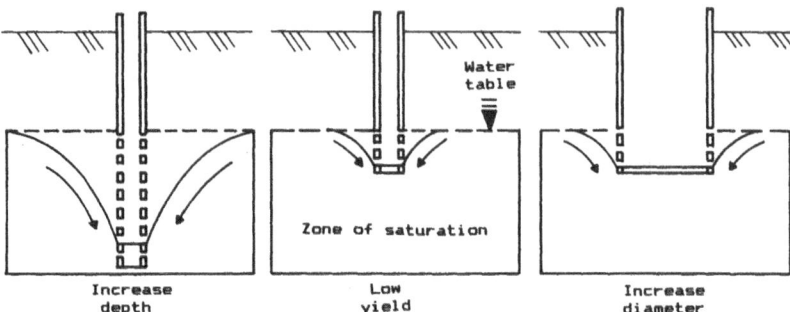

Figure 11.9. Techniques for increasing yield of hand dug wells.

1. increase the depth of the intake area into the zone of saturation, or
2. increase the diameter of the intake area.

In most cases, increasing the depth of the intake area is the most logical method for more yield; however, an aquastat may exist and expansion of the intake area is the only alternative.

Digging a well by hand is a tedious, pick-and-shovel operation. The excess material is hauled to the surface in a pail on a rope either by hand or by a winch with a hand crank, electric motor, or tractor power. A stone or concrete wall must be constructed around the top of the well (wellhead) to keep the sidewalls from caving-in and prevent pollution. Sometimes, the dug well is cased in-place or with precast concrete sections of 50 cm in height by 1 m in diameter. Brick or rock and mortar, concrete blocks, or precast sections should be as well sealed and as watertight as possible to reduce the opportunity of surface water from directly entering the zone of saturation (see Figure 11.7).

A cover or roof should be placed over the wellhead to keep out rainwater and the dug well must have a sanitary seal at the ground surface. The wellhead should have a large apron of at least 2 m width, sloping from the wellhead, for drainage away from the headwall of the dug well. The drainage apron will keep surface runoff from contaminating the water table. Space between the well casing and the surrounding earth should be backfilled with clean sand and the top 3 m plugged with a 1-2 cm coat of waterproof cement grout (1 part cement to 3 parts sand).

At the bottom of the dug well, a caisson or movable liner is constructed which extends from 1 m above the water table, down into the zone of saturation as deep as possible. For example, a caisson 3 m in diameter and 1 m in height can store 7,069 l (7.069 m^3) of water. The caisson method is used because the water-bearing strata is usually unstable, sandy or gravelly material. Generally, the caisson is porous or made with seepage holes to permit rapid recharge into the intake area. When interlocking brick or concrete blocks are used no mortar is added to permit seepage of water into the intake area. If the saturated zone is of consolidated materials of sandstone, fissured rock or hard stone, digging can be continued in the same manner and the wall built up with brick or precast concrete blocks. If the well is hand dug, breathable air must be supplied to the digger by using either bellows, an air compressor, or compressed air tanks. Power-operated pumps may be needed to remove water temporarily as the digging gets into the saturated zone below the water table. The base of the hole should be packed with clean crushed stone to act as a filter for water entering the well.

Bored wells

A bored well is essentially a dug well made with an earth auger instead of a shovel. If the auger is turned by hand, the hole could be between 20-40 cm in diameter. Earth augers work only in consistent soils, such as loams or clay. If there are boulders, or if the sidewalls tend to cave in, a different digging tool is necessary. Power-driven augers can bore wider holes of up to 1 m in diameter. Also, bored wells are somewhat deeper than dug wells, but are rarely more than 30 m deep.

Vitrified tile, steel, and plastic are the lining materials most often used for bored wells. Normally, these casings are perforated where the pipe extends into sand or gravel of the zone of saturation. These perforations receive water from the surrounding strata and should be covered with stainless steel screening to keep silt from entering the discharge line. As with all wells, a bored wellhead must be sealed and protected from rainfall and surface drainage.

Percussion and rotary drilled wells

A common well in the Near East and North Africa is the percussion type well. Cable tool, percussion type drilling rigs (shown in Figure 11.10) are used to beat and punch a hole into the ground. They are mounted on portable derricks with self-contained hoists and use a heavy hammer (bit and stem), a bailer, and a water tank. The hammer is lifted and dropped repeatedly, pounding and crushing its way into the soil and rock strata while water is added judiciously. At a prescribed depth, the bailer (a 3-5 m length of pipe with a valve at its lower end) is lowered into the hole, allowed to fill with the slurry of fractured material, and then removed from the hole. Hammers vary from 20-50 cm in diameter and can weigh more than a metric ton. This method can dig a well at a rate of 10-15 m/day in clay or soft limestone but only 2-3 m/day in hard limestone.

A percussion drill is quick, relatively inexpensive, and dependable and has the advantage of a reciprocating action that fractures whatever is in its path. The hard shock is meant to free water in rock fissures and soft formations near the hole. Bailing gives the driller an indication of the formations the hammer is breaking through. By constant monitoring of the consistency of the slurry and the speed with which water enters the hole, the drillers know when a sufficient supply of groundwater has been reached. Problems for this method arise when

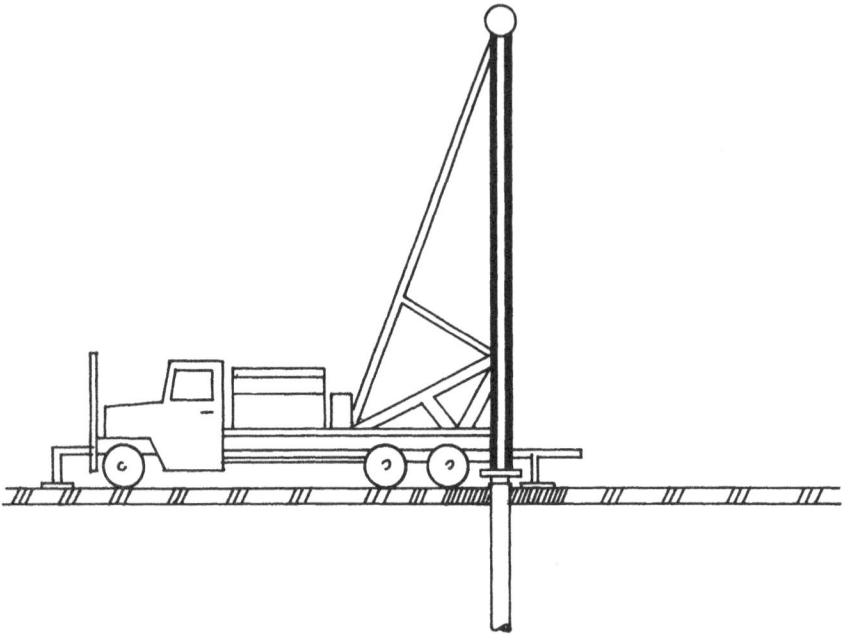

Figure 11.10. Percussion-type well drilling equipment commonly used in the Near East and North Africa.

drilling in heavy plastic clays and dense solid rock, then the rotary drilling method must be applied if further depth is required. Oftentimes, farmers drill a number of wells using the percussion drilling method and connect them to one pump or conveyance system to insure enough water throughout the year and in times of drought. As noted previously, percussion drilled wells can be 5-400 m deep, though most water is drawn from 50-75 m.

Rotary drilling is done with the help of hydraulics. Sometimes water is used as the drilling fluid or sometimes air; but, both air and water can be used. When both are used, this method drills as fast as 20 m/hr and will drive casing. The rotary drilling rig looks similar to the percussion drilling rig (see Figure 11.10).

Rotary drills are fitted with carbide buttons mounted on the drilling bit and attached to the starter rod. They drill into the bedrock about 600 m before the drill bit requires replacement and once into bedrock, the rig works more or less automatically. Where the hole passes through solid rock, no casing is needed. The starter bit is normally 20 cm in diameter to make a hole large enough to accept a casing with 15 cm inside diameter. Before the casing is driven, special Calgon detergent (which disperses clay particles) is pumped down the hole with the drilling water which creates a muddy foam to firm up the sides of the well hole. Once the casing is in place, a 15 cm diameter drill bit is used. When both air and water are used, for each meter depth 4-8 l of drilling water are pumped down the drilling rod, along with 6-18 bars of compressed air. This air pressure blows cuttings and water out of the hole at a speed of 900 m/min even from greater depths.

A flat-faced drive bit is mounted on the end of the starter rod to pound the casing into place. When the casing is firmly installed in the bedrock, drilling continues with a smaller diameter bit. Casings of 10-50 cm in diameter are not unusual for both the percussion and rotary drill techniques. Sections of casing are often connected by threaded couplings, then welded because they must be driven into the ground with great force. The thicker the casing, the stronger the wall. If the well casing extends to the bottom of the well, it can be slotted or screened at its lower end to allow sand-free water to flow into the well during pumping.

Life Expectancy of Wells and Pumps

In general, if an aquifer does not supply a sufficient volume of water to provide the crop water requirement, the well is dug/drilled a little deeper into the aquifer. This changes the economics of the system and a new cost analysis and comparison should be done to estimate the increased costs of operation. Digging wells to the 150 m depth and pumping at the 65 m depth provides a margin of safety for the farmer. In practice, when existing wells are at the 100 m depth, a farmer will drill new wells to the 110 m depth. Moreover, the next farmer may drill a well to the 120 m depth or deeper, increasing the margin of safety.

In this region, wells for irrigation have an adequate supply for about 6 months of the year. Over-pumping can cause wells to be temporarily dry but aquifers are normally rechargeable, a renewable resource. Low-water electrical cut-off switches have been developed to shut down the pumping system whenever the water level gets dangerously low.

In determining the service life expectancy of various pump models, the economic life and not the technical life should be used. Even though a pump manufacturer or supplier claims that their pump will last for 15-20 yr under normal operating conditions, in practice such a life is rare under prevailing service and maintenance found in rural areas. To predict the service life or economic life of a pump is difficult for newly introduced pumps on the market. It is necessary to collect data on yearly repair and maintenance costs for similar kinds of pumps used in the region. Wells are considered excellent if they continue to yield adequate flow when pumped for 25 yr.

Groundwater Pollution

Surface water moves swiftly in streams and rivers and, therefore, pollution can be dispersed quickly. Surface water is also radiated by sunlight and aerated with oxygen and this helps the cleansing action. In contrast, groundwater may move only a few meters per year and sunlight and oxygen in the atmosphere cannot reach it easily. As a result, centuries may be needed for groundwater cleansing.

When rivers are polluted with sewage and used for irrigation as in the Near East and North Africa, the sewage slowly trickles down through the various layers of soil. Some farmers believe that the water is purified of contaminants but many types of chemicals are not filtered out, especially solvents and pesticides which can pass through the soil and into the aquifer. Pollution of groundwater is especially common around population and industrial centers. The best method of controlling groundwater pollution is to control land use and recharge of water into the aquifer.

Surface waters are constantly polluted with silt, algae, decaying leaves, a multitude of salts, chemicals, bacteria, and protozoans. Groundwater, because of contact with soil and rock, is permeated with impure elements like calcium, magnesium, sodium, potassium, gypsum, anhydrite, unhealthy amounts of fluoride, and even arsenic. The well should be constructed carefully to avoid contamination from nearby sewers, cesspools, human and livestock wastes from runoff which can transmit typhoid and hepatitis. Rule-of-thumb distances are presented in Table 11.2 as a guideline for choosing a well site.

Table 11.2. Guidelines for minimum critical distance from sources of contamination for wells.

Source of contamination	Minimum distance (m)
Waste disposal lagoon	100
Cesspools	50
Livestock and poultry yards	40
Manure piles, Privies	40
Seepage pits, Silo pits	50
Milkhouse drain outlets	40
Septic tanks and disposal fields	50
Gravity sewer or drain not pressure tight	25
Pressure-tight gravity sewer or drain	10

Domestic wells: If the well is to be used for domestic purposes, then disinfection should be considered immediately after construction. Most wells in the region serve as multipurpose sources of water for irrigation, recreation, and domestic use. The well should be thoroughly cleaned and sterilized (chlorination) with a strong disinfectant which is allowed to remain in the well for several hours before pumping water for domestic use. A mixture of common household bleach with water (2 l of ordinary bleach added to 35 l of water) to give a chlorine solution is poured into the well, left overnight, and pumped as waste water until the odor of chlorine has disappeared (dependent on the volume of water in the well and the pumping rate). The procedure will kill any bacteria and

Groundwater Supply

parasites that entered the well or pumping system during construction. Disinfecting the well immediately after construction will not keep the well water sterilized for more than a few days if contamination is allowed to re-enter the well after this sterilizing procedure.

Groundwater Abstraction and Delivery

A well's yield depends upon the geological formation it penetrates. Yields of wells pumping from fissures in bedrock are the most difficult to predict. Layered aquifers yield more predictable amounts of water. Fine sand, for example, will yield up to 40 m^3/hr for a properly designed irrigation system; whereas, coarse gravels can yield more than 90 m^3/hr.

Pumping Using Suction

Gravity is the prime mover of water in its liquid state. If an artesian well or spring is located on a hillside, it is possible to conduct the water downhill through aqueducts or tubes, using gravity which develops the hydraulic gradient. The water needed is usually below the surface and must be raised or lifted to a location where it can be transported.

Pumps are machines for lifting water. The term, pump, refers to both the machine which lifts water and the power source, whether a windmill, electric motor, gasoline or diesel engine, or operated by hand. As pumps come in various sizes and types, proper selection and sizing are vital. There are piston pumps, centrifugal pumps, jet pumps, turbine pumps, submersible pumps, plus many other types of wind and hand pumps. Farmers should learn a little about each type before investing in a pumping system.

To move water from a well requires suction. Regardless of the amount of power applied from the pump's motor, the maximum vertical distance that water can be lifted by suction is 450-750 cm. Fundamentally, water is easier to push than to pull upwards though it may not be convenient. Recall that suction pulls and pressure pushes. Normal atmospheric pressure at sea level presses down everywhere at 1 kg/cm^2. Theoretically, suction alone should lift water 10 m at sea level if a perfect vacuum could be created and if there were no friction in the pipe. When a suction pump is drawing air out of a pipe, this action lowers the atmospheric pressure within the pipe. This low-pressure area developed within the pipe is replaced by water from below that is being pushed by the external atmospheric pressure of 1 kg/cm^2.

If the suction pipe has a leak, the vacuum will be reduced or lost and the pump's ability to move water up the pipe is diminished. The pump is said to have lost its prime if the water level in the suction pipe falls all the way back to the level of the water source. Furthermore, lift by suction is reduced as elevation above sea level increases. For example, if the field to be irrigated is at 1,000 m elevation, the suction pump could not lift water higher than 600 cm. A pump's lifting capacity is limited by elevation and water friction in the pipe.

While shallow-well pumps lift water using suction, deep-well pumps are lowered, or parts of them are lowered, into the well itself where they push water. Pushing pumps lift water to heights many times greater than 750 cm because they can develop more power. The line of demarcation between shallow-well and deep-well pumps is, not coincidentally, 750 cm.

Pressure and Head

When water is pushed, this action is termed pressure with units of kilograms. When this pressure is expressed in height, it is called head. The pressure of 1 kg/cm² will produce 10 m of head: conversely, 1 m of head requires 0.01 kg/cm² of pressure. To elevate water 100 m in a tube 1.0 cm², should require a pressure of 10 kg/cm². Only the total vertical distance is important not whether the tube is perfectly vertical or slanted at an angle.

If water does not flow in the tube, this 100 m height of water would be called static head as shown in Figure 11.11. Head below a pump's intake is called static suction lift. Pressure at the point where water leaves a pump is known as static discharge head. Total static head is a combination of static suction lift and static discharge head; that is, the total elevation from the water level in a well to the highest outlet in the field is a well system's total static head.

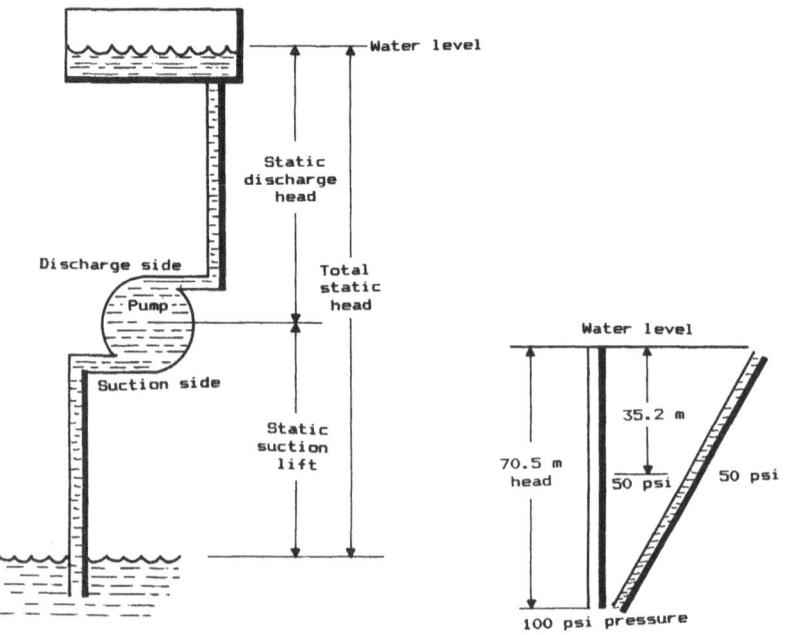

Figure 11.11. Diagrams illustrating terminology relating to head.

Flow and Lift

Water is not stationary in a pumping system and as it moves through pipes or tubing it has flow which is expressed as l/min or m³/hr. Flow resistance in the form of friction inside the pipes causes water to slow down by turbulence at corners, passing through jagged fittings, or from encounters with rough, crusty corrosion. The further that water travels, the greater the friction, hence the greater the head loss, e.g. the smaller the pipe, the more friction resulting in a greater head loss. The greater the flow, the more the pressure will drop. The head loss can be calculated from friction-loss tables that consider pipe diameter, tubing material, age of piping, and fittings. A system's total head is the sum of its total static head and its friction head described as head loss in meters of water per 100 m of pipe. Power from the pump and discharge pressure to a tank or an irrigation system must be enough to push water to a height equivalent to the system's total head.

If the vertical distance between the water table and the highest elevation at the field is 750 cm or less, a suction pump, operating in the 1-3 kg/cm² range, should supply the irrigation system with adequate pressure. This assumes that pipes are properly sized and installed to keep friction to a minimum. If the total static head is slightly greater than 750 cm, a heavy-duty shallow-well pump should be installed which would produce a higher discharge pressure.

When a deep-well pump is pushing water from below the water table, total lift must be computed. Total lift is the well depth added to the highest point of discharge. For example, if the well's depth is 120 m and the highest point of discharge is 5 m, then the pump must be capable of a total lift of 125 m plus friction head losses in the delivery line (estimated to be 10% of the total lift), say 12.5 m. If a standard sprinkler irrigation system is operating at a pressure of 70 psi or 5 kg/cm², the total head needed is about 50 m (70 psi × 70 cm/psi/100 cm/m = 50 m). For this example, the total lift or the delivery head required of the pumping system is 125 m + 12.5 m + 50 m = 187.5 m.

Chapter 12

Water Quality, Irrigation Measurement and Efficiency

Quality of Water for Supplemental Irrigation

Both the quantity and quality of water supply are important for supplemental irrigation. A water supply must be adequate to fulfill anticipated irrigation needs. If water of low quality is applied, then special management practices are needed to maintain optimal crop productivity. Problems that occur from using water of poor quality vary in type and degree of severity. Osmotic effects on the plants, effects on soil infiltration and permeability, as well as specific ion toxicities can be serious. Irrigation water of inferior quality can cause excessive vegetative growth or lodging; or can delay crop maturation because of excessive nutrients in the water supply; or the water can dry and deposit salt residues on fruit or leaves from sprinkling; or cause nutritional disorders by high pH of the water supply.

Table 12.1 gives guidelines for evaluating the quality of the water supply to be used for irrigation. These guidelines are limited to common problems and their effects on crop production. The values are for a wide range of conditions found in irrigated agriculture and criteria incorporate some of the newer concepts in soil-water-plant relations.

Even though rainwater may have the lowest salt content of any water source, it is not completely free of dissolved salts. Salts present in minute amounts in the atmosphere from terrestrial and marine sources are removed and concentrated by rainfall. As oceanic air masses move inland, atmospheric salinity decreases rapidly, with typical salt concentrations of rainfall decreasing from 40 mg/l along the coastline to only 2-3 mg/l in the continental interior.

In general, water quality is not a severe problem for supplemental irrigation since, in rainfall areas of 350 mm, the amount of water demanded for plant growth may be only 60-70 mm of irrigation water. That is, less than 20% of the water for plant growth will be supplied by supplemental irrigation, thus, 350 mm of rainfall is available for leaching any salt accumulated from supplemental irrigation.

E.R. Perrier and A.B. Salkini (eds), Supplemental Irrigation in the Near East and North Africa,
177-189.
© 1991 *ICARDA.*

Table 12.1. Guidelines for interpretation of water quality for irrigation.

Water quality criterion	Degree of problem		
	None	Increasing	Severe
Salinity, EC (dS/m)	< 0.75	0.75 – 3	> 3
Permeability			
1. Low salt water, EC	< 0.5	0.5 – 0.2	< 0.2
2. Sodium hazard, SAR			
Montmorillonitic	< 6	6 – 9	> 9
Illite-Vermiculitic	< 8	8 – 16	> 16
Kaolinite-Sesquioxidic	< 16	16 – 24	> 24
Specific ion toxicity			
Sodium (SAR)	< 3	3 – 9	> 9
Chloride (mol/m^3)	< 4	4 – 10	> 10
Boron (mg/l)	< 0.75	0.75 – 2	> 2
Miscellaneous effects			
Nitrogen (mg/l)	< 5	5 – 30	> 30
Bicarbonate (mol/m^3)	< 1.5	1.5 – 8.5	> 8.5
pH		[Normal range 6.5 to 8.4]	

Determination of Water Quantity for Irrigation

Any conveyance of water forming a free surface is considered an open channel. The channel is usually referred to as a ditch when conveying irrigation water to a cultivated field. The conveyance of water in pipes is used to supply water under pressure for sprinkler and drip/trickle systems. The choice of system to carry water to a cultivated field is usually dependent on the cost and scarcity of water for irrigation. Occasionally, it is dependent upon the level of technology dominant within a given community.

Sprinkler distribution measurement

Wind speed and direction are two characteristics considered detrimental to measurement of the quantity of water applied by sprinkler irrigation. Wind speed which increases with height above a crop or soil surface also effects spray evaporation as well as evaporation from a wetted soil surface. Sometimes these errors can be corrected by scheduling successive irrigations at night when there is a reduction in the wind velocity. Under normal irrigation practices, spray and evaporation losses constitute a rather small percentage of the total applied water. Typical losses are as follows.

Irrigation method	*Spray and evaporation losses*
Hand move sprinkler	6 – 10%
Undertree sprinkler	2 – 3%
Surface	1 – 2%

Instances in which these losses can be noticeably higher are associated with sprinkler irrigation. Hot and dry winds, high sprinkler pressures, and short set durations can, in specific instances, combine to produce evaporation losses greater than 60%.

Measurement of the volume of water applied either by sprinkler or surface irrigation techniques is needed for farmers to know whether or not they are over- or under-irrigating. Meters of various types may answer this question but there are other less expensive and nearly as accurate techniques which can be used in the field. For sprinklers, can catchment can be used: fixed grid systems of cans are used by placing cans on both sides of a lateral or across laterals to evaluate the spacing and efficiency of the sprinkler system. If pressure and discharge data are available then evaluation of gasket leakage or non-uniformity of discharge along the lateral can be done in the field. One simple approach is described below.

1. Place tin cans (10-20 cm diameter top with straight sides are best) at 2 m distances horizontal and perpendicular from the sprinkler head (a minimum of 20 cans are needed) and special care should be taken in the overlap area. Evaporation is retarded by wiping the interior of each can with a cloth moistened with light oil prior to the test.
2. Place leveled open-topped cans at height of crop or close to ground level.
3. Measure diameter of can opening and volume of water collected in can after an irrigation of at least 10 mm. (If plastic cups are used the volume of water captured must be related to depth based on the area of the top rim.)
4. Water should be applied at a rate less than or equal to the infiltration rate of the soil. From these measurements calculate average depth of water applied. If interested, compute distribution uniformity (DU), uniformity coefficient (UC), potential application efficiency (PAE), and actual application efficiency (AAE).

Factors which affect the can catchment uniformity include:
1. wind;
2. sprinkler pressure (pressure differences, elevation and friction);
3. sprinkler spacing;
4. crop interference;
5. nozzle size (different sizes used in the same field); and,
6. sand wear on nozzles and plugging.

Sprinklers must be operated at the proper pressure. A low pressure will create a doughnut (ring-shaped) pattern. The jet of water will not break up and therefore lands in a narrow ring with a radius of 10-12 m away from the sprinkler. At high pressures, there is too much break-up of the spray jet resulting in excessive water application close to the sprinkler and, therefore, susceptibility to wind distortion.

When pressure differences occur sprinklers will have a flow rate variation of 10% if there is a 20% pressure difference. In most hand-move and side-roller systems there is considerably more than a 20% difference between sprinkler pressures. Pressure differences can be minimized by:

1. running laterals downhill when possible;
2. using large pipe diameters;
3. keeping reasonable lateral lengths;
4. using pressure regulators on individual sprinklers; and,
5. using flow control nozzles on individual sprinklers.

Sand wear of nozzles and sprinkler drivers is a major problem in some areas. The sand usually originates in wells. A good well design can minimize sand pumping. However, if a sandy well needs to be used, sand separators can be purchased to remove the sand. They can be installed in the well at the suction of the pump or at ground surface. These separators remove more than 90% of the sand and minimize sand damage. Submersible separators will protect both the pump and the irrigation system.

An important factor when considering both can catchment and distribution is that an irrigation pump supplies water to all of the field, not to just a single lateral of sprinklers. Therefore, the measurements which evaluate a sprinkler system design must be taken throughout the field. Evaluations which only analyze one lateral will give unreliably high uniformity results.

Recently, there has been an effort to reduce the pressure requirements of irrigation systems. The energy bill is directly proportional to the pressure requirement, assuming all other factors remain constant which is not always true. Low pressure sprinkler nozzles have gained popularity in some areas. The objectives of switching from standard to low pressure nozzles are:
1. reduction of sprinkler pressure from 60 psi to 35 psi;
2. achievement of good jet breakup even at these lower pressures; and,
3. achievement of good uniformity at these low pressures.

The reduction of sprinkler pressure by 30% will decrease the actual total pump pressure requirements by different amounts depending upon the total pump lift.

Water source	Total pump pressure reduction for 30% reduction of sprinkler pressure
Canal	23%
Well lift of 30 m	14%
Well lift of 92 m	8%
Well lift of 153 m	5%

When pumping from a canal, the percentage of power savings is appreciable, but it is significantly less with a deep well water supply. There can be potential disadvantages to using low pressure nozzles. Some farmers are satisfied with the nozzles, but others have abandoned them because of problems.
1. Large droplets associated with lower pressures can cause sealing of the soil surface and increased runoff.
2. Lower volume distribution uniformity will always decrease efficiency. The increased hours of operation necessary to offset this lower distribution uniformity may remove any savings in horsepower.

3. Pump efficiency may decrease depending upon the pump and the specific system.

Volume flow rate in ditches

Lined or unlined ditches are used to convey irrigation water to cultivated fields. Special precautions are needed in erodible soils to prevent washouts. Ditches are designed for a capacity equal to the crop water requirements during peak demand and the capacity includes operational losses. Ditches should be designed somewhat larger than immediately thought necessary because with time they become partly clogged with sediment, grass, weeds, or trash. Generally, the farmer uses some device such as a meter, flume, or weir to measure volume flow rates. These devices can measure volume flow rates very accurately but this accuracy may not be needed for all estimates of flow rates. For estimating volume flow rates in ditches, channels, and canals, stream velocity and cross-sectional area (toothpick method) can be used.
1. Place a meter stick (leveled) across the top of the ditch or channel.
2. Measure the depth of water across the ditch or channel at intervals of 2 or 3 cm for small irrigation ditches or wider intervals for larger channels.
3. Calculate the area of each interval and accumulate areas to determine total flow area cross-section.
4. Measure stream velocity by a stream flow meter or estimate the velocity by dropping a floating object (wooden match stick or a toothpick) in the center of a smooth water surface in the ditch or channel. Determine the time required for a floating object to traverse a 2 m interval using a stop watch or observing the sweep second hand of a wrist watch. Calculate the velocity of the object which is now the estimated stream velocity.
5. Calculate the volume flow rate (usually in cm^3/sec) using accumulative area (cm^2) × velocity (cm/sec).
6. Determine the depth of water applied for surface irrigation from the area of the plot, observing units of calculation.

Irrigation Efficiency

The term "water applied" equals the net water supplied specifically to a field. Runoff from a field which is collected and reused on other fields must be subtracted.

Water applied = water onto field − runoff from the field

Applied water values are very important in all of the calculations. The water applied values are often difficult to determine because there may be no flow meters or because a water source may supply several fields. Water may be applied for reasons other than evapotranspiration and leaching. For example, pre-irrigation includes water used for frost protection and weed germination. Some of the water used for frost protection may eventually be stored in the root zone for plant use later in the season. Also, farmers may use light irrigations to germinate weeds, and then disc or spray with an herbicide.

Some crops and some harvesting techniques require a special irrigation to prepare the soil. For example, the soil may be water tamped to make a hard surface prior to shaking almond or olive trees for harvest. Crops such as apples and potatoes may need extra water applications to keep the crop cool. These applications generally exceed the evapotranspiration requirements but are necessary to produce a marketable crop and could be as important as the evapotranspiration requirement.

Preceding a forecast for frost, supplemental irrigation can be beneficial to aid in frost protection for temperatures down to $-9°C$. The freezing point of plant fluids is slightly below $0°C$ and if plant temperatures can be held at $0°C$, frost damage can be prevented. Moisture in the soil increases its capacity to store heat and an irrigation before frost can supply that moisture. After an irrigation, the soil has more heat energy in storage and this heat will be released during the night and can provide protection against light frosts of -2 to $-3°C$. For more severe frosts, a continuous application of water during the hours of frost, wets the plants and as the water freezes on the plant (releasing its latent heat) the temperature cannot go below $0°C$. The principle of frost protection by supplemental irrigation is simple but equipment, management, and wind speed play a significant role.

Water applied by supplemental irrigation has many possible destinations. The water is either lost in the air by evaporation, used by plants in transpiration, stored in the soil root zone, lost as surface runoff, or exits the root zone as deep percolation. Deep percolation can contribute to high water tables, recharge streams and groundwater aquifers, and be either beneficial or harmful. The expression "irrigation efficiency" is defined as the ratio of evapotranspiration to the amount of water applied. Sometimes this value is calculated for large projects, project segments, farmers, cultivated fields, or accurately measured research plots; the magnitude of the measured value is dependent on the data input.

Percent Irrigation efficiency (I_e) is defined as:

$$I_e = \frac{\text{Crop Water Use}}{\text{Water Applied}} \times 100.$$

Low irrigation efficiency is implicit in the definition. Low on-farm irrigation efficiencies do not generally result in a loss of water to an irrigation basin or groundwater unit. The water which runs off or percolates from a field or farm is generally used by another farmer downstream.

The improvement of low on-farm irrigation efficiencies is to avoid the following problems.
1. Excessive irrigation and over-pumping by farmers can cause excess energy use.
2. Deep percolation contributes to fertilizer leaching.

Water Quality, Irrigation Measurement and Efficiency

3. Low irrigation efficiency reduces or lowers crop yields.
4. Water quality can be degraded.

To understand the need for improvement of on-farm irrigation efficiency, an explanation of "beneficial" versus "non-beneficial" use of water in on-farm irrigation is necessary.

Beneficial uses	Non-beneficial uses
Crop transpiration	Weed transpiration
Required leaching of salts	Deep percolation in excess of leaching needs
Special practices	Evaporation from wet soil surfaces
packing soil for harvest	Uncollected surface runoff
weed seed germination	Canal and channel seepage
climate control	Spray losses

To evaluate how irrigation efficiency affects the farmer directly the well system is used where horsepower (Hp) is calculated as follows:

$$Hp = \frac{Q \times h}{273},$$

where Q = pump flow rate in m³/hr, and
h = vertical lift or depth to the water table (m).

Irrigation efficiency on the farm directly affects the horsepower requirement because a low efficiency makes a large flow rate necessary. The annual hours of pumping are also directly affected by irrigation efficiencies. A system which is poorly scheduled will result in excessive hours of operation per year. The following factors affect the annual energy use for irrigation pumping:
1. irrigation efficiency:
    a. irrigation timing (scheduling),
    b. uniformity of water distribution throughout a field,
    c. unrecovered losses, i.e., evaporation and uncollected runoff.
2. pump efficiency.
3. motor efficiency.
4. pressure requirements:
    a. depth to water,
    b. drawdown in a well,
    c. pipe, valve, and column friction,
    d. final discharge pressure, e.g. sprinkler pressure,
    e. elevation change.

Distribution uniformity

Ideally, irrigation methods should apply water uniformly to each plant in a field. Figure 12.1 shows possible destinations of applied water. The distribution uniformity (DU) is defined as a ratio of the minimum amount of water arriving at each plant to the average amount of water. Various formulas have been used to describe DU in slightly different terms, but Figure 12.1 illustrates the concept:

$$DU = \frac{B}{A} \times 100$$

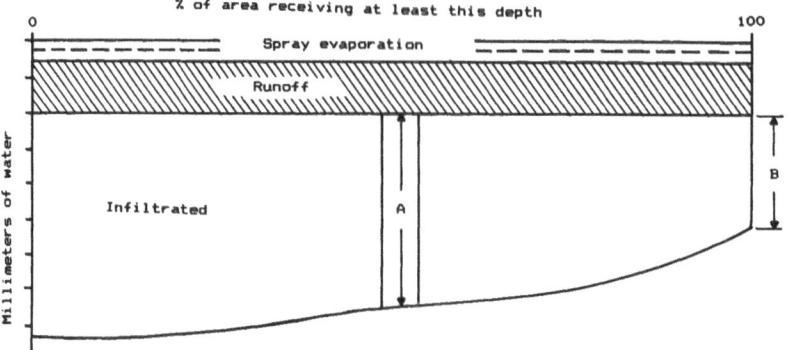

Figure 12.1. Definition of distribution uniformity (DU).

A satisfactorily designed irrigation system will have certain potential distribution uniformities as detailed below.

Irrigation method	"Potential" DU (100 x minimum/ave)
Permanent undertree sprinkler	95
Linear move	93
Sloping furrow	90
Orchard drip	90
Border strip	85
Level furrow	85
Row crop drip	85
Hand move sprinkler	75

Actual distribution uniformities are frequently lower than these potential values. Actual field distribution uniformities depend upon:
1. irrigation system design and installation;
2. suitability of selected method, e.g. furrows are not suitable for variable, sandy soils; and,
3. management of selected method, e.g. pressure regulators adjusted or flow rates used in furrows.

Figure 12.2 illustrates an irrigation in which the quantity of applied water includes estimates of the following:
1. evaporation and spray losses;
2. runoff;
3. root zone storage; and,
4. deep percolation below the root zone.

This diagram specifies a case with perfect irrigation scheduling; the water was shut off exactly at the correct time. No location in the soil profile is under-irrigated, and the area which receives the least amount of water, receives exactly the amount needed to satisfy the soil moisture deficit in the root zone. In this case, a poor DU resulted in deep percolation.

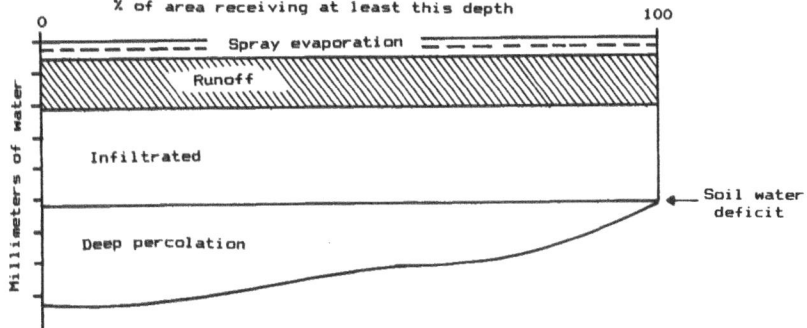

Figure 12.2. Correct timing.

Figure 12.3 illustrates a situation where over-irrigation has occurred. The sprinklers, drip, or furrow set length or amount of time to irrigate was too long. In this case, deep percolation was caused by non-uniformity and poor timing of irrigation.

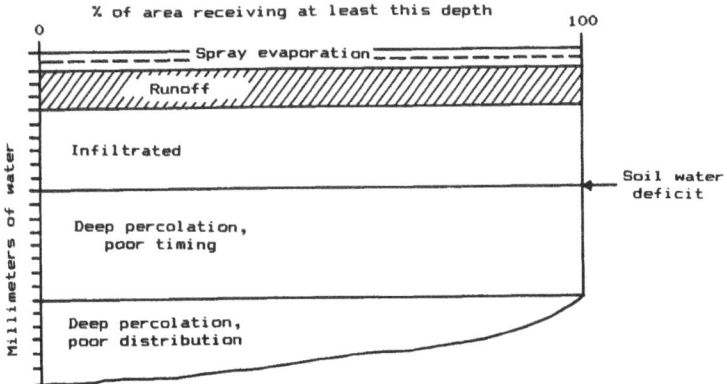

Figure 12.3. Over-irrigation.

Figure 12.4 shows a different concept. Some farmers believe they have no deep percolation losses because the average depth infiltrated equals the soil moisture deficit. This results in both deep percolation and under-irrigation caused by non-uniformity.

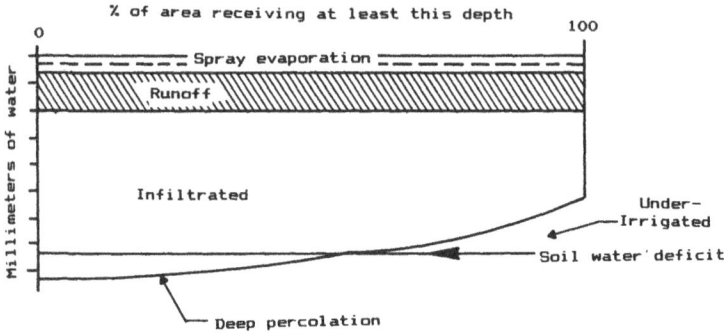

Figure 12.4. Both under- and over-irrigation.

Figure 12.5 illustrates a situation which is possible when surface irrigation is stopped as water reaches the bottom of the field. There is some evaporation loss from the wet soil surface at this time. In this case, nearly 100% irrigation efficiency can be achieved because the field surface has been covered; but, the farmer is under-irrigating the complete field by not replenishing the soil water deficit. All the water will be used by the crop but yields will be unacceptably low at the lower end of the field. The major points this diagram stresses are inadequate runoff and incorrect timing.

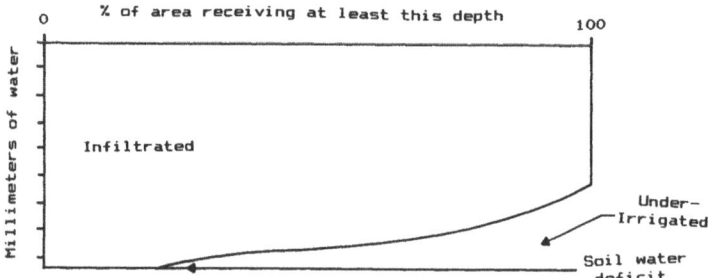

Figure 12.5. 100% irrigation efficiency.

In summary, a good on-farm irrigation efficiency can be obtained by:
1. achieving a good distribution uniformity;
2. proper scheduling of irrigations;
3. collecting and recycling runoff; and,
4. minimizing evaporation and spray losses.

Fertilizer leaching

Water which percolates below the root zone in the soil profile carries dissolved salts. Some leaching is necessary to remove detrimental salts which are brought in by the irrigation water. However, excessive leaching caused by non-uniformity and poor irrigation scheduling can result in heavy fertilizer losses. This fertilizer may enter the groundwater and can cause serious concern in some areas.

Nitrogen fertilizers are particularly susceptible to leaching loss. This is an important energy conservation concern as 1 ton of nitrogen fertilizer requires approximately 42 M BTU's (British thermal units) for manufacture. It has been estimated that only 54% of the applied nitrogen is used by the plants with the remainder lost either by leaching or volatilization.

Micro-Irrigation

Micro-irrigation comprises drip (also called trickle) and micro-spray systems. These systems are characterized by small flow rates at the emitters. The emitters are typically attached to a polyethylene hose of 1.25 cm – 2.5 cm diameter. Micro-irrigation was virtually unknown 20 years ago.

Many misconceptions have arisen about micro-irrigation performance. As with any irrigation system, micro-irrigation has non-uniformities, losses, and potentials for over-irrigation. Understanding the micro-irrigation method is critical for adjusting and maintaining a proper operating condition because of the large amount of sophisticated hardware and moving parts.

The following factors influence the distribution uniformity of micro-irrigation; this is sometimes called emission or emitter uniformity rather than distribution uniformity:
1. clogging;
2. manufacturing variation;
3. aging of materials;
4. pressure variations;
5. varying set times; and,
6. irregular drainage within the soil.

The potential maximum uniformity is estimated to be 90%; in reality, it is considerably lower as indicated by the above factors. The primary factor for drastically lower uniformities is clogging of the small holes in the emitters. The hole sizes range from 0.50-1.52 mm in diameter; so one grain of sand can easily plug them. This problem is so severe that most farmers with drip/trickle systems constantly check, clean, and replace completely plugged emitters.

Clogging of emitters is caused by bacterial growth, dirty source water, and chemical precipitation.
1. Slime bacteria grow inside the hoses and emitters and large fragments can slough off the walls, are carried into the emitter holes, and plug them.

2. Canal water can contain anything from fish to clay. Well water usually contains sand and, often, oil and silt.
3. Carbonates of magnesium and calcium precipitate to form hard white deposits in the emitters.

Clogging can be prevented by a combination of design and maintenance. Emitters should be selected which have large holes and short tubes. Filtration systems must be installed and operated properly in order to clean incoming water. A rigorous and systematic chemical injection program using chlorine and acid controls the build-up of carbonates and bacteria. Hoses and submains need to be flushed by chemical injection on a regular basis.

Manufacturing variability exists which gives differences in flow rate between brand new emitters, all at the same pressure. The smaller the hole in the emitter, generally the higher the variability. Row crop drip tubing is more difficult to manufacture uniformly than orchard emitters because of small hole sizes. Emitters with moving parts can also be difficult to manufacture uniformly. Moving parts tend to have a second problem; their flow characteristics change with time. Manufacturers of micro-irrigation emitters usually guarantee a minimum variability when the emitters are new and over a certain life of the system.

Pressure variations cause the same problems in micro-irrigation systems as in sprinkler systems, although turbulent emitters are more tolerant of pressure changes than others. In an existing system, there are only 3 practical solutions to non-uniformity caused by pressure variations:
1. install pressure regulators at submains or hose entrances;
2. adjust existing pressure regulators so that they all have the same setting; and,
3. change to turbulent emitters.

On sloping topography uneven drainage can be a problem. At low points, some emitters can run continuously and they will drain a complete mainline. This contributes to poor uniformity and possible water logging at low points. Possible solutions to these problems are as follows.
1. Irrigate with longer set times so that mainlines will not have to be drained as often.
2. Install special non-drain valves which are common in golf course installations. These shut off the flow when the pressure drops to a pre-set value.
3. Install pipe loops with air vents in the mainline and submains. The loops act as dams when the water is shut off, and prevent line drainage. Figure 12.6 shows an example of this method.
4. Use turbulent emitters which have constant flow rates over a wide range of pressures.

In many orchards, especially with stone fruit (peaches, plums, nectarines, etc.), farmers plant 5-10 ha blocks of different varieties and continually replant as blocks get old. One drip system can supply many small blocks. There is usually more than one tree spacing among the various blocks. The emitters are often installed with the same number of emitters per tree regardless of the tree

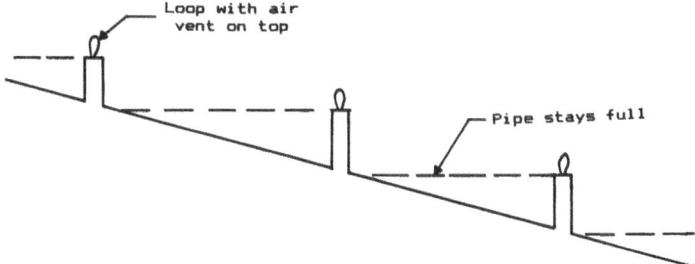

Figure 12.6. Side profile view of submain design.

spacing. Then each block is irrigated the same number of hours. The result is a very different application rate per hectare. For example, there would be a difference of 12% between a 5 m × 5 m tree spacing versus a 5 m × 7 m spacing. The resulting non-uniformity can be solved by adjusting the hours of application to each block.

Drip/trickle irrigation has definite advantages under certain conditions. However, as with all methods, the system must be designed and managed properly to obtain a reasonable irrigation efficiency. For example, it is just as easy to over-irrigate with drip systems as with sprinklers. Applying the water for an extra hour in a 12 hr day will result in an 8% drop in efficiency with both methods.

Chapter 13

Land Leveling and Simplistic Surveying

Farming with supplemental irrigation requires a knowledge of field leveling and surveying. Surveying is the science by which lines, distances, angles, and elevations are established and measured on the earth's surface. All supplemental irrigation methods are made more efficient by some degree of land smoothing or leveling, and several systems require accurate leveling in order to operate. Leveling for supplemental irrigation is different from most other earth-moving jobs because the amount of soil to be moved may not be large, but the depth of cut and the height of fill have to be precise. Natural soil surfaces are generally irregular consisting of elevated areas and depressions. For irrigation basins, this results in pondage of rainfall and interrupted surface runoff. The land surface is reformed to provide a continuous slope through the cultivated soil. There are three main types of land forming.
1. Land leveling creates an almost horizontal soil surface with not more than 0.5% slope in the direction of runoff or cultivation.
2. Land grading provides a continuous sloping plane on the surface of a graded cultivated field.
3. Land smoothing fills localized depressions and removes localized elevations for continuous water flow over a cultivated field.

Equipment

The bulldozer has several advantages in construction of supplemental irrigation systems. It can be used on very rough, stony, or eroded soils and will operate under conditions where other types of equipment cannot function efficiently such as in extremely dry soils. For moving earth short distances, bulldozers give the cheapest cost per cubic meter but costs increase as distance increases. For moving soil over 100 m, an earth mover which can pick up and carry soil using a bottom scraper is more cost effective. Also, for longer hauls, wheels are preferable to track layers. After the soil has been moved, a final smoothing needs a grader or land plane to complete the operation.

Self-propelled road graders permit good results, especially when preparing long graded beds like terraces or contour borders. The road grader, God's gift to water harvesting, is excellent for light or medium work where only small amounts of soil are to be transported. A skilled operator can cut a channel true to grade with little follow-up work required.

To construct large diversion borders or bunds, the area under the ridge should be stripped of shrubs and grass (sod) and disced or cultivated by Canadian. Scarifying or loosening of the foundation surface area for terraces and borders improves the bond between the borders and the soil materials. The depth of scarifying should be limited to 15 cm or less to insure satisfactory blending and compaction of the borders. Compaction by bulldozers or road graders is desirable and often required when fine grained soil materials are predominant. The roughing-in is accomplished with a bulldozer by making three or more "cuts" and "bucks" at right angles to the ridge depending on the size of ridge required. After roughing-in is completed, two rounds or more are made lengthwise to shape the front and back slopes. Sometimes one round is made in the channel to give the diversion its final shape and cross section. If a road grader is used in combination with a bulldozer, the final shaping and compaction of the diversion border can be accomplished with greater precision.

Notwithstanding, the most accurate leveling is achieved by a land plane or even better the automated laser-leveler. The land plane should only be used for the final smoothing after the main movement of earth has been done by a dozer or road grader. The land plane should be over 8 m long with a blade or box blade about two-thirds of the way towards the back. The blade should be tilted forward at the top and not backwards like a dozer blade. The front is hitched to a hydraulic lift on a tractor (75-100 hp or more) which controls the depth of cut and allows the blade to be lifted clear for turning. The rear wheels run on the planed surface. The soil is pulverized as the land plane rolls along in front of the blade.

Lasers are a relatively recent development in land leveling and allow machines to work fast and accurately. A laser produces a fine parallel beam of light which can be used to control grading equipment automatically. The source is mounted on a tripod somewhere in the field. The laser slowly rotates and the beam of light is set so that its plane of rotation is parallel to the gradient required on the land. Towers attached to the cutting blade or large sized box blade on the machine intercept the laser beam and automatically adjust the position of the blade to the correct level. As the machine moves over the high and low points of the soil surface in the field, the blade automatically cuts soil above the desired grade (red-topped stakes) and fills areas below grade (blue-topped stakes).

Laser equipment can be used on 100-125 hp tractors as well as larger equipment such as bulldozers, earth movers (scrapers), road graders, and land planes. Leveling and smoothing can be accomplished four times faster with this method. Furthermore, there is no need to resurvey the land after grading because the operator can be sure that the work has been accurately done when no more soil is being moved.

When irrigating a newly constructed cultivated field for the first time, settlement of the ridges and borders can occur. This results in minor irregularities in the soil surface even on land which has been carefully smoothed and graded. Settlement can cause breaks in the borders which can destroy a good field design and result in excessive soil erosion. If heavy equipment has been used, several trips or passes parallel to the ridge or dike can be extremely helpful for compacting the ridges. During and after the first season's use, it is essential that maintenance be performed on a newly installed supplemental irrigation system to stabilize the soil and prevent any washouts or other types of degradation.

Survey of the Land Surface

A survey of the land surface is needed so that farmers can locate high and low points in fields and identify surface slopes. In the field, points to cut are marked by red-topped stakes or by tying a piece of red cloth or red plastic tape to the stake. If the stake marks a fill, the top of the stake is painted blue. Normally the tractor driver does not know the depth of cut or fill; therefore, an assistant will have to indicate this by a prearranged code.

The instrument to use for surveying is the level. The Dumpy level is very accurate, but expensive and easily damaged and requires trained workers to operate and adjust. The leveling work for supplemental irrigation does not require extreme accuracy because lines are being marked on ordinary rough field surfaces. Three simple levels for surveying, each dependent on the availability of materials, are explained below: the garden hose and A-frame devices do not have to be sighted onto a distant rod, but the water manometer necessitates a measurement rod held by a second person some distance away. These 3 devices do not require delicate handling or highly skilled operators and can be constructed inexpensively using local materials.

Water manometer

For the water manometer, a 30 cm length of smooth soft wood about 4 cm × 2 cm is fitted with two rubber straps (which can be made from a bicycle inner tube) at each end which are bolted or nailed to the board on the same side. A 60 cm length of clear plastic tubing is inserted behind each rubber strap to form a U. A tin can is cut with tin-snips or metal shears to form a V-notch to be attached at the sighting end (see Figure 13.1) and a sliver of metal added at the other end to complete the sight. The block assembly is attached to a pole by two bolts, screws, or nailed carefully. The pole and board should be of a convenient height for sighting.

Water is poured into the U-tube arrangement until the level in each tube is 3 cm from the top. At this point, the tube ends can be fitted with some type of

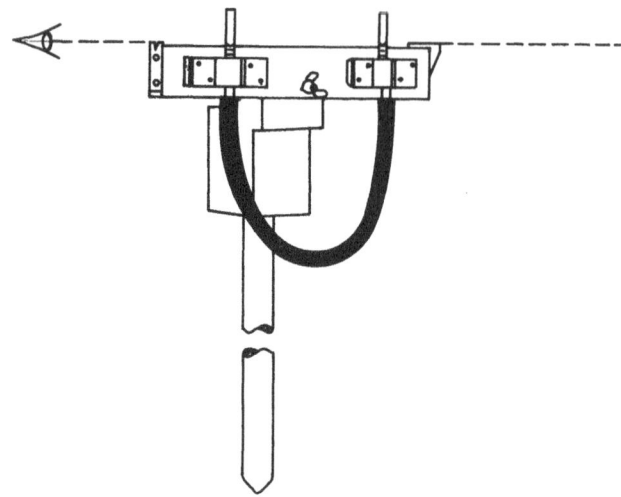

Figure 13.1. Water manometer used as a leveling device.

stopper to prevent loss of water during transport. The mounting pole is pressed firmly on the ground surface and held as vertical as is possible with the leveling device at a convenient eye level.

The device is leveled by removing the two stoppers and adjusting the angle until the two menisci in the tubing are exactly level or at the same height at the top edge of the board. To achieve level, add or remove water in the tubing or slide the tubing up or down in their rubber strap holders.

The survey height-measurement rod can be made from a two-meter board and tape measurements marked on the board with a wide black felt tipped pen. To begin the survey, the rod should be brought to the leveled end of the water manometer sights. The rod is then marked with a pencil or, if a ruler has been added to the rod, the height is recorded with the instrument held level to identify the instrument height. When this is completed, a person with the rod moves a distance of about 10 m, places a card or the forefinger at the mark and is instructed to move either right or left until the location of the horizontal target (card or forefinger) is found in the aiming or sighting notch (level). A stake or mark is placed in or on the soil surface to specify the level location. The person with the rod (rodman) moves another 10 m down the field and the process is repeated. If the level sighting cannot be found, then the instrument must be moved closer to the rodman. The device must be kept upright when changing position to avoid water loss or alternatively the tubing must be stoppered. Also, if a ruler has been attached to the rod, elevations for contours can be made for irrigation, runoff, and drainage ditches.

Within a total distance of 100 m, the instrument is reasonably accurate. With this instrument, cuts and fills can be determined and the estimated depth of each marked on a stake. Other types of manometers can be constructed with

Land Leveling and Simplistic Surveying 195

possibly better sighting techniques. The device as discussed is not too popular but is easy to construct and understand.

A-frame

The A-frame with plumb bob (Figure 13.2) is one of the oldest methods in use for determining a level contour. Two soft wooden boards 6 cm × 2 cm × 2.5 m are laid together at one end and spread apart to 4 m at the other end. The joint can be glued and nailed together, crimping the nails on the backside. Another soft wooden board (6 cm × 2 cm) is glued and nailed to the side pieces to form the horizontal of the A-frame. The so-called feet or legs of the A-frame are cut level after marking using a long straight board for marking. A hook is screwed into the top or apex of the A-frame to attach the string of the plumb line. A plumb line is a line or cord with a weight at one end (plumb bob) to determine the vertical position.

Figure 13.2. A-frame and plumb bob (hammer) used as a leveling device.

The following method is used to mark the horizontal of the "A" of the A-frame to show when the legs of the A-frame are exactly level on the soil surface. Two bricks or stones are placed 4 m apart and the feet of the A-frame are placed upon them. A light mark is made on the horizontal board when the plumb bob has come to rest. The A-frame is then turned around in the opposite direction and placed on the same stones and the procedure repeated. A permanent mark is made halfway between these 2 marks to show when the legs of the A-frame are exactly level.

To set a level contour across a field, one leg of the A-frame is placed at a designated starting point and the other positioned so that the plumb is level with the permanent mark. This position is marked with a stake or flag in or on the soil surface. Then the A-frame is moved to this stake and the procedure repeated. Care must be taken to get the plumb bob as close as possible to the permanent mark. The plumb bob can be affected by physical movement or wind, damping of the plumb bob (excessive movement) is possible by movement of the string line against the horizontal board of the A-frame.

With some experience, this instrument can be quite fast even though each step is limited by the physical width of the span. The taller the A-frame, the more sensitive the instrument becomes to differences in level. Although the A-frame can be operated with only 1 person, another person installing stakes at each step of the A-frame increases the speed of using the device.

Garden hose

The clear plastic garden hose or flexible tube water level method (Figure 13.3), a simple approach, is preferred by Bedouins. Two soft wooden boards 4 cm × 2 cm and 160 cm long have a thin batten or board 1 cm × 1 cm and 1 m long nailed to one side of the wider board. If available, a 1 m tape measure can be carefully glued and tacked to each thin batten or board making certain that the height and position of the tapes are exactly the same. The 0-mark should be at the top end of the thin batten or board.

Clear, see-through, 16 mm internal diameter, PVC tubing or garden hose is excellent for this method. Two ends of a 13 m length of clear or see-through garden hose or plastic tubing are attached to the soft wooden boards by taping or by drilling 1.5 mm holes at 40 cm intervals through the board and securing the hose or tubing against the edge of the board and batten with a soft copper wire or other soft wire. The tube is then slowly filled with water, care being taken to expel all air bubbles, until the water level is about 1 m high in each of the tubes (stands) when they are held together. The ends of the tubes can be fitted with rubber stoppers to prevent water loss during transport.

To begin using the device, the two stands are brought together at the starting point, the stoppers removed and the readings on the measuring tapes taken level at the bottom of each meniscus. If measuring tapes are not available then a pencil mark can be made on each board at the bottom of each meniscus (exactly the same height). The ends of the tubes are then stoppered and the lead person moves one stand in the leveled direction to about 10 m. The stoppers are carefully removed and the lead person moves the stand up or down the slope until the meniscus is at the original mark or elevation. The garden hose is stretched to the 10 m length for estimating distance across the field. Both stands should have the same elevation or menisci at the same level. A stake or mark is placed in or on

Land Leveling and Simplistic Surveying 197

Figure 13.3. Garden hose technique used as a leveling device.

the soil surface and both persons move forward to repeat the operation.

If care is taken, this technique can be the most accurate method of the three presented as well as the fastest to use. With experience, it does not take long to move the stands and settle the water levels in the tubes. Previously de-aired water can be made by boiling water for 20 minutes and allowing it to cool down to air temperature. De-aired water reduces the amount of bubbles in the tube and the level in the stands occurs much faster. When moving the instrument, a procedure to avoid spillage must be adhered to. Being flexible, this leveling device can be used for leveling 2 points not in sight of each other, a common problem in construction.

Measurement of Horizontal Distances

There is a problem when farmers are asked the size of their fields. They usually do not know unless a survey has been done on the farmstead. As an example, during a questionnaire pretest, a farmer in the Mediterranean region was asked by an economist, "What is the size of this field?" After a period of serious thinking and looking at the perimeter of the field we were standing by, the farmer replied, "about two and one-half hours." "No, no," came the response of the economist, "I mean what is the size, you know the area; that is, the aerial extent or surface area of this field?" "Oh, yes, I understand," came the farmer's reply, "but, it is as I've told you, two and one-half hours. That's exactly how much time it takes me to sow the field." If pacing is done as described below, then a more accurate value could have been obtained. For purposes of construction for supplemental irrigation, only pacing and chaining will be examined for measuring horizontal distances.

Pacing

Pacing (Figure 13.4), the measurement of distance based on the length of a human step, may be used for approximate measurement when an error of 2% is permissible (2 m per 100 m). Measurements by pacing are generally valid for terraces, ridges, borders, or diversion layouts as well as preliminary plot or field layout and catchment basins for surface runoff. Measurement by pacing consists of counting the number of steps between two points and multiplying the number by a predetermined pace factor. Pace factors will vary between individuals. Everybody should determine their individual pace factor while walking a natural stride.

Actually, the pace factor for each individual is the average distance in meters per step and is determined best by pacing a measured distance (usually 100 m) several times. It should be paced enough times to make certain the number of paces for the distance does not vary over 2-3 steps. The pace factor then would be the distance in meters divided by the number of paces. The pace factor may vary with the roughness and slope of the ground. Adjustments to length of step should be made to correct these variations. A pedometer can be used to sum the number of steps in a given distance with the measurement corrected by the pace factor.

Land Leveling and Simplistic Surveying

Figure 13.4. Pacing technique for determining field areas.

Some people prefer to use a stride in place of a pace. It consists of about 2 paces, so the stride factor would be roughly two times the pace factor. A stride usually has more error because the longer the line of measurement the more everybody tires and shortens their stride thereby introducing added error.

Chaining

Chaining or taping is a method of measuring horizontal distance with a steel or reinforced plastic tape (Figure 13.5). It is the most common method known and should be used for most measurements in order to attain greater accuracy.

Survey distance is recorded by stations. The distances between full stations is 25 m. When distance is counted as so many stations, this value refers to the number of 25 m lengths, e.g. 6 stations = 150 m. The fractional part of a distance between a full station is called a plus station. Fractions of a meter are indicated by decimals, centimeters, or millimeters depending upon the level of measurement accuracy needed. For example, a point on a line 53.4 m beyond station 10 + 00 is indicated as station 12 + 03.4.

Stakes set along the line are marked with a waterproof crayon or ink markers. Station designation is written on the face of stakes so that as a person walks along the line in the direction of progressive stationing, the station markings are easily seen as each stake is approached.

The following procedure is generally used for chaining a line within a desired accuracy:

Figure 13.5. Chaining or taping technique for measuring horizontal distances.

1. The line to be measured can be a meandering line along a contour border, irrigation ditch, or it can be a straight line in a predetermined direction. For the contour border or irrigation ditch, measurements are taken parallel, or nearly so, to the meandering line. For the straight line, a range pole is set ahead on the line as far as can be seen, or the direction is set by a far tree, shrub, fence post, or other convenient point. This land mark is then used to sight in a straight line from the point of beginning.
2. For purposes of explanation, a straight line is assumed to have been measured and a stake, marked $0+00$, set at the beginning point.
3. The lead chain person takes the 0-end of the tape and advances in the direction of the line to be measured (this could be the level line determined by previous measurements). When the lead person has walked the length of the tape, the rear person calls out "chain" to signal the lead chain person to stop.
4. The rear chain person then sights on the line to be measured and holds the 25 m mark of the tape exactly on the beginning stake. The lead chain person pulls the tape straight and reasonably tight and sets a stake or pin on line exactly at the zero end of the tape.
5. The rear chain person calls out the number of the station (in this case $0+00$) and the lead chain person marks the stake $1+00$ indicating that 1 station has been measured.
6. Both chain persons then move forward along the line to be measured and the rear chain person again calls "chain" when the lead chain person has gone forward 25 m. The line is sighted in, stakes are set and stations marked, each time the rear chain person calls off the station to the lead chain person and the new station is given the next consecutive number.

On slopes, the uphill end of the tape should be held at the soil surface and the rear chain person at the marked station holds the tape level or at least as high as can be reached and uses a plumb bob to confirm vertical (level tape) at the station. At grades over 10%, the chain should be broken into convenient lengths that can be held approximately level using the plumb bob.

Chapter 14

Economics of Supplemental Irrigation

Water for agriculture is virtually synonymous with practical survival of modern agriculture. The day when water was believed to be an inexhaustible supply has long since passed. Historical and current trends of food and agricultural production in the Near East and North Africa indicate a circumstance which is unsatisfactory despite substantial achievements.

Economically feasible alternatives of irrigation technology have been developed which can resolve the farmers' dilemma of production and alleviate or remove constraints to sustainable agriculture. Water scarcity can be crop-specific which emphasizes the importance of contemplating alternative water supplies such as wells, pumpback systems, water harvesting, or intermittent streams. Customized water management to extend these alternative water supplies (temporally and spatially), through supplemental irrigation, will be imperative to future deliberations. Furthermore, a major socio-economic consequence of supplemental irrigation is to interrupt the variable productivity of decreasing production and increasing migration which currently occurs in rural communities with only rainfed farming.

Costs increase and higher technical skills are required commensurate with the degree of an irrigation technology's sophistication and its capability to control water and to reduce labor for operations. Estimates of available capital and labor as well as existing technical skills are essential for identification of possible tradeoffs in the development of supplemental irrigation. For example, surface irrigation is usually less expensive and requires lower technical skills than sprinkler irrigation, which, in turn, is less expensive and requires lower technical skills than drip irrigation. Of course, criteria of design and local labor restrictions may stipulate one type of system in preference to another and yet prevail as economically feasible. Planning for, and selection of, a farm irrigation system only for supplemental irrigation is difficult; therefore, longterm reflection may suggest, to the farmer, that cultivation of high-value summer crops is the most reasonable investment.

So far, most applications of supplemental irrigation in the Near East and North Africa have been small-scale; individual farmers have established and developed systems with limited water sources, principally groundwater.

E.R. Perrier and A.B. Salkini (eds), Supplemental Irrigation in the Near East and North Africa, 203-227.
© 1991 *ICARDA.*

However, increased interest in the technology and enhanced efforts for development in several countries have encouraged many governments to conduct feasibility studies for expanding these small-scale systems to medium or large projects whose basic designs would provide delivery to multiple users for supplemental irrigation of wheat and other winter crops. Based on the results of these feasibility studies, construction of small-scale systems has already started.

Before involvement in new endeavors, several methods of appraisal should be conducted to determine the feasibility of supplemental irrigation. In the past, the focus has been on the engineering aspects involving the physical edifice of irrigation projects, and little attention was paid to the scale of system design or the economic and social consequences of maintenance and operation. More recently, acknowledgement of the crucial impact of socio-economic issues is being accredited for success or failure of supplemental irrigation development.

Feasibility studies for government schemes have followed general but basic principles and techniques of socio-economic appraisal for evaluating medium-to-large scale projects for supplemental irrigation. Effective socio-economic examination of irrigation projects pertains to the viewpoints and interests of the society and the farmers. Financial viability and economic profitability might be of major immediate concern to farmers (and in many cases, is the sole criterion or objective), but their societal objectives are broad and vary from the socio-economic to the psychological and organizational.

Logically, economic appraisal techniques designed for practical assessment of irrigation ventures have to be versatile. These techniques must gauge an individual farmer's criteria for incorporating supplemental irrigation into the farming system and still discern what societal objectives, the farmer's and the nation's, are being encouraged by the development of irrigation technology. To help in isolating these points of reference, techniques for project level analysis are presented in this chapter with classic examples for clarification and ease of future application. Also, a detailed discussion of economic evaluation at the farm level is presented with discussions of record keeping forms for farmers to follow budgetary elements of production using supplemental irrigation. These two approaches gather data from different components of the farming system but are complementary sources for discerning dynamic equilibrium in authority activities of agricultural production for food security.

Economic Appraisal for Supplemental Irrigation Projects

Purpose

Benefit-Cost Analysis (BCA) is project appraisal. BCA is a method to evaluate goals and objectives of planned activities to rank and select those endeavors that contribute the most benefit to society with the least cost. Choices are reached using criteria which best describe the standards of achievement by

Economics of Supplemental Irrigation

calculating benefits and costs of the design and implementation of a proposed project. A publicly funded supplemental irrigation project has two dimensions of concern: a public component, executed by a government agency for mobilization of funds and coordination of construction as well as operation and management of some comprehensive irrigation projects (including construction of necessary physical structures for water sources development, storage facilities for adequate supply, and farm delivery systems); and, the farm site component, managed or controlled by farmers and requiring capital for investment and acceptance and adoption of new farming practices for supplemental irrigation agriculture. Other production costs, a result of increased cropping intensity, and any major changes to farm operations are integrated into a thorough economic appraisal to determine successful introduction of supplemental irrigation.

The purpose of performing benefit-cost analysis of a supplemental irrigation project is to ensure the economic feasibility of proposed objectives predicated on a nation's economy and needs of the population (specifically the farmer). Considering this duality of demands, BCA uses 2 principal approaches:
1. financial analysis, for analyzing the profitability of production to the farmer; and,
2. economic analysis, for establishing economic feasibility and the benefits in comparison to costs of the project to the national economy.

Identification of Relevant Benefits and Costs

BCA is an analytical technique for appraisal of supplemental irrigation projects or other development efforts by weighing benefits acquired from project execution against the accumulated costs of operation and maintenance of the system. BCA can be conducted for project evaluation and selection to reflect the reality of financial, economic, and social objectives. The solutions of analysis can be expressed using different measures: the financial rate of return; the economic internal rate of return; the net present value; and, the benefit-cost ratio.

The beginning step in BCA identifies costs and benefits of implementing supplemental irrigation agriculture using a before/after comparison of the national situation with project and without project. Supplemental irrigation project costs can be grouped into 3 levels:
1. construction and maintenance of the reservoir, conveyance canals, and water courses;
2. development costs in the project area of land leveling, field channels and drains, and improvement of facilities for storage, marketing, and transport of agricultural production; and,
3. costs at the farm including various inputs and services for agricultural production.

Only those costs and benefits that directly result from the project are included, and only incremental changes in benefits and costs between the with and without project situations are considered. For example, if an irrigation dam is already in operation prior to implementation, then construction costs of the dam are considered as sunk costs and not included for calculation in BCA. Cost of water which is allocated to these new ventures would be included but valued on the basis of declining production if new supplemental irrigation deprives other farmers of a water supply.

Public costs include the raw materials and other physical goods which are used in construction, the irrigation system, labor, land removed from production because of the reservoir and conveyance networks, as well as infrastructural costs such as additional roads, services, and administration. Farmer costs include capital investment for equipment (irrigation pumps, pipes and fuel), land development or improvement, additional labor, and farm inputs.

The primary benefit is increased agricultural output. The secondary benefits are stabilized production and farm income. Other peripheral benefits are new employment generated, redistribution of income, increased food self-sufficiency, improvement in land value, income earned by suppliers of agricultural input items and services (backward linkages), and by processors of agricultural outputs and by-products (forward linkages).

Usually, indirect benefits and costs (externalities or secondary effects) are not included in BCA because of the difficulty of identification and measurement. When indirect effects can be quantified and valued in monetary terms, they are included as direct effects in the calculations. Even though indirect effects are not included in calculations, recommendations are to discuss these effects thoroughly during planning and development activities to provide rational estimates of the social, economic, and environmental impact of supplemental irrigation.

Costs in the general category of transfer payments (taxes, tariffs, subsidies, loan receipts and repayments) are included only in financial analysis. These costs are excluded from economic analysis and considered as transfers or claims within the economy but they do not add to or subtract from national income: they constitute redistribution effects only. Finally, depreciation costs are never included in financial analysis or in economic analysis. The inclusion of depreciation costs would constitute double counting since analysis with BCA integrates the stream of benefits and costs over time including maintenance and replacement costs. However, a salvage (or scrap) value for all equipment is included as a benefit during the final year of project implementation.

Financial Analysis

The method has 2 steps: (1) model farm budgets; and, (2) farm level analysis of the benefits and costs of investment. The objectives of financial analysis of supplemental irrigation agriculture are:
1. to assess financial viability of supplemental irrigation from the farmer's viewpoint;

Economics of Supplemental Irrigation

2. to determine if credit will be needed as an incentive to the farmer to adopt supplemental irrigation and, if so, to outline a plan for mobilizing funds;
3. to assess practical administrative and management alternatives;
4. to examine the impact of supplemental irrigation on the population identified as "gainers and losers" then estimate costs and explore mechanisms for potential compensation of losers; and,
5. to provide basic data for Economic Analysis by adjusting results from the Financial Analysis.

Model farm budgets

A model farm budget lists primary production activities affected by potential adoption of supplemental irrigation. The budget details resources used, output produced, and economic returns associated with each farming activity (on-farm as well as off-farm). A standardized model farm budget should be prepared with a defined structure of farm production activities capable of being adjusted for individual localities.

A description of existing farming systems including operation and management (farm structure) should be incorporated into the standardized model farm budget with projections of possible changes in the farm structure as a consequence of supplemental irrigation. In other words, a typical budget would contain incremental changes which occur because of implementation of supplemental irrigation; i.e. changes resulting from with project in contrast to without project activities. This method is referred to as partial budgeting. A discussion of farm-level analysis is presented at the end of this chapter (see Figure 14.1).

Incremental changes require evaluation yearly; therefore, standardized record keeping is essential. Two time-related issues can be exemplified.
1. The rate of adoption of modern technology varies with time; therefore, realistic assumptions of rates of adoption need to be established together with annual estimates of farm area in production using supplemental irrigation.
2. The learning curve is a related issue. New techniques are learned gradually; furthermore, expected increases in output could be lower during the beginning years when gaining experience. Costs and returns during the first few years will reflect this gradual learning and adaptation of a new technology.

Assessment of agricultural inputs and outputs in BCA is based on a general economic principal of opportunity cost, or cost of the second best alternative. If additional farm output reflects the addition of supplemental irrigation and if the produce is sold in the market place, the value of this produce is the current market price. Whereas, the value of labor is the prevailing wage rate if additional labor is needed but not provided by family members. On the other hand, if family members provide the additional labor without affecting production in other on-farm or off-farm activities, then the value of labor for supplemental irrigation would be 0.

In general, current market prices are good indicators of the value of farm inputs and outputs; however, if the current market is not functioning well or if it is distorted, prices may not be reliable indicators. For instance, farmers may be obligated to sell part of their output at a low official price and the remainder at the current market price; therefore, the value of the output will be a weighted average of these two prices. Often adjustment of market prices is necessary to reflect true benefits and costs received by farmers. Similar situations are frequent using current market prices as indicators for a wide range of inputs and outputs. In BCA, these adjusted prices are often referred to as shadow prices which contain indirect transaction costs such as transportation and loading-unloading of farm produce. It is important to specify at what point in space valuation of farm inputs or outputs occur: farm-gate, project boundary of supplemental irrigation, local market, or regional market.

Farm-gate prices are most commonly used in BCA, especially in financial analysis; however, it is important to differentiate farm-gate prices of sales and purchases. For example, two farmers increased barley production with supplemental irrigation: one farmer used a part of the production to feed a small flock of sheep, whereas, the other farmer used the total production to feed a larger flock and, in addition, had to purchase feed from the market. The increased barley output enabled the first farmer to sell more barley in the market place and the second farmer purchased less feed from the market. The farm-gate price of barley is calculated as the market price minus transport costs for the first farmer; but, for the second farmer, the farm-gate price of barley is equal to the market price plus transport cost. In other words, if produce is partially or totally sold in the market, the farm-gate sales price would apply, but if it is totally consumed on the farm then the farm-gate purchase price would be the correct value for the barley. This distinction becomes more important with bulky products, such as hay, when transport costs are high.

Analysis of investments at the farm level

Investment analysis at the farm is a measure of profitability of supplemental irrigation for the farmer over the lifetime of the irrigation system. These future values must be converted to their present values because benefits and costs occur at different points in time and vary from year to year. The distinction between these two values is crucial in BCA and is based on the economic concept of time value of money. For example, if a choice is given between receiving $100 today or in a year's time, the choice would be to receive the money now! Moreover, if the choice was to receive $100 in a year or $95 today, the choice might be to receive $95 today because $95 today is worth more than $100 in a year. In other words, in comparing future and present benefits, future benefits are usually discounted, i.e. they are converted into today's values.

Economics of Supplemental Irrigation

Again, if the choice was payment of $100 next year or another value, $90, paid now and if the $90 was accepted then a conclusion can be reached that the present value of $100 next year would equal $90. The discount rate would equal $100 \times (100 - 90) / 90 = 11\%$.

Calculation of discounted present value, PV, uses the general formula:

$$PV = V_t \frac{1}{(1+r)^t},$$

where, V_t = value in year t;
r = discount rate;
t = time period (year);
$1/(1+r)^t$ = discount factor or PV coefficient.

If present value is calculated using items from the model farm budget (Table 14.1) for an estimated project lifetime of 12 yr, e.g. the average operational life of an irrigation pump, then net annual benefits (NAB) or annual gross benefits minus annual costs can be calculated. The total net benefits over an estimated lifetime equal $11,576 but are worth only $1,423.50 if a 15% discount rate is used to determine the net present value (NPV). However, if a discount rate of 25% is used then NPV = − $560.50.

Table 14.1. Example of partial budget form.

Year	NAB($)	Discount factors 15%	NVP($) at 15%	Discount factors 25%	NVP ($) at 25%
1	−3,443	.870	−2,995.4	.800	−2,754.4
2	−1,334	.756	−1,008.5	.640	− 853.8
3	704	.658	463.2	.512	360.4
4	1,227	.572	701.8	.410	503.1
5	1,436	.497	713.7	.328	471.0
6	1,476	.432	637.6	.262	386.7
7	1,476	.376	555.0	.210	310.0
8	1,476	.327	482.7	.168	248.0
9	1,476	.284	419.2	.134	197.8
10	1,476	.247	364.6	.107	157.9
11	1,476	.215	317.3	.086	126.9
12	4,130	.187	772.3	.069	285.0
	11,576		1,423.5		− 560.5

NPV is a common criterion to assess profitability of supplemental irrigation and represents the total value of net benefits (benefits less costs) discounted at a selected rate (usually the opportunity cost of capital). In the case of financial analysis, the interest rate at which money can be borrowed by the farmer is a good approximation of the opportunity cost of capital. However, actual verification of opportunity cost is not easy to measure because farmers may borrow money from two sources: the banking sector and the local money lenders. Differences often exist between rates charged by the two sectors.

NPV measures total benefits; therefore, project size is reflected in the value as well as the rate of return. NPV can have a positive or negative value; however, for a project to be considered as acceptable, the NPV has to be greater than or equal to zero. A negative NPV indicates that the rate of return is below opportunity costs of the capital invested. In the above example, rate of return to investment is between 15-25%. Comparison of alternative projects with different levels of investment becomes difficult since the measure, NPV, is not size-neutral.

Another measure of profitability, internal rate of return (IRR) is considered to be size-neutral. IRR is defined as the discount rate that gives an NPV = 0 or, stated another way, IRR is a measure of the rate of return to the investment in supplemental irrigation. Calculation of IRR is achieved through trial and error experience and is based on calculations of NPV; whereas, net present value is calculated using increasingly higher discount rates until NPV = 0. Estimation of IRR is done by extrapolating between the negative NPV and the value of NPV closest to 0:

$$IRR = \begin{matrix}\text{Lower} \\ \text{Discount} \\ \text{Rate}\end{matrix} + \begin{matrix}\text{Difference} \\ \text{Between the} \\ \text{Two Rates}\end{matrix} \times \frac{\text{NPV at lower rate}}{\text{Sum of both NPV's}} \text{ (ignoring signs)}.$$

Using the example in Table 14.1:

$$IRR = 15 + 10 \frac{1{,}423.5}{1{,}423.5 + 560.5} = 22.17\%$$

If IRR is greater than the opportunity cost of capital, then the project will probably be profitable. Calculation of IRR enables comparison and ranking of profitability between projects of unequal size.

Another commonly used measure of project profitability is the benefit-cost ratio (BCR). Calculation of BCR is a variation of the calculation of NPV: rather than subtracting costs from gross benefits, BCR = gross benefits/costs. Benefits and costs are discounted and BCR calculations take the ratio of the present value of gross benefits to costs. A project is considered profitable if BCR is greater than or equal to 1 at a selected discount rate.

Economic Analysis

Economic analysis is an appraisal of the economic feasibility of an activity to contribute towards attaining the national objectives of increasing national incomes, promoting a country's economic growth, diminishing poverty and improving income distribution, when protecting the environment. Social cost-benefit analysis, a method of BCA, attempts to incorporate some or all of these

Economics of Supplemental Irrigation

objectives (often conflicting) into calculations of project worthiness. These methods remain controversial and involve obvious predicaments when applied in practice. Standard BCA methods (used by World Bank and other international donor agencies) focus on a single objective, to increase national income. Other objectives or concerns are implicated in the analysis, but are discussed separately and do not enter the calculation of benefits and costs.

Economic analysis requires 3 steps for calculation. The first step takes values from the model farm budget and adjusts these values to reflect true economic benefits and costs of inputs (resources) and outputs (production). In the second step, aggregation of benefits and costs (including public investment) are estimated for the total farm area receiving supplemental irrigation. In the final step, measures of worthiness are decided from the discounted value of benefits and costs to confirm the economic feasibility of supplemental irrigation.

Estimation of shadow prices

Shadow prices in BCA are the market prices of inputs and outputs adjusted to reflect the true value to the economy. In estimating shadow prices, all transfer payments are removed. For instance, if fertilizers are sold at a subsidized price to farmers, then their true value to the economy would not include the subsidy.

In many developing countries, foreign exchange rates are fixed by the government and the domestic currency is usually over-valued. As a consequence, imports become cheaper than their true value and exports become more expensive. A pivotal element in economic analysis is estimation of a shadow exchange rate (SER). Estimation of SER is also based on the concept of opportunity cost. Therefore, if an active parallel market, or black market, for foreign exchange exists, then the true value of a product is the price received if sold in that parallel market. Note, however, that the market rate can be distorted because all public and many private transactions are made at an official rate (often several official rates operate at the same time); whereas, the market rate reflects only the remaining private transactions. In the case of illegal black markets, the market rate includes a significant risk premium and sometimes a monopolistic profit margin imposed by the foreign exchange dealers. Therefore, as a rule-of-thumb, the market rate is considered as the upper limit for setting the true value of foreign exchange, while the official rate represents the lower limit, with SER lying somewhere in between.

If values of benefits and costs are expressed in foreign currencies, US dollars, then the value of traded goods is equal to their border price or, for imported goods, the price of cargo, insurance, and freight (CIF), and for exports, the price of free on board (FOB). If, on the other hand, benefits and costs are expressed in local currency, then the border price of traded goods is converted to local currency using SER. Traded goods are not limited to imports and exports, they also refer to inputs used by supplemental irrigation agriculture which would have been exported (diverted exports), and to surplus production (outputs) that supplants agricultural goods currently imported (import substitutes). Indirectly traded goods are locally produced goods but with a sizable import component, such as in the case of tractors assembled from imported parts. The value of indirectly traded goods should be partially adjusted by SER using a weighted average reflecting the relative contribution of imported and domestic components.

Generally, non-traded goods refer to water, land, and labor resources not traded in the international market, but these can also include bulky and highly perishable products, with costs of transport too high for profitable international trade. In addition, are those products that could be traded but are not because of government bans or excessive export duties. If benefits and costs are expressed in US dollars then, using SER, the value of non-traded goods should be converted to dollars.

The valuation of land in BCA depends largely on the form of ownership and whether this would be affected by implementation of supplemental irrigation. For example, if land already in rainfed production is converted to supplemental irrigation and does not require new land to be brought into cultivation then the value of the land is considered in the calculation of incremental changes of benefits and costs of production between the with project and without project situations. If, on the other hand, implementation of supplemental irrigation involves changes in land ownership, i.e. if land is purchased as part of the project, then the new land is valued in terms of the potential for earning income based on the land's rental value or purchase price. Since construction of permanent structures, e.g. irrigation/drainage canals and wells for supplemental irrigation, increases land value, the recommendation is to include the increased land value as a benefit (salvage value) in the final year as determined by estimated lifetimes of system components.

Estimation of the shadow price of labor, shadow wage rate, depends on 3 categories of labor: family labor and skilled or unskilled hired labor. The value of family labor is not directly included in analysis because the opportunity cost is implicitly considered in the incremental changes of output between the with and without situations. The assumption that a functioning market exists for skilled labor suggests there is no need to adjust the market wage rate because it adequately reflects the economic value of skilled labor. The valuation of unskilled labor requires adjustment to reflect the existence of seasonal unemployment or underemployment. The wage rate paid for unskilled labor during peak season is multiplied by the number of days of the period of peak labor, which equals the annual economic wage. This method assumes that unskilled labor is unemployed during the off-season and may not be a realistic measure; therefore, it is preferable to include potential wages earned during the off-season in estimating annual economic wage.

Measures of project worth

Total benefits and costs are estimated by extrapolating the computed values from the model farm budget to the area with supplemental irrigation when these have been adjusted to reflect their true economic value with the inclusion of public costs of construction, maintenance, and administration. Since benefits and costs occur at different points in time, they should be converted into their present value (NPV) applying the method discussed under financial analysis.

Choice of a discount rate is the main difference in the discounting procedure between financial and economic analyses. As previously discussed, the appropriate discount rate used in financial analysis is the interest rate paid by the farmer for borrowed money. In economic analysis, the discount rate reflects the true opportunity cost of capital to the national economy. If the supplemental irrigation project is funded by a loan from an international aid agency, then the interest rate is a good approximation of the opportunity cost of capital. Contrarily, if the project is funded from the general development budget of a national government (typically limited funds in most developing countries) then investments allocated for supplemental irrigation would come at the expense of other development efforts. The opportunity cost of capital is defined as the forgone returns from the second best alternative for investing national development capital; therefore, the appropriate discount rate would be the expected rate of return from capital investments in alternative development projects.

In practice, however, the above approach predicated on opportunity cost has obvious obstacles. As a rule-of-thumb, the interest rate that a government pays to borrow money in the international financial market represents a reasonable basis for estimating the discount rate. Some BCA methods recommend the inclusion of an inflation premium in the discount rate, especially if benefits and costs are calculated in local currency. An inflation premium is not always necessary because general inflation of benefits and costs is expected to increase at a more or less equivalent rate. But if prices of specific items are expected to increase faster than the prevailing inflation rate, e.g. the price of fuel, then these expected price increases should be reflected in future benefits and costs.

After selection of a discount rate, then calculation of NPV or BCR is possible using the general formula presented for financial analysis. Similarly, the discount rate can be used as a criterion to rank alternatives weighed against comparisons of the IRR for implementation of supplemental irrigation. Applying all the methods of calculation presented, a preliminary decision can be made of the economic feasibility of implementing supplemental irrigation. Note, however, that the criteria for determining economic feasibility constitute only the degree of profitability for proceeding with implementation of the project. Other considerations of equal importance should contribute to the final decision of project selection. Socio-economic, political, and environmental impacts of supplemental irrigation, in addition to economic feasibility, are integral to the economic appraisal process.

An example of calculations for benefit-cost ratio

The BCR is calculated from the formula:

$$BCR = \frac{NAB}{AC},$$

where NAB is net annual benefits and AC is annual cost.

NAB are the net gains in the value of agricultural outputs from operation of supplemental irrigation, i.e. the net gains are the equivalent of NAB after irrigation minus NAB before establishment and operation (with project and without project situations). NAB is calculated taking the total gross benefits (value of agricultural products) minus the total costs of production.

Estimation of benefits uses data from activities of cropping patterns, cropping intensities, crop yields and by-products, and prices of inputs and outputs. The data should include the values before and after implementation of supplemental irrigation for the project area (the incremental changes). Table 14.2 presents the data for regions of Syria which receive 250-350 mm average annual rainfall, showing changes in the cultivated land allocated to specific crops before and after the introduction of supplemental irrigation.

Table 14.2. Cropping patterns, intensities, and yields before and after supplemental irrigation.

Crop	Land allocation (%)		Yield (t/ha)	
	before	after	before	after
Wheat	27.7	43.0	0.5-1.5	2.5- 3.5
Barley	22.0	—	0.5-1.5	—
Food legumes	4.1	1.4	0.7-1.0	2.0- 3.5
Cotton	—	33.5	—	3.0- 4.5
Maize	—	8.0	—	3.0- 5.0
Forage crop/vegetable	—	24.2	3.0-5.0	30.0-50.0
Other crops	8.2	14.9	0.5-1.5	1.5- 3.5
Total	62.0	125.0		
Fallow	28.0	—		

From data of cropping intensity and cropping patterns, the total cultivated land can be estimated for each crop. The total area sown to each crop (before supplemental irrigation) is multiplied by yield levels (before supplemental irrigation) then by current market prices to determine gross benefits (GAB) before project implementation. Table 14.3 shows a GAB of US$ 40.17 M before supplemental irrigation (1). The same procedure is completed to estimate input requirements and costs of production (US$ 26.7 M). By subtracting production cost from GAB, a net benefit of US$ 13.47 M is obtained in the project area

Economics of Supplemental Irrigation

before project implementation. The total procedure is completed for the project area after supplemental irrigation (2), and the corresponding values (M US$) are 210.4 GAB, 98.9 for production costs, and 111.9 for net benefits. Subtraction of net benefits, (2) minus (1), gives NAB of US$ 98.4 M due to the establishment and operation of the project. Table 14.3 shows a summary of estimations of the net value of output with and without supplemental irrigation. However, it should be noted that the data include only aggregate values with no details on how these numbers were obtained.

Table 14.3. Estimated benefits and costs of supplemental irrigation (in millions of US$).

Categories of benefits/costs	Supplemental irrigation		Differences due to Suppl. irrig. (2)–(1)
	(1) Before	(2) After	
a. Gross benefits of agricultural production			
1. Crop outputs	35.40	191.90	
2. Livestock	2.75	14.70	
3. By-products	2.02	3.80	
Total gross benefits	40.17	210.40	170.23
b. Costs of agricultural production			
4. Seeds	2.50	11.50	
5. Fertilizer/pesticides	2.50	31.80	
6. Machinery (annual) maintenance/operation	8.60	18.10	
7. Land rent and tax	2.70	10.30	
8. Land leveling/field channels	—	4.00	
9. Labor (human)	10.40	22.80	
Total costs	26.70	98.50	71.80
NET BENEFITS	13.47	111.90	NAB 98.43

So far, the NAB of the project is estimated at US$ 98.43 M. Before calculating BCR, annual cost has to be estimated for the irrigation project to determine the expense of establishing and operating the system. These data items include interest on the capital invested, depreciation calculations, and administrative costs. Table 14.4 presents the total cost of capital and Table 14.5 shows estimation of annual costs; Table 14.6 combines the results of Tables 14.3, 14.4, and 14.5, to summarize the results of BCR for the example.

If 15% of the stored water in the dam were used for non-agricultural purposes such as industry or household then the cost of the dam could be discounted by an amount equal to 15% (194.0 × 0.15 = 29.1 and 474.0 – 29.1 = 444.9). Therefore, estimated total capital costs of supplemental irrigation for agricultural purposes would be reduced to US$ 444.9 M. This value is used to produce the interest on capital cost and depreciation cost in Table 14.5.

Table 14.4. Estimated capital costs for supplemental irrigation.

Capital item	Total cost (US$ M)
Dam and adjunct works	194.0
Canal systems	223.0
Distribution systems	43.0
Marketing and storage facilities	5.0
Transport and communication networks	2.0
Estimated total capital costs	474.0

Table 14.5. Annual costs of establishment and operation of supplemental irrigation.

Item	Annual cost (US$ M)
Interest on capital, 10%	44.50
Depreciation, 2%	8.90
Administration, $20/ha	2.00*
Total annual costs	55.4

* Assuming that total area of the project is 100,000 ha.

Table 14.6. Summary results of BCR for the example project.

BCR = NAB/AC = 99.43/55.40 = 1.786 (US$ M)	
Net annual benefits (see Table 14.3)	99.43
Annual cost (see Table 14.5)	55.40
BCR (Benefit-Cost Ratio) for SI (US$ M)	1.79

Sensitivity and Risk Analysis

Cost-benefit analysis involves making some assumptions about a number of parameters which can ultimately determine actual benefits and costs of supplemental irrigation. By definition, present investment only generates revenue in the future and returns (gain or loss) on investments are inherently uncertain.

Two principal causes are defined to explain uncertainty: data deficiencies (also referred to as risk) are distinguished from the uncertainty which emanates from unexpected events. Two techniques can be used to examine uncertainties of production: sensitivity analysis for the uncertainty of unexpected events and risk analysis for identifying data deficiencies.

Sensitivity analysis identifies the likely effects of assumed value changes of important parameters which can assure the success or failure of supplemental irrigation. This is useful information to demonstrate the range over which returns can change the value of any variable. In sensitivity analysis, an estimation is made of the extent that NPV or IRR is affected because of changes in the values of one of the chosen variables. The variables are manipulated, one at a time, holding all other variables to their best estimates. The variables are:
1. land area irrigated;
2. time required to achieve implementation and size of yields when supplemental irrigation is fully operational; and,
3. the costs of construction, operation, and maintenance.

Risk analysis incorporates selection of the range of possible values for each major variable and an estimation of the probability of each value occurring. The purpose of risk analysis is to avoid narrow judgments of evaluation (optimistic, pessimistic, or best). This is accomplished by analyzing, throughout calculations, the complete range possible for each variable and a measure of the likelihood of each value being within this range. At each step of risk analysis, these judgments are reviewed in various combinations in conjunction with the variables themselves. In risk analysis, a single point estimate (or the best estimate) is replaced by a probability distribution of alternatives.

The major advantage of risk analysis is that this method gives the most likely outcome (or average outcome) and the extent of risk of supplemental irrigation. Risk analysis can also examine when to undertake a marginal project, how to handle a project with inherent uncertainty, how to choose the best combination of specifications in a single project, and, how to identify a project with only minimal information.

Farm Level Economic Procedures

Farm Level

At the farm level, the farmer makes a decision to invest money in irrigation, e.g. to dig a well to utilize groundwater, based on the economic feasibility of such an investment including expected costs and returns, changes in the cropping patterns and intensity, and the economics of crop production. Other concerns that might be considered are the impact of the new investment on employment and income (on-farm and off-farm) and other socio-economic components of the farming system.

A reliable source of water and an adequate supply are the primary elements for the farmer to contemplate to make a preliminary assessment to adopt supplemental irrigation. When water resources are available for supplemental irrigation, the farmer would then need to compare the profitability of rainfed farming in contrast with supplemental irrigation agriculture for winter crops to ascertain the extra benefits obtained against extra costs incurred due to irrigation. Farmers with larger, more complex farms could require additional deliberations to furnish comparative data to assess the diverse management associated with the magnitude of operation for different crop production and land conditions. The small farmer can determine the comparative profitability of dividing a limited water supply between winter crops with supplemental irrigation and summer crops with conventional irrigation.

To achieve this appraisal, farmers can employ 3 different methods of budgeting to match different management and assessment purposes; whole farm cash flow budget, farm resource budget, and partial budgeting. The procedures focus on the small farmer who possesses restricted inputs of capital and operational funds. To evaluate the annual farming endeavor, annual cost estimates of pertinent items related to production as well as the actual costs of production should be established on a routine basis:
1. water cost including permits, tax assessments and fees, and district costs;
2. annual fixed costs of owning the irrigation system, including depreciation and interest on investment;
3. energy costs for operation;
4. costs for operation, maintenance, and repairs including the labor;
5. taxes and insurance; and,
6. actual costs of crop production.

Small Farm Budgeting

A farm resource budget is a systematic collection of relevant costs and returns associated with a specific crop, supplemental irrigation, livestock, and other farm activities. The small farm is emphasized because large farming systems more closely resemble a commercial enterprise and by necessity must keep records and maintain a budget to meter the flow of cash and capital. Small farmers tend to separate the business details of family farms in their heads and may, in fact, mistrust the accountant's ledger, viewing it as a constraint to an independent life style. For small farmers operating at subsistence levels, farm resource budgeting can be beyond their grasp: feeding the household is enough!

When surplus production demands a market economy, then farm resource budgeting takes on a greater meaning for these farmers, i.e. cash flow becomes critical. Without current information, the farmer can make serious errors in assessing the economic consequences of available alternatives. Resource budgets do not remain static (the same from year to year) and mechanisms to maintain current input costs and product prices can be synthesized to accurately predict production costs and potential revenue (for both levels: country and farmer).

Individual cost items reflect the cropping patterns and production practices for selected crops; therefore, these items change for each resource budget and inclusion of variable costs and fixed costs are common to farm resource budgets. To illustrate these variable values, Table 14.7, a resource budget for supplemental irrigation of spring wheat, presents items of seed costs and costs for manually sowing wheat. Variable costs for day-to-day operations change

Table 14.7. Farm resource budget of cost and return per hectare of spring wheat production with supplemental irrigation.

Item	Unit	Number of units	Price or cost/unit ($)	Total income or cost ($)
INCOME				
Wheat grain	kg	3,500	0.0684	239.40
Wheat straw	kg	8,870	0.0224	198.69
Total income				$438.09
VARIABLE COSTS				
Purchase inputs:				
Seed	kg	140	0.0728	10.19
Urea (46% N)	kg	66	0.0864	5.70
Super phosphate (15.5% P)	kg	15	0.0368	0.55
Production practices:				
Plowing	ha	1	10.29	10.29
Smoothing	ha	1	4.19	4.19
Sowing	Man hr	6	0.57	3.42
Weeding	B&G hr[a]	24	0.29	6.96
Fertilizer application	Man hr	8	0.57	4.56
Irrigation				
Water pumping	hr	40	0.44	17.60
Water application	Man hr	40	0.57	22.80
Ditch maintenance	Man hr	24	0.57	13.68
Harvesting and transportation:				
Harvesting	M&W hr[b]	72	0.48	34.29
Threshing	Mach. hr	7	5.71	40.00
Winnowing	kg	18	1.77	31.89
Loading	Man hr	19	0.57	10.86
Transporting	kg	2,000	0.0038	7.62
Total variable costs				224.60
Return over variable costs				213.49
FIXED COSTS				
Land rent	Month	3	11.43	34.29
Land tax	Month	3	1.58	4.74
Management charge	Month	3	1.90	5.70
Total fixed costs				44.73
Grand total costs				$269.33
Return above all costs				$168.76

[a] B & G = boys and girls.
[b] M & W = men and women.

with the production level throughout a given time period. With the adaptation of supplemental irrigation, the production level of wheat increases, the farmer may purchase more seed, increase the cultivated area, use more labor, or apply more water and fertilizer. These costs vary with the rate at which they become exhausted in the production of crops or livestock. If variable costs are not recovered with adequate yields, other things being equal, that particular crop or product should not be produced. In a sense, variable costs could be avoided because they are not incurred if production is not undertaken.

The farm resource budget includes fixed costs which are aligned to ownership or overhead expenses because they are difficult to allocate to specific farm resource activities or agricultural production. Fixed costs are not related to level of production and continue to accrue with or without production. These costs include depreciation, interest, and taxes. To minimize economic losses, farmers must receive returns sufficient to at least cover all variable costs and make some contribution to fixed costs in the short term. Land rent and a management overhead charge are classified as fixed costs (see Table 14.7). The example is for a Syrian farmer who owns a 16 ha farm, has 1 wife and 3 children of employable age, and uses supplemental irrigation. This farmer applied 1 irrigation to assist germination following the sowing of Durum spring wheat and applied 4 supplemental irrigations when rainfall essentially ceased in March.

Table 14.8 provides information to identify monthly requirements of labor and water use for wheat crop production. The total monthly requirements for labor and water can be estimated from the relevant table to help farmers select activities and plan labor and water management. By correlating the input requirements and availability, an estimation can be made each month of the labor supply and availability of water, thereby identifying times of surplus supply or when shortages could occur. If a particular month is indicated to have shortages, the farmer can plan to hire additional labor and to procure additional water, or contrarily, to choose alternative production practices. The

Table 14.8. Monthly labor and supplemental irrigation water requirements per hectare of spring wheat.

Month	Labor requirements (hr)			Water requirements (m^3)
	Man	Woman	Boy/Girl	
Nov	22	8	3	415
Dec	3		6	
Jan	2	4	14	
Feb	11	4	7	
Mar	12		3	367
Apr	22		10	1,242
May	12		4	654
Jun	26	18		
Season	98	34	47	2,678

Economics of Supplemental Irrigation 221

crux of the matter rests with the farmer having adequate, timely information to rationally choose production alternatives before crisis: if financial or other constraints prevent farmers from obtaining the necessary inputs, different combinations of farm activities must be considered. Although budgets for labor and supplemental irrigation are simpler than cash flow budgets, they are indispensable for the prosperity of the farm.

Farm resource budgets can be used by individual farmers to evaluate existing farm practices and identify where problems may exist. They target where farm costs, production, or returns appear to be out of line with the farmer's planned analysis of management and production. These methods routinely followed can lend clarification to specialists, such as engineers, economists, agronomists, and sociologists, and explain why certain factors of production may reduce yields or diminish family income.

Cash Flow Budget

Partial budgets delineate the economic benefits derived from improvements in water use efficiency, but do not specify the capability to pay off debts incurred in making improvements. A cash flow budget prepared in one form or another is critical to the farmer's decision-making in modern agriculture. A budget that shows the sources, amounts, and timing of cash income and expenses is especially relevant to decisions concerning investments for supplemental irrigation.

The cash flow budget contains 2 tasks:
1. to identify the actual cash flow during the preceding year by documenting monthly income and expenditures for each of the last 12 months; and,
2. to project the next year's cash flow, the more difficult task.

The actual cash flow for the preceding year can be prepared from existing farm records or, when just beginning, a farmer can use the best estimate. To project for next year's cash flow, this estimate should be guided by historical results and crop and livestock plans for the future. When credit is necessary for major investments (installing lined head ditches, irrigation wells, land leveling, pumps, or new land purchases) and pay-off of the principal extends over a period of more than one year, then expenditures and incomes have to be projected over a longer time period for more accurate estimates of capability to repay indebtedness.

A standard worksheet can be developed and modified for local situations to provide area farmers with a step-by-step procedure to construct the cash flow budget. A standard form can be developed with general categories for comparative appraisal among farms but with entries self-explanatory, thereby allowing farmers the opportunity to personalize their own particular characteristics of cash flow. At the top of each column the sequence of adding and subtracting can be identified as follows:

Column 1. cash inflow minus (−) outflow excluding credit;
       2. − scheduled debt payments;
       3. + capital purchase loans;
       4. = uncommitted cash flow;
       5. + new short-term debt;
       6. − additional debt payments;
       7. = cash flow: this month;
       8. + balance from last month; and,
       9. = cash flow: balance.

Partial Budgets

Farmers continually face choices between 3 alternatives; choices which involve the following questions which can be evaluated with partial budgeting.
1. Should the farmer purchase a new diesel water pump or continue to rebuild and repair the old one?
2. Should the farmer buy plastic tubing for water conveyance or maintain open ditches in the traditional manner?
3. Should the farmer produce wheat with the traditional rainfed system or use supplemental irrigation?

Partial budgeting is an organized method for appraising the effect a proposed change could have on the net returns of the farm. A typical partial budget has 7 salient parts which are shown in the example of a partial budget form in Figure 14.1. Partial budgeting is only as accurate as the data used and erroneous data can lead to serious mistakes of judgment. A positive net change suggests that to implement the proposed plan could be all right but to proceed with an alternate plan could be the wiser choice. A negative net change implies that to proceed with the proposed change would not be profitable.

The 4 main steps in constructing a partial budget are:
1. describe the change contemplated;
2. list details of contemplated change and conditions:
    a. key information to estimate the impact of the potential change; and,
    b. the assumptions made of future events impinging on the proposed change;
3. prepare an economic analysis to show the net effect of any changes on income; and,
4. itemize all factors which are difficult to allot a monetary value.

There are 7 key parts of the partial budget.
1. A listing of additional costs that will be incurred if a new enterprise or factor is introduced. For example, identify and add all costs that will increase if the decision is made to introduce a new diesel powered irrigation pump. Fixed costs include an allowance for depreciation, interest on borrowed funds, taxes, shelter, etc. Variable costs include fuel, oil, grease, repairs, and labor. For the example, added annual costs would be $2,000 during the expected life of the pump.

Economics of Supplemental Irrigation

```
                    PARTIAL BUDGET FORM

STEP 1.  Proposed  change:  Replace animal powered water wheel
                            with a diesel powered water pump.

STEP 2.  Key information and assumptions:
              1.  More irrigation water will be available.
              2.  Fuel costs will increase at about the same
                  rate as the past 5 years.
              3.  Crop prices will be about the same as the past
                  5 years.
              4.

STEP 3.  Economic analysis:
```

ITEMS THAT REDUCE NET INCOME

Part 1. Added costs:
Variable costs: fuel, grease, repairs, oil, etc. $ 1,500

Fixed costs: interest on loan, depreciation, etc.   500

Total added costs $ 2,000

Part 2. Reduced returns:

All crops & livestock produced when waterwheel $ 0.0
was used can be produced using a diesel pump
(Milk & meat may be reduced if animals used
for power are sold)

Total reduced returns $ 0.0

Part 3. Total added costs and reduced returns $ 2,000

ITEMS THAT INCREASE NET INCOME

Part 4. Added Returns:

Wheat (better yield) $ 500
Forage (better yield)   150
Maize (better yield)   300
Vegetables (increased cultivated area)   800

Total added returns $ 1,650

Part 5. Reduced costs:

Draft animal feed $ 500
Waterwheel repair   100
Labor   420

Total reduced costs $ 1,020

Part 6. Total added returns and reduced costs $ 2,670

Part 7. Net change in farm income
        (part 6 minus part 3): .... $ 670

STEP 4. Non-monetary considerations:

Diesel pump repair parts may be difficult to obtain.
Farmers are more familiar with animals than with
machinery.

Figure 14.1. Example of partial budget form.

2. The identity of the receipts-returns which would reduce because they no longer would be received after the proposed change. For the example, returns might not reduce because the previous level of production would be maintained with a combination of products made possible by the new method of lifting water. Therefore, the total reduced returns are zero but, if a certain crop or product could no longer be produced after installation of the diesel pump, the sacrificed income would equal the value of the product not produced.
3. A summation of estimated total additional costs from part 1 and reduced returns from part 2 with a value of $2,000.
4. A list of all expected additional returns if the production factor is introduced or new enterprise is implemented. In the example, it is the total value of additional crops or livestock produced because of the diesel pump delivering additional water and permitting more timely irrigations. Assumptions are made of increased yields or the production of higher value crops to increase annual returns by $1,650 during the pump's useful life.
5. Estimates of all costs eliminated if an existing enterprise is discontinued or production factor is replaced. When the animal powered water wheel is replaced, a reduction would occur in the number of draft animals to feed; repairs for the irrigation wheel would be eliminated; and, a possible reduction in labor for irrigation could occur which gives savings in costs of $1,020.
6. Total gains, i.e. additional returns plus the reduced costs of part 5. For the example, these are $ 2,670, annually, for the useful life of the pump.
7. The net change in income acquired from the potential new enterprise or production input. The change is calculated by subtracting the negative impacts of part 3 from the total gains of part 6. If the results of part 7 show income could be significantly increased through implementation of the proposed change, then action can be taken to achieve the change. If part 7 indicates income would be reduced or there might be indiscernible positive impact, the farmer seeing the risk in making the change, can alter the decision before action is taken.

Identification of potential benefits or costs is less disruptive to sustainable production if done before money or labor is committed to an unprofitable project. Introduction of the pump for irrigation shows a benefit of $2,670; but, before a final decision is made, the farmer still should make a sensitivity analysis of the proposed change by testing various levels of crop yields, prices, or costs to see the total effect on net changes in farm income. Before decisions are finalized, non-monetary considerations have to be carefully examined.

Discounting and Depreciation

When farmers weigh the potential payoff of investments for a new well, sprinkler irrigation equipment, land leveling, or lining head ditches, they expect these will eventually pay returns equal to the original value plus reasonable earnings. In consequence, the time value of money must be considered; both the cost and income of an investment must be weighed over time. If money is borrowed to finance a project and repayment made at different periods through time, certain questions arise. What is the present value of the future payments? What is the present value of the future income that will be earned during the useful life of the investment item?

An associated problem is evaluating the net results of two alternative investment possibilities, e.g. a new well versus installing sprinkler irrigation equipment. Alternatives might have different earnings within a time period, different amounts of time during which the investment is useful in producing farm income, and different arrangements for loan repayment: these variations tend to complicate comparisons. Discounting is an economic tool which facilitates comparing investment alternatives when future income or costs are a factor.

Depreciation of components of an irrigation system is calculated on the expected life of the component. Table 14.9 presents an overview of expected life values for several components of an irrigation system. Variability in expected life can occur for many of these components because of contrasts in maintenance practices or quality of repairs as well as the time of use for any particular piece of equipment.

Wells and Pumping

When wells and pumping equipment are not readily available, correct scheduling of irrigation applications and maintenance of an efficient irrigation system become increasingly crucial factors in the conservation of water resources and efficient use of energy. The cost of supplemental irrigation water is influenced by the source of water, surface or groundwater, distance of conveyance, and height of pumping. The information in Table 14.10 provides a detailed estimate of the cost of production of groundwater.

The average cost ($US) of digging a well in Tel Dhaman sub-district, Syria, is $4,423; whereas, at Mare'a, Syria, the average cost is $3,077. The cost for the engine and pump is $6,410-7,692. The plastic pipe to convey water from the well to the field costs $5.64/m at Tel Dhaman and $3.85-5.13/m at Mare'a. Investment to dig a well is a high risk for farmers in these areas. There is no guarantee that an adequate supply of water will be found at an economic depth and credit from the Agricultural Bank is not available until a well has been dug and water found. Only then can credit be obtained to buy pumping and conveyance equipment.

Wells in both sub-districts are individually owned. Tel Dhaman water is exclusively used by the well owner although a few farmers help each other when water supplies are low. In Mare'a sub-district, water is used by the well owner but farmers with good wells occasionally sell water to neighbors at a rate of

Table 14.9. Estimates of depreciation for irrigation equipment components.

Component	Depreciation (1,000 hr)	Period (yr)	Annual maintenance and repairs (% initial cost)
Wells and casings		20-30	0.5-1.5
Pumping plant			
structure		20-40	0.5-1.5
pump, vertical turbine			
bowls	16-20	8-10	5-7
column, etc.	32-40	16-20	3-5
pump, centrifugal	32-50	16-25	3-5
power transmission			
gear head	30-36	15-20	5-7
V-belt	6	3	5-7
flat belt, rubber & fabric	10	5	5-7
flat belt, leather	20	10	5-7
power source			
electric motor	50-70	25-35	1.5-2.5
diesel engine	28	14	5-8
gasoline engine			
air cooled	8	4	6-9
water cooled	18	9	5-8
propane engine	28	14	4-7
Open farm ditches (permanent)		20-25	1-2
Concrete structures		20-40	0.5-1.0
Pipe			
PVC buried		40	0.25-0.75
aluminium, gated, surface		10-20	1.5-2.5
steel, waterworks, buried		40	0.25-0.50
steel, coated & lined, buried		40	0.25-0.50
steel, coated surface		10-12	1.5-2.5
steel, galvanized, surface		15	1.0-2.0
steel, coated & lined, surface		20-25	1.0-2.0
wood, buried		20	0.75-1.25
aluminium, sprinkler use, surface		15	1.5-2.5
reinforced plastic mortar, buried		40	0.25-0.50
plastic, drip/trickle, surface		10	1.5-2.5
Sprinkler heads		8	5-8
Drip/trickle emitters		8	5-8
Drip/trickle filters		12-15	6-9
Mechanical move sprinklers		12-16	5-8
Continuous move sprinklers		10-15	5-8

Table 14.10. Estimated production costs for groundwater in Aleppo province.

Parameter		Tel Dhaman (US$)		Mare'a (US$)	
I.	Well, 150 m deep:				
	1. Drilling	@17.95 $/m,	2,692	@8.97 $/m,	1,346
	2. Casing	@ 5.13 $/m,	769	@5.13 $/m,	769
	3. Pump shaft	@ 6.41 $/m,	962	@6.41 $/m,	962
		Total:	4,423		3,077
II.	Pumping units:				
	1. Pump		4,103		
	2. Motor		3,590		
		Total:	7,693		
III.	Fuel + oil + labor		2,564		
	Maintenance		897		
		Total:	3,461		
IV.	Water pumped/yr: 3,740 hr of pumping/yr × 40 m³/hr = 149,600 m³/yr (3,740 hr = 170 days × 22 hr/day)				
V.	Total annual cost: Assume life of well = 25 yr; life of pumping unit = 10 yr; interest = 8%				
	Interest and depreciation, well		544		379
	Interest and depreciation, pumping unit		872		872
	Operation and maintenance including labor		3,670		3,670
		Total:	5,086		4,921
VI.	Cost/m³ of water: Tel Dhaman = $ 5,086/149,600 = $0.034/m³; Mare'a = $ 4,921/149,600 = $0.033/m³				

$10.26/hr for wells of 4 in flow, and $6.40-7.70/hr for wells of 3 in flow. Otherwise, well owners of Mare'a may receive one-third of the produce in payment of a season's irrigation of a neighbor's plot.

The topography in the two sub-districts is rolling and uneven which makes water conveyance and distribution difficult for irrigation. Since land levelers are not used, fields have to be subdivided into small plots to simplify irrigation of the entire field. Normally, one irrigation is applied during December before planting winter crops with a second irrigation applied between March and April. Farmers of the Mare'a sub-district do not usually apply a pre-sowing irrigation for winter crops but wait to sow after the first rains when the first 20-40 cm of the soil surface are at field capacity.

Chapter 15

Evaluation of Supplemental Irrigation

Evaluation procedure interfaces with chapter 2 where the activities of research design and methodology were discussed and chapters 3-14 where specific theories and methods delineate the techniques of supplemental irrigation. This procedure lends a structure to monitor and assess the components of supplemental irrigation as they link agricultural production to the total socio-economic system. Evaluation becomes a continuing process for affirming the results from basic research using verification studies and on-farm applications as they integrate within technology transfer for sustained agricultural productivity.

The peculiar diversity of agricultural industries and the intensity of competition for resources have hindered elaboration of a database for support of decision making in water resource management. Included in this diversity is the variation in type of family unit, the small and large scale farming system, a juxtaposition of traditional with contemporary farming practices, a lack of formal training, spatial dissimilarities, and variability of production. Evaluation measures the effectiveness of operations and identifies the most efficient mechanisms to link research with planning and development. These measures of objective fulfillment and job performance can then be used as feedback or a critique of the existing system and, if needed, changes or modifications can be introduced.

The method promotes regulated monitoring and reviewing to assess the performance level of supplemental irrigation and of other system components through consistent data collection and comparison of itemized data sets. Evaluation of irrigation technology emphasizes action and activity adding to the conceptual framework and integrating the analytical results into an ideal model or decision field. The focus is on modernized agriculture but with emphasis on authority of organization and stipulated management responsibilities fundamental to sustainable agriculture.

Cyclical Process for Data Acquisition

The cyclical process is structured to investigate management of risk by delineating procedures and processes for testing the working hypothesis that supplemental irrigation can sustain agricultural productivity in the Near East and North Africa. The process promotes a self-study of indigenous farming systems designed to discover and to quantify the combining endeavors of farmers, extension workers, scientists and engineers, and policymakers. The resolution method of evaluation, employing computer technology, accommodating data in an ideal model and supported by a conceptual framework of modernized agriculture, reinforces predictive methodology through classification of data and information according to the 4 categories of decisional caveat.

1. Point-in-time net results are verified discrete data and topical units of information. They are the current fact net or the characteristics and attributes of supplemental irrigation agriculture and other components of the system.
2. Period-in-time net results are the management data, descriptively grouped data and information for precise purposes of decisional authority.
3. Automatic notification of resource need indicates system imbalance and forecasts supply and demand for resource mobilization to restore or maintain dynamic equilibrium.
4. Statistical analyses are precise manipulations of information and data for determining analytical relationships and predictions for on-going system intervention and involvement.

Evaluation of supplemental irrigation is systemized to match the goals and objectives of production. The goals describe the purposes and actions of the technology and the objectives delineate and quantify the techniques and practices. In unison, goals and objectives provide the means of establishing criteria necessary and sufficient to determine rates of change and to explain trends of accomplishments. The degree of precision of quantifiable objectives constitutes an appraisal of the performance level of each component and relative contribution to the efficiency and effectiveness of irrigation technology. Data results of evaluation summarize the progress of goal attainment and rate of achievement of objective fulfillment. The total evaluational venture demonstrates the impact of development programs of supplemental irrigation which best fit specific national or localized situations.

Application of the Working Hypothesis

Evaluation methods allow for a multi-directional process for monitoring and reviewing a total system of activity and effort that contributes to private needs and public accountability as well as to scientific knowledge. The method includes 8 stages of activity, each stage becoming progressively more precise with increasing specificity of data or information. The baseline data collection and analysis is designed to characterize existing boundaries of an agricultural system and serve as benchmarks for future development. The data provide the basis for determining system needs leading to: program response for new projects; progress monitoring and project assessment; and intensive research review for response to supply and demand. The method provides for system analysis using feedback to modify or correct baseline data if the initial measurements were insufficient or if periodic audit is deemed critical for verification of the working hypothesis.

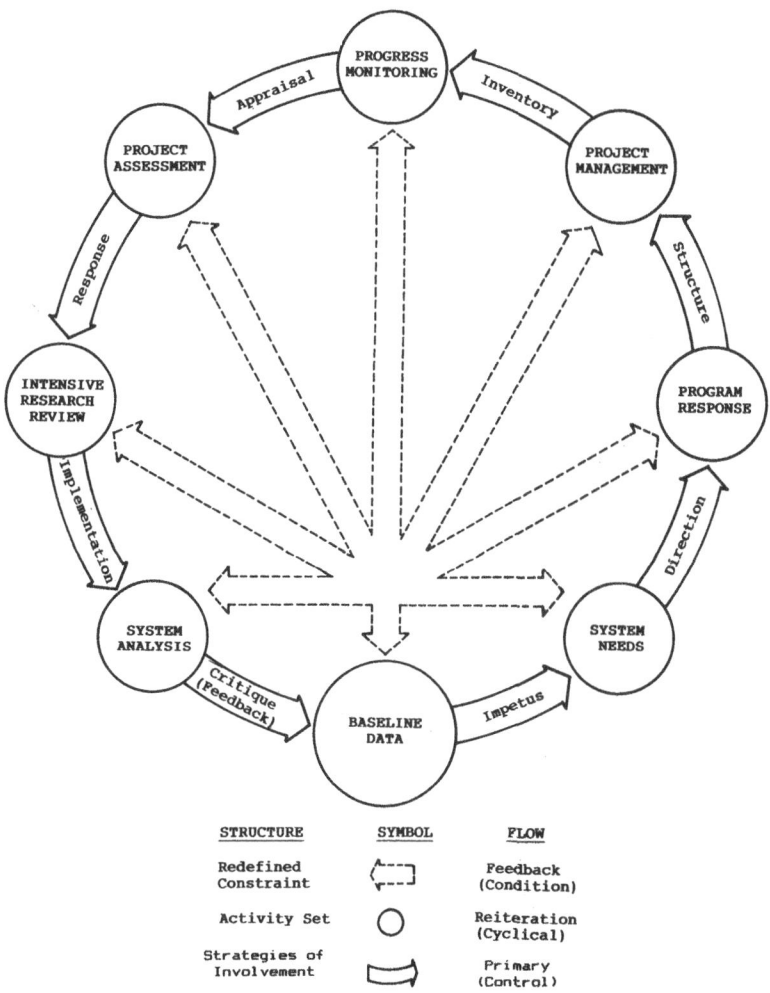

Figure 15.1. Resolution process method of evaluation.

Resolution Method for Evaluation

Figure 15.1 presents a schematic of the cyclical process of the Resolution Method of Evaluation. Results can be categorized for rational resource mobilization and revenue allocation before crisis occurs. The performance of the agricultural production system can be determined using the ideal model (see Figure 16.1) as the standard for comparison.

Strategies of Involvement

The resolution method permits access at any stage of data collection to modify the system of agricultural production without proceeding through the complete cycle. The level of complexity of data rests with the urgency of maintaining system integrity which can be achieved by selection of a strategy of involvement. The magnitude of perceived crisis defines the point of intervention and the appropriate response for resolving the crisis, i.e. the strategy of involvement.

A critique is made of a defined area of concern using existing data and information for determination of current agricultural productivity. Review of the pertinent literature of agricultural science leads to a general conclusion that a demographic explosion is occurring, that the current population desires a better living standard than previous generations, and that degradation of natural resources is an unintended consequence of trying to achieve a better life style.

Baseline data furnishes a descriptive profile and identifies organizational capability, operational constraints, and resource capacity of the agricultural system, the impetus for change. Under the existing rainfed farming system, production is usually limited to subsistence farming with low crop intensity whose corollary is unstable yields and a high yield gap ratio. In consequence, the food supply is inadequate for population requirements and results in famine and/or poor nutrition.

Employing the results of baseline data, system needs are ranked to delineate the constraints to planning and development. This furnishes the direction to proceed or denotes the urgency of the problem for which to design a structure for program response. In semi-arid and arid regions, water resources are scarce, distribution is variable, and supply is erratic. There is a potential to increase yields and stabilize production with supplemental irrigation and sustain production with planned management.

An inventory of alternative technologies is formulated from development activities using quantified objectives and project management. The implementation of supplemental irrigation using these alternative technologies corroborates that the potential exists for a surplus food supply.

Project monitoring is the dynamic phase of program auditing which generates an appraisal of the day-to-day fulfillment of objectives through a routine documentation of the relative success of supplemental irrigation. A possible consequence of the technology of supplemental irrigation is higher cropping intensities which support differing farming practices, changing cropping patterns, effective land use, and above all the more efficient application of water.

Project assessment establishes trends, factors of change, and correlated relations as validation of the effects of program goals and objectives that are directly attributed to supplemental irrigation. Sustainability of basic food crop production becomes a reality and agronomic diversification becomes an actuality.

Intensive research review analyzes the comparative contribution of each activity of agronomic production as related to supplemental irrigation and to introduce a measure of effectiveness. Analysis interrelates the multi-directional relationships of farming systems comparing supplemental irrigation to other components. These results are then used to refine or modify activities for precipitating changes. With these data, action can be taken to initiate the structure and framework for a viable market economy which cannot only promote self-sufficiency but also bolster the choice for autonomy in agricultural productivity.

The impact of supplemental irrigation on the agricultural system can be fully evaluated with system analysis. The result of the evaluation contributes new knowledge towards planning and development as well as a revised comprehensive critique of a country's capability and capacity to maintain productivity using irrigation technology. The ultimate consequence is to disrupt the rhythm of production expressed in low yields, inadequate food supply, diminishing resources, famine, and low nutrition. Multivariate analysis can demonstrate divergent and similar system characteristics, e.g. resources and constraints, prevalent in local, regional, national, and international agricultural systems.

Survey Methods, General Information and Forms

In this section, forms and data items to evaluate socio-economic and technical factors are presented for the varying requirements of monitoring, auditing, assessing, and evaluating supplemental irrigation. These can be adapted to suit any information system with a reminder that these should be pre-tested to insure validity and reliability of results.

When contemplating implementation of supplemental irrigation, current values for selected variables should be compiled to identify the optimal supplemental irrigation method for desired crop production. After implementation of an irrigation system, annual records of complete water use and pumping should be documented at the farm level as well as a comprehensive analysis performed by local extension personnel for comparison of the local farmers' production to the production of regional/national agriculture. Although water use and evapotranspiration vary from year to year, major problems and approximate efficiencies can be identified through consistent record keeping. Profits are reduced when poor practices are pursued such as excessive pre-irrigations, gross over-irrigation, or malfunctioning equipment but these can be rapidly identified when baseline data is established and routine records are kept.

Survey Questionnaire for Baseline Data

Instructions to Administer Questionnaire

A precoded questionnaire has been designed to assist in obtaining information for baseline data collection. It is designed to permit comparative analysis and locate similar and dissimilar regions within a country where supplemental irrigation is practiced as well as provide comparability across national boundaries. This interaction with the various countries involved, can lead to an in-depth understanding of both the need and potential for development of supplemental irrigation in the region.

The survey questionnaire of socio-economic variables of the farming system must be pre-tested to ensure validity and reliability of results. Pre-testing is required to perfect the questionnaire into a survey instrument that gives valid measurements over space and time. Pre-testing involves administering the questionnaire in the field, preferably in locations with characteristics similar to those of the proposed study area. The pre-test examines that a statement's position in the questionnaire does not introduce bias and that each time the same question is asked the meaning remains constant and therefore the distribution of responses are consistent. Statistical analysis must be performed on the pre-test questionnaires to identify problem areas, e.g. non-randomness of responses. When pre-testing has been accomplished, the survey instrument can be likened to a calibrated meter stick that measures exactly 100.00 cm each time it is used. If the meter stick measured 99.50 cm and 102.00 cm the next, it is equivalent to a questionnaire which has not been pre-tested; the results cannot be relied upon nor are the data valid.

The example as presented is not a self-administered questionnaire but is designed for an interviewer to ask the questions and record the responses. This enables the interviewer to discuss and explain each question to the respondent in the context of the questionnaire. Where diverse expertise exists within a district office, it may be useful to interview several professionals at the same time to complete a questionnaire. Administration of the questionnaire may require 4-6 hr depending upon the interest of the respondents or clients; it may be desirable to divide the session into 2-3 parts. Responses to the questions should not be hurried or rushed as the client may require time to consider the most appropriate answer. Some questions may elude a response from the client and, in this instance, the use of secondary data sources may be necessary to complete the questionnaire.

The completed questionnaires will be statistically analyzed; therefore, it is

Evaluation of Supplemental Irrigation

essential that all questions be completed. However, do not add responses to questions which do not apply, e.g. if water harvesting does not exist in the district, then these questions should be left blank during pre-test and removed from the questionnaire. For statistical comparability, terminology must have the same meaning for all respondents in all countries. For this reason, some terms are defined within the questionnaire and the following list of definitions should be used as a point of reference.

1. Uniform agro-ecological zone is defined as a homogeneous region having the same non-varying form of agricultural science. These zones are primarily concerned with the amount of rainfall and with the inter-relationship or patterns of soils, vegetation, and farming practices within a designated geographical boundary.
2. District is defined as the smallest administrative organization for farm extension services.
3. Demography is defined as a collection of quantitative data for the study of human populations with particular reference to size and density as well as distribution.
4. Household is defined as those persons who dwell under the same roof and compose a family.
5. Infrastructure is defined as the permanent installation or framework required for management of a farming system.
6. Potable water is defined as non-contaminated or suitable for human drinking.
7. Fragmented refers to a land parcel or agricultural field which is detached from the major farm unit.
8. Tenure is defined as the act or right of possessing landed property.
9. Livestock is defined as farm animals kept for use and profit, e.g. sheep, goats, cattle, horses, or donkeys.
10. Constraints are defined as unnatural confinements to decision making, limitations of activities, or deficiencies of resources in the existing farming system which can be restructured or corrected.

Most questions are self-explanatory; however, two-dimensional questions using rank order may require further explanation. These questions are also pre-coded to simplify statistical analysis. For example, in question 29, district personnel are asked to respond to land use management problems and the interviewer must enter the appropriate rank order number in the blank provided. There are 3 farming systems presented in this question: rainfed, supplemental irrigation (sup. irri.) and water harvesting (water harv.). The problems associated with each farming system must be rank ordered in accordance to priority. If, for example, the client does not have water harvesting farming in the district, no responses are entered for this category. Responses are entered only for rainfed and supplemental irrigation farming systems within the district.

During the pre-test, insert blank sheets at the end of the questionnaire to write additional comments and observations which will help in data interpretation for development of the survey instrument.

Supplemental Irrigation Systems
District Level Questionnaire for the Near East and North Africa

<u>Survey of Uniform Agro-Ecological Areas</u>

Questionnaire I. D. _____
Title of Principle Participant _____
Date of Interview _____
Name of Interviewer _____

Country _____
Province or
Governorate _____
District _____

Add any unit used that is not listed. Give conversion of all units to metric equivalent, e.g., 0.39 inches = 1 centimeter. Final measurements will be converted to **metric units**.

Land Unit (LU) _____ _____ Hectares
Distance Unit (DU) _____ _____ Meters
Weight Unit (WU) _____ _____ Grams
Volume Unit (VU) _____ _____ Liters
Yield Unit, Grain (YG) _____ _____ kg/ha
Yield Unit, Straw (YS) _____ _____ kg/ha
Monetary Unit (MU) _____ _____ $ US

GENERAL DEMOGRAPHIC INFORMATION FOR THE DISTRICT

1. What is the total population? .... _____
2. What is the total number of villages? .... _____
3. What is the total number of households? ...... _____
4. Average number of members per household? ..... _____
5. What is the ratio of male to female in the household? _____
6. Average number of members per household
   in the following age groups? .... less than 5 years _____
                                        5 thru 9 years _____
                                        10 thru 14 years _____
                                        15 thru 65 years _____
                                        more than 65 years _____
7a. What is the number of weather stations? .... _____
 b. Are daily data available for: rainfall? .... Yes ( ) 1
                                                      No ( ) 2
                              pan evaporation? .... Yes ( ) 1
                                                      No ( ) 2
                              air temperature? .... Yes ( ) 1
                                                      No ( ) 2
                             relative humidity? .... Yes ( ) 1
                                                      No ( ) 2
                              solar radiation? .... Yes ( ) 1
                                                      No ( ) 2

DEFINITIONS OF FARMING SYSTEMS
Management of crop production under natural rainfall, i.e., no added water, is termed <u>rainfed</u>.
****
In an area where a crop can be grown by natural rainfall alone but additional water stabilizes and improves yield, this management technique is termed <u>supplemental irrigation</u>.
****
Collection of natural rainfall from a modified or treated area to either maximize or minimize runoff whichever method is selected for a specific site, this management technique is termed <u>water harvesting</u>.

INFRASTRUCTURE/RURAL SUPPORT SERVICES IN THE DISTRICT

8a. How many schools are there? .... _____
 b. How many schools at each level? ..... grades 1 thru 3 _____
                                          grades 4 thru 6 _____
                                          grades 7 thru 9 _____
                                          grades 10 thru 12 _____
9a. Do both males and females attend school? .... Yes ( ) 1
                                                         No ( ) 2
 b. If Yes, do they attend school the same number of years?
                                                         Yes ( ) 1
                                                         No ( ) 2
10a. What percentage of farmers are illiterate? .... _____ %
  b. Is illiteracy of farmers: .... increasing? ( ) 1
                                                decreasing? ( ) 2
                                                unchanged? ( ) 3

To rank order responses, list by number where one (1) is the
highest occurrence, with the next highest occurrence as
two (2), then three (3), etc.

---

11. What is the average education level for farmers?
                    (check one choice for each farming system)
                               rainfed   sup. irri.   water harv.
             illiterate   ( ) 1   ( ) 1   ( ) 1
    literate, no certificate   ( ) 2   ( ) 2   ( ) 2
         primary certificate   ( ) 3   ( ) 3   ( ) 3
    intermediate certificate   ( ) 4   ( ) 4   ( ) 4
       secondary certificate   ( ) 5   ( ) 5   ( ) 5
    other _____   ( ) 6   ( ) 6   ( ) 6
12. How many centers (bureaus) are there for:... credit? _____
                                      agricultural extension? _____
                                             veterinary care? _____
                                                medical care? _____
                             other (specify) _____? _____
13a. How many cooperatives are there? .... _____
  b. How many of each service type are there?
                       specify type _____ _____

14. Percentage of farmers enrolled in cooperatives? ... _____ %
15. What are the sources for credit?
             (rank order according to priority, 1 thru 6)
                                agricultural bank .... _____
                                     cooperatives .... _____
                            other government sources .... _____
                                          private .... _____
                                           family .... _____
                      other (specify) _____ .... _____
16. Is credit adequate for farmers' needs? .... Yes ( ) 1
                                                         No ( ) 2
17a. How many villages have electrical power? .... _____
  b. What is major source of electricity? ...... wind power   ( ) 1
                                        government service   ( ) 2
                                          village generator ( ) 3
                                          private generator ( ) 4
                                                solar power ( ) 5
                    other (specify) _____   ( ) 6
18. What is the primary source of household fuel for:
                                 (check one choice for each)
              heating?   cooking?
             wood/shrub ( ) 1   wood/shrub ( ) 1
               oil/gas ( ) 2   oil/gas ( ) 2
           crop residue ( ) 3   crop residue ( ) 3
           animal waste ( ) 4   animal waste ( ) 4
    other _____   ( ) 5   _____   ( ) 5

19a. How many villages have potable water
     for domestic uses? .... _____
  b. How many have pressure systems? .... _____
20a. Are there public water stands (faucets)? .... Yes ( ) 1
                                                         No ( ) 2
  b. If No, major source of household water is ... cisterns ( ) 1
                                                private well ( ) 2
                                                  check dam ( ) 3
             other (specify) _____   ( ) 4
  c. If Yes, do farmers buy water for domestic use? ... Yes ( ) 1
                                                         No ( ) 2
21a. Where are the main markets for:
                           agricultural inputs? _____
                           agricultural outputs? _____

b. What is the average distance to these markets?
agricultural inputs _____DU
agricultural outputs _____DU
22. Are secondary roads for motor transport: ... seasonal? ( ) 1
all year? ( ) 2
23a. What is the condition of secondary roads? .. excellent ( ) 1
good ( ) 2
fair ( ) 3
poor ( ) 4
bad ( ) 5
b. Major type of secondary road surface? .... asphalt ( ) 1
gravel ( ) 2
dirt ( ) 3
other (specify) _____ ( ) 4
23c. Are secondary roads adequate for farmers' needs? Yes ( ) 1
No ( ) 2
24. Is there refrigerated transport for farmer use? ... Yes ( ) 1
No ( ) 2
25. Is transport to major market considered expensive? Yes ( ) 1
No ( ) 2

LAND AREA AND LAND USE IN THE DISTRICT

26a. What is the total land area? .... _____LU
b. What is the total area for non-arable lands of:
trees/desert shrubs? _____LU
steppe and pastures? _____LU
rocky and sandy? _____LU
marshes and lakes? _____LU
development and roads? _____LU
c. What is the total area for arable lands of:
not in crop production? _____LU
rainfed? _____LU
fallow? _____LU
supplemental irrigation? _____LU
water harvesting? _____LU
27. What is the total number of farms? ....
28a. Percentage of farms that are fragmented? ... rainfed _____%
supplemental irrigation _____%
water harvesting _____%
b. Average number of fragments per farm? .... rainfed _____
supplemental irrigation _____
water harvesting _____
29. What are the problems with management of farm units?
(rank order 1 thru 7 for each farming system)
rainfed   sup. irri.   water harv.
fragmentation   _____   _____   _____
inadequate size   _____   _____   _____
not enough water   _____   _____   _____
too large a land area _____   _____   _____
access in winter   _____   _____   _____
multiple ownership _____   _____   _____
other _____
30. What percentage of farmers use: .... rainfed farming? _____%
supplemental irrigation? _____%
water harvesting? _____%
31. Percentage of farms having the following type
of land tenure? .... individual ownership _____%
joint ownership (family) _____%
agrarian reform _____%
leased/rented _____%
sharecropping _____%
other (specify) _____
32. What is the average size of farms for: ... rainfed? _____LU
supplemental irrigation? _____LU
water harvesting? _____LU
33. What is the minimum size and number of farms for:
rainfed? supplemental irrigation? water harvesting?
number   _____   _____   _____
LU
34. What is the maximum size and number of farms for:
rainfed? supplemental irrigation? water harvesting?
number   _____   _____   _____
LU

SOIL CHARACTERISTICS IN THE DISTRICT

35a. Are specific soils preferred for field crops? .... Yes ( ) 1
No ( ) 2
b. If Yes, which soil type do farmers prefer?
(check one choice for each farming system)
rainfed   sup. irri.   water harv.
heavy (clay)   ( ) 1   ( ) 1   ( ) 1
slightly heavy (clay loam) ( ) 2   ( ) 2   ( ) 2
moderate (loam)   ( ) 3   ( ) 3   ( ) 3

Evaluation of Supplemental Irrigation

```
           moderately light (silt loam) ( ) 4      ( ) 4      ( ) 4
           light (sandy loam)            ( ) 5      ( ) 5      ( ) 5
           very light (loamy sand)       ( ) 6      ( ) 6      ( ) 6
36. Are shallow soils predominant?
             rainfed          supplemental irrigation       water harvesting
          Yes ( ) 1              Yes ( ) 1                    Yes ( ) 1
          No  ( ) 2              No  ( ) 2                    No  ( ) 2

37. What is the average depth of major soils?
                    (check one choice for each farming system)
                                  rainfed    sup. irri.   water harv.
              Less than 50cm      ( ) 1       ( ) 1        ( ) 1
              51cm to 75cm        ( ) 2       ( ) 2        ( ) 2
              76cm to 1m          ( ) 3       ( ) 3        ( ) 3
              More than 1m        ( ) 4       ( ) 4        ( ) 4
38. What are the predominant soil characteristics?
     (check one choice each, a thru g, for each farming system)
                                  rainfed    sup. irri.   water harv.
    a. color: ........... red     ( ) 1       ( ) 1        ( ) 1
                         brown    ( ) 2       ( ) 2        ( ) 2
                      tan/sand    ( ) 3       ( ) 3        ( ) 3
                         black    ( ) 4       ( ) 4        ( ) 4
                        yellow    ( ) 5       ( ) 5        ( ) 5
    b. slope: ......... level     ( ) 1       ( ) 1        ( ) 1
                      moderate    ( ) 2       ( ) 2        ( ) 2
              slightly sloping    ( ) 3       ( ) 3        ( ) 3
                  steep slopes    ( ) 4       ( ) 4        ( ) 4
    c. stoniness:   stone-free    ( ) 1       ( ) 1        ( ) 1
                 slightly-stony   ( ) 2       ( ) 2        ( ) 2
                         stony    ( ) 3       ( ) 3        ( ) 3
                    very stony    ( ) 4       ( ) 4        ( ) 4
    d. salinity: ........ none    ( ) 1       ( ) 1        ( ) 1
                           low    ( ) 2       ( ) 2        ( ) 2
                        saline    ( ) 3       ( ) 3        ( ) 3
    e. infiltration: excessive    ( ) 1       ( ) 1        ( ) 1
                      moderate    ( ) 2       ( ) 2        ( ) 2
                          slow    ( ) 3       ( ) 3        ( ) 3
                       limited    ( ) 4       ( ) 4        ( ) 4
    f. crusting: ........ none    ( ) 1       ( ) 1        ( ) 1
                          thin    ( ) 2       ( ) 2        ( ) 2
                      moderate    ( ) 3       ( ) 3        ( ) 3
                          hard    ( ) 4       ( ) 4        ( ) 4
    g. tilth: ........... good    ( ) 1       ( ) 1        ( ) 1
                          fair    ( ) 2       ( ) 2        ( ) 2
                          poor    ( ) 3       ( ) 3        ( ) 3
```

WATER SUPPLY AND MANAGEMENT IN THE DISTRICT

39. What is the average annual rainfall .... _____mm
40a. Are weather and farm reports available on radio
     or television to assist in farm planning? .... Yes ( ) 1
                                                        No ( ) 2

40b. If Yes, do farmers use these reports to estimate:
              (rank order 1 thru 6 for each farming system)
                               rainfed   sup. irri.   water harv.
           planting time?   _____   _____   _____
     approach of severe storms? _____ _____   _____
     approach of killing frosts? _____ _____   _____
          drought conditions? _____   _____   _____
               harvest time?   _____   _____   _____
     other _____?_____   _____   _____

41a. What is the source of water for supplemental irrigation?
         (rank order according to priority, 1 thru 7)
                                       well .... _____
                                    cistern .... _____
                            canal/galleries .... _____
                         intermittent streams .... _____
                                 check dams .... _____
                           water harvesting .... _____
                            other (specify) .... _____
   b. What is the land area for supplemental irrigation by the
      source ranked as first priority? .... _____LU
42. What is the water quality for supplemental irrigation
    by the source ranked as first priority in question 41a?
            suitable for all uses .... ( ) 1
              (0 - 250 ppm or EC = 0.4 mmhos/cm salt)
           reduced yields for most crops .... ( ) 2
              (251 - 600 ppm or EC = 0.5 - 0.8 mmhos/cm salt)
        yields from salt tolerant plants only .... ( ) 3
              (601 - 1000 ppm or EC = 0.9 - 1.6 mmhos/cm salt)
        management skills needed for yield production ... ( ) 4
              (greater than 1000 ppm or EC > 1.6 mmhos/cm salt)
43. What is the maximum intensity of rain storms ? _____mm/hr

44. What is the average length of time for maximum
    intensity rain storms? .... _____ hr
45. How many storms per year are of: average intensity? _____
                                       maximum intensity? _____
                                       minimum intensity? _____
46a. Do farmers buy water for supplemental irrigation?   Yes { } 1
                                                         No { } 2
   b. If yes, what is the cost per land unit for:
                                  one irrigation? _____ MU/LU
                          volume of water delivered? _____ MU/LU
                             total cropping season? _____ MU/LU
                          annual delivery of water? _____ MU/LU
                        other (specify) _____? _____ MU/LU

```
flow is usually given in one of three units, "inches" or
   "cubic meters per hour" or "liters per second";
                    specify unit
```

47a. What is the average depth to water table? ...... _____ DU
   b. What is the average pumping depth of wells? .... _____ DU
   c. What is the average flow of water from the pump at the
      average pumping depth? .... _____
48. How many productive wells are in the district? ... _____
49. How many wells have gone out of production? ..... _____
50a. How many new wells were saline or salty? .... _____
   b. How many new wells were "dry holes"? .... _____
51. How many wells are owned by: .... individual farmers? _____
                                     cooperative society? _____
                                          farmer groups? _____
                                       government agency? _____
                        other (specify) _____? _____
52. What types of pump engines are used?
            (rank order according to priority, 1 thru 5)
                                            windmill .... _____
                                              diesel .... _____
                                            electric .... _____
                                            gasoline .... _____
                        other (specify) _____ .... _____
53a. Is there a seasonal decline in pumping capacity?   Yes { } 1
                                                        No { } 2
53b. If Yes, how long does decline last? .... _____
54. What criterion do farmers use when deciding to dig a well?
            (rank order according to priority, 1 thru 5)
                                  to increase farm income .... _____
                                    household requirement .... _____
                                          livestock need .... _____
                                      yield stabilization .... _____
                        other (specify) _____ .... _____
55. What are average costs for installing a well?
      digging (drilling) ...... from _____ MU to _____ MU
      casing .... from _____ MU to _____ MU
      pumping shaft .... from _____ MU to _____ MU
      pump shelter .... _____ MU
56a. What are average costs for buying a pump and power unit?
                        pump .... _____ MU
                        engine .... _____ MU
            installation fee .... _____ MU
56b. When deciding to buy a new pump, what are the constraints?
            (rank order according to priority, 1 thru 7)
                            initial capital investment .... _____
                                    cost of energy/fuel .... _____
                            cost of maintenance/repair .... _____
                         availability of parts/service .... _____
                                    credit availability .... _____
                            requirement of skilled labor .... _____
                        other (specify) _____ .... _____
57. What is the annual operating cost
      of a well? .... energy/fuel _____ MU
                                              lubrication _____ MU
                                    maintenance and repair _____ MU
                                            skilled labor _____ MU
58. What is the volume of water pumped per year? ...... _____ VU
   a. Is well water used exclusively by owners? .... Yes { } 1
                                                        No { } 2
   b. If No, how is water shared? .. _____

SUPPLEMENTAL IRRIGATION IN THE DISTRICT

59. What is the total land area under
      supplemental irrigation using: ..... surface flow? _____ LU
                                             sprinklers? _____ LU
                                           drip/trickle? _____ LU

Evaluation of Supplemental Irrigation 241

```
                      other (specify) _____?  _____LU
60.  In general, is pumping needed to deliver water
     for supplemental irrigation? ...................... Yes ( )  1
                                                          No  ( )  2
61a. Is land leveled for supplemental irrigation? ...... Yes ( )  1
                                                          No  ( )  2
  b. If yes, how is land leveling usually done?
                     (rank order according to priority, 1 thru 7)
                                          hand equipment ....  _____
                               animal traction equipment ....  _____
                                     tractor attachments ....  _____
                                   grader or land planes ....  _____
                          bulldozers and earth movers    ....  _____
                                          auto levelers ....  _____
                 other (specify) _____  ....
62.  How many supplemental irrigations are applied to each crop?
                                    RAINFALL SEASON
```

	Normal	Below Normal	Above Normal
CROP	NUMBER OF SUPPLEMENTAL IRRIGATIONS		
wheat	_____	_____	_____
barley	_____	_____	_____
faba bean	_____	_____	_____
lentil	_____	_____	_____
chickpea	_____	_____	_____
forages	_____	_____	_____

```
other _____
63a. Do farmers apply a presowing supplemental irrigation
     for winter crops? ................................. Yes ( )  1
                                                         No  ( )  2
  b. If Yes, percent farmers using presowing irrigation? _____ %
  c. If Yes, how much water is usually applied? ........        mm
64.  If presowing irrigation is not applied, do farmers apply a
     supplemental irrigation immediately after sowing?   Yes ( )  1
                                                         No  ( )  2
65.  What is the percentage of supplemental irrigation
     systems using: ................................ basin? _____ %
                                            surface flooding? _____ %
                                                      furrow? _____ %
                                                   sprinkler? _____ %
                                        drip/trickle, surface? _____ %
              other (specify) _____?
66.  What criterion is used to schedule supplemental irrigation?
                     (rank order according to priority, 1 thru 6)
                                          weather factors ....  _____
                                  water supply availability ....  _____
                                        fixed time periods ....  _____
                                     neighbor's scheduling ....  _____
                                       condition of plants ....  _____
                  other (specify) _____ ....
67a. Is surface or subsurface drainage provided for supplemental
     irrigation? ....................................... Yes ( )  1
                                                         No  ( )  2
  b. If Yes, major type of drainage used? ........... tile ( )  1
                                              open ditches ( )  2
                                             recharge wells ( )  3
                                                 pump-out ( )  4
                  other (specify) _____     ( )  5
68a. Do farmers tend to over-irrigate crops? .......... Yes ( )  1
                                                         No  ( )  2
  b. If Yes, do they have problems with:
                     (rank order according to priority, 1 thru 6)
                                                   runoff? ....  _____
                                            water-logging? ....  _____
                                          deep percolation? ....  _____
                                           salt build-up? ....  _____
                                      fertilizer leaching? ....  _____
                  other (specify) _____? ....  _____
```

WATER HARVESTING IN THE DISTRICT

```
69.  Do farmers use water harvesting
     to collect rainfall? ............................. Yes ( )  1
                                                         No  ( )  2
70.  Are dew and mist harvested as a source
     of additional water? ............................. Yes ( )  1
                                                         No  ( )  2
71.  What is major method used for water harvesting?
                     (rank order according to priority, 1 thru 8)
                               conservation bench terrace ....  _____
                                       desert strip farming ....  _____
                                 tied-ridges/micro-catchment ....  _____
                                  contour ridges or bunding ....  _____
                                         check dams/tanks ....  _____
```

```
                            catchment basin/direct use  ....  _____
                                   floodwater farming  ....  _____
              other (specify)_____  ....  _____
72. Do water storage facilities have silt traps? ...... Yes ( ) 1
                                                         No ( ) 2
73. How is collected water used?
              (rank order according to priority, 1 thru 7)
                                          field crops  ....  _____
                                           tree crops  ....  _____
                                              pasture  ....  _____
                                   household, drinking ....  _____
                                   household, general  ....  _____
                                            livestock  ....  _____
              other (specify)_____  ....  _____
74. Do water harvesting facilities have problems with:
              (rank order according to priority, 1 thru 7)
                                          maintenance  ....  _____
                                   reliable collection ....  _____
                                              erosion  ....  _____
                                        water quality  ....  _____
                                               health  ....  _____
                                 cultural acceptability ....  _____
              other (specify)_____?  ....  _____
75. Do farmers who use water harvesting have field problems with:
              (rank order according to priority, 1 thru 6)
                                             runoff?   ....  _____
                                       water-logging?  ....  _____
                                     deep percolation? ....  _____
                                        salt build-up? ....  _____
                                  fertilizer leaching? ....  _____
              other (specify)_____?  ....  _____
76a. Is land leveling done for water harvesting? ...... Yes ( ) 1
                                                         No ( ) 2
  b. If Yes, how is land leveling done?
              (rank order according to priority, 1 thru 7)
                                        hand equipment ....  _____
                              animal traction equipment ....  _____
                                    tractor attachments ....  _____
                                  grader or land planes ....  _____
                              bulldozers and earth movers ....  _____
                                         auto levelers  ....  _____
              other (specify)_____  ....  _____
77. Are the soils used in the rainfall catchment areas of a
    water harvesting facility: ............. impermeable? ( ) 1
                                           crust-forming? ( ) 2
                                          poorly drained? ( ) 3
              other (specify)_____?  ( ) 4
78. What is the major water harvesting technique used for
    collecting domestic water? ....... roof top catchment ( ) 1
                                  ground/surface water catchment ( ) 2
                                              open tank ( ) 3

              CROP PRODUCTION IN THE DISTRICT

79. What are the three (3) major winter crops
    for each farming system?
                        rainfed       sup. irri.      water harv.
           crop-1 ....  _____       _____      _____
           crop-2 ....  _____       _____      _____
           crop-3 ....  _____       _____      _____
80. What criterion is used to determine the size of crop area
    for rainfed farming for a particular year?
              (rank order according to priority, 1 thru 7)
                      total rainfall up to planting time ....  _____
              number of intensive storms before planting ....  _____
                             plant entire cultivated area ....  _____
                     depth of water in soil at planting time ....  _____
                                 crop rotation requirement ....  _____
              amount of seed available at planting time ....  _____
              other (specify)_____  ....  _____
81. What criterion is used to determine the size of crop area
    for supplemental irrigation for a particular year?
              (rank order according to priority, 1 thru 8)
                      total rainfall up to planting time ....  _____
                             plant entire cultivated area ....  _____
                     depth of water in soil at planting time ....  _____
              amount of seed available at planting time ....  _____
                                 crop rotation requirement ....  _____
                              water available for irrigation ....  _____
                                cost of water for irrigation ....  _____
              other (specify)_____  ....  _____
82. What criterion is used to determine the size of crop area
    for water harvesting for a particular year?
              (rank order according to priority, 1 thru 7)
                      total rainfall up to planting time ....  _____
```

Evaluation of Supplemental Irrigation

```
                number of intensive storms before planting ....  _____
                          plant entire cultivated area ....  _____
                  depth of water in soil at planting time ....  _____
                amount of seed available at planting time ....  _____
                               crop rotation requirement ....  _____
          other (specify)_____
```

83. Do farmers sow (plant) in dry soil?
    rainfed   supplemental irrigation   water harvesting
    Yes ( ) 1   Yes ( ) 1   Yes ( ) 1
    No ( ) 2   No ( ) 2   No ( ) 2

84a. Do farmers sow winter crops after first rains? ... Yes ( ) 1
                                                        No ( ) 2

84b. If Yes, how much rainfall is considered enough
     before sowing? .... rainfed   _____ mm
                                                supplemental irrigation _____ mm
                                                water harvesting   _____ mm

85. What is the major sowing method for winter crops?
                               rainfed   sup. irri.   water harv.
           broadcast ....   ( ) 1   ( ) 1   ( ) 1
           seed drill ....   ( ) 2   ( ) 2   ( ) 2
             spinner ....   ( ) 3   ( ) 3   ( ) 3
    other _____   ( ) 4   ( ) 4   ( ) 4

86. For **rainfed farming**, what is the major sowing date,
    planting depth, and seeding rate for the following crops?
                    date   planting depth, cm   seed rate, WU/LU
                              manual   drilled   manual   drilled
         wheat   _____   _____   _____   _____   _____
         barley   _____   _____   _____   _____   _____
         faba bean _____   _____   _____   _____   _____
         lentil   _____   _____   _____   _____   _____
         chickpea   _____   _____   _____   _____   _____
         forages   _____   _____   _____   _____   _____
     major crop-1   _____   _____   _____   _____   _____
     major crop-2   _____   _____   _____   _____   _____
     major crop-3   _____   _____   _____   _____   _____

87. For **supplemental irrigation**, what is the major sowing date,
    planting depth, and seeding rate for the following crops?
                    date   planting depth, cm   seed rate, WU/LU
                              manual   drilled   manual   drilled
         wheat   _____   _____   _____   _____   _____
         barley   _____   _____   _____   _____   _____
         faba bean _____   _____   _____   _____   _____
         lentil   _____   _____   _____   _____   _____
         chickpea   _____   _____   _____   _____   _____
         forages   _____   _____   _____   _____   _____
     major crop-1   _____   _____   _____   _____   _____
     major crop-2   _____   _____   _____   _____   _____
     major crop-3   _____   _____   _____   _____   _____

88. For **water harvesting**, what is the major sowing date,
    planting depth, and seeding rate for the following crops?
                    date   planting depth, cm   seed rate, WU/LU
                              manual   drilled   manual   drilled
         wheat   _____   _____   _____   _____   _____
         barley   _____   _____   _____   _____   _____
         faba bean _____   _____   _____   _____   _____
         lentil   _____   _____   _____   _____   _____
         chickpea   _____   _____   _____   _____   _____
         forages   _____   _____   _____   _____   _____
     major crop-1   _____   _____   _____   _____   _____
     major crop-2   _____   _____   _____   _____   _____
     major crop-3   _____   _____   _____   _____   _____

89. What is the major source of seed? .... farm storage   ( ) 1
                                              government supply ( ) 2
                                              commercial   ( ) 3
                    other (specify) _____   ( ) 4

90. What is the land area allocated to the following crops?
                          rainfed   sup. irri.   water harv.
         barley ....   _____ LU   _____ LU   _____ LU
         wheat   ....   _____ LU   _____ LU   _____ LU
         faba Bean .... _____ LU   _____ LU   _____ LU
         lentil ....   _____ LU   _____ LU   _____ LU
         chickpea ....   _____ LU   _____ LU   _____ LU
         forages ....   _____ LU   _____ LU   _____ LU
         tree crops .... _____ LU   _____ LU   _____ LU
         vegetables .... _____ LU   _____ LU   _____ LU
         summer crops .... _____ LU   _____ LU   _____ LU

91a. Do farmers use fertilizer in cereal production?
     rainfed   supplemental irrigation   water harvesting
     Yes ( ) 1   Yes ( ) 1   Yes ( ) 1
     No ( ) 2   No ( ) 2   No ( ) 2

  b. If Yes, the major method of applying fertilizer is by:
                               rainfed   sup. irri.   water harv.
         broadcasting? ....   ( ) 1   ( ) 1   ( ) 1
              banding? ....   ( ) 2   ( ) 2   ( ) 2
    other _____?   ....   ( ) 3   ( ) 3   ( ) 3

91c. If Yes, average fertilizer application rates (WU/LU)?
         rainfed   sup. irri.   water harv.
  nitrogen .... _____   _____   _____
  P_2O_5 .... _____   _____   _____
  organic .... _____   _____   _____

92a. Is it difficult to buy fertilizer? .... Yes ( ) 1
                                                        No ( ) 2
  b. If Yes, what are the reasons? .... market availability ( ) 1
                                                      cost ( ) 2
                                         transport distance ( ) 3
                                      governmental control ( ) 4
                              other (specify) _____ ( ) 5

93a. Percent farmers who apply fertilizer? ...... rainfed _____ %
                                       supplemental irrigation _____ %
                                            water harvesting _____ %
  b. Is the application of fertilizer:
         rainfed   sup. irri.   water harv.
  increasing? .... ( ) 1   ( ) 1   ( ) 1
  decreasing? .... ( ) 2   ( ) 2   ( ) 2
  no change? .... ( ) 3   ( ) 3   ( ) 3

94. How many cultivations are usually done for wheat?
         rainfed   sup. irri.   water harv.
  number ..
  timing of cultivations :: 1 _____   1 _____   1 _____
                         :: 2 _____   2 _____   2 _____
                         :: 3 _____   3 _____   3 _____
  equipment used :: 1 _____   1 _____   1 _____
                 :: 2 _____   2 _____   2 _____
                 :: 3 _____   3 _____   3 _____

95. How many cultivations are usually done for barley?
         rainfed   sup. irri.   water harv.
  number ..
  timing of cultivations :: 1 _____   1 _____   1 _____
                         :: 2 _____   2 _____   2 _____
  equipment used :: 1 _____   1 _____   1 _____
                 :: 2 _____   2 _____   2 _____

96a. What is principal crop rotation?
         rainfed   sup. irri.   water harv.
  continuous cropping .... ( ) 1   ( ) 1   ( ) 1
  no specific rotation .... ( ) 2   ( ) 2   ( ) 2
  two-course rotation .... ( ) 3   ( ) 3   ( ) 3
  three-course rotation .... ( ) 4   ( ) 4   ( ) 4
  other _____ .... ( ) 5   ( ) 5   ( ) 5
  b. If two-course rotation, what is sequence for?
  rainfed .... _____/_____
  supplemental irrigation ... _____/_____
  water harvesting .... _____/_____
  c. If three-course rotation, what is sequence for?
  rainfed .... _____/_____/_____
  supplemental
  irrigation: _____/_____/_____
  water
  harvesting: _____/_____/_____

97a. What is the percentage of farmers using principal rotation?
         rainfed   sup. irri.   water harv.
          _____ %   _____ %   _____ %
  b. Why do they use this rotation?
         (rank order 1 thru 6 for each farming system)
         rainfed   sup. irri.   water harv.
  maintain yield ..   _____   _____   _____
  weed and pest control .. _____   _____   _____
  soil fertility ..   _____   _____   _____
  livestock needs ..   _____   _____   _____
  tradition and habit .. _____   _____   _____
  other ..   _____   _____   _____

98. What are the major weeds during each season?
  spring   summer   autumn   winter
  _____   _____   _____   _____
  _____   _____   _____   _____

99. Percentage of farmers who use the following methods
  of weed control on winter crops?
         rainfed   sup. irri.   water harv.
  herbicides ..   _____ %   _____ %   _____ %
  manual weeding ..   _____ %   _____ %   _____ %
  mechanical cultivation .. _____ % _____ %   _____ %
  other _____ ..   _____ %   _____ %   _____ %

100a. Is the application of herbicides on winter crops:
         rainfed   sup. irri.   water harv.
  increasing? .... ( ) 1   ( ) 1   ( ) 1
  decreasing? .... ( ) 2   ( ) 2   ( ) 2
  no change? .... ( ) 3   ( ) 3   ( ) 3
  b. If no herbicides are used on winter crops,
  what is estimated yield loss (YG/LU)?
     rainfed   supplemental irrigation   water harvesting

Evaluation of Supplemental Irrigation

101. Are seeds for winter crops dusted or fumigated?   Yes ( ) 1
                                                        No ( ) 2
102. What are the major diseases on: the major insects on:
     wheat? _____   wheat? _____
     barley? _____   barley? _____
     faba bean? _____   faba bean? _____
     lentil? _____   lentil? _____
     chickpea? _____   chickpea? _____
     forages? _____   forages? _____

103a. Percentage of farmers who use pesticides on winter crops?
                         rainfed   sup. irri.   water harv.
                         _____%   _____%   _____%
   b. If no pesticides are used on winter crops,
      what is estimated yield loss (YG/LU)?
                         rainfed   sup. irri.   water harv.

104. What minor pests (rodents, birds, etc.) affect winter crops?
           pest   crop

105. What is the normal harvest date for:
                         rainfed   sup. irri.   water harv.
          wheat?   .... _____   _____   _____
          barley? .... _____   _____   _____
          faba bean? .. _____   _____   _____
          lentil? .... _____   _____   _____
          chickpea? ... _____   _____   _____
          forage? .... _____   _____   _____
          major crop-1? ._____   _____   _____
          major crop-2? ._____   _____   _____
          major crop-3? ._____   _____   _____
     other _____?

106a. What is the major method of harvesting cereals?
                         rainfed   sup. irri.   water harv.
     combine harvester .... ( ) 1   ( ) 1   ( ) 1
                manually .... ( ) 2   ( ) 2   ( ) 2
     manually and thresher .. ( ) 3   ( ) 3   ( ) 3
                   other .... ( ) 4   ( ) 4   ( ) 4
   b. Percent farmers who harvest mechanically ... rainfed   _____%
                                                   supplemental irrigation _____%
                                                   water harvesting _____%

107a. What is the level of grain yield (YG/LU) for wheat?
                         rainfed   sup. irri.   water harv.
       average yield .... _____   _____   _____
       minimum yield .... _____   _____   _____
       maximum yield .... _____   _____   _____
   b. Percent farmers who obtain these grain yields for wheat?
                         rainfed   sup. irri.   water harv.
       average yield .... _____%   _____%   _____%
       minimum yield .... _____%   _____%   _____%
       maximum yield .... _____%   _____%   _____%
107c. What is the level of straw yield (YS/LU) for wheat?
                         rainfed   sup. irri.   water harv.
       average yield .... _____   _____   _____
       minimum yield .... _____   _____   _____
       maximum yield .... _____   _____   _____

108a. What is the level of grain yield (YG/LU) for barley?
                         rainfed   sup. irri.   water harv.
       average yield .... _____   _____   _____
       minimum yield .... _____   _____   _____
       maximum yield .... _____   _____   _____
   b. Percent farmers who obtain these grain yields for barley?
                         rainfed   sup. irri.   water harv.
       average yield .... _____%   _____%   _____%
       minimum yield .... _____%   _____%   _____%
       maximum yield .... _____%   _____%   _____%
   c. What is the level of straw yield (YS/LU) for barley?
                         rainfed   sup. irri.   water harv.
       average yield .... _____   _____   _____
       minimum yield .... _____   _____   _____
       maximum yield .... _____   _____   _____

109. What is the most important agronomic factor to increase
     cereal production?
                (rank order 1 thru 9 for each farming system)
                         rainfed   sup. irri.   water harv.
       time of seeding .. _____   _____   _____
       seedbed preparation ._____   _____   _____
            seeding rate .. _____ _____   _____
            seed variety .. _____ _____   _____
       irrigation after sowing .._____ _____ _____
         pesticide/herbicide .. _____ _____   _____
         nitrogen fertilizer .. _____ _____   _____
         phosphorus fertilizer .. _____ _____   _____
              other _____

110. What are the major conflicts between cereal production and

other kinds of crop production?
(rank order 1 thru 8 for each farming system)
```
                                    rainfed    sup. irri.   water harv.
            cropping sequence       _____    _____      _____
     timing of soil preparation     _____    _____      _____
            harvest operations      _____    _____      _____
            threshing operations    _____    _____      _____
            pest/disease control    _____    _____      _____
            labor requirement       _____    _____      _____
         equipment availability     _____    _____      _____
```

LIVESTOCK IN THE DISTRICT

111. Do most farmers have livestock?
```
     rainfed           supplemental irrigation       water harvesting
     Yes ( ) 1         Yes ( ) 1                     Yes ( ) 1
     No  ( ) 2         No  ( ) 2                     No  ( ) 2
```
112. What is the number of: rainfed   sup. irri.   water harv.
```
              sheep?  ....    _____      _____       _____
              goats?  ....    _____      _____       _____
              cattle? ....    _____      _____       _____
              camels? ....    _____      _____       _____
              horses? ....    _____      _____       _____
              donkeys? ....   _____      _____       _____
     other _____?  ....     _____      _____       _____
```
113. Do variations in feed supplies occur?
```
     rainfed           supplemental irrigation       water harvesting
     Yes ( ) 1         Yes ( ) 1                     Yes ( ) 1
     No  ( ) 2         No  ( ) 2                     No  ( ) 2
```
114a. Are crop residues grazed by livestock?
```
     rainfed           supplemental irrigation       water harvesting
     Yes ( ) 1         Yes ( ) 1                     Yes ( ) 1
     No  ( ) 2         No  ( ) 2                     No  ( ) 2
```
b. If Yes, are residues grazed by livestock:
```
                              rainfed      sup. irri.    water harv.
     inside district?  ....   ( ) 1        ( ) 1         ( ) 1
     outside district? ....   ( ) 2        ( ) 2         ( ) 2
     other _____?  ....     ( ) 3        ( ) 3         ( ) 3
```
114c. If No, how are crop residues disposed of?
(rank order 1 thru 7 for each farming system)
```
                              rainfed      sup. irri.    water harv.
              baled     ....  _____      _____       _____
              sacked    ....  _____      _____       _____
     removed by trailer ....  _____      _____       _____
     left in field, mulch ... _____      _____       _____
              burned, field.. _____      _____       _____
              burned, fuel .. _____      _____       _____
     other _____ ....       _____      _____       _____
```
115. What is the average cost for grazing (MU/LU):
```
                              rainfed      sup. irri.    water harv.
     pasture and forage? .... _____      _____       _____
     hay and straw?      .... _____      _____       _____
     cereal crops?       .... _____      _____       _____
     other _____?      .... _____      _____       _____
```
116a. Do farmers plant feed for livestock?
```
     rainfed           supplemental irrigation       water harvesting
     Yes ( ) 1         Yes ( ) 1                     Yes ( ) 1
     No  ( ) 2         No  ( ) 2                     No  ( ) 2
```
b. If yes, what is the major type of feed produced?
(rank order 1 thru 4 for each farming system)
```
                              rainfed      sup. irri.    water harv.
     pasture and forage ....  _____      _____       _____
     hay and straw      ....  _____      _____       _____
     cereal grains      ....  _____      _____       _____
     others _____     ....  _____      _____       _____
```
117. Percent contribution of feed type to total requirement?
```
                              rainfed      sup. irri.    water harv.
     cereal straw    ....     _____%     _____%      _____%
     legume hay      ....     _____%     _____%      _____%
     crop residues   ....     _____%     _____%      _____%
     concentrates    ....     _____%     _____%      _____%
     natural pastures ....    _____%     _____%      _____%
     other _____   ....     _____%     _____%      _____%
```
118. What percent of feed is purchased from off-farm sources?
```
                              rainfed      sup. irri.    water harv.
     cereals straw   ....     _____%     _____%      _____%
     legume hay      ....     _____%     _____%      _____%
     crop residues   ....     _____%     _____%      _____%
     concentrates    ....     _____%     _____%      _____%
     natural pastures ....    _____%     _____%      _____%
     other _____   ....     _____%     _____%      _____%
```
119a. Are flocks of the district moved? .... Yes ( ) 1
                                                        No ( ) 2
b. If Yes, when do moves occur? .... rainfed _____
                                supplemental irrigation _____

Evaluation of Supplemental Irrigation

119c. If Yes, where is the major location that flocks are moved?
(rank order 1 thru 4 for each farming system)

	rainfed	sup. irri.	water harv.
to the steppe	____	____	____
to wetter areas	____	____	____
remain in-place?	____	____	____
other _____?	____	____	____

SOCIO-ECONOMIC FACTORS IN THE DISTRICT

120. How are winter crops marketed?
(mark first (1) and second (2) choice according to priority)

	government	private
wheat	( )	( )
barley	( )	( )
faba bean	( )	( )
lentil	( )	( )
chickpea	( )	( )
forages	( )	( )
major crop-1	( )	( )
major crop-2	( )	( )
major crop-3	( )	( )
other (specify) _____	( )	( )

121. For the three major winter crops produced for **rainfed farming**, what are the following **total costs**:

	crop-1	crop-2	crop-3
names of major crops			
a. seed (MU/WU)			
b. fertilizer:			
Nitrogen (MU/WU)			
P_2O_5 (MU/WU)			
organic (MU/WU)			
other			
c. pesticide (MU/VU)			
d. pesticide rate: (VU/LU)			
e. herbicide (MU/VU)			
f. herbicide rate: (VU/LU)			
g. seedbed preparation: (MU/LU)			
h. sowing (MU/LU)			
i. weeding (MU/LU)			
j. harvest (MU/LU)			
k. threshing (MU/WU)			
l. transportation (MU/WU)			
m. storage (MU/WU)			
n. average yield (WU/LU)			
o. minimum yield (WU/LU)			
p. maximum yield (WU/LU)			
q. current market price: (MU/WU)			

122. For the three major winter crops produced for **supplemental irrigation farming**, what are the following **total costs**:

	crop-1	crop-2	crop-3
names of major crops			
a. seed (MU/WU)			
b. fertilizer:			
Nitrogen (MU/WU)			
P_2O_5 (MU/WU)			
organic (MU/WU)			
other			
c. pesticide (MU/VU)			
d. pesticide rate: (VU/LU)			
e. herbicide (MU/VU)			
f. herbicide rate: (VU/LU)			
g. seedbed preparation: (MU/LU)			
h. sowing (MU/LU)			
i. weeding (MU/LU)			
j. irrigation (MU/LU)			
k. harvest (MU/LU)			
l. threshing (MU/WU)			
m. transportation (MU/WU)			
n. storage (MU/WU)			
o. average yield (WU/LU)			
p. minimum yield (WU/LU)			
q. maximum yield (WU/LU)			
r. current market price: (MU/WU)			

123. For the three major winter crops produced for **water harvesting farming**, what are the following **total costs**:

	crop-1	crop-2	crop-3
names of major crops			
a. seed (MU/WU)	.. _____	_____	_____
b. fertilizer:			
Nitrogen (MU/WU)	.. _____	_____	_____
P_2O_5 (MU/WU)	.. _____	_____	_____
organic (MU/WU)	.. _____	_____	_____
other	.. _____	_____	_____
c. pesticide (MU/VU)	.. _____	_____	_____
d. pesticide rate: (VU/LU)	.. _____	_____	_____
e. herbicide (MU/VU)	.. _____	_____	_____
f. herbicide rate: (VU/LU)	.. _____	_____	_____
g. seedbed preparation: (MU/LU)	.. _____	_____	_____
h. sowing (MU/LU)	.. _____	_____	_____
i. weeding (MU/LU)	.. _____	_____	_____
j. harvest (MU/LU)	.. _____	_____	_____
k. threshing (MU/WU)	.. _____	_____	_____
l. transportation (MU/WU)	.. _____	_____	_____
m. storage (MU/WU)	.. _____	_____	_____
n. average yield (WU/LU)	.. _____	_____	_____
o. minimum yield (WU/LU)	.. _____	_____	_____
p. maximum yield (WU/LU)	.. _____	_____	_____
q. current market price: (MU/WU)	.. _____	_____	_____

124. What percentage of total farm income is from:

	rainfed	sup. irri.	water harv.
winter crops?	_____%	_____%	_____%
tree crops?	_____%	_____%	_____%
vegetable crops?	_____%	_____%	_____%
livestock?	_____%	_____%	_____%
other _____?	_____%	_____%	_____%

125a. What percentage of farmer's total income is from:

	rainfed	sup. irri.	water harv.
farm production?	_____%	_____%	_____%
off-farm, agriculture?	_____%	_____%	_____%
non-agriculture?	_____%	_____%	_____%

b. What percentage of farmers rely on farm production for total income? ....
   rainfed _____%
   supplemental irrigation _____%
   water harvesting _____%

126. What are the major agronomic constraints to farmers' income?
     (rank order 1 thru 9 for each farming system)

	rainfed	sup. irri.	water harv.
seedbed preparation	_____	_____	_____
fertilization	_____	_____	_____
sowing	_____	_____	_____
cultivation	_____	_____	_____
weeding	_____	_____	_____
pest/disease control	_____	_____	_____
harvesting	_____	_____	_____
storage	_____	_____	_____
other _____	_____	_____	_____

127. What are the financial arrangements for sharecropping?
   rainfed _____
   supplemental irrigation _____
   water harvesting _____

128. What percentage of hired labor comes from:

	rainfed	sup. irri.	water harv.
within district?	_____%	_____%	_____%
outside district?	_____%	_____%	_____%
outside country?	_____%	_____%	_____%

129. What percentage of labor requirements are provided by:

	rainfed	sup. irri.	water harv.
family?	_____%	_____%	_____%
hired labor?	_____%	_____%	_____%
sharecropping?	_____%	_____%	_____%
free labor?	_____%	_____%	_____%
mechanization?	_____%	_____%	_____%
animal power?	_____%	_____%	_____%
other _____?	_____%	_____%	_____%

130a. When using hired labor, what is the major problem?
      (for a and b, check one choice for each farming system)

	rainfed	sup. irri.	water harv.
no problem	( ) 1	( ) 1	( ) 1
timing conflict	( ) 2	( ) 2	( ) 2
labor supply	( ) 3	( ) 3	( ) 3
cost	( ) 4	( ) 4	( ) 4
other _____	( ) 5	( ) 5	( ) 5

130b. When using hired labor, what is the major timing conflict?

	rainfed	sup. irri.	water harv.
seedbed preparation	( ) 1	( ) 1	( ) 1
sowing	( ) 2	( ) 2	( ) 2

Evaluation of Supplemental Irrigation

```
                      weeding ....  ( ) 3       ( ) 3       ( ) 3
                   harvesting ....  ( ) 4       ( ) 4       ( ) 4
         other _____...  ( ) 5       ( ) 5       ( ) 5
131. Percentage age distribution of hired labor used?
                              less than 15 years ....  _____ %
                              15 thru 30 years ....    _____ %
                              31 thru 65 years ....    _____ %
                              more than 65 years ....  _____ %
```

```
╔══════════════════════════════════════════════════════════════╗
║                         DEFINITIONS                          ║
║ All movements in physical space with the assumption more or  ║
║ less implicit that a change of residence or domicile is      ║
║                  involved is termed migration.               ║
║                         **********                           ║
║ Those migrants who move their activity but not their "usual" ║
║             place of residence are termed temporary.         ║
║                         **********                           ║
║ Those migrants who move their domicile and their activity    ║
║                    are termed permanent.                     ║
╚══════════════════════════════════════════════════════════════╝
```

132a. What type of migration occurs?
(in a, b, and c, rank order 1 thru 4 for each farming system)
```
                                rainfed    sup. irri.   water harv.
             no migration ....  _____    _____    _____
      temporary migration ....  _____    _____    _____
      permanent migration ....  _____    _____    _____
      other _____...  _____    _____    _____
```
b. To where does temporary migration go?
```
                                rainfed    sup. irri.   water harv.
      neighboring district ....  _____    _____    _____
             urban centers ....  _____    _____    _____
           outside country ....  _____    _____    _____
      other _____....  _____    _____    _____
```
c. To where does permanent migration go?
```
                                rainfed    sup. irri.   water harv.
      neighboring district ....  _____    _____    _____
             urban centers ....  _____    _____    _____
           outside country ....  _____    _____    _____
      other _____....  _____    _____    _____
```
133. What are the major reasons for migration?
(rank order according to priority, 1 thru 7)
```
                                    temporary         permanent
              no work ....          _____         _____
    unsatisfactory work ....        _____         _____
    new work opportunities ....     _____         _____
    seek education/training ....    _____         _____
    family commitments ....         _____         _____
           social feuds ....        _____         _____
    other (specify) _____...   _____         _____
```
134a. At what time of year does temporary migration occur?
(rank order 1 thru 4 for each farming system)
```
                                rainfed    sup. irri.   water harv.
                winter ....     _____    _____    _____
                spring ....     _____    _____    _____
                summer ....     _____    _____    _____
                autumn ....     _____    _____    _____
```
134b. How long do most temporary migrants remain away?
(check one choice for each farming system)
```
                                   rainfed    sup. irri.   water harv.
       less than 3 months          ( ) 1      ( ) 1        ( ) 1
       between 3 and 6 months      ( ) 2      ( ) 2        ( ) 2
       between 6 and 9 months      ( ) 3      ( ) 3        ( ) 3
       between 9 and 12 months     ( ) 4      ( ) 4        ( ) 4
       more than 12 months         ( ) 5      ( ) 5        ( ) 5
```
135. During the last five (5) years has the level of migration:
```
                                rainfed    sup. irri.   water harv.
              increased? ....   ( ) 1      ( ) 1        ( ) 1
              decreased? ....   ( ) 2      ( ) 2        ( ) 2
              no change? ....   ( ) 3      ( ) 3        ( ) 3
```
136a. Do permanent migrants still have farms?
```
        rainfed          supplemental irrigation     water harvesting
      Yes ( ) 1               Yes ( ) 1               Yes ( ) 1
      No  ( ) 2               No  ( ) 2               No  ( ) 2
```
b. If Yes, do they send money to invest in farms?
```
        rainfed          supplemental irrigation     water harvesting
      Yes ( ) 1               Yes ( ) 1               Yes ( ) 1
      No  ( ) 2               No  ( ) 2               No  ( ) 2
```
c. If Yes, who currently operates farm?
(rank order 1 thru 4 for each farming system)
```
                                rainfed    sup. irri.   water harv.
                family ....     _____    _____    _____
               tenants ....     _____    _____    _____
```

```
                 sharecroppers ....  _____      _____      _____
    other_____          ....  _____      _____      _____
137a. How many years has the following farm equipment been used?
                          rainfed      sup. irri.   water harv.
               tractor  _____yrs    _____yrs    _____yrs
    combine harvester   _____yrs    _____yrs    _____yrs
          seed drills   _____yrs    _____yrs    _____yrs
   fertilizer spinners  _____yrs    _____yrs    _____yrs
 pest and weed sprayers _____yrs    _____yrs    _____yrs
 other _____        _____yrs    _____yrs    _____yrs
137b. What percent of the following groups own tractor equipment?
                          rainfed      sup. irri.   water harv.
               farmers ....  _____%     _____%     _____%
      machine operators ....  _____%     _____%     _____%
          cooperatives ....  _____%     _____%     _____%
    other_____         ....  _____%     _____%     _____%
138. What are the major constraints to the farmer for management
     of supplemental irrigation?
                (rank order according to priority, 1 thru 7)
                             level of literacy   ....  _____
                             farming knowledge   ....  _____
                           irrigation training   ....  _____
                            years of experience  ....  _____
                               farmer's attitude ....  _____
                             tradition and habit ....  _____
           other (specify) _____ ....  _____
```

Technical Data Collection

When assessing an irrigation system, certain features should be considered and investigated before selection of an irrigation method and purchase of equipment. The initial data set should be retained and the data items integrated into the record keeping system for continuous evaluation. Some features require daily monitoring while others need only periodic assessment or auditing. The following categories are suggested for assessment and example forms are included:

1. power requirement,
2. initial investment,
3. availability and cost of energy and fuel,
4. depreciation of equipment,
5. dependability of method,
6. portability requirement of method,
7. maintenance and convenience of operation, and
8. labor availability and skill level.

Evaluation of Supplemental Irrigation

Pump and power requirement

1. Make, model _____
2. Cubic centimeters displacement _____
3. Stroke in centimeters _____
4. Rpm @ design load _____
5. Piston speed @ design load _____
6. Type of power conversion _____
7. Speed rate _____
8. Electric motor _____ hp; _____ phase; _____ rpm _____

Irrigation pump specifications

1. Make, model, and size _____
2. Impeller diameter _____
3. RPM @ design load _____
4. Efficiency _____ %
5. Required brake horsepower @ design load _____
6. Shutoff head _____ meters _____
7. Required brake horsepower @ minimum design load _____
8. Rpm @ minimum load _____

Figure 15.2. Power plant characteristics and irrigation pump specifications.

Data items	Field number and area (ha)					Design standards
	1	2	3	4	5	
1. Application rate (mm/hr)						
2. Time/set (hr)						
3. Settings/day						
4. Days of operation/interval						
5. Method's capacity (m^3/hr) (preliminary)						

Figure 15.3. Form for estimating irrigation capacity.

Energy and fuel requirements: To assist in cost evaluation of energy for power plants to pump irrigation water, Table 15.1 presents estimated fuel requirements for shallow lifts. Record the type of energy and the unit cost on farm and update on a regular basis.

Table 15.1. Estimated fuel requirements for pumping plant.

Flow of water (m³/hr)	Lift (m)	Horsepower (hp)	Fuel required (l/hr)	
			Diesel	Gasoline
100	20	7.5	2.7	3.5
	50	18.5	6.2	8.5
	70	26.0	9.0	11.7
150	20	11	3.7	5.2
	50	28	9.5	13.0
	70	39	13.5	18.2
200	20	15	5.2	6.7
	50	37	12.5	16.5
	70	52	17.7	23.5
250	20	19	6.5	8.5
	50	46.5	16.0	21.0
	70	65	22.2	20.2

Discharge pressure and lift: If there is a pressure at the pump discharge, add the number of meters shown in Table 15.2 to the distance the water must be lifted when pumping (see Table 15.1).

Table 15.2. Relationship between discharge pressure and lift.

Discharge pressure (bars)	Equivalent "lift" (m)
1	10.3
2	20.7
3	31.0
4	41.4
5	51.7
6	62.0
7	72.4

Horsepower requirements: To compute the theoretical horsepower (100% efficiency) required to lift different quantities of water to different elevations compute:

$$hp = \frac{Q \times H}{273},$$

Evaluation of Supplemental Irrigation

where hp = horsepower,
  Q = discharge (m³/hr),
  H = vertical lift (m).

Table 15.3 presents the required horsepower for 50% efficiency of theoretical hp (100%). These values are used for estimating required horsepower under normal field conditions. To determine the horsepower required for lifting a stream of water of any size a given elevation multiply the observed value for 100 m³/hr by the output selected. For example, to lift a flow of 100 m³/hr from a depth of 60 m would require 44.0 hp. Therefore, 200 m³/hr would require about 88.0 hp (87.9 hp) for a 60 m lift.

Table 15.3. Horsepower required to lift different quantities of water to elevations of 10-90 m.

Discharge		Horsepower (hp) required for elevations (m)								
m³/hr	l/sec	10	20	30	40	50	60	70	80	90
10	2.78	0.7	1.5	2.2	2.9	3.7	4.4	5.1	5.9	6.6
20	5.56	1.5	2.9	4.4	5.9	7.3	8.8	10.3	11.7	13.2
30	8.34	2.2	4.4	6.6	8.8	11.0	13.2	15.4	17.6	19.8
40	11.1	2.9	5.9	8.8	11.7	14.7	17.6	20.5	23.4	26.4
50	13.9	3.7	7.3	11.0	14.7	18.3	22.0	25.6	29.3	33.0
60	16.7	4.4	8.8	13.2	17.6	22.0	26.4	30.8	35.2	39.6
70	19.5	5.1	10.3	15.4	20.5	25.6	30.8	35.9	41.0	46.2
80	22.2	5.9	11.7	17.6	23.4	29.3	35.2	41.0	46.9	52.7
90	25.0	6.6	13.2	19.8	26.4	33.0	39.6	46.2	52.4	59.3
100	27.8	7.3	14.7	22.0	29.3	36.6	44.0	51.3	58.6	65.9
125	34.8	9.2	18.3	27.5	36.6	45.8	54.9	64.1	73.3	82.4
150	41.7	11.0	22.0	33.0	44.0	54.9	65.9	76.9	87.9	98.9
175	48.6	12.8	25.6	38.5	51.3	64.1	76.9	98.7	102.6	115.4
200	55.6	14.7	29.3	44.0	58.6	73.3	87.9	102.6	117.2	131.9
250	69.5	18.3	36.6	54.9	73.3	91.6	109.9	128.2	146.5	164.8
300	83.4	22.0	44.0	65.9	87.9	109.9	139.9	153.8	175.8	197.8

Costs for power plant operation: To determine costs of pumping irrigation water, the following factors should be included (Figure 15.4):
1. interest on capital investment, i.e. on first cost;
2. taxes;
3. depreciation on irrigation equipment;
4. power (fuel and lubricating oils);
5. water;
6. repair, operation, and maintenance; and,
7. labor.

Cost of power plant: $

    Well...
    Pump ....
    Engine (installed) ....
    Shelter...
                                                     Subtotal No. 1

Annual fixed costs:

    Interest on subtotal No. 1 @ ____ % ....
    Taxes (estimated) ....
    Depreciation on engine @ ____ %............................
    Depreciation on pump @ ____ % ....
    Depreciation on well and shelter @____ % ....
                                                       Subtotal No. 2

Annual operating costs (_____ hr operation/yr)

    Fuel (_____ l @ _____ $/l) ....
    Lubrication @ _____ $/100 hr ....
    Tax on fuel & lubrication @ ____ %..........................
    Repairs @ _____ $/100 hr..................................
    Attendance @ _____ $/100 hr ....
                                                       Subtotal No. 3

                      Sum of subtotal No. 2 and No. 3

Total cost per m³ of water.......................................
Operating cost per m³ of water..................................

Figure 15.4. Cost sheet for pumping plant operation.

Energy cost: The energy cost from a well includes all pumping costs of the deep well pumps and any booster pumps used. If return flow pumps are used, those energy costs should be apportioned among the various water supplies. The energy cost should only include energy consumed on the farm itself; energy costs incurred by the irrigation authorities in conveying water to the farm are not included.

Cost of energy used can estimate the benefit of minimizing pressure losses in the system, e.g. pressure throttling with a gate valve to reduce volume flow. The%/kw-hr is the unit price paid to the electrical power authority.

Water requirement and annual cost determinations

To calculate the water requirement:

$$Q \text{ (m}^3\text{/hr)} = \frac{\text{area (ha)} \times \text{precipitation (mm)} \times 10}{\text{days of water cycle} \times \text{daily operating hours}}.$$

Evaluation of Supplemental Irrigation

For example, if the required amount of water, Q, for sprinkling an area of 10 ha were assigned a precipitation value of 25 mm and a watering cycle of 14 days operating at 10 hr/day, the water requirement, Q, would be:

$$\frac{10 \times 25 \times 10}{14 \times 10} = 17.9 \text{ m}^3/\text{hr}.$$

Water supply: Semi-arid and arid regions may adjust irrigation practices on the basis of reliable information concerning alternative water supplies. This information should be collected for each source of supply. Figure 15.5 is an example of the data items necessary for evaluation.

---

1. Source _____
2. Estimated or measured quantity _____ m³/hr
3. Quality of water _____
   (good, fair, poor: verify by test)
4. Delivery schedule _____
5. Seasonal variation in quantity _____ m³/hr to _____ m³/hr
6. Pressure available @ source _____ bars
7. Well data: total depth _____ m; diam _____ cm

    water bearing material _____

    water found @ _____ m to _____ m and

    _____ m to _____ m

    -

    test: no of hours _____ by _____

    static W.L.: @ _____ m

    drawdown: _____ m @ _____ m³/hr

    _____ m @ _____ m³/hr

    _____ m @ _____ m³/hr

Figure 15.5. Water supply.

Test pumping of wells: Test pumping is probably the most accurate means to measure yield of a well. Water is pumped from the well for several hours, i.e. pumping is recommended for 24 hr, at a rate higher than the calculated daily demand. Using this technique, the pumping characteristics of the well can be determined. Every hour, pumped water is collected in containers, usually for 5 min. If 95 l are collected during this time (19 l/min), this number multiplied by 12 will be the well's per hour yield (95 × 60/5 = 1,140 l/hr). Averaging these values over several hours will give an accurate accounting of the pumping rate.

Drawdown for shallow wells is measured using rulers or floats. In a deeper well, drawdown, an indication of the recharge capacity of a well, can be measured with a float and weight or a pressure or vacuum gauge. A column of water 70.2 cm high will generate 0.45 kg/cm^2 of pressure. The calibrated pressure difference in a well, before and after pumping, multiplied by 70.2 cm is equal to drawdown.

Drawdown can be checked if there is room to drop a weighted float on a string into the well as shown in Figure 15.6. Proceed as follows:
1. lower a weighted float on a string until it reaches the water level;
2. tie a knot in the string at the water surface level before starting the pump;
3. repeat steps 1 and 2 at end of the pumping period; and,
4. measure the difference between the upper and lower knots.

The distance between the 2 knots represents the drawdown during the pumping period. If your well is too small in diameter to check drawdown with a string and float, you can use another type of water-level indicator.

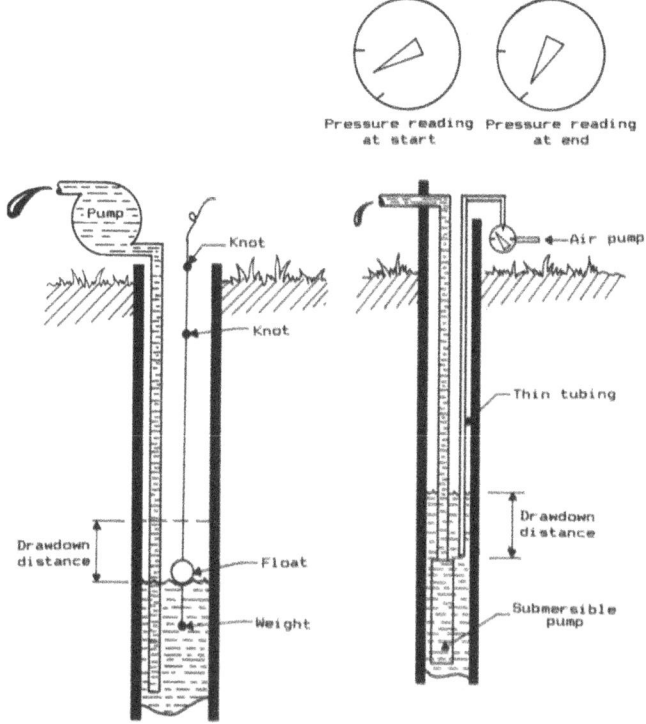

Figure 15.6. Techniques for measuring drawdown in wells.

The pressure gauge technique (see Figure 15.6) consists of plastic or copper tubing, about 0.3-0.5 cm in diameter, attached to a low pressure gauge, 15-30 psi (1-2 kg/cm^2), and an air pump. The end of the tube should not be closer than

Evaluation of Supplemental Irrigation

50 cm above the intake of the pump and never below the intake. If the end of the tube is too close to the intake of the submersible pump, the pressure gauge reading will be incorrect (too low) because it will measure a portion of the pump suction. If the pressure gauge pointer returns to 0, there may be a small leak in the tube or at the connections. Check for leaks and then replenish the air until the pressure gauge pointer remains constant. Proceed as follows:

1. drop the open end of the tube into the well to a position below the water level;
2. pump air into the tube until the pressure gauge reading remains constant;
3. record pressure gauge reading (for example, 15 psi);
4. start the pump;
5. check pressure gauge reading at end of pumping but while the pump is still running (for example, 10 psi);
6. subtract the lower reading from the higher reading (for example, 5 psi); and,
7. if the gauge reads in psi (lbs/in^2), then multiply the reading by 2.3.

A column of water that is 2.3 ft (5.84 cm) high will develop a pressure of 1 psi. Consequently, by multiplying 2.3 × 5 psi (example from step 6), the total number of feet of drawdown is obtained.

reading, start of pumping . . . .   15 psi
reading, end of pumping . . . .   10 psi
                  difference . . . .   5 psi

5 psi × 2.3 (ft/lb)   = 11.5 ft drawdown
11.5 ft × 30.5 cm/ft   = 351 cm of drawdown

Another technique uses a sink stopper on a string lowered into the well; when the stopper comes in contact with the water surface, it makes a popping sound. The pop sound is at the water surface with the length of the line plus stopper equaling the depth to water table. The concave portion of the sink stopper makes the pop when coming in contact with the water surface and the sound travels up the well casing. If surface noise drowns out the sound, a bare electrical wire can be made as a switch to turn on a low voltage flashlight bulb or electrical meter when in contact with the water surface (water is a good conductor). The electric wire is marked for each meter of length, put on a large fishing reel, and lowered into the well with a lead weight at the bottom to keep the line taut.

Water quality: The average electrical conductivity, EC, should be measured for the seasonal water supply. The units are mmhos/cm or deci-siemens/m (1 dS/m = 1 mmhos/cm). Typical values range from 0.3-1.5 mmhos/cm.

Water delivery: The total water applied (m^3) should be measured on a seasonal basis. The water applied equals the net water supplied specifically to a field. Runoff which is collected and reused on other fields must be subtracted.

Water Applied = Water onto field − Reused runoff from field

The applied water values are important in all calculations. If these cannot be determined or estimated reasonably well, results of the evaluation will be unreliable. Values for total water applied are often difficult to determine because there may be no flow meters available or a water source may supply several fields simultaneously (see Chapter 12).

In cases of obvious gross over-irrigation, e.g. a farmer applies 240 mm when only 90 mm is needed, reasonable estimates would provide information to identify the problem of scheduling and water control. The effectiveness of an evaluation would be minimal without information of the total water supply. Reasonable estimates cannot be made of the potential saving of water and energy from field measurements of a single irrigation event. The total water applied should be estimated subtracting water lost to evapotranspiration and leaching (m^3). This value is not in addition to values previously discussed. Some water may be used for special purposes and the extent of use should be identified. Some of the water used for these special purposes may eventually be stored in the root zone for evapotranspiration and leaching later.

1. Excessive pre-irrigation (m^3) should be measured; it can be a major source of deep percolation on many farms.
2. Water used for frost protection (m^3) should be measured. The principle of frost protection by supplemental irrigation is simple but equipment, management, and wind speed play a significant role.
3. Irrigation for weed germination (m^3) should be measured. Farmers may use a light irrigation to germinate weeds and then disc or spray with a herbicide.
4. Irrigation for soil conditioning (m^3) should be measured. Some crops or harvesting techniques require a special irrigation to prepare the soil. For example, the soil may be water-tamped to make a hard surface prior to shaking almond trees for harvest.
5. Irrigation for temperature reduction (m^3) should be measured. Crops such as apples and potatoes may need extra water applications to keep them cool. These applications may be necessary to produce a marketable crop.

Water cost (excluding pumping): If an irrigation authority charges $60/ha and the annual water delivered per hectare is 4,000 m^3, i.e. 400 mm = 0.4 m, 1 ha = 10,000 m^2, then 0.4 m × 10,000 m^2 = 4,000 m^3. The value entered for the irrigation authority cost should be $0.015/$m^3$, i.e. $60/4,000 = $0.015/$m^3$. Many farmers do not assign a water cost to well water but this cost should be determined for each m^3/yr to allow for equipment maintenance.

Irrigation operations

1. The irrigation interval (days) should be recorded during the peak evapotranspiration as well as the gross depth of water applied per irrigation, GDWAI (mm), during the peak evapotranspiration period. The gross depth of water applied includes all losses. One formula which can be used is:

Evaluation of Supplemental Irrigation

$$\text{GDWAI (mm)} = \frac{\text{flow rate (m}^3/\text{min)} \times 60{,}000 \times \text{application time (hr)}}{\text{field area (m}^2)},$$

2. Conveyance losses between source and field should be measured to determine percent gross water applied. This value may be negligible, as in the case of a pipeline system, or 10-40% on a long unlined ditch. No specific procedure exists for estimating the conveyance losses. In the case of an unlined ditch, a flume or weir should be set up at the water source and another flume or weir set up where the water supply reaches the pipeline or individual furrows (see Chapter 10). The difference in flow rates between the 2 flumes or weirs will give conveyance loss which is then converted to a percent.
3. Distribution uniformity (%DU) should be measured for each irrigation system. Individual methods may have slightly different names for DU, such as emission uniformity (EU) for micro-irrigation. This measure is a parameter of the irrigation system rather than the value of distribution uniformity down a single furrow or from can catchment on a single lateral of sprinklers.
4. Measure the amount of field runoff (% of runoff that reaches field) which has not been reused on another field.

Surface runoff: The surface runoff is the total runoff from all sets evaluated and any spills that might occur at the head ditch. Irrigation runoff includes water from the ends of the furrows and any spills that might occur at the field head ditch. Runoff collected and reused on the farm is not a loss to the system. For evaluation of surface runoff, the following questions (with yes or no responses) and measurements should be recorded.
1. Is there irrigation water runoff? (Y/N)
2. Is the runoff collected for reuse on the farm? (Y/N)
3. Measure runoff not collected for reuse and calculate the percent of water applied.
4. Are border strip ends blocked? (Y/N)
5. Is water prevented from draining freely from the border strips? (Y/N)

Annual irrigation requirement and system performance: Figures 15.7 and 15.8 should be completed on an annual basis for each farm unit. These records should be maintained from year to year for comparative purposes.

Soil data: Basic soils data should be completed on each field to ensure efficient irrigation management. Figure 15.9 presents the data items for collection. Additional information which should be collected follows.
1. Record the percent of cropped area for each soil type.
2. Determine the available water capacity (mm) in the total active root zone.
3. Measure the depth of the active root zone for the crop on the specific soil type. Actual root zone depths may be affected by conditions of poor drainage or hard pans.

	Field number and area (ha)					Design standards
Data items	1	2	3	4	5	
1. Net seasonal req'd moisture (mm)						
2. Estimated effective stored moisture (mm)						
3. Estimated effective rainfall (dry yr)						
4. Estimated net irrigation requirement (mm)						
5. Gross irrigation requirement (mm)						
6. Maximum number of irrigations req'd						
7. Minimum number of irrigations req'd						

Figure 15.7. Annual irrigation requirements.

Irrigation efficiency (%):

Irrigation water beneficially used (consumptive use or crop evapotranspiration) × 100 / irrigation applied = _____ :

Water requirements and use (mm):
   Water applied, irrigation:
   Effective rain (Chapter 3):
   Total water applied:
   Water used, evapotranspiration + leaching:
                other beneficial use:
                losses, non-beneficial:
   Total water used:
Under-irrigation (evapotranspiration + leaching):

Annual energy costs + water costs for the field: _____ $
Annual water used: _____ $
Annual water losses;
   uncollected runoff: _____ $
   conveyance: _____ $
   deep percolation, non-uniformity: _____ $
   deep percolation, scheduling: _____ $
   spray drift and evaporation: _____ $
                        Total value of losses: _____ $
                        Total annual costs: _____ $

Figure 15.8. Irrigation system performance.

Evaluation of Supplemental Irrigation

4. Measure the available water capacities (AWC) of the various soil horizons within the active root zone.

These variables can be estimated by examining the soil texture at various depths from an augered hole. For example, assume a wheat crop with an estimated 150 cm active root zone. A hole is augered in soil no. 1 and the texture is noted for each depth of the horizon. The water holding capacity (WHC) is estimated or the soil moisture characteristic curve is determined in a laboratory from a soil sample. For this example, AWC is 450 mm × 145 mm/1,000 mm/m = 65 mm. Continuing down the horizon gives the following values:

Depth (mm)	Texture	Est. WHC (mm/m)	AWC (mm)
0 – 450	Silt loam	145	65
450 – 760	Loam	133	41
760 – 1070	Clay loam	167	52
1070 – 1520	Loam	133	60
		Total AWC of soil no. 1 =	218

	Field number and area (ha)					Design standards
Data items	1	2	3	4	5	
1. Surface texture						
2. Subsoil texture						
3. Effective depth of soil (cm)						
4. Moisture-holding capacity (mm/m)						
5. Intake rate (mm/hr)						
6. Soil limitations (depth, hardpan, gravel, fertility, drainage, etc.)						

Figure 15.9. Basic soils data.

Evaluation of Specific Methods of Irrigation

Forms are presented for field evaluation of 4 methods of irrigation. These are notes or guidelines for evaluating the efficiency and effectiveness of the design, maintenance, and operation of specific irrigation systems. Data should be routinely collected.

Furrow Irrigation

The furrow method is effective when water flows uniformly along the run and if water movement within the soil profile is horizontal into the plant root zone, i.e. not a sandy soil with a high infiltration rate. Compared to other surface irrigation methods, there is less exposed water surface, and therefore, less evaporative loss from furrows; the risk of puddling in clayey soils is reduced, and equipment can work in the field sooner after the end of water application. For each field, identify the type of conveyance system used to deliver water from source of supply to the field. A system is considered low pressure if it has a pressure of 1 bar (15 psi) or less. A high pressure system has a pressure of greater than 1 bar.
1. Select the type of furrow design for the existing field: sloping, level, or sloping with level or blocked ends.
2. Identify the type of conveyance which delivers water to field, for example, head ditch (siphons, cuts, or tubes), low pressure pipe, or high pressure pipe.

Pumping plant measurement: Enter the total loss across the manual valve. Any pressure drop or reduction because of a throttled manual valve is considered a loss even though there may be a logical reason for regulating the pressure or flow. An automatic control valve is designed to provide slow opening and closing, pressure regulation, or check irrigation operation. Enter the total loss across the valve, i.e. the difference between the inlet and outlet pressures of the valve. Pressure losses greater than 1/3rd bar (5 psi) are considered excessive.
1. Measure pressure loss across throttled manual valves (bar or psi).
2. Measure pressure loss across automatic control valves (bar or psi).

Field observations: Field observations should be answered with a yes or no response. Crop height differences are indications of non-uniform water application.
1. Are there differences in crop height along the furrow? (Y/N)
2. Is there poor surface drainage at the lower end of the furrows? (Y/N)
3. Within the same furrow set, is water advancing at different rates (uneven furrow advance)? (Y/N)
4. Does variation in water delivery flow rate go beyond the irrigator's easy control? (Y/N)
5. Is the infiltration rate limiting (water penetration problems) at end of season? (Y/N)

Set times: A maximum of 4 set times are used to describe a typical 12-24 hr irrigation cycle, including changes made as a result of starting additional furrows. For example, a 10-11 hr set is used during the daylight hours and a 12-14 hr set is used during the night time hours. The total hours for water to travel the entire length of furrow includes travel time through level sections at the end of the furrow. The same data items are collected for each irrigation set.

Evaluation of Supplemental Irrigation

1. Record total time (hr) water is applied to the set.
2. Record total time (hr) that water is applied to the furrow.
3. Record time (hr) required to reach end of furrow.
4. Compute average distribution uniformity (%DU).
5. What percent of the furrow length is affected by blocked ends or leveled sections?
   For example, the total furrow length is 50 m, the length of the section where water stands because of the level or blocked end section is 5 m. The percent of length affected equals 5/50 × 100 = 10%.
6. What is the average time (hr) water stands in the level or blocked section of the furrow?

Techniques to determine when to stop water application: More than 1 method can be used to determine when to shut the water off. The method used should be noted.
1. Measure water applied minus measured runoff? (Y/N)
2. Measure water applied minus estimated runoff? (Y/N)
3. Probe soil during irrigation for depth of water penetration? (Y/N)
4. Soil moisture measurements after the first set? (Y/N)
5. Estimate soil moisture, no measurement? (Y/N)
6. Measure applied water with runoff returned to the same set? (Y/N)
7. When beds receive water uniformly across field? (Y/N)
8. When water has reached the lower end of the furrow? (Y/N)
9. At the end of a 12 hr or 24 hr set? (Y/N)

Assessment of irrigation problems in the field, rate from 0 to 3: This is an assessment of problems observed in the field while conducting the evaluation. Assign the most suitable rating to the specific problem:

0 – no problem or no visible effect to method,
1 – visible problem but little effect to method,
2 – definite problem, affects performance of method, or
3 – severe problem, dysfunctional to performance of method.

Problem 1, furrow operation: High and low spots are indications of non-uniform advance rates; water remains standing in part of the furrow after irrigation; and, there can be large variations in the top width of the open water surface along the furrow.

    High and low spots in the furrow _____.

Problem 2, furrow erosion: Indicated by clarity of the water, vertical cuts on sides of furrow, and deposition of eroded material in furrow or drain ditches at end of field.

    Erosion at head of furrow _____.

Erosion along furrow _____.

Erosion at end of furrow or field _____.

Buried pipe risers are operating at full capacity with riser valve in the full open position _____.

Excessive pressure caused by minor valve or gate change results in large variations of flow rate with soil erosion at pipe outlet _____.

Furrows flowing at full capacity, i.e. water is at or near top of furrow _____.

Furrows do not maintain their shape during irrigation due to instability of the soil structure _____.

Cracks or lack of furrow capacity allow water to crossover to adjacent furrows when irrigating alternate furrows _____.

Graded Border Strip Irrigation

Description of graded border strip method:
For each field, identify the type of conveyance system used to deliver water from source of supply to the field. A system is considered low pressure if it has a pressure of 1 bar (15 psi) or less. A high pressure system has a pressure of greater than 1 bar.
   Identify the type of conveyance which delivers water to field, for example, head ditch (siphons, cuts, or tubes), low pressure pipe, or high pressure pipe.

Pumping plant measurement: Enter the total loss across the manual valve. Any pressure drop or reduction because of a throttled manual valve is considered a loss even though there may be a logical reason for regulating the pressure or flow. An automatic control valve is designed to provide slow opening and closing, pressure regulation, or check irrigation operation. Enter the total loss across the valve, i.e. the difference between the inlet and outlet pressures of the valve. Pressure losses greater than 1/3rd bar (5 psi) are considered excessive.
1. Measure pressure loss across throttled manual valves (bar or psi).
2. Measure pressure loss across automatic control valves (bar or psi).

Field observations: Field observations should be answered with a yes or no response. Crop height differences are indications of non-uniform water application.
1. Are there differences in crop height along the border strip? (Y/N)
2. Is there poor surface drainage at the lower end of the border strip? (Y/N)

Evaluation of Supplemental Irrigation 265

3. Within the same border strip, is water advancing at different rates? (Y/N)
4. Does variation in water delivery flow rate go beyond the irrigator's easy control? (Y/N)
5. Is the infiltration rate limiting (water penetration problems) at end of season? (Y/N)

Set times: A maximum of 2 set times can be used to describe 1 complete irrigation cycle. Time measurements are cumulative time in minutes. Time 0 for all points is when the set begins, i.e. irrigation has started. The advance time and recession time are to be measured at 5 locations along the border strip: at 0%, 25%, 50%, 75%, and 100% of the border strip length. The data items are to be collected for each set. Data for a second set should be collected to determine if day-time and night-time sets are significantly different.
1. The total length of border strip should be noted.

Location of measurement as % of total length				
0%	25%	50%	75%	100%

2. Denote the time (min) irrigation water reaches each of the 5 locations (advance time).
3. Record time (min) when water no longer remains on soil surface at each of the 5 locations (recession time).
4. Record time (min) that water is no longer being applied to the border strip.

Techniques to determine when to stop water application: More than 1 method can be used to determine when to shut the water off. Select the method that most nearly describes the strategy used to decide when to turn off the irrigation water.
1. Measure water applied minus measured runoff? (Y/N)
2. Measure water applied minus estimated runoff? (Y/N)
3. Probe soil during irrigation for depth of water penetration? (Y/N)
4. Soil moisture measurements after the first set? (Y/N)
5. Estimate soil moisture, no measurement? (Y/N)
6. Measure applied water with runoff returned to the same set? (Y/N)
7. When basins receive water uniformly across field? (Y/N)
8. When water has reached the lower end of the basin? (Y/N)
9. At the end of a 12 hr or 24 hr set? (Y/N)

Assessment of irrigation problems in the field, rate from 0 to 3: This is an assessment of problems observed in the field while conducting the evaluation. Assign the most suitable rating to the specific problem:
0 – no problem or no visible effect to method,
1 – visible problem but little effect to method,
2 – definite problem, affects performance of method, or
3 – severe problem, dysfunctional to performance of method.

Problem 1, border strip operation: High and low spots or uneven grade are indications of non-uniform advance rates in the border strip; water remains standing in part of the basin after irrigation.

High and low spots in the border strip _____.

Excessive cross slope in the border strip _____.

Problem 2, border strip erosion: Indicated by clarity of the water, and deposition of eroded material in border strip or drain ditches at end of field.

Erosion at head of border strip _____.

Erosion along border strip _____.

Erosion at end of border strip or field _____.

Buried pipe risers are operating at full capacity with riser valve in the full open position _____.

Excessive pressure caused by minor valve or gate change results in large variations of flow rate with soil erosion at pipe outlet _____.

Border strip flowing at full capacity, i.e. water is at or near top of border _____.

Cracks or lack of border height allow water to crossover to adjacent border strips _____.

Hand Move or Side-Roller Sprinkler Irrigation

Description of sprinkler method: Typical pipe diameters for hand move sprinklers are 5, 8, and 10 cm (2, 3, and 4 in) and pipe diameters for side-roller systems are 10 or 12 cm (4 or 5 in). Sprinklers for most hand move systems have a 12 m spacing, and for side-roller systems have a 14 m spacing. The riser height is actually the height of the nozzle above the ground surface (standard = 1 m). Frequently, sprinkler risers are not tall enough to clear the mature crop. To measure volume flow rates, a flow meter can be placed at turnouts at the water source or at a farmer's well to enable the farmer to measure flow onto the field. Flow rates can be measured using examples given in Chapter 12. The following items can be used in data collection for evaluation of irrigation system characteristics.
1. Record the lateral pipe diameter (cm or in) _____.
2. Record the sprinkler spacing (12 or 14 m, 30 or 40 ft) on the lateral _____.

Evaluation of Supplemental Irrigation

3. Enter the spacing (m) or the move distance between laterals _____.
4. Measure the height (m) of sprinkler nozzle above the ground surface _____.
5. Record the set duration (hr) for the time a sprinkler lateral operates at 1 location (typical values are 12 hr or 24 hr) _____.
6. Are laterals perpendicular to prevailing winds, i.e. does the wind blow down rather than across the lateral? (Y/N)
7. What is the maximum plant height (m) expected when the crop reaches maturity or the maximum plant height expected during irrigation operations _____.
8. Is there a flow meter to measure water volume for irrigation application? (Y/N)
9. Itemize the planned irrigation operations:
   a. Surface area coverage in _____ days,
   b. No. of moves _____,
   c. Method used to move system: manual labor _____
      owner only _____
      mechanical _____
      other type _____
10. Prepare a diagram and data profile for the main line design and include data to estimate the system capacity. Figures 15.10, 15.11, and 15.12 provide data items and example forms for the data collection.

---

1. Final application rate _____ mm/hr
2. Spacing _____ m. (line) × _____ (move)
3. Sprinkler: model _____ ,
   nozzle _____ mm × _____ mm,
   _____ m³/hr @ _____ bars (ave)
4. Lateral:
   total length _____ m,
   _____ m of _____ cm,
   _____ m of _____ cm
   no. of sprinklers _____
5. Head loss in lateral _____ bars
6. Pressure @ head of lateral _____ bars _____ m.
7. Total lateral capacity _____ m³/hr
8. No. of laterals operating _____
9. Total system capacity _____ m³/hr
10. Hectares covered/set _____
11. Hectares/day _____

Figure 15.10. Sprinkler system specifications.

		Pipe size and length (cm)				Head loss (m) for trial pipe combination			
I.	Trials					1	2	3	4
	1						/////	/////	/////
	2					/////		/////	/////
	3					/////	/////		/////
	4					/////	/////	/////	
II.	Pressure @ lateral or gated inlet (m)								
III.	Max. elevation difference (m)								
IV.	Total suction lift (m)								
V.	Minor losses and fittings (m)								
VI.	Total dynamic head (m)								
VII.	Water horsepower required								

Figure 15.11. Total dynamic head for main line combinations and horsepower required.

	Head loss (m) for trial pipe combination			
Data items	1	2	3	4
1. Approx. pipe combination cost				
2. Annual fixed cost				
3. Annual fixed cost difference				
4. Water horsepower difference				
5. Cost/water horsepower/season				
6. Annual power cost difference				

Figure 15.12. Economic pipe size for sprinkler irrigation method.

Field observations: Field observations should be answered with the appropriate response. Sprinklers should be tall enough to clear mature plant height. Crop height differences are indications of non-uniform water application. Nozzle plugging can be determined by shutting a lateral line and opening the end plug to see what is in the last joint of a pipe.

Evaluation of Supplemental Irrigation 269

1. Is there crop interference with sprinkler pattern? (Y/N)
2. What percentage of the sprinklers are not rotating? _____
3. What percentage of water is leaking from the system, i.e. from the mainline, end caps, or lateral gaskets, and is difficult to quantify? _____
4. What is the cause of sprinkler nozzle plugging?
   a. no plugging _____
   b. sand or gravel only _____
   c. aquatic growth (fish, algae) _____
   d. mixture of sand and aquatic growth _____
   e. other cause _____
5. What percentage of the sprinklers are plugged? _____
   The example includes all fully plugged sprinklers and considers the following situation.

   300 sprinklers in the field
   2 sprinklers are fully plugged (2 × 1)
   10 sprinklers are ½ plugged (10 × 0.5)
   15 sprinklers are ⅓ plugged (15 × 0.33)

   $$\frac{(2 \times 1) + (10 \times 0.5) + (15 \times 0.33)}{300} \times 100 = 4\% \text{ are plugged}$$

6. Is sand wear noticeable on the sprinkler spoons or nozzles? (Y/N)
7. Are alternate sets used (placing laterals in intermediate positions every other irrigation)? (Y/N)

Can catchment data: Cans must be placed just above the canopy and the openings must be absolutely horizontal. If a can tips over, estimate the catch based upon an average of the surrounding catches. Cans are placed always at 3 m spacings parallel to the lateral.

Move distance	Can spacing perpendicular to lateral
9, 12, 15, 18	3.0 m
11	2.7 m
14	3.0 m
18	3.4 m

Enter the total volume (ml) caught in cans. A zero value means no water fell into the can.

			(4th column used only for 12 m spacing.)
L33:	L34:	L35:	L36:
L29:	L30:	L31:	L32:
L25:	L26:	L27:	L28:
L21:	L22:	L23:	L24:
L17:	L18:	L19:	L20:
L13:	L14:	L15:	L16:
L9:	L10:	L11:	L12:
L5:	L6:	L7:	L8:
L1:	L2:	L3:	L4:
R1:	R2:	R3:	R4:
R5:	R6:	R7:	R8:
R9:	R10:	R11:	R12:
R13:	R14:	R15:	R16:
R17:	R18:	R19:	R20:
R21:	R22:	R23:	R24:
R25:	R26:	R27:	R28:
R29:	R30:	R31:	R32:
R33:	R34:	R35:	R36:

Runoff problems: Runoff problems can be caused by several factors, including low pressures, large droplet sizes, and high application rates. However, water quality is a frequent, little recognized contributor. Values entered here should only include water which is not used somewhere else on the farm. Runoff is difficult to estimate and may require setting up a flume.
1. Are there runoff or infiltration problems? (Y/N)
2. What percentage of the applied water runs off the field? _____

Equipment assessment, sprinkler method

Nozzles and regulators: Special nozzles and pressure regulators are relatively inexpensive to install and can drastically affect system performance. If they are incorrectly installed, they can decrease performance. A low pressure nozzle is one especially designed to break up the water jet at lower-than-normal sprinkler pressures. They may look like a regular nozzle at first glance; however, instead of a standard straight bore, they will have rectangular, square, triangular, hex, or other bore designs. Flow control nozzles take the place of regular nozzles and have a flexible washer which becomes smaller as the sprinkler pressure increases to maintain a constant flow rate over a wide range of pressures. If the system has flow control nozzles, do not place the pitot tube of the pressure gauge into the nozzle.

Evaluation of Supplemental Irrigation 271

Flow control bases look like a threaded coupling at the base of a sprinkler. They may be made of metal or plastic, but will have a flexible washer inside which performs the same function as a flow control nozzle. Usually they are stamped with the nominal flow rate and are not the same as pressure regulators. Pressure regulators are 8-15 cm in length, and are found directly at the base of sprinklers. They have a spring assembly inside and are meant to maintain a constant discharge pressure as long as the incoming pressure is greater than the pressure at the regulator. Usually the nominal discharge pressure is stamped on the regulator.

1. Does the system have low pressure nozzles? (Y/N)
2. Does the system have flow control nozzles? (Y/N)
3. Does the system have flow control bases? Y/N)
4. Are pressure regulators at the sprinkler base? (Y/N)

Filtration: Tubular screens are popular on canal installations. The perforated inside part of the screen is stainless steel; the outside part is typically galvanized steel. Media filters are also called sand tanks. Sand separators are common on well installations. They are usually installed above ground next to the pump but there are submersible types installed on the suction side of a deep well pump.

Figure 15.13 shows an example of a hydrocyclone device or separator designed to spin turbid water at high speeds. Solids are thrown outward against the sides of the separation chamber while clear water stays near the center.

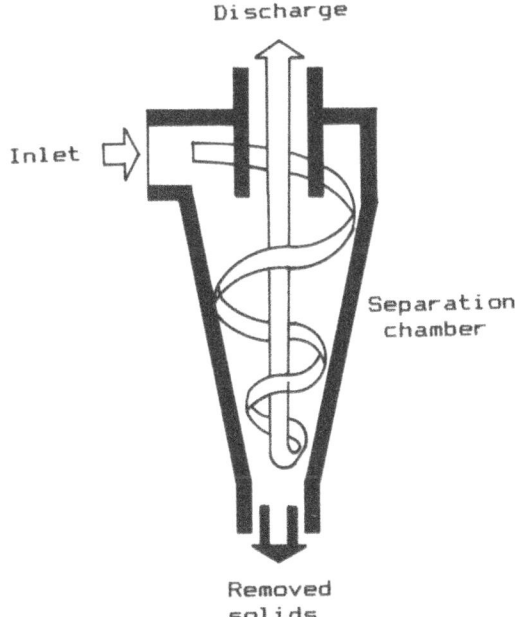

Figure 15.13. Hydrocyclonic separator centrifuge for separation of solids by centrifugal force.

There is a wide variety of overflow screen designs to remove weed seeds, canal trash, etc., which would be included in this category.
1. Is there filtration prior to mainline intake? (Y/N)
2. Is there a perforated stainless steel screen? (Y/N)
3. Are media filters used in the system? (Y/N)
4. Have sand separators been installed? (Y/N)
5. Are there other types of screen used, e.g. overflow, ring? (Y/N)

Pressure and flow measurements: Pressures and flows should be measured on at least 3 laterals on each side of the mainline. This gives a total of 6 laterals if laterals are on both sides of the mainline. Data for the first and end or last laterals along the mainline must always be measured using the first sprinkler (inlet end, I/S) and last sprinkler of each lateral (downstream end, D/S).

Pressure measurements are made with the standard pressure gauge and pitot tube device. The volume caught for each sprinkler to obtain flow measurement is an important data item and, if the water collection containers are too small, can be the least accurate value. All water from the sprinkler head should be diverted into a large container for at least 2 min to minimize measurement error. Error can also be reduced by using large graduated cylinders (over 2,000 ml) to measure the water collected in the container. It is best to take at least 2 measurements for each sprinkler per location and enter the average volume caught in the example form below.

Check to be certain that sprinkler nozzles are not plugged before measurements are made. The form below can be used to list the required data in the appropriate columns for evaluation. Use decimal equivalents for the manufactured nozzle size in data tabulation.

Location/ Lateral	Nozzle Diameter (in)	Pressure (bar or psi)	Volume Caught (ml)	Time (sec)
I/S No. 1				
D/S No. 1				
I/S No. 2				
D/S No. 2				
I/S No. 3				
D/S No. 3				
I/S No. 4				
D/S No. 4				
I/S No. 5				
D/S No. 5				
I/S No. 6				
D/S No. 6				

Excess pressure losses: A typical 400 m line of hand move sprinkler system will have 44 rotating sprinklers. If a valve opening tee is used to supply laterals on both sides of the mainline, this number can total 88-89. Many irrigators have a

Evaluation of Supplemental Irrigation 273

butterfly or gate valve at the pump which they use to restrict the flow if the pump pressure is too high. An automatic control valve is designed to provide slow opening, slow closing, pressure regulation, or pressure check functions. The manufacturer's recommended minimum operating pressure is not a readily apparent value. This value will depend upon the type of drive (wedge vs. spoon) and bearing assembly. Refer to the manufacturer's literature for sprinkler standards most common in the area.

1. Number of sprinklers supplied by 1 valve opener _____.
2. Riser valve diameter (typical size, 8 and 10 cm or 3 and 4 in) _____.
3. Is there a throttled manual valve at the pump? (Y/N)
4. For the throttled manual valve, what is the pressure loss (bar or psi) across the valve? _____.
5. Is there an automatic control valve at the pump? (Y/N)
6. For the automatic control valve, what is the pressure loss (bar or psi) across the valve? _____.
7. What is the manufacturer's recommended minimum operating pressure (bar or psi) of the sprinkler and nozzle? _____.

Drip/Trickle Irrigation

Description of the drip/trickle method: Evaluation of a drip/trickle irrigation system should follow the same or similar data items as presented for the sprinkler irrigation system. A typical drip/trickle irrigation system includes the following elements: mainline, e.g. aluminum or steel pipe; submain, e.g. polyvinyl chloride pipe (PVC), or polyethylene tube (PE); lateral line (PE tubes); and, other attachments such as, valves, pressure regulators, filters, pressure gauges, vacuum breakers, and drain valves or plugs.

1. What is the age (yr) of the system? _____.
2. Do you have infiltration (water penetration) problems? (Y/N)
3. Do the fields have undulating topography? (Y/N)
4. Estimate change in elevation (cm) from filter to highest point in field, " + " if going uphill from filter _____.

Pumping plant measurements

1. What is the pressure (bar or psi) at the pump discharge pipe at ground level, not the total head (TH) at the pump? _____
2. What is the pressure (bar or psi) into the mainline taken immediately before the mainline goes into the ground or across the field, downstream from filters and control valves? _____
3. What are the optional pressure valves if pressure differences in the system are excessive? _____
4. What is the total head loss (bar or psi) from the filters? _____
5. What is the total head loss (bar or psi) from the pump control valve? _____

6. What is the total head loss (bar or psi) from throttled manual valves? _____.

A reasonable head or pressure loss depends upon the type of filter and control valves. For example, an automatic control valve should have about a 1/3 bar (5 psi) pressure drop or loss. Other standard losses would be 0.07 bar (1 psi) for a screen, 0.3 bar (5 psi) for a media filter, and 0.4-0.53 bar (6-8 psi) for a sand separator.

Field observations: Field observations should be answered with the appropriate response. On sloping ground some emitters continue to flow for several hours after water is shut off. A reasonable estimate should be made of the percentage of emitters that continue to flow.
1. What is the time (min) that emitters continue to flow after water has been shut off? _____
2. What is the percentage of emitters that continue to flow after water has been shut off? _____
3. What is irrigation duration (hr/block)? _____
4. What is the flushing time (sec) to get clear water from the lowest, most distant pipe? _____

Using a watch with a second hand, determine the time required for water to turn clear after the pipe is flushed. The system must be running when the pipe is flushed.

Emitter flow measurements: One manifold in a typical, average location should be used for testing. A manifold is a submain, at the ground surface or buried underground, which supplies water to the emitters directly. Within the test, there should be a pressure difference between emitters of less than or equal to 0.07 bar (1 psi). If a group of 15 emitters along 1 pipe has a pressure difference of more than 0.07 bar (1 psi), test emitters from 2 adjacent submains. Water should be collected in a can or graduated cylinder from individual emitters for at least 3 min each.

Select a manifold (submain) near the field inlet, so that pressures can be adjusted easily for Test No. 2. These measurements will give data of the variation in flow between emitters at the same pressure caused by manufacturing variability and plugging. Locate Test No. 1 at middle of the submain closest to inlet.

1. Test No. 1 (min):
    a. average emitter pressure (bar or psi) _____
    b. enter the volume (ml) collected below:

1:	6:	11:
2:	7:	12:
3:	8:	13:
4:	9:	14:
5:	10:	15:

Evaluation of Supplemental Irrigation

Test No. 2 is done at the same location as Test No. 1 but at a different pressure (0.67-1.00 bar or 10-15 psi higher or lower).

The results from Test No. 2 combined with Test No. 1 indicate how flow rates respond to changes in pressure. If an average emitter pressure was 1.7-2 bar or 25-30 psi in Test No. 1, reduce the pressure to 0.67-1.3 bar or 10-20 psi by restricting the plastic pipe or throttling the inlet valve. If an average emitter pressure was 0.67-1 bar or 10-15 psi in Test No. 1, try to increase the pressure (temporarily) for Test No. 2 by adjusting pressure regulators. In any case, the pressure difference between Test No. 1 and Test No. 2 should be from 50-100%.

2. Test No. 2 (min):
   a. average emitter pressure (bar or psi) _____
   b. enter the volume (ml) collected below:
      1:        6:   11:
      2:        7:   12:
      3:        8:   13:
      4:        9:   14:
      5:        10:   15:

Field pressure measurements: Water must be flowing through the plastic pipes when all measurements are made. Data measurements should be made on the submain closest to the pump or inlet. In the case of a system with only a single submain (manifold), there will be data only for 1 manifold. On most systems, pipes run in 2 directions from a manifold and the end pressures on both pipes should be measured to verify that they have the same inlet pressure.

Record the number of pressure regulation valves, manual or automatic, in series after the pump station and en route to the pipe with the lowest inlet pressure. A regulator valve is used to adjust pressure which includes pipe pressure regulators but does not include electrically actuated on-off valves which provide no pressure regulation. However, an electrically actuated valve could be included if the flow control stem has been adjusted to develop a specific downstream pressure. Before taking pressure measurements across an individual automatic valve, set the valve adjustment to be completely open. After taking pressure readings across the valve, readjust opening to the original pressure setting.

1. What is the pipe inlet pressure (bar or psi)? _____
2. What is the downstream end pressure (bar or psi) on 1st pipe? _____
3. What is the downstream end pressure (bar or psi) on 2nd pipe? _____

The following measurements are made on the most distant pipe from the submain inlet.

4. What is the pipe inlet pressure (bar or psi)? _____
5. What is the downstream end pressure (bar or psi) on 1st pipe? _____
6. What is the downstream end pressure (bar or psi) on 2nd pipe? _____

If there are multiple submains, repeat the first 6 measurements.

7. What is the difference between inlet and discharge pressures at the valves?
   a. valve 1 (bar or psi) _____
   b. valve 2 (bar or psi) _____
   c. valve 3 (bar or psi) _____
8. What is the average drop in pressure or head loss (bar or psi) across pipe entrance screens in the field? _____

Dirty pipe screen washers or field screens can be a major maintenance problem and are the cause of large drops in pressure or head losses. The measurement of this pressure drop generally requires installation of a regulator valve upstream from the screen.

Equipment assessment, drip/trickle method

Emitters: Plugging of the emitters is considered the major maintenance problem with the drip/trickle method. An in-line flow meter can indicate when emitter flow is reduced, e.g. when the flow is reduced by 10% then plugged emitters should be cleaned or replaced. Ideally, emitter flow rates should match the soil infiltration rate; however, in practice this usually results in a greater probability of plugging; therefore, some degree of compromise is required. Most emitters or microsprayers are rated as delivering a certain flow rate at a specific pressure (1 bar or 15 psi).
1. Emitter manufacturer _____
2. Model _____
3. Nominal flow rate (G = gph or L = lph) _____
4. What is the emitter path type?
   Long smooth path _____
   Pressure compensating _____
   Vortex _____
   Orifice _____
   Tortuous (turbulent) path _____
   Multiple flexible orifice _____

Filtration: An automatic flush valve is controlled by a timer or a pressure differential sensor. For a combination media and screen filtration station, the media filter is considered the primary filter. Some systems have more than one filter type and most micro-irrigation systems have multiple filtration systems. Pipe screen washers and small screens at pipe inlets are not considered filters.
1. Is the automatic flush valve on the primary filter?   (Y/N)
2. Which type of filter does the system have?
   a. tubular screen _____
   b. overflow screen _____
   c. media filter _____
   d. sand, centrifugal or separator _____
   e. other type of screen (e.g. filtomat, ring type) _____

Evaluation of Supplemental Irrigation

Valving: Some systems have several stages of pressure regulation. Pressure regulators can be either automatic or manually adjusted valves, or both. The path of flow in the plastic pipes must be followed from pump to emitters recording where pressure adjustments could be made. Drip/trickle systems often have one or more automatic valves which are located either before or after the filters near the pump and filter station. Standard check valves or slow opening-closing valves without pressure regulator features are not considered pressure control valves. Submain pressures are frequently adjusted with manual gate valves. One pressure regulator is commonly used per 3-5 plastic pipes.
1. How many automatic pressure control valves are near the filter and pump? _____
2. Is there a throttled manual valve at the pump? (Y/N)
3. Are submain pressures regulated individually? (Y/N)
4. Are pipe pressures regulated individually? (Y/N)
5. Is there a flow meter in the system? (Y/N)

Chemical injection technique: The most common chemical injector is a pressure differential tank. The throttling valve action must serve no other purpose than the chemical injector. Sometimes a device is placed across a gate valve which throttles the valve to force water through the injector. Note whether injector is downstream of filter or upstream of filter.
1. What is the location (upstream or downstream) of injector with respect to filter? _____
2. Does system use a throttling valve in the main line specifically for chemical injection? (Y/N)
3. Is chemical injection possible at a constant rate? (Y/N)

Servicing drip/trickle irrigation systems

1. How often (yearly, monthly, other) is chlorine, acid, etc. injected? _____
2. How often (yearly, monthly, other) are pipe laterals flushed? _____
3. How often is automatic flushing of pipe ends used? _____

Assessment of irrigation problems in the field, rate from 0 to 3: This is an assessment of problems observed in the field while conducting the evaluation. Assign the most suitable rating to the specific problem:
0 – no problem or no visible effect to method,
1 – visible problem but little effect to method,
2 – definite problem, affects performance of method, or
3 – severe problem, dysfunctional to performance of method.

Problem 1, contaminants: Use knee-high nylon stockings to catch the water as it flows from the end of the pipe outlet. After some experience of data collection on different fields, a scale can be developed for comparative analysis of the results.

1. Total amount of material caught in nylon sock when flushing hoses _____

2. Rate amount of: Sand _____
                     Clay _____
           Bacteria/Algae _____

Problem 2, emitter plugging: For this data item, remove 3 emitters at each end of the pipes, inspect and assess the severity of plugging. Before removing the emitters or micro-sprinklers, obtain enough spare parts (spare emitters, plastic pipe pieces, and couplings for various pipe sizes) to repair any damage which may occur. Carbonate precipitates can be detected by putting a drop of acid (hydrochloric is recommended) onto an emitter. If the white, brownish deposits or sludges fizz, you have a positive indication of a precipitate.
1. Rate amount of:
             Sand _____
          Bacteria _____
           Insects _____
     Plastic parts _____
2. Rate amount of precipitate (bubbles with acid drop) _____

Problem 3, abnormal emitter flow: Runoff is a major problem in some fields with micro-irrigation. If the runoff is not collected in a sump and reused an estimate should be recorded.
1. Note and rate cracked hoses, barbed or snagged tube leaks, oversized wet spots, surface runoff, etc. _____
2. Runoff from the field of irrigation water _____
3. Rate the subsurface wetted area per emitter, to a depth of 50 cm below the soil surface (cm^2) _____

Chapter 16

Introduction to Technology Transfer

What is the Future for Supplemental Irrigation?

For national programs working to establish sustainable agriculture in a deficit moisture area, supplemental irrigation can be a blessing. In Iran, Iraq, Jordan, Libya, and Morocco, analogous experiences can be found of increased yields and stabilized production applying supplemental irrigation – the blessings of augmented rainfall. Why? At high elevations in Turkey, for example, the average yield is 1.7 t/ha from rainfed spring wheat in the Konya Province of the Central Anatolia Region; but, it increases to 3.75 t/ha with supplemental irrigation. For rainfed barley, the average yield is 1.7 t/ha but is 4.0 t/ha with supplemental irrigation. In both cases, to achieve more than double the yields, a mere 20 mm of water was added to aid germination in the fall and, later in the spring, an application of 30 mm was added.

In most countries of the Near East and North Africa, an increasing food deficit between domestic production and consumption has been witnessed in the last two decades. Syria was a net exporter of basic food commodities (wheat) and feed stuff (barley) in the 1960s. In the 1970s, Syria still reached self-sufficiency in barley production but was only 72% self-sufficient in wheat production. In the 1980s, production lost ground again, Syria slipped to 84% self-sufficiency in barley and to 60% in wheat. These circumstances are critical; but, many countries in the region find themselves in an even worse situation. If current levels of productivity continue to the year 2000, not only in Syria but in most countries of the region, the deficit in wheat production will be 55% of domestic needs, putting serious pressure on national resources. The ups and downs of the availability of food on the world market have been caused by climatic fluctuations but they mask a dangerous underlying trend. In most regions of the world, land availability for agricultural purposes has diminished or ceased. Expansion to new frontiers is not an alternative; future increases in productivity will necessarily rely on increasing yields per cultivated hectare instead of bringing new land into production. In the 1970s, 1 ha of arable land maintained an average of 2.6 persons: in the 2000's, this single hectare will be expected to sustain 4 persons. The pressing need to raise agricultural production persists in every developing country. Conceiving a system of sustainable agriculture invites actions that cut across a range of vested interests which involves more than a simple marshaling of science and technology to attain a single objective. Three major causes have been identified which underlay the problem of food deficiencies.

1. High rates of population growth: controlling population growth rates is difficult because this is related to considerations of ethical or religious beliefs which take a long time to change (requiring long term objectives and comprehensive policy).
2. Low and fluctuating yields from rainfed production (less than 1.5 t/ha for wheat): expansion of rainfed wheat land is limited because most land suitable for wheat is already in production.
3. A decrease in land area sown to wheat in preference to other crops, e.g. a decrease in Syria from 1.6 M ha in the 1960s to 1.2 M ha in the 1980s: the genuine potential for closing the food gap is to increase yields and stabilize production on the lands which are now under rainfed cultivation or which have a potential for production with management of the average annual rainfall.

Notwithstanding, efforts are being made to increase yields and stabilize production with rainfed farming but these have been neutralized by the harsh environment of the region with its erratic rainfall patterns. Moreover, only modest increases can be realized in yields of rainfed crops with reliance on current technologies of plant breeding and agronomy. On the other hand, integration of supplemental irrigation into rainfed farming has generated higher yields (about 200% above the yield of rainfed cultivation). Application of supplemental irrigation or water harvesting techniques offers definite possibilities for increasing productivity on the vast areas of rainfed land to realize self-sufficiency and to eventually provide food security in the basic food and feed commodities.

Technology transfer executed as experimental development is measured activity which engenders processes and procedures to incorporate new scientific knowledge into existing farming experience, practices, and products. This activates an amalgam of basic, applied, and evaluation research methods mixed thoroughly with political realities and social ideals until boundaries of activities and actions can become blurred. Technology transfer becomes calculated, deliberate social change. A theoretical perspective for supplemental irrigation combines the basic scientific theories for water, soil, and plant associated with climate and incorporates the philosophy of time and space. From these theories which have been developed into methodologies, combined with conservation of resources and balanced with the environment, a routine for social engineering can evolve to successfully implement the technology of supplemental irrigation.

Endorsing this perspective, farmers who follow traditional practices are not perceived as being ill; they do not require a diagnosis or a remedy as assumed by the medical or diagnostic models so prevalently used in earlier agricultural development methodologies. On the contrary, they are responding to centuries of experience gained over generations to minimize the risk, to lessen the gamble of farming. Rainfall is "God given", for the traditional farmer, and land is a symbol of wealth, a guarantee of survival. In contrast, for modern farmers who are, themselves, a country's human resource, water and land become agricultural resources and are the means with which they can increase production and sustain productivity on a given land unit with an appropriate water supply by using warranted inputs and their creative management.

The master key to unlock downward spiraling yields is constructive risk anticipation of agricultural productivity. Technical verification of supplemental irrigation is not enough to achieve and sustain an impact on food security. These tasks require comprehensive planning. Technical imperatives define only what is possible, not what is necessary – that which can be done, not what must be done. The distinction between possibility and necessity can be lost on many contemporary observers who often lack imagination and social vision. Planning maintains the integrity of scientific precision but consolidates social vision and individual imagination with technical possibility. This combination ascribes responsibility of access to information and procedures of operation collectively with shared standards of record keeping. Elements must be quantified; comparative values can and must be assigned for appraisal of benefits and costs of modern agricultural production.

Responsibility of management entails flexibility not only in decisional capability but especially in ease of information retrieval for refinement of future performance. Experience and opinion can then derive from a much wider participatory audience than is ordinarily observed, with results disseminated to varying levels of management. Data collection occurs on a more continuous time frame than is usually designated with the population being repeatedly sampled for intermediate analysis to test whether or not the components of supplemental irrigation technology are in tune with current agricultural practices.

Successful technology transfer encompasses the technical advances of supplemental irrigation integrated with supportive national investments for implementation of related farming practices. These investments are planned to coincide with development and expansion of improved networks of marketing as well as production disposal and distribution facilities. Experimental development of supplemental irrigation, at the farmer level, should be targeted toward small-scale water management systems which lessen the effect of technology transfer and diminish agro-economic constraints. Thereby, farmers, gaining experience but with reduced risk, can produce sufficiently for local demand and dispose of surplus production with continued reliance (initially) on the country's existing infrastructure. At the same time, the government can gain experience from this different technology with all the inherent ramifications from the focal point of the total agricultural system.

Decision Field, an "Ideal Model" for Technology Transfer

The decision field (Figure 16.1) is graphically designed as a grouping of information categories. It allocates the contact of other system components with the individual farmer. The accessibility of resources for intervention, e.g. institutional, technical, economic, social, commercial, and financial, are more readily discerned to modify farming practices. For development purposes of supplemental irrigation (and for discussion in this chapter), entry for proposed implementation will focus on water, therefore entering from the technical point of intervention with emphasis on the level of management, development parameters, and resource exploitation. Resource exploitation acknowledges that water is the principal resource for regions of water scarcity with the selected strategy of involvement dependent upon the required precision of data and critical need.

What the format does allow is a means to view associated activities which are affected by water resource management using supplemental irrigation technology. The previous chapters provide the technical requisites for implementing and operating supplemental irrigation. The chapters which follow offer examples of supplemental irrigation research and data and information of national scientists to determine the possibility of integrating these research findings into their country's agricultural system.

The decision field is an ideal model presented as a stochastic matrix of elements into which data of deterministic units of decisional action and operational activity can be introduced, integrated with research findings, and empirically tested. The decision field is founded on the principles of irrigation technology with system components ranked by management levels, arranged into relative points of intervention, and ordered by alternative strategies of involvement. Comprehension of systems theory helps in understanding the ideal model which displays the interrelationships of components and also the independence of each category of the catenated procedures of planning and development.

The arrangement of the model prescribes a structure for defining and correlating data analysis and a classification scheme to determine system needs, the development rate possible, and the capability of the system to respond. Through analyses using the decision field "best available data" is compiled for estimating the success of integrating technology into the contemporary agricultural system. The arrangement distributes information for direct access to a single category memory for simple monitoring. These results identify what can be absorbed by the population-at-large within the existing socio-economic structure and institutional organization.

Points of Intervention for Technology Transfer

The points of intervention can be grouped into 6 divisions: institutional, technical, economic, social, commercial, and financial.

Introduction to Technology Transfer

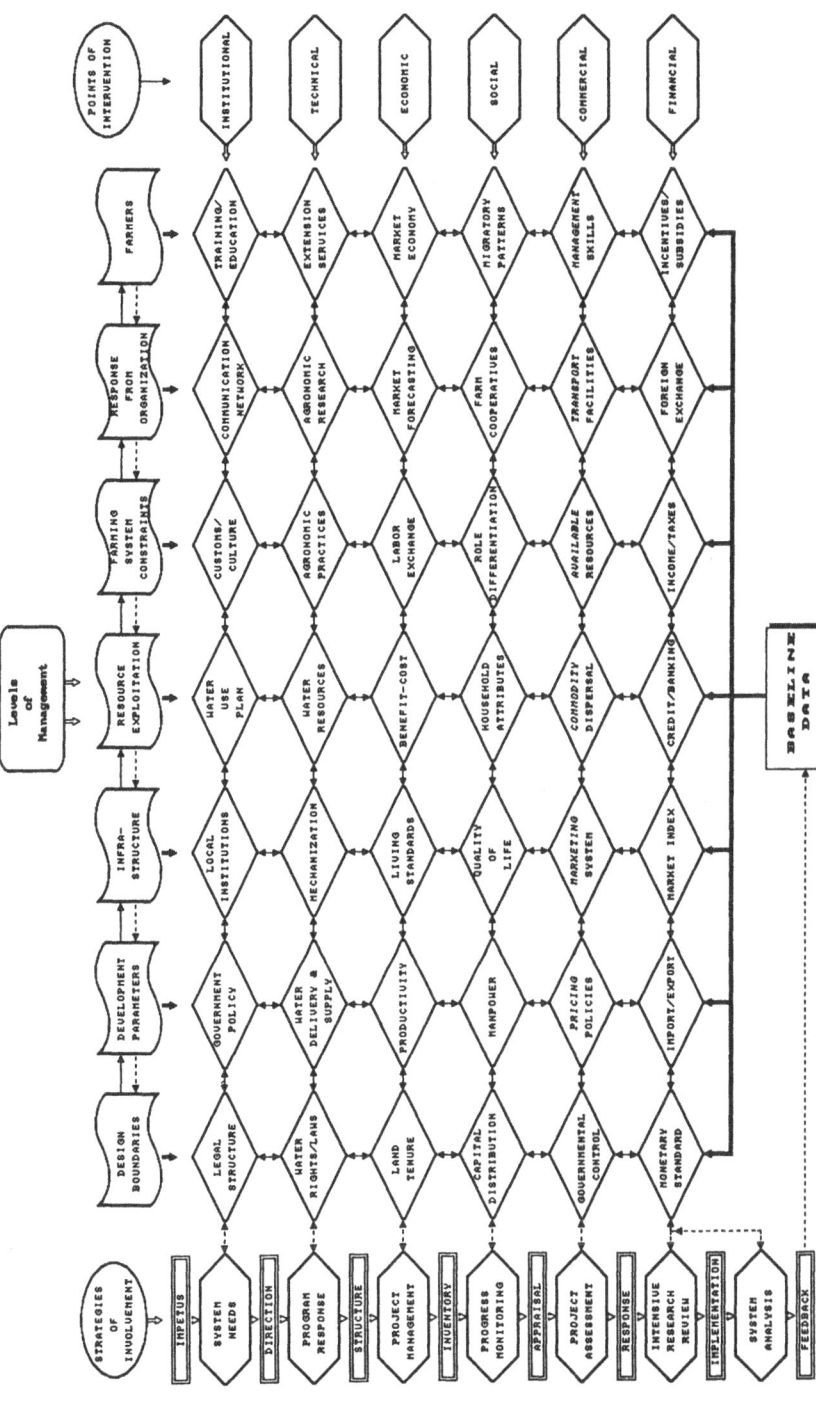

Figure 16.1. A flow chart describing an ideal model for transfer of supplemental irrigation technology, illustrating the categories of information for a data bank and authority (levels of management), the process of evaluation (strategies of involvement), and requisite points of intervention.

Institutional intervention: The socio-cultural patterns and established structures of the country in which technology transfer for supplemental irrigation will take place.

A nation's legal structure establishes the basic stipulations of implementation and outlines the regulations governing supplemental irrigation agriculture. Competent activation of the technology enlists endorsement of existing government policies or requires formulation of new policy. Farmers of traditional farming systems depend on local associations for sustenance during change or crisis; therefore, cooperatives should be established and viable for the period of transition. Re-establishing production equilibrium to food deficient nations of the Near East and North Africa relies on rational water use plans for the best allocation of water resources: enough water for production, well managed by individual farmers but with adequate regulation by governmental authorities for prediction of supply and demand with a goal of the least volume of waste possible.

Customs and culture confine attitudes to an articulated sphere; therefore, cultural background and custom indigenous to the farming community discriminates the variety of agronomic practices introduced at the start. Exposure to and exchange of new information and sharing of experiences rely heavily on communication networks. In the countries of the region, measures must be found to mobilize methods and facilities for training and educating high numbers of illiterate farmers.

Technical intervention: The technical factors which relate to irrigation technology and are involved in development of modern irrigated agriculture.

Access to water resources resides in water rights and laws. If historical water rights are dysfunctional to development of supplemental irrigation technology, modifications or changes to these rights and laws have to be drafted with equity to adjust to the needs of modern agriculture. Effectual scheduling of irrigation and water conservation is pivotal to the farmer using supplemental irrigation for the first time. A thriving, vigorous and profitable farming enterprise derives from the chosen means of water delivery and supply. Establishment of delivery and supply has to entail the least capital outlay in the short term but offer minimal maintenance in the long term. Group affiliation coalesces with satisfactory conversion to mechanization. The sharing of equipment and exchange of skills and knowledge among a community of farmers distributes the burden of applying new techniques of farming. Paramount to exploitation of water resources is determination of strategic reserves and regular supplies, and then, execution of responsible blueprints of use and renewability.

Intensity of constraints to developing supplemental irrigation determines the difficulty of instructing farmers in, and establishment of, the agronomic practices associated with supplemental irrigation. National research agencies bear the responsibility for technical verification of technologies for local and regional application and appraisal of economic and financial profitability to identify, nationally, the scale and scope of efforts. Instruction, explanation,

Introduction to Technology Transfer

guidance for adapting the technology, and an overwhelming quantity of empathy for the farmers are all essential; this avocation demands dedicated, resilient extension services.

Economic intervention: The constituents that bolster the total economy to use existing resources during modernization of agriculture applying irrigation technology.

The realm and extent of development is guided by prevailing land tenure. The levels of interest and investment demonstrated by the farmer are curtailed by restrictive elements of tenure. Consecutive and incremental development of other system components has to occur concurrently with supplemental irrigation technology to ameliorate productivity. Achievement of anticipated production means anticipating need and accounting for availability of the various elements that go into crop production. Increased production and additional income, consequences of modern agriculture, contribute to the improvement and enrichment of living standards for farmers and their households. These must be forthcoming for the farmer to consider that the additional physical effort necessitated by supplemental irrigation is worthwhile. Capable benefit-cost computations of alternatives of irrigation technology have to occur including an appraisal of benefits and costs of research studies which devised the technologies.

Intensity of experience and education furnishes the tenacity of the labor exchange. Actual demands for labor will increase but the skill levels can be remarkably different from traditional farming needs. Reliable market forecasting fortifies the farmer's crop selection and anticipated yields as well as refining the government's capacity to better plan allocation of resources. To guarantee tolerable returns from diversified cropping, farmers require a market economy that has the complexity to disperse their increased scale of production but steady enough to manipulate supply and demand for maximum benefit with minimum cost.

Social intervention: The categories for facilitation of social adaptiveness entail societal implications of proposed investments as well as affecting living patterns and the issue of environmental quality.

Capital acquisition and distribution is an indication of a country's capacity for modernization of irrigated agriculture. The scope for realistic elaboration of the technology is furnished by the ability and initiative inherent in the society. The speed of establishment of the technology and accuracy of implementation will rest with human resources. Adequacy and diversity of public services and consumer goods strengthen the quality of life in the rural community. Recognition of the viability of the infrastructure to fulfill public demands for services and goods to bolster the quality of life alleviates public dissatisfaction. Constructive application of household attributes to dissimilar farming patterns and demands are required to foster participation of household members in the expanded tasks of production.

The crisis and urgency of basic food deprivation in the region prompts reassignment of role differentiation for achieving the unaccustomed tasks incumbent to modern irrigation. Auspicious delivery of agricultural inputs and services from agriculture cooperatives helps alleviate risk to the farmer in the adoption of irrigation technology. The cooperative has proven an adequate mechanism for distribution of goods and services necessary for affecting reliable adaptation to new practices. Investments and returns from irrigation technology have to be enough to provide occupations, community services, and rewards to the rural community to diminish migratory patterns. Size and movement of population have to settle into routine predictable patterns to support development efforts.

Commercial intervention: Conceived arrangements of production supplies and disposal facilities of agricultural enterprise to promote and sustain the productivity of supplemental irrigation agriculture.

Governmental controls affect operational substance of irrigation systems. Regulation and control is an essential feature of a complex technological society in which the public costs associated with the production and use of goods and services must be balanced against the social benefits. But these controls should be used sparingly to avoid inhibiting farmer initiative. Encouragement to participate in the farming practices synonymous with modern irrigation evolves from pricing policies; the farmer must make a profit. Introduction of supplemental irrigation technology means strengthening the marketing system and the parallel production services for ease of acquisition of inputs and disposal of produce. The government will become more involved in the sequence of marketing of surplus production and entry into national distribution networks with a focus on international markets. Pressures increase markedly for fulfillment of the farmer's supply and demand.

Time and space take on renewed meanings with supplemental irrigation when scheduling and volume of application revolve around stages of plant growth. Impartial scheduled allocation of the available resources to farmers in transition removes a major constraint to crop production. Transportation facilities and an interlinked roadway network which is adequately maintained is imperative to reinforce maximal disposal of production from farmer to consumer. Coordination of a variety of transportation methods will help to move agricultural supplies and produce to a wider market. Traditional farming experiences must be interpreted and used for bridging the gap between traditional farming practices and modern management skills. Recognition of their limitations and qualifications is an important step towards helping farmers adapt to the requirements of new practices and techniques.

Financial intervention: For identification of fiscal effects of proposed changes on various components of the system with the introduction of modern irrigated agriculture.

The stamina of the implemented technology relies on fiscal accountability of the monetary standard. The reliability of a nation's banking system and stability of national currency reflects on probable sustainability of development efforts. Investment for the expansion and diversification of agriculture asserts a potent structure for an import/export market. Reasonable tariffs and taxes should be considered which minimize interference with market processes. Establishment of irrigation technology interposes production planning with greater reliance on the market index. Funds have to be apportioned for investment and credit for execution of the objectives of supplemental irrigation.

Farmers cannot be asked to bear the burden of agricultural expansion without credit available at fair costs. They require support from community investments to further their efforts for increasing production. Equitable distribution of taxes and the receipt of adequate revenues and income from investments contribute heavily to total productivity. No sector of the society should be asked to carry a larger share than another. Foreign exchange expands the opportunity of increased rewards for the farmer and the country. Participation in the foreign exchange broadens opportunities for building capital and increases a country's reserves. The amplitude of incentives and subsidies, as well as ease of acquisition by the farmer, generates expeditious transfer of technology.

Linkage to Strategies of Involvement

The strategies of involvement discriminate the pattern of effect and hierarchy of response to alleviate constraints or mobilize resources for maintaining and re-establishing equilibrium to the program area of supplemental irrigation under stress. Each strategy evolves a dimension of activity: impetus, direction, structure, inventory, appraisal, response, implementation, and feedback. Correlated to these activities is action which can be taken: baseline data, system needs, program response, project management, progress monitoring, project assessment, intensive research review, and system analysis. Choice of a strategy depends upon the crisis discerned, the relative urgency of the situation, the decisional position (in the matrix) of the problem, and the available resources and capital to support alternative solutions.

For credible results, interdisciplinary concern and contribution should be integral to technology transfer. Participation is not only a social principle; it is an essential requirement for equitable planning, technology transfer, and equally important, lends an understanding of the human condition. Of critical concern in formulating responses is the ability to quantify specific contributions and to articulate the perceptions of organizational objectives and goals, i.e. the contribution of individuals. Conscious comparisons and adaptations from differing disciplines introduces creative forces for manageable change within agriculture. An open exchange of expertise, experience, knowledge, opinion, and ideas alleviates any tendency for an imbalance of input resulting from differing authority.

Activities for technology transfer of supplemental irrigation

An effective information format (see Figure 16.1) creates a data bank that obtains the information necessary to mobilize the authority for leadership to maintain operational control of system activity. Manipulation of the data and information of system action, activity, and performance reflects the levels of management required, narrows the choice of essential points of intervention, defines the primary strategy for involvement, and identifies resources and constraints available for response. This process is effective and efficient for investigating, planning, and developing supplemental irrigation although providing for comprehensive decision making through coordinated administration. It builds a framework for controlling the availability of resources and the flow of economic support as reflected from the conditioning of the system by farmers in response to decentralized delivery of programs and services of supplemental irrigation.

Baseline data analysis, impetus for action: The first step in data and information acquisition, baseline data provides a descriptive profile of the existing system. After analysis of these data, an elementary picture of the state of agriculture in the national context can be drawn which provides insight to constraints of production and an overview of the available resources. The broad categories designated for describing farming systems in each country give the impetus for designing program guidelines:
1. general demographic information;
2. land area and land use;
3. infrastructure/rural support services;
4. water management and supply;
5. water harvesting farming;
6. supplemental irrigation agriculture;
7. climate characteristics;
8. soil characteristics;
9. crop characteristics; and,
10. socio-economic factors.

System needs, definition of constraints: Referring to the descriptive profile of the country, goals can be stated based on parameters set forth by national agricultural policy. The constraints to agriculture outline the needs of the production system and provide direction to proceed in development activities. Objectives are stated precisely with quantitative measures of effort included for continuous monitoring to support assessment, appraisal, and evaluation of relative task performance and fulfillment of objectives. A comprehensive plan delineates proposed action to be taken:
1. goals and objectives;
2. policy parameters; and,
3. government plan for development.

Program response, action characterization:
After preliminary findings of the existing farming systems, the environmental and agro-ecological conditions for the sites selected for agricultural production can be defined and correlated to structure future ventures of increasing production. As stressed in the chapters on technical aspects of supplemental irrigation, understanding these areas of concern is critical to the enterprise of sustainable agriculture. In preparing for the development and implementation of supplemental irrigation, 6 categories of data collection and analysis are required to characterize the country's environmental setting for agricultural production:
1. collect, analyze, and map data sets of weather and climate;
2. itemize water resources; then, map and quantify their geohydrological properties;
3. establish hydrological characteristics as well as soil physical and chemical properties;
4. calculate water balance relations and evaluate with probability analysis;
5. select irrigation method and drainage design most feasible technically and economically; and,
6. estimate water requirements of certain crops for supplemental irrigation and determine comparative profitability of production.

Project management, activity selection: Executing change in production practices of existing farming systems stipulates actual activities based on the stated objectives. The scope and scale of activity evolve from an inventory of possible alternatives. The inventory is compared to standards based on the state-of-the-art technology and classified using the categories outlined in the ideal model. Actual activities are then selected for development and implementation:
1. delineate the socio-economic profile with particular attention to the country's critical resources: water, land, and human.
2. identify major constraints to altering production systems; then, prioritize and rank according to ease of resolving the constraint;
3. design water management systems based on available agricultural resources;
4. describe target population and select farmer sample to receive technology transfer of supplemental irrigation; and,
5. interpret data results using standards of success with referral to the decision field.

Progress monitoring, measures of performance: In the early stages of effort, general measures of performance derive from current literature describing contemporary practices of modern agriculture. The objectives further define these measures with the quantitative values to denote when objectives are fulfilled as planned. Specific criteria are necessary for concrete measurement of degrees of performance. Periodic appraisal of field application is needed for assurance that technology transfer is taking place. Criteria are established for

selecting which variables to measure and the instruments are designed and tested to evaluate the pace of varying activities to verify program congruence to the comprehensive policy, national plan, goals, and objectives. The selection of variables is achieved by:
1. designation of criteria for monitoring, assessment, appraisal, and evaluation of the progress of development;
2. review of policy and guidelines for mobilization of resources and funds of the development plan; and,
3. coordination of farm level data collection, research field plot designs for verification, and demonstration trials for observation and comparison with farmer's fields.

Project assessment, trends and factors of change: A thorough assessment of activities contributing to the success of technology transfer is critical for reproducibility of efforts. A descriptive analysis of the data presents results for comparative purposes. These results isolate the factors which contributed towards changing the rainfed farming systems with supplemental irrigation to achieve increased yields. Trends are indicated that can support additional efforts or which suggest that objectives were adequately fulfilled. Moreover, the results promote a response of intensive research review to partition variance for improvement of planning precision:
1. preparation of analytical summaries for descriptive project profiles; and,
2. outline the objectives describing the fulfillment of each as a direct consequence of the planned effort as well as unintended consequences which may have occurred.

Intensive research review, effectiveness of effort: Precision of effort has to be maintained with the least cost and greatest achievement. Intensive research review integrates data collected from various components of the system for analytical comparison of efficient and effective strategies of involvement to establish sustainable agriculture with increased production. Furthermore, it evaluates the contribution of supplemental irrigation technology towards this achievement. This review of the implementation of supplemental irrigation technology allows a realignment of ideas, assumptions, alternatives, and resources for maintenance of food self-sufficiency with a possibility of food security:
1. linkage of the ideal model, logic of inquiry, activity paradigm of authority, resolution method of evaluation and the theories and methodologies of supplemental irrigation technology, to planning, design, research, and development of program activities to assure food security through increased agricultural productivity; and,
2. integration of policy, procedure, process, and technology of program response and project implementation: locally, regionally, nationally, and internationally.

System analysis, impact of activities: Data results are categorized in the decision field to ascertain what changes have occurred in the various elements of the matrix. Conclusions can be reached from these data to retain the goals and objectives or refine, alter, or reject the original plan. If care were taken in the initial planning and design as well as consistent record keeping followed, only adjustments and refinements to original planning should be required. Relying on feedback, agriculture can be compared with other components of the national system, e.g. agriculture, industry, or urban development, for determination of resource allocation and investment funds by:
1. interpretation to determine future programs of basic research and definition of system need for development.
2. contribution to "new knowledge" and reliability of efforts for generalization of results of technology transfer for application to other locations.

Chapter 17 is a presentation of research in supplemental irrigation developed from theoretical considerations for efficient and effective water use. Chapters 18 through 29 describe and evaluate existing supplemental irrigation systems within countries of the Near East and North Africa with limited data sets and a defined time constraint. The more comprehensive the data, the more accurate is the evaluation of a country's capacity to profitably develop supplemental irrigation and implement technology transfer.

Chapter 17

Verification of Supplemental Irrigation of Spring Wheat

About 6,000 years ago, supplemental irrigation started in the uplands and foothill regions near many small rivers and streams in the Near East. These efforts were devoted to the production of cereal grains, mainly barley and wheat. Even now these traditional practices of supplemental irrigation continue to be indigenous to the region. With a stable water supply, modern agricultural technology exists to increase yields; for example, the world's record wheat yield on a farmer's field is 14.1 t/ha accomplished in Washington State, USA, using conventional irrigation (Kirkham and Kanemasu, 1983). To achieve maximum yield in this region, a wheat crop requires at least 450 mm of water to produce more than 5.0 t/ha.

This chapter is the linkage between irrigation techniques, agronomic practices, and socio-economic theory organized to implement the systems approach for technology transfer. The objectives of the research activities were to improve supplemental irrigation techniques, to determine plant varieties and levels of fertility that respond to supplemental irrigation, to estimate consumptive use, to determine irrigation scheduling requirements under local conditions to ensure water use for an economic increase in yields, and to stabilize crop production. Data are presented verifying research approaches for scheduling supplemental irrigation and the amount of water required for economic production.

Agronomists and agricultural engineers concentrate on measurements of soil water availability to schedule irrigation and tend to disregard the characteristics of plant growth; whereas, scientists who propose economic concepts, focus on plant indicators to schedule irrigation at critical growth stages where "limited water or deficit irrigation" become primary techniques for scheduling (Shipley, 1977; Marttin and van Brocklin, 1985). The phenology of the crop helps to define the schedule and the volume of water to apply. The chapter integrates soil water availability and plant indicators with climate, deploying water balance methods (see Chapter 3) to accentuate the economic dimension of effectively scheduling irrigation to maximize production.

E.R. Perrier and A.B. Salkini (eds), Supplemental Irrigation in the Near East and North Africa,
293-313.
© 1991 *ICARDA.*

Characteristics of Plant Growth for Spring Wheat

Wheat is usually planted during November and December in the Near East and North Africa. If the air and soil are dry, the exposed spring wheat plants suffer little and can readily survive winter temperatures which do not fall below $-10°C$. For the most satisfactory growth and development of grain, a cool, moist growing season followed by a bright, dry and warm ripening period of 6-8 weeks with mean temperatures of 18-19°C is best.

Germination of wheat in a moist soil takes 2-5 days with supplemental irrigation but emergence may not occur until later if the soil is cold and dry. Wheat plants are extremely resistant to drought during the first stages of growth. Plants which germinate during mid-November with a soil profile at field capacity enter the winter with good vegetative cover and resistance to low winter temperatures. Three to four weeks are needed after germination to establish adequate vegetative vigor before growth is slowed by conditions of the approaching winter.

After the arrival of spring, the first internode is visible or culm elongation occurs (jointing). At this stage of growth, the rate of water use begins to accelerate. Near 1 March, jointing is followed by the boot stage where the tip of the flag leaf becomes visible (Kirkham and Kanemasu, 1983; Finkel, 1983). Heading or the emergence of inflorescence is a sensitive stage for available soil moisture and occurs in most wheat varieties from mid-March to April. Heading is followed by anthesis (flowering) which is the most sensitive growth stage for soil water. When anthesis is complete, near 1 May which coincides with diminishing rainfall in the Near East and North Africa, the plant demand for available soil moisture rapidly declines. Irrigation should be discontinued when wheat grains have completed the soft dough growth stage.

Severe water deficiency during the grain filling stage which follows anthesis can result in shriveled grains. Peak vegetative development occurs from mid-February through April corresponding to high plant water demand. Evapotranspiration is about 50% of pan evaporation from mid-February to early March and reaches 80-110% within the vegetative development stage in April and early May, then diminishes to about 20% during maturation from late May through June.

Design for Supplemental Irrigation Research, Tel Hayda, Syria

To investigate the concepts of supplemental irrigation that required verification, the following studies were implemented in 1986 and 1987. A split block statistical design used 4 replications with the supplemental irrigation treatments as the main plots:
1. rainfed (no irrigation), I_0;
2. irrigated to replenish the total water requirement, I_1;

3. irrigated to replenish 2/3rds of water requirement, I_2; and,
4. irrigated to replenish 1/3rd of water requirement, I_3;

where the water requirement was computed at 50% of the plant available soil moisture (see Addendum 2A). The subplots had different levels of nitrogen in 1986 (none, 70, 140, 210 kg/ha) and wheat varieties in 1987 (Bread wheat, Cham IV and Mexipak; and, Durum wheat, Sebou and Cham I). At planting time, phosphorus, P_2O_5, was broadcast and incorporated at a rate of 80 kg/ha, and nitrogen (Urea) was applied at a rate of 40 kg/ha on all nitrogen treatments. The remaining quantities of nitrogen (30, 100, and 170 kg/ha) were applied before tillering in mid-February.

Immediately after sowing and fertilization, because rainfall was inadequate, an irrigation for germination, 30 mm in 1985 and 20 mm in 1986, was applied to the total plot area in 1985 but excluded the rainfed plots in 1986. Scheduling of supplemental irrigations was estimated by the water balance method employing the class A evaporation pan with measurements of soil moisture using either soil samples or the neutron soil moisture method for verification.

During the 1987-88 and 1988-89 seasons a line-source sprinkler system was used (see Addendum 2A) and the supplemental irrigation treatments were:
1. rainfed (no irrigation), I_0;
2. irrigate to replenish 1/5th (20%) of water balance requirement, I_1;
3. irrigate to replenish 2/5ths (40%) of water balance requirement, I_2;
4. irrigate to replenish 3/5ths (60%) of water balance requirement, I_3;
5. irrigate to replenish 4/5ths (80%) of water balance requirement, I_4; and,
6. irrigate to replenish total (100%) water balance requirement, I_5.

Spring wheat varieties Cham I (Durum), V_1, and Cham IV (Bread), V_2, were sown with a drill at 125 kg/ha. Four levels of nitrogen (Urea) were applied: N_0 = none; N_1 = 50 kg-N/ha; N_2 = 100 kg-N/ha and, N_3 = 150 kg-N/ha. Nitrogen was applied at the rate of 30 kg/ha at sowing time and the remaining quantities of nitrogen (20, 70 and 120 kg/ha) were applied before tillering, during February. Phosphorus (P_2O_5) was broadcast at a rate of 100 kg/ha at sowing time.

Weather Data for Tel Hadya, 1985-89

Seasonal rainfall at Tel Hadya for September through May, 1985-86 was 315.1 mm and for the 1986-87 season was 316 mm. Although the quantity of rainfall was similar for both seasons, the distribution of rainfall was different. For 1987-88, rainfall was 503.6 mm, a wet season, but during 1988-89, rainfall virtually stopped near the end of December with only 234.4 mm for the season. Table 17.1 presents the quantity of water applied for all 4 seasons where the supplemental irrigations were added to the seasonal rainfall to determine total amount of the water applied in each treatment. Using gated pipe, 2 irrigations were added in the spring of 1986 and 2 irrigations in the spring of 1987. Using the sprinkler line-source, 2 irrigations were applied in the spring of 1988 and 3 irrigations in the spring of 1989. The total water requirement treatment, I_1, for supplemental irrigation received 705.1 mm of water in 1985-86 and 456 mm in 1986-87. The same treatment with the line-source, I_5, received 692.6 mm in 1987-88 and 417.8 mm in 1988-89. Treatments, I_1 and I_5, were irrigated when 50% of the available water was removed from the root zone; the wheat was under no plant-water stress throughout the growing season.

Table 17.1. Quantity of water applied and rainfall for the supplemental irrigation treatments for the 1985-86, 1986-87, 1987-88, and 1988-89 growing seasons at Tel Hadya.

Irrigation levels	Rainfall (mm)	Water added at germination (mm)	Supp. irrig. (mm)	Total added (mm)
1985-86				
Rainfed	315.1	30	0	345.1
1/3rd replenishment	315.1	30	120	465.1
2/3rds replenishment	315.1	30	240	585.1
Total requirement	315.1	30	360	705.1
1986-87				
Rainfed	316	0	0	316
1/3rd replenishment	316	20	40	376
2/3rds replenishment	316	20	80	416
Total requirement	316	20	120	456
1987-88				
Rainfed	503.6	0	0	503.6
0.2 W. Bal.	503.6	0	33	536.6
0.4 W. Bal.	503.6	0	75	578.6
0.6 W. Bal.	503.6	0	104	607.6
0.8 W. Bal.	503.6	0	140	643.6
1.0 W. Bal.	503.6	0	189	692.6
1988-89				
Rainfed	234.4	0	0.0	234.4
0.2 W. Bal.	234.4	0	31.5	265.9
0.4 W. Bal.	234.4	0	67.1	301.5
0.6 W. Bal.	234.4	0	104.9	339.3
0.8 W. Bal.	234.4	0	141.3	375.7
1.0 W. Bal.	234.4	0	183.4	417.8

Figure 17.1 shows the monthly rainfall variation by season. The 1987-88 rainfall was higher and better distributed than rainfall of the other seasons. The rapid cut-off of rainfall after December for the 1988-89 season shows the effect of drought. Normally, rainfall declines rapidly by mid-March with no effective rain in mid-April or May. By mid-March, the growth stage of wheat is heading or in anthesis and supplemental irrigation is essential to stabilize and increase yields. High rainfalls in the second week of February, 1986, refilled the soil profile to the 1.05 m depth and the low rainfalls of May aided the grain filling

process; however, supplemental irrigation was required. During 1987, rainfall effectively ceased by 1 April and during 1989 rainfall ceased at the end of December and, as would be expected, wheat yields were reduced under rainfed conditions.

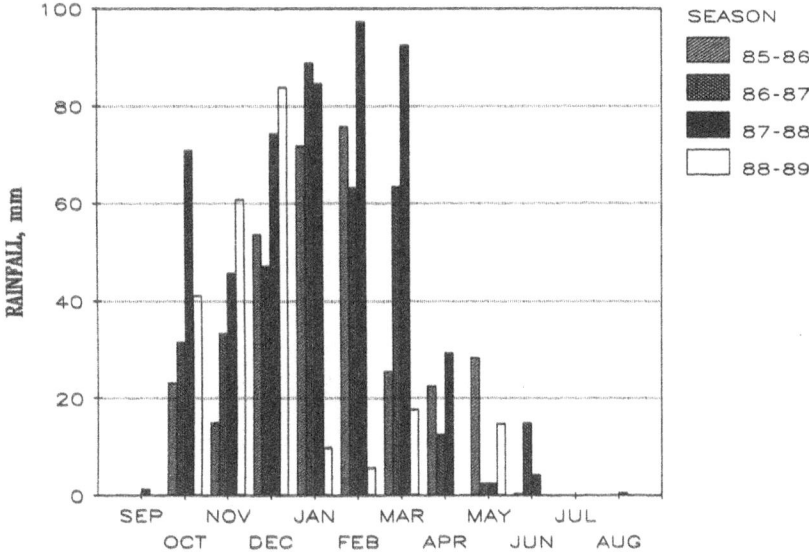

Figure 17.1. Monthly rainfall, 1985-89, Tel Hadya.

Figure 17.2 shows the weekly mean temperature starting on the first week of January and ending on the last week of May. From sowing time in mid-November to mid-December the temperatures were comparable for all 4 seasons. Only during 1988-89 did cold temperatures persist from mid-December through February which suppressed vegetative growth. A rapid rise in temperature occurred near the first week of March, 1989, and continued with high temperatures, accelerating senescence, until harvest time in the last week of May. Other seasons had relatively normal temperature increases except for a cold spell during 1987 where the temperature remained about 5° cooler than the other seasons until the last week of March then started to rise the first week of April. These low temperatures in 1987 delayed heading and anthesis for about a month as compared to 1986.

The vegetative stage shows a lag of about 17 days when comparing 1985-86 with 1986-87 (Table 17.2). Of more importance, a comparison of the heading stage of 1986 with 1987 identified a lag of 17 days. Comparing the 2 years, a lag of 15 days was indicated for the time of starting anthesis but a 10 day lag when completed. These data follow similar intervals as expressed in the weekly temperature data for the 4 years (Figure 17.2).

Research data show that for readily available soil moisture, the most sensitive stages of growth for wheat are boot (shooting), heading, and anthesis; however,

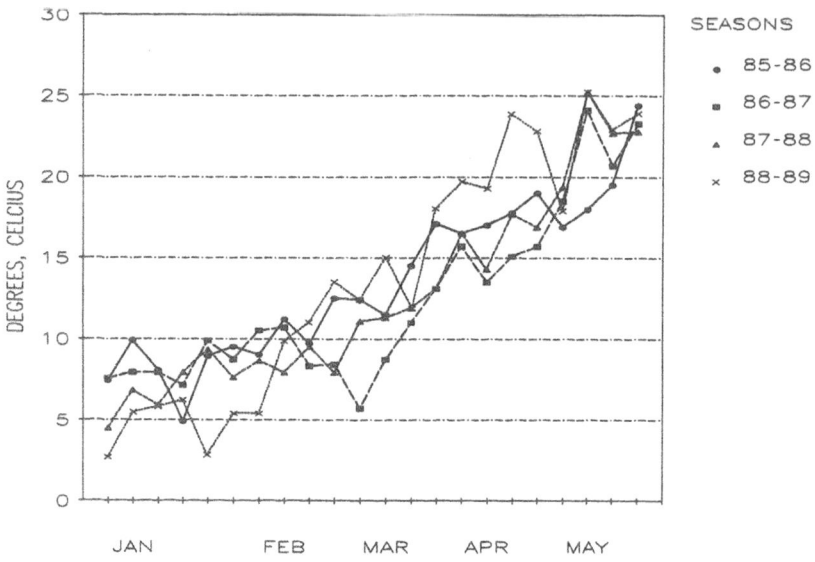

Figure 17.2 Weekly mean temperature from the first week of January to the end of May, 1986-89, Tel Hadya.

Table 17.2. Dates of growth stages for spring wheat at Tel Hadya, 1986-87.

Year	Stage of growth			
	Sowing	Vegetative	Heading	Anthesis
1986	1/12/85	10/01	17/03	3/04
1987	21/11/86	26/01	3/04	18/04
1988	19/11/87	7/01	10/04	27/04
1989	15/11/88	22/01	1/04	8/04

the different stages of rooting and the period of grain filling (milk, soft dough to hard dough) are of lesser importance. Therefore, where possible and for the most economic use of limited water supplies, supplemental irrigation should be scheduled at the moisture sensitive stages of plant growth.

Results of Supplemental Irrigation Studies

Research studies, Tel Hadya, 1985-89

The crop coefficient or K_c curves (Figure 17.3), show that the boot stage occurs at the initial change of slope or acceleration of the curve; heading starts when the curve is nearing its peak; and, anthesis occurs at the peak. Although rooting occurs during the increasing steep portion of the K_c curve, the effect of available soil moisture on root development is of lesser importance if soil moisture is available. After anthesis, the K_c curve declines rapidly because maturation and the process of grain filling and hardening is performed without a high soil moisture demand. The K_c curve for the 1988-89 season shows a distinct change from the previous 3 growing seasons. Here 2 effects occurred; first the cold spring reduced vegetative growth, then the rapid change to higher temperatures and a normal complete vegetative development (full cover) did not occur.

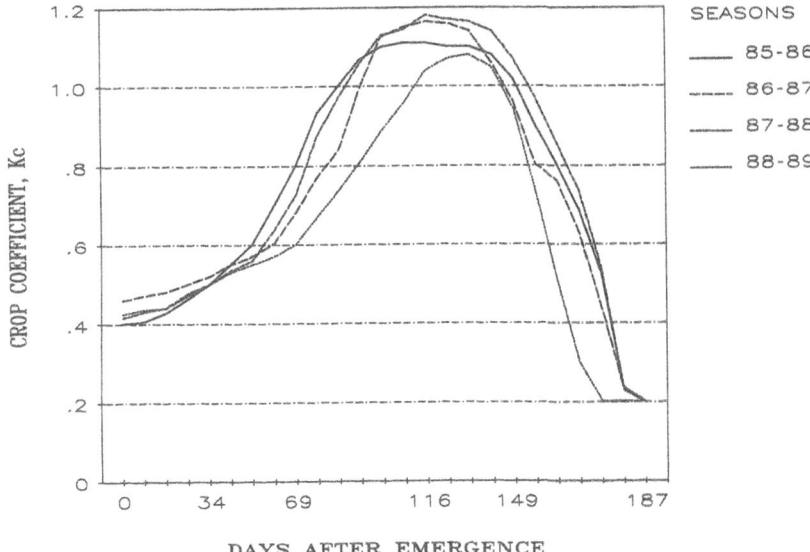

Figure 17.3. Crop coefficients for spring wheat, Tel Hadya.

In 1986, irrigation was needed on 23 March during heading and on 16 April near the end of anthesis. These irrigations were the most critical as supported by length of grain head and grain yield data. Two additional irrigations were applied, the first on 3 May when the wheat was in the soft dough stage and the second on 18 May during the hard dough stage. In 1987, 2 irrigations were needed, 1 on 20 April near the end of anthesis and 1 on 7 May when the wheat was in the milk stage. Two irrigations were applied in 1988, 1 on 30 April at the beginning of the milk stage and 1 on 12 May during the soft dough stage. In 1989, 3 irrigations were applied, 1 on 4 February during a sluggish vegetative stage, then 1 on 16 March, and 1 on 19 April just before the milk stage. In 1989, further irrigations were scheduled but strong wind conditions did not permit usage of the line-source sprinkler system.

The active root depth as determined by moisture extraction patterns using the neutron probe is shown in Figure 17.4. The final active root depth is fixed by the soil depth in the various fields. The effect of supplemental irrigation on active root depth could not be determined as during the 4 seasons of measurement the

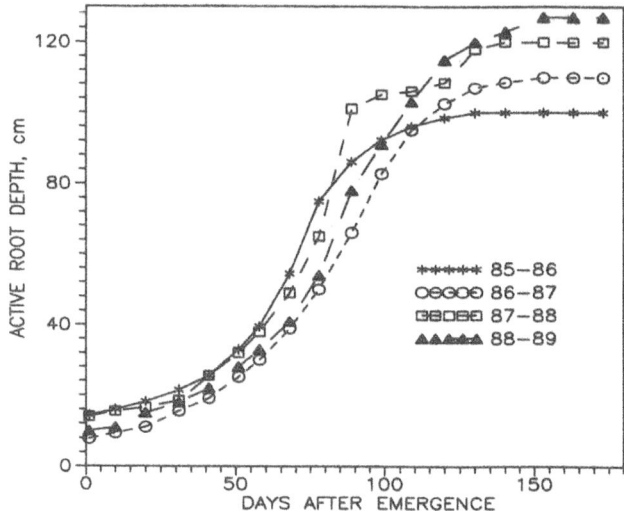

Figure 17.4. Active root depth in relation to days after emergence for spring wheat.

soil profiles were at field capacity by the first week of January. In general, the active roots are 1 m deep by the last week of February.

Plant heights for the 4 seasons comparing rainfed and water replenishment treatments either I_3 or I_5 are shown in Figure 17.5. The data show that without supplemental irrigation or nitrogen fertilizer applications, wheat heights are depressed; however, there was a response to low levels of nitrogen and low quantities of supplemental irrigation. These data would suggest that the wheat

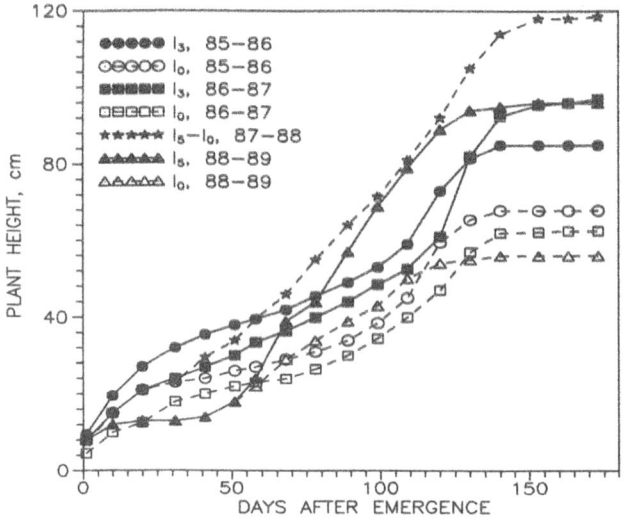

Figure 17.5. Plant height measurements for rainfed and total water replenishment treatments, I_3 or I_5, for 4 seasons.

varieties tested are excellent for rainfed agriculture and do not respond well to increased usage of water and nitrogen. No lodging of the plants was observed with high levels of nitrogen and supplemental irrigation.

Supplemental irrigation increases plant height by at least 40 cm. The comparison of rainfed data from 1985-86 and 1986-87 seasons demonstrates that a 30 mm supplemental irrigation of wheat in 1985 at sowing may have increased plant height by as much as 10 cm. There was a significant response in plant height, 89 cm, to low levels of nitrogen and low quantities of supplemental irrigation. The 1987 data for plant height on 4 wheat varieties is similar to the data presented for 1986; however, Cham I reached 99 cm in 1987, 10 cm taller than in 1986. Cooler temperatures in 1987 were a factor in extending the vegetative growth period for an additional month when compared to 1986.

During the 1987-88 season rainfed and supplemental irrigation treatments had the same plant height measurements. Also, plant height showed no measurable differences with nitrogen fertilizer. Differences in plant height data are attributed to the distribution of seasonal temperatures, plant available water throughout the season, and soil depth. Scheduling of supplemental irrigation was more important to increasing plant height than the quantity of water applied. Water available at sowing time either by rainfall or by the addition of a light supplemental irrigation was the most important.

Figure 17.6 shows the relation between active root depth and the plant height for spring wheat at Tel Hadya. The curves show a definite separation as to rainfed, I_0, and the 100% water replenishment treatments for supplemental irrigation, I_3 and I_5. The plant height can be used (with caution) to estimate the active root depth which will permit the calculation of the root zone moisture

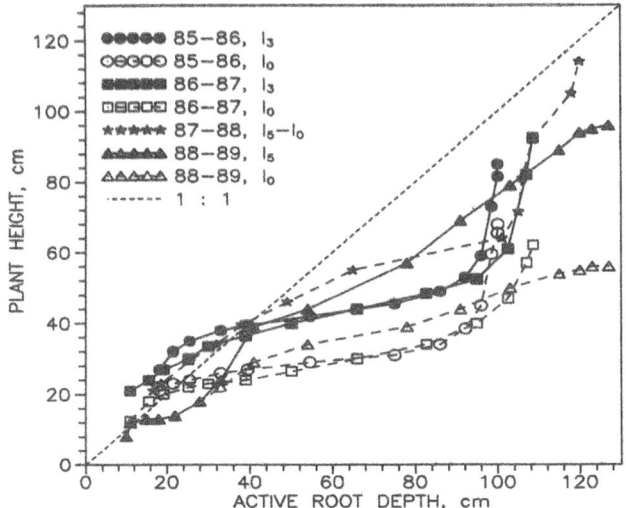

Figure 17.6. Plant height in relation to active root depth of spring wheat, Tel Hadya.

content, RZM, in the water balance equation. Because of the high rainfall during the early portion of 1987-88 season (see Figure 17.1), there was no difference between the curves for rainfed and supplemental irrigation. This type of information permits farmers within the region an indicator from which to calculate the water balance equation for irrigation. The flat portion of the curves takes place during the early stages of plant growth; however, as the active root depth approaches its maximum, at about 90 cm, then the plant height accelerates to its maximum. Both the active root depth and plant height are maximized immediately following anthesis.

Results of data analyses for the 4 seasons showed that wheat has similar responses to treatments of supplemental irrigation and additions of nitrogen. Grain yield of Cham I, 1985-86, increased 311% from 2.10 t/ha to as much as 8.64 t/ha (Table 17.3); in 1986-87, yield increased 218% from 1.7-5.4 t/ha (Table 17.4); in 1987-88, a wet year, yield increased 36% from 5.04-6.87 t/ha (Table 17.5); and, in 1988-89, a drought year with severe temperature changes, yield increased 450% from 0.66-3.63 t/ha (Table 17.6). For the first 3 seasons, grain yields maximized with 50-70 kg of N/ha and supplemental irrigation treatments I_3 or I_5; however, for the drought season of 1988-89, the best yield response was with a nitrogen treatment of 100 kg of N/ha. At the higher levels of the water balance treatments, the plants were never under a stressed condition. However, Cham I and Cham IV, which are excellent rainfed varieties (neglecting the poorer drought response of Cham I), showed that growth under some stressful conditions is not limiting to grain yield or straw weight. These data emphasize that timing of irrigation is an important factor

Table 17.3. Data for 1985-86 showing the effect of nitrogen and supplemental irrigation treatments, SIT, on grain yield, Tel Hadya.

Irrigation levels	Grain yield (t/ha)				Irrigation means [b]
	Levels of nitrogen (kg/ha) [a]				
	None	70	140	120	
Rainfed	2.10	3.56	2.74	3.46	2.97
0.33 W. Bal.	3.37	8.64	5.03	6.23	5.82
0.66 W. Bal.	4.26	7.10	7.18	6.50	6.26
1.00 W. Bal.	3.53	6.36	6.10	6.17	5.54
Nitrogen means [c]	3.32	6.41	5.26	5.59	

Level of significant difference, LSD:
[a] LSD, 5% among SIT for same nitrogen level = 2.14
              among SIT for different nitrogen levels = 2.25
[b] LSD, 5% = 1.07
[c] LSD, 5% = 1.28

Analysis of variance, ANOVA:
1. Levels of nitrogen: significant (F = 10.81, df = 3/6)
2. SIT: significant (F = 15.7, df = 3/9)

Table 17.4. Data for 1986-87 showing the effect of supplemental irrigation treatments and wheat variety on grain yield, Tel Hadya.

Irrigation levels	Grain yield (t/ha)				Irrigation means [b]
	Wheat varieties [a]				
	Bread		Durum		
	Cham IV	Mexipak	Sebou	Cham 1	
Rainfed	1.61	1.23	1.77	1.78	1.60
0.33 W. Bal.	4.69	3.41	5.17	5.35	4.66
0.66 W. Bal.	4.36	2.71	4.13	4.90	4.03
1.00 W. Bal.	4.01	2.85	4.55	4.75	4.04
Variety means [c]	3.67	2.55	3.91	4.10	

[a] LSD, 5% among SIT for same variety = 1.09
    among SIT for different varieties = 1.15
[b] LSD, 5% = 0.60
[c] LSD, 5% = 0.47

ANOVA: 1. SIT: significant (F = 82.4, df = 3/9)
       2. Wheat varieties: significant (F = 16.8, df = 3/6)

Table 17.5. Data for 1987-88 showing the effect of supplemental irrigation treatments and wheat variety on grain yield, Tel Hadya.

Irrigation levels	Grain yield (t/ha)		Irrigation means [b]
	Wheat varieties [a]		
	Bread	Durum	
	Cham IV	Cham 1	
Rainfed	5.02	5.04	5.03
1/5 W. Bal.	5.56	5.66	5.61
2/5 W. Bal.	6.10	6.44	6.27
3/5 W. Bal.	6.43	6.86	6.64
4/5 W. Bal.	6.20	6.76	6.48
1.0 W. Bal.	6.81	6.87	6.84
Variety [c]	6.02	6.27	

[a] LSD, 5% among SIT for same variety = 0.54
    among SIT for different varieties = 0.58
[b] LSD, 5% = 0.38
[c] LSD, 5% = 0.33

ANOVA: 1. SIT: significant (F = 9.0, df = 5/120)
       2. Wheat variety: significant (F = 31.4, df = 1/18)
       3. Fertilizer: significant (F = 2.01, df 15/120)

Table 17.6. Data for 1988-89 showing the effect of supplemental irrigation treatments and wheat variety on grain yield, Tel Hadya.

Irrigation levels	Grain yield (t/ha)		Irrigation means [b]
	Wheat varieties [a]		
	Bread Cham IV	Durum Cham 1	
Rainfed	0.81	0.66	0.74
1/5 W. Bal.	0.96	0.83	0.90
2/5 W. Bal.	1.65	1.47	1.56
3/5 W. Bal.	2.37	2.12	2.25
4/5 W. Bal.	3.11	2.92	3.01
1.0 W. Bal.	4.04	3.63	3.83
Variety [c]	2.15	1.94	

[a] LSD, 5% among SIT for same variety = 0.12
  among SIT for different varieties = 0.17
[b] LSD, 5% = 0.08
[c] LSD, 5% = 0.05

ANOVA: 1. SIT: significant (F = 1626.2, df = 5/120)
2. Wheat variety: significant (F = 164.8, df = 1/18)
3. Fertilizer: significant (F = 24.7, df = 3/18)
4. Interaction of fertilizer × SIT: significant (F = 7.1, df = 15/120)
5. Interaction of variety × fertilizer: significant (F = 10.1, df = 3/18)

for increasing grain yield. Other wheat varieties in 1986-87 showed a similar response to supplemental irrigation but the increases in yield were not as high.

The components of yield (Table 17.7), show that tiller survival in terms of head number/ha, spikelet number (head length), and kernel weight (1,000 grain weight) all contributed to the increased yield. In the 1985-86 season, supplemental irrigation was required before anthesis but irrigation occurred after anthesis in 1986-87. Supplemental irrigation had a significant effect on the length of grain head in 1985-86. Cham I had a longer head length in 1986-87 but the effect of supplemental irrigation was not significant for any variety. If supplemental irrigation is required before anthesis and during head development, there is a considerable increase in head length.

Weight/1,000 grains was significantly different for both supplemental irrigation and variety with bread wheat having smaller kernels than Durum. The bread wheat, Cham IV, had the lightest grain under rainfed conditions and the Durum wheat, Sebou, had the heaviest. With supplemental irrigation in 1986, only 34.7% of the yield increase of Cham I was attributable to weight/1,000 grains while, in 1987, it accounted for 71.7%. An additional 27.7% is due to head length (15.4%) and number of heads/ha (12.3%). The weight/1,000 grains for Sebou is attributed to a yield increase of 86.9%;

Verification of Supplemental Irrigation of Spring Wheat

Table 17.7. Plant components contributing to yield formation.

Variety	Heads/ha × 10⁶		Head length (cm)		Weight/1000 grains (gm)		Straw wt. (t/ha)	
	RF	SI	RF	SI	RF	SI	RF	SI
1985-86								
Cham I	6.08	11.84	4.40	5.40	30.95	41.68	4.95	8.87
1986-87								
Cham IV	6.70	9.00	7.73	8.20	23.35	36.95	4.40	7.59
Mexipak	4.58	7.13	8.23	9.15	27.18	37.98	4.05	7.44
Sebou	5.86	7.65	7.00	7.18	27.40	51.20	5.51	9.07
Cham I	6.75	7.58	6.38	7.38	28.05	48.15	4.84	8.60
1987-88								
Cham IV	4.29	5.67	7.66	7.93	27.49	34.73	8.38	9.56
Cham I	3.82	4.24	7.21	7.78	33.25	42.27	8.85	9.66
1988-89								
Cham IV	3.75	4.92	6.38	7.29	23.65	33.62	2.86	5.08
Cham I	2.56	3.31	6.11	6.50	25.47	38.25	3.09	5.57

whereas, an additional 46.3% was attributable to head length (15.7%) and number of heads/ha (30.6%). These percentages indicate that Sebou is a more efficient wheat plant than Cham I which supports other evidence that increases in yield using supplemental irrigation are related to increases in all plant components of yield formation.

Straw weight is a critical component of wheat production because straw weight can be livestock feed with a cash value almost the same as that for grain and can be as important to farmers as grain weight. Supplemental irrigation has a significant effect on straw and nearly doubles yield for some varieties. However, the Durum spring wheat varieties produced significantly more straw weight than the Bread wheat varieties.

Water use efficiency, WUE, as outlined by Bolton (1981) and Cooper (1983) is computed using the following equation:

$$\text{WUE} = \frac{\text{Biological yield (kg/ha)}}{\text{Total water applied (mm)}},$$

where biological yield is the sum of grain and straw weight in kg/ha and total water applied is the sum of precipitation and supplemental irrigation in mm. Figure 17.7 shows clearly the benefit in WUE for limited amounts of supplemental irrigation. The 1/3rd and 2/3rds replenishment treatments were significantly different from the rainfed and total water requirement treatments. The 1/3rd replenishment treatment was significantly different from the other treatments which confirms earlier results that the 1/3rd replenishment is the most efficient supplemental irrigation treatment: the least amount of water is

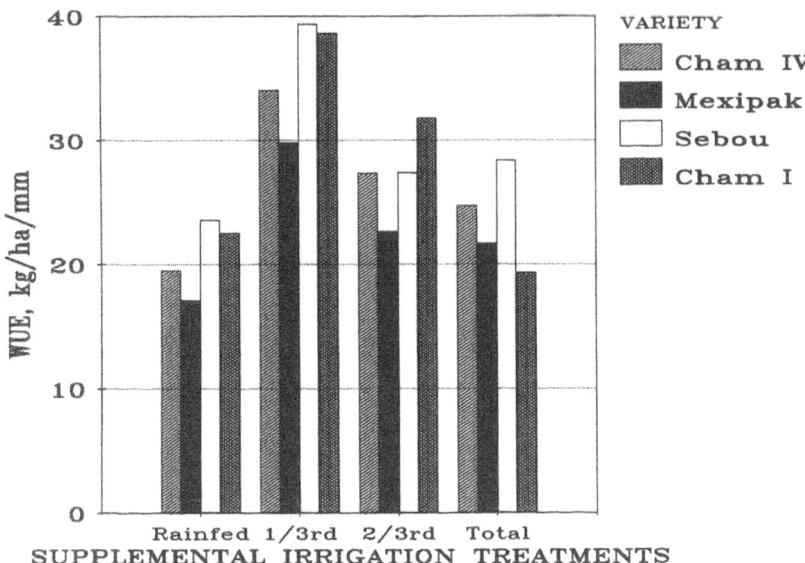

Figure 17.7. Water use efficiency for supplemental irrigation treatments and spring wheat varieties, 1985-86, Tel Hadya.

applied but irrigation is scheduled at the critical growth stage, 50% available soil moisture of plant water stress.

If the response to supplemental irrigation is calculated as

$$WUE_{irr} = \frac{\text{Increase in biological yield}}{\text{Irrigation water applied}},$$

Cham I produced 81 kg/ha/mm in treatment I_3 during 1985-86 and 199 kg/ha/mm in 1986-87. Some interesting relations were obtained for the wet season 1987-88 and the drought season of 1988-89 as shown in Table 17.8.

Table 17.8. WUE_{irr}, response to supplemental irrigation treatments for Cham I and Cham IV varieties, 1987-89, Tel Hadya.

Supplemental irrigation treatment	WUE_{irr} (kg/ha/mm)			
	1987-88		1988-89	
	Cham I	Cham IV	Cham I	Cham IV
I_1	30.6	18.8	9.0	17.4
I_2	27.6	27.2	14.3	23.3
I_3	25.1	25.0	20.6	26.6
I_4	19.0	17.1	26.8	27.5
I_5	16.3	14.7	29.7	29.7

Rainfall was sufficient to produce at least 5 t/ha for both varieties of spring wheat in 1987-88 and applications of only light amounts of supplemental irrigation were sufficient to increase the WUE_{irr} to high values. For example, Cham I produced 9,120 kg/ha of straw and 5,040 kg/ha of grain in treatment I_0 and 9,510 kg/ha of straw and 5,660 kg/ha of grain in treatment I_1. The biological yield difference = 14,160 − 15,170 = 1,010 kg/ha; therefore, the WUE_{irr} = 1,010/33 = 30.6 kg/ha/mm. In 1988-89, Cham I had nearly the same WUE_{irr} response in the higher irrigation treatments of I_4 and I_5 as Cham IV; but, Cham IV was better able to respond to low levels of irrigation as shown in treatments I_1, I_2, and I_3. The data show that Cham IV gives higher responses to lower amounts of irrigation than does Cham I during seasons of drought.

Research managed on-farm trials, 1986-89

Rainfall and water added using supplemental irrigation for 3 seasons at Breda and Mare'a, Syria, are shown in Table 17.9. No water was added for germination during 1987-88 but 20 mm was added in 1986-87 and 30 mm in 1985-86. Also, for 1985-86, a germination irrigation was added to all treatments including rainfed. At the Mare'a site, difficulty was experienced with farmer cooperation during 1985-86 and essentially all supplemental irrigation treatments for this site were the same.

Table 17.9. Quantity of water added to the supplemental irrigation treatments for 1985-86, 1986-87, and 1987-88, Breda and Mare'a, Syria.

Irrigation levels	Rainfall (mm)		Supp. irrig. (mm)		Total added (mm)	
	Breda	Mare'a	Breda	Mare'a	Breda	Mare'a
1985-86						
Rainfed	220	356	30	30	250	386
0.33 W. Bal.	220	356	90	210	310	566
0.66 W. Bal.	220	356	150	210	370	566
1.00 W. Bal.	220	356	210	210	430	566
1986-87						
Rainfed	225.8	359.4	0	0	225.8	259.4
0.33 W. Bal.	225.8	359.4	80	80	305.8	419.4
0.66 W. Bal.	225.8	359.4	140	140	365.8	499.4
1.00 W. Bal.	225.8	359.4	200	200	425.8	559.4
1987-88						
Rainfed	416.6	503.9	0	0	416.6	503.9
0.33 W. Bal.	416.6	503.9	88	98	504.6	601.9
0.66 W. Bal.	416.6	503.9	122	197	538.6	700.9
1.00 W. Bal.	416.6	503.9	265	295	681.6	798.9

For the 3 seasons, grain yields with supplemental irrigation appear to be in the region of 5-6 t/ha; whereas without supplemental irrigation, yields fall to within 1-3 t/ha and are highly dependent on rainfall (Tables 17.10, 17.11, and 17.12). The low grain yields during 1986-87 were attributed to management problems and to the use of the Sebou variety.

Table 17.10. Effect of supplemental irrigation treatments and location on grain yield of spring wheat (Cham I), 1985-86, Breda and Mare'a.

Irrigation levels	Grain yield (t/ha)		
	Location[a]		Irrigation means[b]
	Breda	Mare'a	
Rainfed	2.63	4.84	3.74
0.33 W. Bal.	5.55	8.01	6.78
0.66 W. Bal.	6.42	7.24	6.83
1.00 W. Bal.	4.02	8.69	6.36
Location means	4.66	7.19	

[a] LSD, 5% for same location and different SIT = 1.40
[b] LSD, 5% = 0.99

ANOVA: SIT: significant (F = 20.8, df = 3/12)

Table 17.11. Effect of supplemental irrigation treatments and location on grain yield of spring wheat (Sebou), 1986-87, Breda and Mare'a.

Irrigation levels	Grain yield (t/ha)		
	Location[a]		Irrigation means[b]
	Breda	Mare'a	
Rainfed	0.38	2.33	1.35
0.33 W. Bal.	1.21	3.16	2.19
0.66 W. Bal.	1.45	3.44	2.44
1.00 W. Bal.	1.50	3.02	2.26
Location means[c]	1.13	2.99	

[a] LSD, 5% for same location and different SIT = 0.68
[b] LSD, 5% = 0.40
[c] LSD, 5% = 0.48

ANOVA: 1. Location: significant (F = 147.9, df = 1/3)
       2. SIT: significant (F = 12,2, df = 3/18)

Figure 17.12. Effect of supplemental irrigation treatments' level of nitrogen on grain yield of spring wheat (Cham I), 1987-88, Breda and Mare'a.

Irrigation levels	Grain yield (t/ha)				Irrigation means
	Fertilizer level[a]				
	F_0	F_1	F_2	F_3	
Rainfed	3.30	3.04	3.98	4.16	3.62
0.33 W. Bal.	3.41	4.01	4.84	5.32	4.39
0.66 W. Bal.	3.98	4.97	4.79	5.01	4.69
1.00 W. Bal.	3.34	4.06	5.92	5.45	4.69
Variety means[c]	3.51	4.02	4.88	4.98	

[a] LSD, 5% for same fertility level and different SIT = 1.28
[b] LSD, 5% = 0.64
[c] LSD, 5% = 1.08

ANOVA: 1. Location: significant (F = 42.1, df = 1/3)
2. Fertility: significant (F = 8.58, df = 3/3)
3. SIT: significant (F = 5.83, df = 3/12)

Economics of Supplemental Irrigation of Spring Wheat

The economics of supplemental irrigation compares system costs with the market value of wheat grain, straw, and stubble in the short-term. To decide whether or not to use supplemental irrigation, the farmer must have immediate appraisal of local market values and expected profits on estimated returns. If an irrigation would cost more than the expected gain, then no additional water should be applied. An economic evaluation is computed on yield obtained from 1986-87 research studies to update water cost estimates, to assess supplemental irrigation and its impact on revenue, to predict the net benefits after irrigation costs, and to describe recommendations for optimal profitability. The comparison is made between the rainfed and 1/3rd replenishment treatments. Soumi (1987) reports data on water cost and profitability of supplemental irrigation in Syria which are comparable to results and trends for the Tel Hadya region.

Water cost estimates are a conservative appraisal of costs associated with delivery and supply which are based on the following measurements and calculations where:
1. total annual cost of water pumped = 39,500 SL;
2. annual pumping hours = 170 days × 22 hr/day = 3,740 SL;
3. average pumping capacity = 31 m³/hr (pumping test survey of 220 wells in Aleppo Province, Syria conducted by MAAR); and,
4. water pumped/yr = 3,740 hr × 31 m³/hr = 115,940 m³/yr.

The cost of water pumped is 0.34 SL/m³. The cost of supplemental irrigation for the season at 600 m³/hr for the 1/3rd replenishment treatment is 204 SL/hr (600 × 0.34).

The output prices for spring wheat yields are given as:

grain = 2.5 SL/kg or 2,500 SL/t for bread wheat;
 = 2.6 SL/kg or 2,600 SL/t for durum wheat; and,
straw = 0.85 SL/kg or 850 SL/t.

Partial budget analysis is a simple economic technique used to evaluate economic superiority of different treatments tested in a research trial. The results from the analysis are shown in Figure 17.8. Data from the supplemental irrigation study showed that the 1/3rd replenishment treatment was more efficient than the other treatments. Therefore, only data from the rainfed and 1/3rd replenishment treatments will be compared in the partial budget analysis.

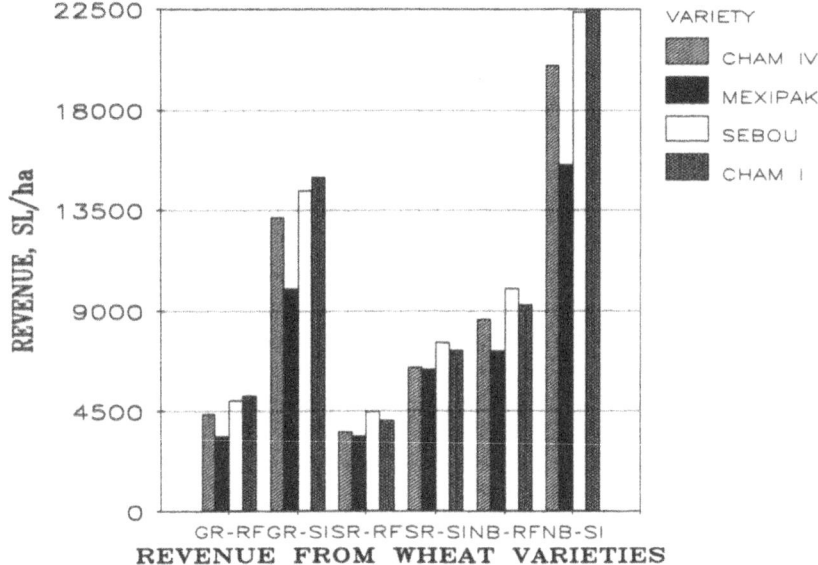

Figure 17.8. Partial budget analysis comparing grain, straw and net benefit revenue derived from rainfed and the 1/3rd replenishment supplemental irrigation treatment for the 4 spring wheat varieties.

Durum wheat varieties, Sebou and Cham I, produced higher revenues from grain and straw with higher net benefits than did the Bread wheat varieties, Cham IV and Mexipak, under both rainfed and supplemental irrigation treatments. Sebou and Cham I which have similar net benefits are at the top of the superiority rank order, followed by Cham IV which has a reduced net benefit of 16% of Sebou's net benefit from the rainfed treatment and 14% lower for the 1/3rd replenishment treatment. Mexipak, at the bottom of the

Verification of Supplemental Irrigation of Spring Wheat

superiority rank order, produced a net benefit 30% lower than Sebou for the rainfed treatment and 31% lower than Sebou for the 1/3rd replenishment treatment.

Grain yields increased 190% and straw yields increased 73% with a gross benefit for grain and straw of 136% and a net benefit of 124%. Whereas 1 m^3 of precipitation produced 0.56 kg of grain and 1.5 kg of straw, equivalent to 2.50 SL, applications of of 1 m^3 added by supplemental irrigation resulted in yield increases of 5.65 kg of grain and 1.5 kg of straw, equivalent to 19.3 SL.

Rate of return for capital investment in supplemental irrigation of wheat is exceptionally high at 5,670% for 1986-87 and 2,270% for 1985-86. Comparing extra gross benefits produced by 1 m^3 of supplemental irrigation, i.e. 19.3 SL, with water cost estimate of 0.34 SL/m^3, indicates the immense profitability that farmers can expect from a supplemental irrigation of wheat. Conservative estimates of water cost do not exceed 0.50 SL/ha for deep pumping. Supplemental irrigation from other sources costs even less than deep pumping (Soumi, 1987).

The supplemental irrigation requirement of the 1/3rd replenishment treatment of the 1985-86 trials, 150 mm or 1,500 m^3/ha, although representing a decrease of irrigation profitability when compared to 1986-87 trials, is still feasible because the trend did not change. According to 1985-86 results, 1 m^3 of supplemental irrigation increased yield by 2.26 kg of grain and 2.31 kg of straw, equivalent to 7.72 SL, which is higher than the conservative estimates of water cost.

Discussion and Conclusions

The comparison of both seasons shows the effect of weather on plant growth and production. The warmer season of 1985-86 significantly increased grain yield; however, it had a negative effect on straw yield and was a constraint to plant height. The 0 nitrogen treatment in the 1985-86 trials headed nearly a week before the treatments which received various levels of nitrogen. The cooler 1986-87 season delayed heading for all treatments; however, rainfed treatments reached maturity nearly 2 weeks before the treatments which received supplemental irrigation.

In all measurements for the Cham I variety of Durum spring wheat, the 1/3rd replenishment treatment for supplemental irrigation produced the same yield as higher amounts of water application, i.e. 2/3rds and total water requirement treatments. The Durum wheat varieties produced higher grain and straw yields with higher values of water use efficiency than did the two varieties of Bread wheats. The 30 mm of water added to all treatments in 1985-86 to ensure germination contributed to increased yields for the rainfed treatments. These data emphasize that scheduling of supplemental irrigation is as important as quantity of water applied. Water applied at sowing time for early germination is an important factor to plant height.

The lower level of nitrogen (70 kg/ha) produced the same yield as increased levels, 140 and 210 kg/ha. When phosphorus is not limiting and when nitrogen is applied as 30 kg/ha at sowing time and 40 kg/ha before tillering further additions of nitrogen may not be beneficial. Measurements of plant height, head length, straw yield, and weight/1,000 grains showed similar results to those mentioned above.

These data show that supplemental irrigation scheduling for spring wheat is effective using water balance techniques with measurements of rainfall, pan evaporation and estimates of rooting depth and crop coefficient curves for the region. The standard guidelines should be followed for scheduling irrigations to replenish the water in the root zone when 50% of the available water has been used. The technique is an accurate indicator for scheduling irrigation but the quantity of water to be applied can be reduced to 1/3rd replenishment in the root zone with a higher water use efficiency. In the Near East and North Africa, spring wheat usually completes the vegetative and tillering stages of plant growth well before the end of effective rainfall, and well-timed light applications of supplemental irrigation if needed at heading, anthesis, or milk stage can insure an increased and stabilized yield.

The margin of return from supplemental irrigation is higher than the cost of capital investment and operation. The small volume of water, 60 mm or 600 m^3/ha, applied as supplemental irrigation with 20 mm at germination, 20 mm during anthesis, and 20 mm during the milk stage had a significant effect on yield when compared with yields of rainfed wheat. The net benefits derived from these increased yields with these low volumes of water applied reflects the impact supplemental irrigation can have on the production economics of rainfed wheat. In the long-term, supplemental irrigation research and policy development must reflect that the shortage of water is critical to agricultural production because population density is increasing exponentially. Supplemental irrigation is one method of managing this escalating shortage by applying only sufficient water to improve and stabilize yield to provide an adequate profit margin.

References

Bolton, F.E., 1981, "Optimizing the Use of Water and Nitrogen through Soil and Crop Management," pp. 231-248, *In* Soil Water and Nitrogen in Mediterranean-type Environments, Eds. J. Monteith and C. Webb, Martinus Nijhoff/Dr. W. Junk Publs., The Hague.

Cooper, P.J.M., 1983, "Crop Management in Rainfed Agriculture with Special Reference to Water Use Efficiency," Proc. 17th Coll. Int. Potash Inst., Bern, pp. 63-79.

Doorenbos, J., and W. O. Pruitt, 1984, "Guidelines for predicting crop water requirements," FAO Irrigation and Drainage Paper No. 24, FAO/UNDP, Rome.

Finkel, H. J., 1983, "Irrigation of Cereal Crops," pp. 159-189, *In* CRC Handbook of Irrigation Technology: Volume II, Ed. H. J. Finkel, CRC Press, Inc., Boca Raton, Florida.

Kirkham, M. B., and E. T. Kanemasu, 1983, "Wheat," pp 481-520, *In* Crop-Water Relation, Eds. I. D. Teare and M. M. Peet, John Wiley & Sons, New York.

Marttin, D., and J. van Brocklin, 1985, "The Risk and Return with Deficit Irrigation," ASAE No. 85-2594, 28 pp.

Perrier, E.R., 1986A, "Small Scale Water Harvesting Techniques," USDA/USAID Workshop on Soil, Water, and Crop/Livestock Mgmt. Systems for Rainfed Agr. in the Near East Region, Amman, Jordan.

Shalhevet, J., A. Mantell, H. Bielorai, and D. Shimshi, 1976, "Irrigation of Field and Orchard Crops under Semi-arid Conditions," Inter. Irrigation Inf. Center, Ottawa, Canada.

Shipley, J.L., 1977, "Scheduling Irrigations with Limited Water," 5th Annual High Plains Grain Conference, Bushland, Texas.

Soumis, G., 1987, "Potentials for supplemental irrigation in SAR," paper prepared for ICARDA/FAO Regional Seminar, Rabat, Morocco.

Chapter 18

Irrigation of Cereals in Algeria

LARBI BAGHDALI

This report contains an overall review of supplemental irrigation of cereals in Algeria and an outline of the prospects of development in that field. The analysis focuses on cereal production because of its importance to the country's agricultural economy and the increase in yields that supplemental irrigation would permit.

This chapter is based on: talks with experts from the Ministry of Agriculture; surveys carried out in sample areas; and, analysis of studies and statistics on agriculture and water resources in Algeria. The chapter is divided into 4 parts:
1. characteristics of field conditions;
2. main characteristics of Algerian agriculture;
3. the role of irrigation in Algerian agriculture and existing methods of supplemental irrigation; and,
4. developmental prospects for supplemental irrigation of cereals.

Characteristics of Field Conditions

Agro-Climatic Zones

Algeria, unlike other North African countries, is a very large country with a total land area of 2,380,000 km^2. Its morphological and bio-climatic characteristics permit a variety of landscapes and agricultural systems. Between the humid areas in the north where intensive cultivation prevails and the arid platform of oases in the south lies a mountain range, the Atlas Tellian – Atlas Saharian. This area is intersected by plains where the agricultural activity differs from east to west. Figure 18.1 presents a map of Algeria.

The type of agriculture is essentially determined by the irregularity of the precipitation and land topography. The seasonal and geographical distribution of precipitation is not uniform and can lead to marked regional diversities. The threat of drought in almost every region compromises cereal production as a consequence of a dry fall, a dry spring, or, as in humid and sub-humid areas,

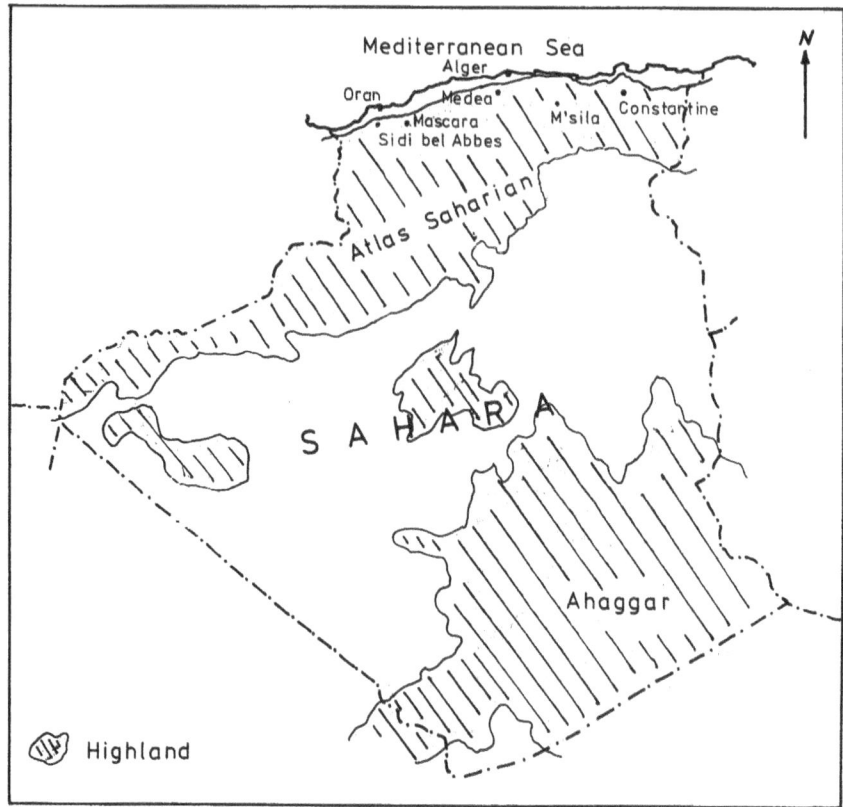

Figure 18.1. Map of Algeria.

late rainfall. The combination of morphological and climatic factors, divides the north of the country into 3 large agro-climatic zones:

Humid zone

The humid zone includes the coastal and sub-coastal regions from the center of the country eastwards. Average annual rainfall exceeds 600 mm and is relatively well distributed and favorable for runoff collection over the greater portion of the year. Pedoclimatic characteristics provide this area with considerable hydro-agricultural potential, particularly in the great plains of Mitidja in the center of Annaba in the east. However, even with these relatively sufficient rainfalls, when compared with the rest of the country, irrigation is essential in this zone for crop production during the summer season.

Semi-arid zone

The semi-arid zone possesses a number of distinctive areas including the Tell Plains and the northern border of high plains. The average annual rainfall varies between 350-600 mm but distribution is irregular. Two distinctive farming systems can be identified: in the west on the Oranie and the Cheliff Plains, mixed farming is practiced, especially irrigated arboriculture (citrus fruits in particular) and viticulture. In the east on the high plains, dryland production of cereals prevails. Under dry conditions in the east, average cereal yields are insubstantial and production is irregular from one year to the next. These yields could be improved with the addition of supplemental irrigation to compensate for erratic seasonal rainfall.

Sub-arid zone

The sub-arid zone has low rainfall and is agro-pastoral with hilly terrain. The largest quantity of rainfall runoff is at flood time only and the excess water is used for irrigation. Average annual rainfall varies between 200-350 mm with continental temperatures prevailing, 20-30 days of frost and more than 40 days of sirocco winds per year. Soils exhibit crusting which limits agricultural potential. Cereal crop production is a high risk, especially the growing of barley associated with extensive livestock breeding of sheep.

Water Resources

On the whole, water resources in Algeria are limited and extremely variable in space and in time. Wadis have seasonal and irregular annual flows with violent and abundant flooding but with low water levels. However, some of these wadis are perennial and maintain water flows in summer. This is the case in Chellif and Soumman where large watershed reservoirs are drained. On the other hand, high plains wadis are usually dry and flow only in flood time.

Average annual runoff at the outlet of watershed reservoirs is estimated at 12.4 B m^3 particularly in the northern part of the country to the north of the Saharan Atlas. The distribution of this runoff varies and constitutes a constraint to the hydro-agricultural improvement of the lands. The humid area contributes 63% of Algeria's total water resources in rainfall runoff of 7.8 B m^3; the semi-arid area contributes 26% with 3.3 B m^3 of runoff; and, the sub-arid area constitutes only 6% with 0.7 B m^3 of total runoff recorded in the north of the country.

On a regional scale, there is also an imbalance in surface water distribution with the west getting only 18% of the total runoff (2.2 B m^3) or 32% of the total surface water in the north. The east of the country gets 44% of the total runoff (5.5 B m^3) or 19% of this surface water. Watershed reservoirs of the sub-arid area represent 42% of the total surface water. Surface water resources are now being stored in large dams and reservoirs. Their volume is maintained at a regular level equal to 1.4 B m³ or 11% of the average annual rainfall runoff.

As for groundwater, in the north of the country its potential amounts to 1.8 B m³. The exploitation of these resources amounts to 80%.

Algeria is a country with difficult agro-climatic conditions. The potential of farming lands is limited and the annual rainfall is irregularly distributed. For winter cultivation, supplemental irrigation would contribute dramatically to the improvement of yields. Likewise, summer cultivation is not satisfactory unless irrigated.

Main Characteristics of Algerian Agriculture

Land Potential

The useful agricultural surface (UAS) amounts to 7,500,000 ha, i.e. 3% of the total land area or 20% of the country's northern region. This percentage cannot be easily expanded because of natural constraints. Furthermore, the UAS per capita is decreasing, from 0.75 ha in 1963 to 0.40 ha in 1980. By the year 2000, arable land per capita will dwindle to 0.20 ha. Table 18.1 shows the distribution of the UAS in each agro-climatic zone.

Table 18.1. Distribution of arable land by agro-climatic zone.

Classification	Annual rainfall (mm)	UAS (ha)	(% of area)
Humid area	600	400,000	5.3
Semi-arid	350-600	5,100,000	68.0
Sub-arid	200-350	1,100,000	14.7
Mountains / oases		900,000	12.0
Total Useful Agricultural Surface (UAS)		7,500,000	100.0

Land use

Table 18.2 presents data on land use under rainfed conditions and with irrigation. Cereal production covers an average 3,300,000 ha, i.e. 44% of the total useful agricultural surface and 74% of the total arable land. When land use distribution is viewed geographically, the semi-arid zone accounts for 75% of the nation's land sown to cereals.

Table 18.2. Land use in Algeria.

Crop	Rainfed (ha)	Irrigated (ha)	Total (ha)	% Area
Cereals	3,293,000	7,000	3,300,000	44
Artificial fodder	725,000	25,000	750,000	10
Vegetables	163,000	95,000	258,000	3.4
Dry pulses	152,000	—	152,000	2
Industrial crops	18,000	8,000	26,000	0.34
Other crops	—	2,000	2,000	0.02
Fallow lands	2,290,000	—	2,290,000	30.5
Total[a]	6,641,000	137,000	6,778,000	90.26
Natural meadows	28,000	—	28,000	0.37
Arboriculture	337,000	81,000	418,000	5.5
Vineyard	211,000	2,000	213,000	2.8
Date palms	—	62,000	62,000	0.8
Total[b]	576,000	145,000	721,000	9.47
Total UAS land	7,218,000	282,000	7,500,000	100.00

[a] Land suitable for cultivation
[b] Land under permanent cultivation

Development of Agricultural Production

Agriculture accounts for 7-8% of the gross domestic production in the national economy. Table 18.3 shows the planning schedule for development of agricultural production.

Cereal production

Even though national yields have been improving since the 1985 season, average yields are still low. The improvement of yields reflects the implementation of a large scale intensification program of cereal production using several cultivation methods based on mechanization, fertilizers, and intensive pest control in addition to the introduction of high yielding varieties.

Table 18.3. Comparison of actual yields and planned production goals (q/ha), 1967-86.

Plans	3 yr 1967-69	4 yr 1970-73	4 yr 1974-77	5 yr 1980-84	1 yr 1985	1 yr 1986
Planned annual production	20	24.3	24	30	30	30
Annual ave. production	18.6	19.3	18.9	17	29.2	24
Ave. yield	6.5	6	6.2	6	9	7.5

Other plant and animal products

Overall agricultural production (Tables 18.4 and 18.5) is improving due to the use of new production techniques and actions aimed at stimulating production efforts.

Table 18.4. Production of non-cereal crops, 1980-86.

Crop	Average production (t)		
	1980-84	1985	1986
Dry pulses	37,000	46,000	65,000
Dry fodder	695,000	849,000	1,260,000
Vegetables	1,311,800	1,957,000	2,500,000
Industrial crops	146,720	186,000	220,000
Aboriculture	610,000	615,000	662,000

Table 18.5. Production of livestock, 1980-86.

Product	Average production (t)		
	1980-84	1985	1986
Red meat	146,500	160,500	170,000
White meat	121,250	163,000	210,000
Milk	717,600	713,000	750,000
Eggs	27,690	83,750	110,000

The Role of Irrigation in Agriculture

The hydro-agricultural sector constitutes a major contribution to agricultural production; this is demonstrated in that 40% of the total production of the country is produced by irrigated agriculture which ensures food security by mitigating climatic hazards. Irrigated lands in the north of the country extend over nearly 300,000 ha or 4% of the UAS not including the area under micro-irrigation which is still not well known. The hydro-agricultural sector has not improved very much since the sixties. However, some important progress has been made towards replanning the infrastructure of irrigation and improving existing drainage.

Structure of Irrigation Planning

Irrigation is subdivided into 2 systems: large surface application with large isolated equipment, and the small or medium irrigation systems. The large systems deliver irrigation to large land areas averaging 90,000 ha, most often with water supplied from a barrage (dam). The management, improvement, and maintenance of the system together with the delivery networks and conveyance equipment is the responsibility of 4 offices of irrigation in the Ministry of Hydraulics.

Small and medium systems are linked to local water resources such as wadi runoff water, groundwater supplies, mountain reservoirs, small dams, or flood water spreading. Usually flood water is used for supplemental irrigation of winter cereals but it can be used for more important cropping in years with above average rainfall. The management of the irrigation systems and their infrastructure, depending on their relative size, are the responsibility of either the Office of Wilaya, farmer operators, or farmer irrigation cooperatives, with the assistance of the Extension Services.

Irrigation Methods

Due to the climatic conditions, irrigation is necessary for all summer cultivation and contributes substantially to the improvement of winter yields. The most widespread irrigation method is to apply water from April to October to those crops grown during the summer with almost no rainfall. Consequently, the cultivation of vegetables in seasonal rotation needs large quantities of water (from 5,000 to 10,000 m^3/ha depending upon the efficiency of application). In addition to summer irrigation, in the winter rainy season, supplemental irrigation supplies plant water requirements during temporary dry periods when the absence of rainfall could affect the yield.

Supplemental Irrigation

Supplemental irrigation is applied to winter cereals, wheat and barley, in the High Plains in the eastern region and more particularly in the area of Hodua. The method is based on flood water distribution which is an uncertain practice because supply is closely linked to rainfall distribution and amount. Table 18.6 presents the existing area sown to winter cereals and cultivated using supplemental irrigation.

Table 18.6. Land area cultivated using supplemental irrigation, 1966-72.

Year	Land area (ha)
1966-72	100,000
1973-76	40,000
1977-83	7,000
1984-85	25,000

Lands irrigated by flood water distribution amounted to 100,000 ha during 1966-72. Since that time the area has decreased to as little as 7,000 ha between 1977-83. This method of supplemental irrigation is gaining interest nowadays, coinciding with the measures taken to revitalize the agricultural sector. In other respects, for example in the Cheliff Valley, supplemental irrigation of winter cereals has almost disappeared.

Flood water distribution in the Hodna

The climate of the Hodna area is essentially sub-arid as prevails in the agro-pastoral zones. The practice of cereal cultivation in association with livestock production of sheep depends mainly on barley. Production relies on techniques of extensive cultivation and is closely linked to the probability of receiving flood water using supplemental irrigation for distribution. Yields vary according to actual annual rainfall amounts and distribution patterns; therefore production sustainability is linked to annual rainfall.

In most cases, rainfall runoff diversion systems are designed on a traditional concept of direct distribution to the cultivated soil or into "gabions" to fall downwards in cascades along the wadis which regulate the flood water flow into storage for application with supplemental irrigation. This water is then conveyed to segments of the cultivated areas through systems of "seguias". In some cases, modern concepts have been incorporated which allow better management and greater control of the flood water.

In Hodna, cultivation of barley requires 3,500 m^3/ha of water if sown before December. For cereals sown later, the vegetation extends through May and therefore requires additional moisture after cessation of the rainy season. The exact quantity of additional water is not known for flood water distribution because of the perceived difficulty of measuring the application rate using supplemental irrigation. Farmers estimate the quantity of water at 2,000 m^3 which is applied to the cereal twice a year (October and March or April). The difference in amounts (3,500 m^3 and 2,000m^3 could be explained by the fact that farmers try to create large water reserves in the soil to compensate for lack of water in the interval between the two irrigation periods.

The actual quantity of water applied to cultivated fields varies according to annual rainfall. The rainy season of March and April can have an extreme effect on yields. This period coincides with a critical growth stage which is severely affected by deficient water. In times of scarcity, cereals are grazed or used as hay but rarely are harvested as grain.

The production of cereals in this area follows a 2-year rotation, cereal – fallow, a rotation of limited value when associated with seed varieties, fertilizer application, and some supplemental irrigation. Average annual cereal production is characterized by medium yields of about 5 q/ha.

Growing of cereals in the Middle Cheliff area

The Middle Cheliff Plain is located in the Tell. It is bounded in the north by the Dahra Mountains and in the south by the Ouarsenis Massif. Despite its proximity to the sea, its climate is remarkable and characterized by an average annual rainfall of 420 mm. Eighty five percent of precipitation is concentrated between October and April. Soil preparation can be done in August but, more frequently, is done after the first rainy days of October to facilitate penetration of tillage equipment and to destroy the early emergence of weeds.

In the Middle Cheliff, cereals are grown on 8,000 ha which is equivalent to one third of the area; they are grown under rainfed conditions even on farms which are equipped with irrigation systems. Durum wheat is the major grain produced with a certain homogeneity of farming practices. Wheat is seeded when the soil is wet. The best time for sowing is the first 2 weeks of November; however, this period can extend until the end of December in dry seasons. Seeds are broadcast or drilled with the chosen varieties restricted to traditional strains better adapted to local conditions but with scant potential for economic production. For many years, the Mexican varieties have been used but they have not adapted well to the heavy soils or to the dry period during April/May. Organic manure is, in general, weak and is applied by hand or with a centrifugal spreader but only a few treatments are given. Harvesting operations begin on 1 June and end by the middle of July. Harvesting is done mechanically.

Yields are low with average yields of 7q/ha in 1984. Low yields are attributed to the method of cereal cultivation combined with the climate which curtails the regular supply of water to the plant, disallowing a sufficient quantity for growth without stress. Continuation of existing practices without introduction of supplemental irrigation for cereal production restricts the use of high yielding varieties and other agronomic inputs. The primary limiting factor is water for intensified cereal production in the Middle Cheliff area: a stable, adequate water supply is necessary, efficiently applied with supplemental irrigation.

Development Prospects for Supplemental Irrigation

Due to the large land area under cereal production (74% of the land area cultivated annually) and to the importance of cereals in food consumption (180 kg/inhabitant/yr), cultivation of cereals constitutes an important activity in Algerian agriculture and has a strategic stake in the national economy.

Analysis of cereal yields over a period of years shows that yields have remained static. Annual production does not fulfill the needs of the population; consequently, continual increases (varying according to the annual production levels) are required in imports of cereals, totaling 50-70% of the total national consumption of wheat. Imports of cereal grains for human consumption alone amount to 30% of total national imports.

In view of these facts, large-scale operations for intensification of cereal production have been set up in the last few years. These operations essentially aim at improving the technical aspects of production by controlling cultural methods of cultivation. Target yield levels have been established for the various agro-climatic zones: 16 q/ha in areas with high potential and 14 q/ha on land with lesser potential. Due to these intensification efforts during the 1985 season, production reached an absolute record of 30 M q. In 1986, production reached only 24 M q because of unfavorable rainfall in the east. Production in 1987 reached 20 M q but production levels suffered, in the west, from a severe deficiency of rainfall during March and April.

The cyclical progress of production shows the necessity of consolidating and supporting comprehensive intensification programs. Programs aimed at the improvement of various factors related to soil and plants are still inadequate and need to be followed up by further action in order to get one of the more crucial factors of production, i.e. water, under better control. For this reason supplemental irrigation of winter cereals is considered vital wherever the potential water supply is sufficient to guarantee stable production of cereals.

The analysis of regional agro-climatic characteristics as well as potential water supply permits identification of the necessary elements associated with the form and size of future programs concerning the supplemental irrigation of cereals.

Areas for Intensive Cultivation

Intensive production of cereal under rainfed conditions takes place in the Tell Plains in the west, in the High Plains in the center (El-Sersou), and in the east. These areas produce 70% of the average annual production of cereals. Despite the favorable amounts of rainfall in these regions, farmers are confronted with the risk of seasonal deficiencies for crops during March and April. This period corresponds to a critical growth stage in the cereal's vegetative cycle. Supplemental irrigation would allow the management of rainfall deficiencies through application of small amounts of water to every hectare. However, regional distribution of water resources imposes certain constraints to the development of infrastructures for individual supplemental irrigation systems.

Water resource availability in the east is favorable for large-scale distribution on the High Plains using collective irrigation equipment and includes inter-basin transfers for certain areas. In the west on the Tell Plains, the solution tends towards more localized water resources being used on established small farms.

From an agronomic perspective, supplemental irrigation could consolidate efforts for intensive production and progressively generalize into the use of high yielding seed varieties. Nonetheless, in both cases, the introduction of supplemental irrigation will require research experimentation, extension activities to popularize the results, and technological transfer to encourage farmers to adapt supplemental irrigation and the recommended methods of cereal production and to abandon other alternative crops of higher profitability.

Cereal production in the humid areas would tend to decrease and the intensive crop rotations recommended for these areas could be implemented during the dry season.

Areas for Extensive Cultivation

In regions of extensive cereal cultivation, flood water distribution methods need to be sustained and developed especially in the Hodna area where large tracts of land are being irrigated even in good rainfall years. With this method of irrigation, cereal production must be confined to favorable soil conditions benefiting from rainfall runoff through adequate hydraulic works. However, it must be kept in mind that flood water distribution is closely linked to rainfall and that the decrease in yields in low rainfall years cannot be avoided.

In addition to the constraint of variable annual rainfall, even if water transfer from the northern basins were technically possible, hydraulic water works would require large capital investment and therefore this option would be dependent on the economic profitability of production.

Conclusions

Supplemental irrigation methods in Algeria are limited except for traditional practices of flood water distribution in the sub-arid zone and in the Cheliff area which uses irregular rainfall runoff.

Agro-climatic conditions in the production areas of the north make supplemental irrigation an essential part of the policy for intensification of cereal production. Water availability throughout the country, and in the areas of intensive production in particular, allows the introduction of this appropriate technique. However, irregularity of water resources in the various regions implies a need for some inter-basin transfers of water thus requiring large capital investment.

Moreover, in order to attain significant objectives, it would be advisable to define production systems which are adapted to each region, particularly those based on selection of seed varieties and appropriate cultivation methods. There is a need to account for regional characteristics of production, to perform local field verification of the experimental results, and to ascertain the feasibility of farmers adapting the techniques for intensive cereal production. Supplemental irrigation could play a significant role in increasing cereal yields thus enabling better food security for Algeria.

Chapter 19

Supplemental Irrigation Systems in Cyprus

V. C. KRENTOS

This study was undertaken within the context of ICARDA's and FAO's cooperation to develop a regional perspective on the efficacy of supplemental irrigation for increasing and stabilizing field crop production in the Near East and North Africa. Although supplemental irrigation is confined to only a few areas in Cyprus, this technique is known to increase significantly the yields of field crops such as wheat and barley. In one selected district, supplemental irrigation has been practiced by farmers for many centuries from generation to generation by virtue of communal and individual water rights and is interwoven with social tradition and farming practice.

In this chapter, the natural and environmental features of Cyprus are described; next, the land and water resources and the infrastructure at the national level are described; then, these aspects are projected onto the district level to contrast existing farming systems in 3 selected districts with contemporary techniques and future prospects of supplemental irrigation.

The National Perspective

Geographical, Morphological, and Geological Features

Cyprus has an area of 9,251 km^2 and is the third largest island in the Mediterranean Sea (after Sicily and Sardinia) situated between 34°33' to 35°41' North and 32°20' to 34°35' East, and located 70 km from southern Turkey, 100 km west of Syria, and 370 km north of Egypt.

The island has, transversely from east to west, the high dome-shaped Troodos Massif in the southwest rising to 1,951 m and the long narrow Kyrenia Range, with highest elevation of 900 m, extending along the northern coast. Between these two mountain ranges the capital, Nicosia, is situated on the central plain with low relief and maximum elevations of 180 m. Along the seaward side of these mountain systems stretch narrow alluvial coastal plains. The bulk of the agricultural land is on the central plain and littoral strip.

The Troodos Massif consists of igneous rocks and the Kyrenia Range is of

E.R. Perrier and A.B. Salkini (eds), Supplemental Irrigation in the Near East and North Africa, 327-365.
© 1991 ICARDA.

hard crystalline limestone. Over half the central plain is composed of Middle Miocene and post Middle Miocene calcareous marine sediments, marls and limestones; the remainder is covered by Pleistocene calcareous or non-calcareous deposits and by recent alluvium in certain low-lying areas. The coastal strips are Upper Miocene to Upper Pliocene limestones and marls, with the remainder covered by Pleistocene calcareous or non-calcareous deposits and by recent calcareous alluvium (Panayiotou, 1983).

Middle Miocene soft bedded limestones and chalks extend from east to west along the southern coast over a distance of 145 km. This land is frequently dissected by deep narrow valleys formed from the action of water flowing to the sea and constitutes the southern flanks of the Troodos Massif.

Climate

The geographical position of Cyprus in the north-eastern corner of the Mediterranean Sea imparts to the island an intense Mediterranean climate of hot almost rainless summers from mid-May to mid-September and rainy rather changeable winters from mid-November to mid-March; these are separated by a short autumn and a short spring with both periods characterized by rapidly changing weather conditions. Topography has a pronounced effect on temperature: with rising elevation, a decrease of about 5°C occurs for every 1,000 m and, from the influence of the sea, cooler summers and warmer winters ensue along the coastline.

Meteorological features

The synoptic features of the lower atmosphere affecting the weather and climate in the Mediterranean basin are closely associated with the motion and development of the great pressure systems of the Atlantic Ocean, Eurasia, and Africa. The upper atmosphere is characterized by the jet streams, Arctic and sub-tropical (Meteorological Service, 1986). These features of the lower and the upper atmosphere define the thermodynamics of the air masses which create and affect the patterns of weather in the Mediterranean zone. The Troodos Massif in the southwest central part of the island and, to a lesser extent, the narrow Kyrenia Range in the north have an important effect on the meteorology of Cyprus. These morphological features affect precipitation, temperature, and wind pattern.

Fronts and depressions approach from the west with desert depressions from the southwest. During the winter, cold air streams come from the north and, during the summer, hot dry air masses affect Cyprus from the surrounding desert areas of North Africa and Arabia (Hadjioannou, 1987). On an average, the island can experience 35 unstable weather systems from October to May.

During the summer period, the island is under the influence of a shallow trough of low pressure extending from the great continental depression over southwest Asia which results in high temperatures and almost cloudless skies. In the winter, Cyprus is influenced by frequent unstable weather systems which cross the Mediterranean Sea from west to east. These systems give rise to periods of disturbed weather lasting from 1-3 days and resulting in most of the annual precipitation. Spring and autumn are transitional seasons with long periods of settled weather.

Precipitation

Average annual precipitation is 477 mm; however, totals can vary from 182 mm (1972-73) to 759 mm (1968-69). Monthly average precipitation values (mm) for 30 yr, 1951-80, are:

Jan	Feb	Mar	Apr	May	Jun	Jul	Aug	Sep	Oct	Nov	Dec
99	69	57	30	18	6	2	3	5	32	49	107

Statistical analysis indicates a trend of decreasing precipitation during the last 30 years (Meteorological Service, 1986). Most of the annual precipitation falls during the winter months of November to March, about 80% of the seasonal total; whereas, in the summer, rainfall is negligible (less than 5%).

The orographic effect on the distribution of precipitation is pronounced as is shown in Figure 19.1. The average annual precipitation on the windward side of the Troodos (south-western slopes) increases from 450 mm to nearly 1,000 mm at the highest elevation. On the leeward slopes, precipitation decreases steadily northwards and eastwards to 300-350 mm in the central plain and the south-eastern parts of the island. The average monthly precipitation (mm) in 5 zones demonstrates this effect:

Month	Central Mesaoria	Southeastern Coastal	Southern Coastal	Western Coastal	Troodos Range
November	25	35	45	50	70
December	70	90	110	110	200
January	60	70	100	100	200
February	40	50	70	70	150

The average annual number of rainy days (more than 0.2 mm of rainfall) varies from 40 days in the south-east to 80 days in the Troodos Massif. Snowfall on the ground above 1,000 m in the Troodos Massif occurs from the beginning of December to mid-April. At elevations above 1,725 m, the average annual number of snowdays is 36.

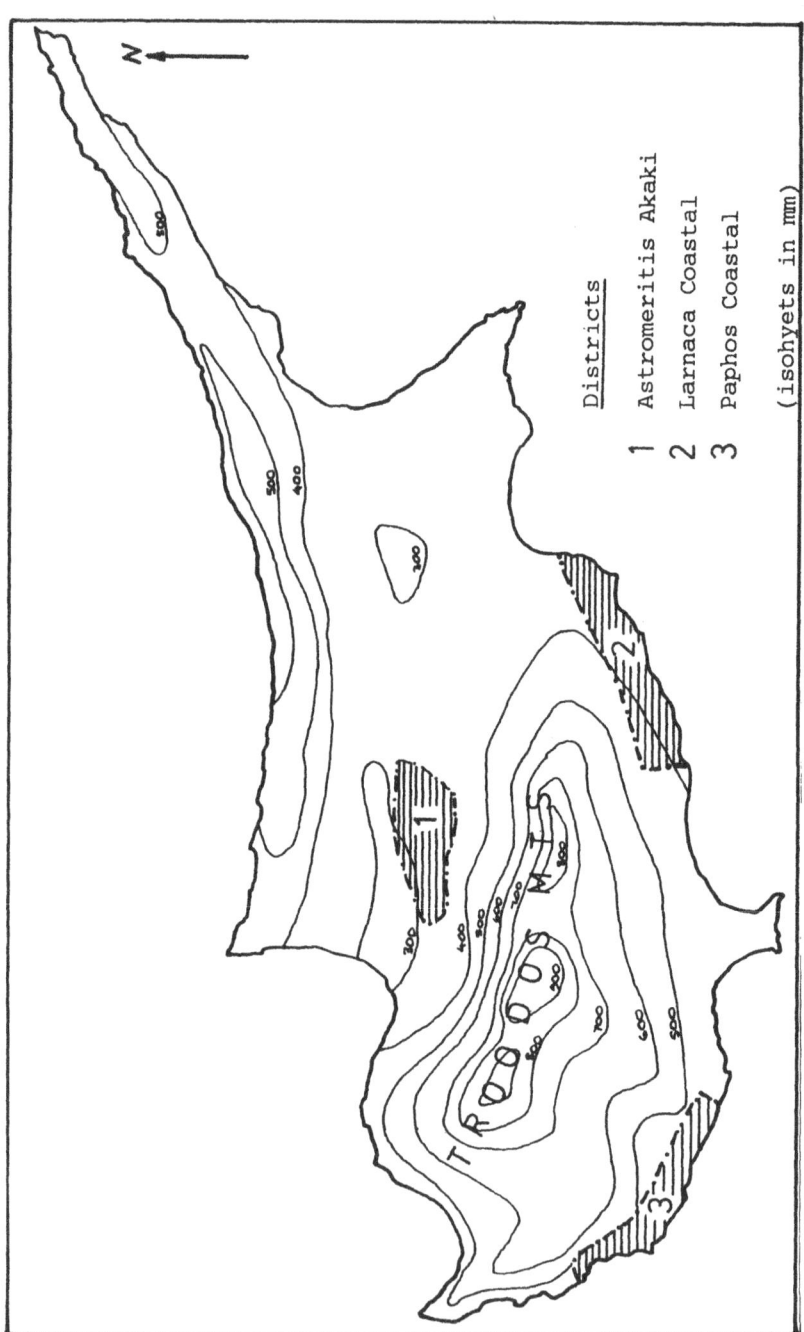

Figure 19.1. Map of Cyprus showing precipitation and district locations.

Temperature

The mean daily temperature in July and August ranges between 29°C on the central plain to 22°C on the Troodos Massif with the average maximum temperatures of 36°C and 27°C respectively. In January, the mean daily temperature on the central plain is 10°C with an average minimum temperature of 5°C and, at the higher elevations of the Troodos, a mean daily temperature of 3°C with an average minimum value of 0°C. The seasonal difference between mid-summer and mid-winter is quite large, 18°C inland and 14°C on the coast. Large differences in the summer, especially inland, are observed between day maximum and night minimum temperatures. These differences reach 10°C in the lower lands and 6°C in the mountains, increasing in the summer to 16°C on the central plain and 9-12°C elsewhere.

Ground frost is frequent but rarely severe. The average number of days with frost per year varies from 3-10 in the coastal areas and from 20-40 in the inland plains. The mean soil temperature at the 10 cm depth in the lowlands is 10°C in January and 33°C in July. At the 1 m depth, the corresponding temperatures are 14°C in January and 28°C in July. On a typical July day, the temperature at the soil surface is 15°C at dawn and 60°C at 1400 hrs; at the 5 cm depth, the corresponding values are 24°C and 42°C; and, at the 50 cm soil depth there is no significant diurnal variation of temperature.

Sunshine duration and radiation

Cyprus enjoys a very sunny climate. In low-lying areas, the average duration of bright sunshine is 75% of the time that the sun is above the horizon. Over the summer period, there is an average of 11.5 hrs/day of bright sunshine reducing to 5.5 hrs/day in the cloudiest winter months of December and January. Theoretical values of the mean sunshine duration vary from 14.5 hrs in June to 9.8 hrs in December.

Solar radiation is abundant. The long sunshine duration and the transparency of the atmosphere facilitate direct and diffuse solar radiation (principally, global solar radiation) to reach the ground surface. Global solar radiation on a horizontal surface varies from about 2 kWh/m² per day in mid-winter to about 7 kWh/m² per day in mid-summer (Meteorological Service, 1986).

Relative humidity

Distance from the coast and elevation above sea level have an effect on the relative humidity of the air. In winter, relative humidity during daylight and during darkness varies from 65-95%. Near midday in the summer, humidity on the central plain is usually 30% but can be as low as 15%. Daily variations in relative humidity are smaller on the coast than the inland plains because of sea breeze effects.

Evaporation

Intense solar radiation, a relatively dry atmosphere, and stirring winds result in high evaporation rates, especially during the warm months. Evaporation measurements (Class A Pan) reach average annual values of 1,800-2,000 mm in low-lying areas and 1,300 mm at the higher elevations of the Troodos (Meteorological Service, 1986). In January, the average pan evaporation varies from 25-75 mm and, in July, varies from 220-330 mm. In April, pan evaporation values reach 130 mm in the central plain and 140-160 mm in the coastal areas.

Land Resources and Land Use

Land use patterns

Although agricultural land constitutes 63.6% of the total area, only half is under cultivation since a sizeable portion is left uncultivated because of marginal fertility, erosion, and steep relief (Ministry of Agric. and Natural Res., 1984). Categories of land use are shown in Table 19.1.

Table 19.1. Land use in Cyprus.

Land use	Land area (ha)		
	Irrigated	Rainfed	Total
Temporary crops	15.053	51,578	66,631
Permanent crops	12,094	30,052	42,146
Fallow	3,603	12,669	16,272
Grazing	NIL	4,469	4,469
Forest	NIL	4,871	4,871
Uncultivated	NIL	35,315	35,315
Non-Productive	NIL	8,353	8,353
Totals	30,750	147,307	178,057

Source: (Dept. of Statistics and Research, 1985)

Land enumerated as irrigated from various citations represents only 17.3% of the total cultivable land. Table 19.2 shows the irrigated land sown to major crop categories and sources of irrigation water. Rainfed agriculture occupies the major portion of land resources but its contribution to the gross crop production does not exceed 45% in an average year.

Table 19.2. Irrigated area by water source and major crop category.

Source of irrigation	Temporary crops (ha)	Permanent crops (ha)	Fallow (ha)	Total (ha)
Borehole/well	11,104	6,847	1,924	19,870
Dam	2,510	2,640	1,020	6,170
River	1,128	1,736	425	3,289
Springs	311	876	234	1,421
Totals	15,053	12,094	3,603	30,750

Land under fallow has decreased from 15% of the total cultivable land in 1977 to 9% in 1985. The decreasing practice of fallow is a consequence of a subsidy paid to encourage production of forage and hay for livestock feed in regions where cereal and livestock production is integrated in a mixed farming system.

Soils

With the exception of those soils developed from igneous rocks of the Troodos Massif, all other soils are calcareous. The majority contain free calcium carbonate and have a predominantly Ca-saturated clay and a pH always above 7.5 and usually above 8.0 as determined from a 1:5 soil to water suspension. Soils endure an intense impact from the arid climate and diverse topography, both of which seem to have acted dominantly on parent materials as the principal soil forming processes. The shallowness of the soil profile and the lack of well defined diagnostic horizons bear evidence that large soil moisture deficits created from low rainfall and high temperatures have arrested the rate of soil formation usual in humid climates.

Land tenure

Land tenure is the outcome of historical interactions of social, economic, and political forces. Predominant is private ownership of numerous small, fragmented agricultural holdings. Size and fragmentation of farm units are serious technical and institutional constraints to agricultural development. The trend is towards increasing the number of holdings because of patterns of inheritance and, in consequence, a decrease in size of farm plots. These data are depicted in Table 19.3, showing plot size and number and reduction in size of holding during a decade.

Whereas the average size of holding decreased from 4.6 ha to 3.8 ha from 1977-85, the number of plots per holding also decreased from 6.4 ha to 5.2 ha

Table 19.3. Size and fragmentation of land holdings.

Census year	Cultivated area (ha)	No. of plots	No. of holdings	Average holding (ha)	Plot size (ha)	No. plots/ holding
1977	201,093	281,690	43,807	4.6	0.7	6.4
1985	178,058	245,135	47,248	3.8	0.7	5.2

Source: (Dept. of Statistics and Research, 1987)

from 1977-85. In comparison, the percentage of holdings with 1 parcel increased from 20-26.5% from 1977-85. Similarly, the number of holdings of 6 to more than 16 plots decreased from 21.3% in 1977 to 14.6% in 1985.

Associated with fragmentation, dispersal of plots, and diminishing plot size is the irregular shapes of plots randomly laid out with little relevance to land relief and contour lines. This irregularity of plot layout makes irrigation or land improvement costly while dispersal and fragmentation of the plots makes access to plots difficult.

In 1960, sharecropping was practiced on only 3.5% of the cultivated land. Now, this system of land tenure is practically extinct. Tenancy with short term leases of 1-3 yr was the system on 23.5% of the cultivated land in 1977 and still exists on 26.7% in 1985. Table 19.4 presents land tenure as enumerated in the 1985 Agricultural Census (Dept. of Statistics and Research, 1987).

Table 19.4. Land holdings by type of tenure, 1985.

Type of tenure	Number of holdings	Land area (ha)	Holdings (%)	Land area (%)
Owned land only	37,331	105,614	79.0	59.3
Rented land only	4,455	22,134	9.4	12.4
Owned & rented	5,462	50,310	11.6	28.3
Totals	47,248	178,058	100.0	100.0

Source: (Dept. of Statistics and Research, 1987)

Land consolidation

Land consolidation is the only measure to overcome fragmentation of holdings at the ownership level. The Land Consolidation Law supports the administrative and technical structure to implement land consolidation. The philosophy of the Law is to obtain voluntary participation in a consolidation scheme. Measures include the consolidation of ownership of fragmented plots into compact holdings, the expropriation of ownership of holdings below the minimum size, enlargement of land ownership, landscape renovation, and the construction of a modern farm road network serving all new plots.

Since the Law's enactment, many schemes have been implemented successfully in areas irrigated from dams as well as locations on hilly terrains. Observations of these schemes show that the absolute number of owners was reduced 27-46%; absentee ownership was reduced an average of 37%; the number of plots per ownership by 38%; and, the absolute number of plots by 63%. The average size of plot was increased 126% and the length of the overall road network increased 250% (Land Consolidation Authority, 1979).

Water Resources and Use

The semi-arid character of the island's climate typified by unreliable rainfall patterns often results in major crop failures, especially of cereals and food legumes which form the bulk of rainfed production. In addition, there are wide annual (as well as regional) fluctuations in rainfall distribution at the critical stages of plant growth in March and April. The available surface and groundwater resources are dependent exclusively on rainfall. The total volume of precipitation in an average year is 4,600 M m^3 which is disseminated as surface runoff, sub-surface infiltration, evapotranspiration, and direct surface evaporation (UNDP/FAO, 1969). The total volume of annual rainfall is estimated at:

	M m^3/year	%
Total precipitation	4,600	100
Evapotranspiration and direct evaporation:		
non-cultivated	2,944	64
cultivated	920	20
Utilization from dams, boreholes, springs	414	9
Surface loss to the sea	322	7

The island has 39 watersheds but no perennial river flow. At most, rivers flow until April or May and the major source is groundwater. This water supply is extracted using thousands of boreholes and shallow wells with the result of over-exploitation of all major coastal aquifers beyond natural replenishment.

On most rivers and streams there are numerous private water rights. *Ab antiquo* water rights are "property rights acquired since times immemorial": they may be written or registered or, in many cases, neither. However, they are fully recognized in the Constitution and in the Law. Ten to 15 percent of the total water supply is estimated to come from these sources with nearly all water encumbered by private rights. Through these rights, individuals, groups of people, a village, or groups of villages are granted the exclusive right or privilege to divert water from rivers and streams through river intakes or chain-of-wells for the irrigation of adjacent or neighboring fields. Consequently, *ab antiquo* water rights can legally block or delay the development of major waterworks in a watershed.

Water use patterns and policy

Thirteen percent of the water supply is utilized for domestic and industrial purposes. Between 1967-71, 36% of the total development expenditure was spent on domestic and industrial water supplies. By the end of 1970, all villages had piped water and, in 1982, 92% of the villages or 99% of the population had domestic water.

By far the major volume of water is for agriculture. The arid climate makes rainfed production a precarious enterprise. Irrigated agriculture is a major input into production, furnishing higher and stabilized productivity and, today, is the key factor supporting development. In the past, irrigation practices wasted water and water use efficiencies did not exceed 50%. In 1965 the Water Use Improvement Project was initiated to promote adoption of improved practices and scheduling for crop water requirements to conserve water resources at the farm level. Under this scheme, the government provided financial and technical assistance to individual farmers as incentives to adopt more efficient irrigation techniques such as sprinklers and drip irrigation. Introduction of these methods has increased water use efficiencies to more than 75% (Konteatis, 1967).

Concurrently, the formation of statutory irrigation divisions encouraged collective and more efficient use of water. Irrigation divisions have been established on state-owned boreholes allocated to farmers for collective irrigation where they had been users of dam water and where surface water from springs, rivers, or chain-of-wells had been used collectively through tradition or *ab antiquo* rights. The official policy stresses the improvement of irrigation efficiencies, the minimization of waste, and control of aquifer exploitation.

Structure and Development of Agriculture

Several crops are grown depending on the climate and the soils. All crops found in the temperate zone grow on the island; some subtropical crops such as avocado and bananas are cultivated successfully in locations with a suitable microclimate. The largest compact area of arable land lies in the Central Mesaoria Plain, where wheat and barley have been grown under rainfed conditions. Food legumes, forages, and vegetables are grown in other regions. Olives and almonds are distributed throughout the island and produced under rainfed conditions. Citrus is confined to the coastal areas where the mild microclimate is particularly suitable and water is available for irrigation.

The Troodos Massif has forest but some valleys are planted in compact areas to deciduous fruit trees and vines in the southern foothills on hillsides of white chalk or soft igneous soils. Carobs thrive on the seaward slopes of the island and at altitudes below 350 m.

Livestock is distributed in all regions but large commercial enterprises of poultry and swine are located near urban centers. Forestry and fisheries remain of minor importance to the national gross agricultural output.

Crop production

Two distinct types of farming charcterize crop production: an intensive crop production on irrigated land (citrus, bananas, table grapes, vegetables, etc.) and an extensive traditional system of rainfed farming (cereals, forages, olives, carobs, grapes, and almonds). Table 19.5 shows the contribution from the broader agricultural sector (Dept. of Stat. and Res., 1986). Both types of crop production, intensive and extensive, are market oriented with practically no subsistence agriculture. Whereas irrigated cropping involves intensive use of land, labor and water rainfed farming makes extensive use of land with underemployment of labor and fluctuation of production due to erratic weather conditions.

Table 19.5. Value of agricultural production, 1985.

Production category	Output/input	
	(1000 Cyprus pounds)	(%)
A. Gross output	197,960	100.0
1. Crop production	108,261	54.8
2. Livestock	64,631	32.6
3. Forestry	1,725	0.9
4. Fishing	4,803	2.4
5. Hunting	1,115	0.6
6. Ancillary production	14,814	7.5
7. Other output	2,372	1.2
B. Intermediate input Feeds, seeds, fertilizers, etc.	86,968	
C. Value added (A–B) at market prices	110,992	

Source: (Agric. Stat. and Res. No. 17, 1986)

In Table 19.6, the 2 types of farming systems are compared in terms of land area, production, and gross output. The level of production of barley during the period 1983-85 fluctuated from 62,000 t in 1983 to 102,800 t in 1985 (a good season for cereals). Wheat takes second place to barley production with an average annual yield of 9,000 t. The stability of durum wheat yields is attributed to the practice of supplemental irrigation (Table 19.7).

With the exception of 2,660 ha of cereals and forage crops which receive supplemental irrigation, the remaining hectares are rainfed. Food legumes, faba beans, cowpeas, and small amounts of chickpeas and lentils are grown on

2,400 ha annually. The overall cropping pattern and value of production is shown in Table 19.8.

Table 19.6. Comparison of irrigated and rainfed production.

Type of cropping	Average values	
	1981-82	1984-85
Irrigated crops		
Area (ha)	26,074	28,057
Production (1,000 CL*)	51,971	65,413
Gross Output (CL/ha)	1,920	2,331
Total Crop Production (%)	60.5	60.4
Rainfed crops		
Area (ha)	125,414	126,749
Production (1,000 CL)	33,922	42,847
Gross Output (CL/ha)	270	338
Total Crop Production (%)	39.5	39.6

* CL = Cyprus pounds
Source: (Dept. Stat. and Res. No. 17, 1986)

Table 19.7. Cereal grain production, 1983-85.

Crop/year	Area (ha)	Production (t)	Ave. yield (t/ha)	Producer's price (1,000 Cyprus pounds)	Value
Wheat					
1983	7,627	9,100	1.2	111.0	1,010
1984	6,021	9,000	1.5	116.0	1,044
1985	6,021	9,000	1.5	121.0	1,089
Barley					
1983	43,485	62,000	1.4	96.1	5,958
1984	46,429	83,250	1.8	101.1	8,416
1985	50,175	102,800	2.1	106.1	10,907

Source: (1985 Annual Rpt., Dept. Agric., 1986)

Table 19.8. Crop area, yield, and production value.

Crop	Land area (ha)	Yield (t/ha)	Value of production – producer's price (1,000 Cyprus pounds)
Cereals			
Wheat	6,000	9,000	1,089
Barley	50,000	102,800	10,907
Oats	3,700	400	96
Straw	—	80,800	2,820

Table 19.8. (Continued).

Crop	Land area (ha)	Yield (t/ha)	Value of production – producer's price (1,000 Cyprus pounds)
Food legumes			
Broadbeans, fresh	1,730	1,500	336
dry		2,250	886
Cowpeas, fresh	400	1,360	471
dry		250	286
Chickpeas	160	140	55
Lentils	130	90	34
Forage crops	27,000	—	5,771
Industrial crops	1,120	—	1,770
Vegetables	12,420	—	32,715
Flowers	40	—	2,045
Tree crops and fruits	62,800	—	40,173
Other tree crops			
Olives	6,250	11,500	7,845
Carobs	7,320	7,600	931
Totals	161,620	—	108,260

Source: (Dept. of Stat. and Res. No. 17, 1986)

Livestock production

Livestock production constitutes one third of the total gross agricultural output with about 30% of the holdings engaged in animal husbandry of cattle, sheep, and goats. Poultry and swine are concentrated in large commercial units. Cyprus is self-sufficient in pork, poultry meat, and eggs but a net importer of beef, lamb, and dairy products. In 1978-80, 52% of beef, 20% of mutton, and 32% of dairy products were imported for domestic consumption. More than 50% of the cows' milk produced is handled and distributed by the pasteurizing industry while the rest is absorbed by cheese, yoghurt, and ice cream. Sheep' and goats' milk is used mainly in halloumi cheese.

Total head of sheep, cattle, and goats with production figures are given in Table 19.9. Poultry, swine, and other livestock are valued at another 30 M Cyprus pounds. Ruminant production represents 53.7% of the total annual livestock production: sheep contribute 21%; goats, 15%; and cattle, an additional 17%.

For successful livestock development, two factors are of paramount importance, namely, balanced nutrition and high standards of animal health. In addition, for sheep, goats, and cattle, adequate supplies of straw and other roughages must be readily available. By promoting mixed farming in rainfed areas, forage production has been increased. Implementation of the Mixed

Table 19.9. Livestock (ruminant) population and production, 1985.

Animal/ product	Total number of animals	Quantity (t)	Value – producer's prices (1,000 Cyprus pounds)
Sheep	276,300		
meat		4,200	7,049
milk		20,500	5,945
wool		450	198
Goats	173,300		
meat		3,230	5,564
milk		23,400	3,744
Cattle			
meat		2,750	3,575
milk		63,500	6,896
Other products (manure, hides, skins)			1,733
Total value			34,704

Source: (Dept. of Stat. and Res. No. 17, 1986)

Farming Project in 1967 introduced forage crops into fallow, improved pasture land, and promoted the substitution of forage crops on marginal cereal land.

Concurrently, forage conservation is encouraged by the Haymaking Project which has gathered momentum slowly from a mere 50 ha, 20 years ago, to almost 15,000 ha in 1985. The Pasture Development Project on a communal basis with appropriate management supports the development and improvement of poor marginal land. In 1986, 20 communal grazing areas totaling 523 ha were established with the participation of 246 livestock keepers owning 27,200 animals. A high standard of animal health is ensured in diagnosing, curing, and preventing animal diseases by well-organized veterinary services.

Institutional Framework

Planning

Planning evolves through preparation of 5-year medium term plans implemented with an annual development budget. The structure for agriculture includes the Central Planning Commission (CPC) with overall responsibility for preparing and implementing development plans; Planning Committees are the second tier of authority; and, the Planning Bureau is the economic, administrative arm of these two bodies (Planning Bureau, 1987). A coordinating committee, comprising the Director-General of the Ministry and the Directors of the departments and services of the Ministry, is under the chairmanship of the Minister of Agriculture and Natural Resources and oversees the broader agricultural sector. The function of this committee is to monitor the implementation of approved projects and programs, to introduce approved policy measures, and to prepare studies, plans, legislation, and policy.

Research and extension

The Agricultural Research Institute was established in 1962 as a cooperative project between the Government of Cyprus and the United Nations Special Fund. The Institute under the Ministry of Agriculture and Natural Resources is entrusted to perform applied agricultural research. The basic objectives of the Institute's programs are to provide viable answers to present and foreseeable problems facing crop and livestock production. In parallel purpose, the research programs focus on introducing scientific and technological innovations adaptable to local conditions which would enhance the potential for agricultural development (Cyprus Agric. Inst. Annual Rpts., 1980-86). The Institute performs a wide spectrum of applied research in agronomy, horticulture, plant protection, soil and water use, animal production, and agriculture economics. The Institute's contribution to the agricultural sector has been substantial.

The Department of Agriculture is the central extension agency and is responsible for the implementation of development projects in crop and livestock production and for the transfer of new technologies to farmers. Close and constant liaison and cooperation between research and the extension service ensures that critical problems receive adequate attention and investigation. The interaction of research and extension with other departments and services of the Ministry is of fundamental importance in coordinating agricultural planning and development. Agricultural extension operates on a "beat" which is the smallest entity, serving an average of 1,500 farmers and 13.5 villages. An agronomist is the leader of a beat and is assisted by specialists from district headquarters when required.

Crop insurance

The Agricultural Insurance Law established the Agricultural Insurance Organization, a non-profit body. A reserve fund was created from farmers' contributions calculated from their annual production. These contributions, matched by an equal amount from government funds, are used to provide indemnity in cases of natural disasters. The contributory scheme is compulsory and the insurance covers many crops (including wheat and barley) which are damaged by drought, hail, or rust (Agric. Ins. Organ., 1986). An annual premium of 3% for wheat and 2.5% for barley is calculated using the current value of grain delivered to the Grain Commission.

Table 19.10 shows the levels of indemnities from 1983-86 paid for cereals damaged by drought, hail, and rust. In 1983, a year of severe drought, the Agricultural Insurance Organization paid 1.25 M Cyprus Pounds in damages; in 1986, farmers received 1.2 M Cyprus pounds in damages from drought. Barley grain production in 1983 was 60% of the 1985 yields (a relatively good year) and, in 1986 was 65% of the 1985 yields. Recently, the scheme has been extended to include forage crops with the value of insurance premium based on barley.

Table 19.10. Indemnities paid for cereals (1,000 Cyprus pounds).

Cause of damage	Amount paid (1,000 Cyprus pounds)			
	1983	1984	1985	1986
Drought	1,244	371	34	1,131
Hail	—	15	23	54
Rust	4	3	—	—
Total	1,248	389	57	1,185

Source: (Agric. Insurance Organization, Annual Report, 1986)

Subsidies

Subsidy schemes for agriculture constitute important policy measures which promote agricultural development and can be seen as effective instruments in resource allocation for attaining structural changes and technological advances. Most direct subsidies are paid for cereals and vine products as shown in Table 19.11. Direct grain subsidy for locally produced cereal is small when compared to total cereal subsidies. The scale paid for subsidies fluctuates according to international prices because more than 67% of the grain is purchased by the Cyprus Grain Commission (Cyprus Grain Comm. Annual Rpt., 1986). The results of a recent World Bank study showed that livestock production and the consumer indirectly benefit more from such subsidies than does the cereal grower; also this subsidy, in addition to being a serious burden to the national economy, may actually serve as a deterrent to more efficient use of scarce resources.

Projects on crop and livestock production benefiting from state creditor subsidies include soil conservation works, establishment of livestock units, promotion of haymaking, and pasture improvement. Irrigation water from dams is sold to farmers at subsidized prices; however, smaller waterworks such as boreholes, small dams, and distribution networks are financed jointly by the state and the farmers involved. The state finances two thirds of the construction

Supplemental Irrigation Systems in Cyprus 343

Table 19.11. Direct subsidies for cereals, vine products, and hay.

Crop	Amount paid (1,000 Cyprus pounds)			
	1983	1984	1985	1986
Cereals	25,530	31,486	20,575	9,282
Vine products	5,914	6,734	6,514	6,958
Hay	212	213	226	117
Total	31,656	38,433	27,315	16,355

Source: (Planning Bureau, Central Planning Comm., Dept. of Stat. and Res.)

costs and also furnishes credit to the beneficiaries at subsidized interest rates for the remaining one third.

Credit

Medium and long term credit is an indispensable means for successful project implementation for development. The bulk of this agricultural credit is provided by the State through the Loan Commissioners and the Cooperative movement; the private sector is involved only marginally. Commercial banks and private financial institutions do extend short term loans to agricultural enterprises.

The village cooperative credit societies provide most of the short term credit required by farmers, with the maximum loan size depending on the financial standing of the local credit society. Most villages have a Cooperative Credit Society with the majority of farmers as members (Mavrommatis, 1987).

The District Perspective: Characterization of 3 Districts

The selection of 3 districts was achieved after close scrutiny of 6 agro-economic regions (Philippides and Papayiannis, 1983); then, data was collected by questionnaire and the results analyzed. Districts I, II, and III (see Figure 19.1) fulfill the requirements of the study based on climatological characteristics and, in particular, on annual and mean monthly rainfall; the agro-economic nature of cropping patterns and farming systems; and, the existing infrastructure for applying supplemental irrigation to increase and stabilize yields of cereals and other field crops. Districts I and III are examples which, traditionally, have practiced supplemental irrigation; whereas, District II is an example of cereal-forage-livestock integration with the potential for using treated sewage effluent (in the future) for supplemental irrigation.

Climatic Aspects

District I, Astromeritis-Akaki, lies in the central plain, 15-20 km west of Nicosia, and on the leeward slopes of the Troodos Massif. The orographic effect of this location is to decrease average annual precipitation to 300 mm, the lowest in Cyprus.

District II, Larnaca Coastal, is a narrow coastal strip west of the port town of Larnaca and has an average annual precipitation of 358 mm. Over 80% of this total occurs, during most winter seasons, between November and April. The stabilized yields of barley grain during the last 6 years give evidence of a favorable rainfall distribution. The average pan evaporation is 1,764 mm and the calculated evapotranspiration, ET (Penman), is 1,254 mm.

District III, Paphos Coastal, is a narrow plain bound on the south by the sea with an area 35 km long and 2 km wide located along the southwestern coast. With a major dam, canal, and distribution network, irrigated agriculture is practiced on 50% of the land. Average annual rainfall is 430 mm and is favorably distributed. Supplemental irrigations of 50-100 mm would increase yields only to a limited extent. The microclimate is particularly mild, favoring sub-tropical fruits such as bananas and avocado. Crop water requirements are similar to those in District II and other coastal areas. Table 19.12 compares mean monthly precipitation for the 3 districts.

Table 19.12. Mean monthly rainfall, 1951-80.

District	Jan	Feb	Mar	Apr	May	Jun (mm)	Jul	Aug	Sep	Oct	Nov	Dec	Annual total (mm)
I	60	40	42	22	12	4	1	2	3	30	31	58	305
II	76	67	33	18	9	2	0	1	1	20	33	98	358
III	94	64	47	24	11	1	0	0	2	29	52	106	430

Source: (Meterological Service, Nicosia.)

Rainfall distribution in District I is unfavorable for winter field crops. Crops have insufficient rainfall to satisfy their moisture requirements throughout the growing season. Rainfed production is vulnerable, in particular, during critical plant growth stages of March and April when evapotranspiration totals are more than 200 mm. During this period, supplemental irrigation could exert the most impact on grain yields of cereals and food legumes, literally increasing from no production to yields of 4-5 tons/ha. The average relative humidity in district I is slightly lower than in districts II and III.

In absolute terms, air temperatures (Table 19.13) and pan evaporation values (Table 19.14) are similar in all 3 districts although the general climate in the coastal districts is substantially modified by the sea when compared to the arid character of climate in District I.

Table 19.13. Mean daily temperature.

	Jan	Feb	Mar	Apr	May	Jun	Jul	Aug	Sep	Oct	Nov	Dec	Annual average (°C)
						(°C)							
District I													
Maximum	15.6	15.0	18.4	22.9	27.7	31.5	33.7	33.7	31.6	26.8	21.0	17.6	24.6
Minimum	6.7	5.8	7.4	10.4	14.3	18.1	20.7	21.7	18.7	15.1	11.3	8.2	13.2
Mean daily	11.1	10.4	12.9	16.6	21.0	24.9	27.2	27.7	25.1	21.1	16.2	12.9	18.9
District II													
Maximum	16.4	16.1	18.5	21.8	26.4	28.9	31.2	31.7	30.2	27.1	22.6	18.4	24.1
Minimum	6.7	5.4	6.7	9.4	13.8	16.3	19.0	19.6	17.5	14.2	12.4	8.3	12.4
Mean daily	11.5	10.7	12.6	15.6	20.1	22.6	25.1	25.6	23.8	20.7	17.5	13.3	18.3
District III													
Maximum	16.6	16.7	18.6	21.9	26.1	29.3	31.3	31.3	29.5	26.9	22.6	18.5	24.1
Minimum	8.7	8.3	9.7	12.1	15.2	18.5	20.9	21.2	19.5	16.9	13.3	10.1	14.5
Mean daily	12.7	12.5	14.2	17.0	20.7	23.9	26.3	26.2	24.5	21.9	18.0	14.3	19.3

Source: (Meteorological Service, Nicosia)

Table 19.14. Average monthly pan evaporation and potential evapotranspiration.

	Jan	Feb	Mar	Apr	May	Jun	Jul	Aug	Sep	Oct	Nov	Dec	Annual total (mm)
						(mm)							
District I													
Pan	55	57	97	169	175	239	250	223	176	99	66	53	1,659
P.E.	33	41	70	111	149	173	183	171	126	88	48	31	1,224
District II													
Pan	55	68	104	150	173	214	228	221	176	139	76	60	1,764
P.E.	37	51	78	126	145	167	182	176	130	102	55	35	1,254
District III													
Pan	76	69	96	128	186	220	240	217	183	145	113	83	1,766
P.E.	42	50	78	110	145	165	181	163	127	99	63	42	1,265

Source: (Meteorological Service, Nicosia)

Farming Practices

Rainfed farming, perennially irrigated agriculture, and livestock production are practiced as integrated activities. With the data available, to isolate completely one production type from another is difficult because the farming systems are, to varying degrees, interdependent. One striking feature of the farming structure is the high incidence of combined part-time farming with part-time (or even full-time) off-farm employment (Ansell, et al., 1984). Table 19.15 shows the findings of the 1985 Agricultural Census (Dept. of Stat. and Res., 1987), describing the full-time equivalent of persons engaged in agriculture and on-farm employment, off-farm agriculture, and employment in non-agricultural sectors.

Table 19.15. Employment of farmers (holders) and family members in agricultural and non-agricultural sectors.

	Number of farmers and family members	Total weeks worked		
		On-farm (%)	Off-farm (%)	Non-agric (%)
District I	1,527	32	3	65
District II	1,153	48	2	50
District III	1,954	35	2	63

Source: (Dept. of Stat. and Res., 1987)

Sixty five percent of the land in District I is under rainfed crops (mainly cereals) with the remaining 35% under seasonal or perennial irrigation. In District II, 81% of the land is rainfed and 19% under irrigation. In District III, 54% is rainfed and 46% is irrigated. Development goals support full use of available water resources from the dam and associated canal and distribution networks; the trend is towards increasing the proportion of crops using supplemental irrigation to augment rainfall.

Revenues from livestock production (sheep, goats, and dairy cattle) is 60% of the total farm income and contributes a sizeable share to the total output of District II. About 20% of the livestock population is integrated into a cereal-forage crop farming system.

Cropping Patterns

Supplemental irrigation of field crops has been practiced traditionally for many centuries in District I. A wide variety of crops can be grown throughout the year; however, rainfed winter field crops are confined to cereals, forages, and food legumes. When vegetable and tree crops (olives and citrus) were integrated into the farming system, priority for use of river flows from February to May was shifted to irrigation of these higher valued crops. Similarly, in District III, the use of dam water is for perennially irrigated vegetables and fruit trees (citrus, bananas, avocados, and table grapes) and the produce is exported. Rainfed farming of cereals is used in the rotation system. In District II, rainfed field crops of barley, forages, and wheat predominate while irrigated agriculture is concentrated in a few villages and accounts for less than 20% of the total land area. Cropping patterns for major crops are presented in Table 19.16.

Table 19.16. Cropping patterns.

Crop	Farming system					
	District I (ha)		District II (ha)		District III (ha)	
	rainfed	irrigated	rainfed	irrigated	rainfed	irrigated
Barley	3,707	700	3,892	—	943	82
Wheat	60	164	249	—	172	53
Food legumes	45	401	6	8	8	250
Forage	411	130	1,415	42	544	181
Vegetables	—	800	—	390	—	577
Tree crops	60	636	331	302	1,200	1,338

Source: (Dept. of Stat. and Res., 1987)

Mechanization

Introduction of farm machinery, tractors, implements, and combine harvesters in the late forties has fostered the development of a high degree of mechanization. Farmers' adoption of mechanized farming practices is an indication of their response to technological innovation. The degree of mechanization in each of the districts is shown in Table 19.17.

Table 19.17. Levels of farm mechanization as indicators of farmers' adoption of technological innovation.

Indicator	District		
	I	II	III
Number of tractors	749	465	788
Holdings per tractor	2.9	2.9	4.0
Cultivated area/tractor (ha)	11.2	15.2	8.7
Number of combine harvesters	154	180	208
Cereal/combine harvester (ha)	3.2	2.9	5.9
Holdings/transport vehicle	5.0	14.7	8.1
Number of forage harvesters	30	23	6
Forage/hay harvester (ha)	108	99	89

Source: (Dept. of Stat. and Res., Nicosia)

Fertilizer Use

Use of fertilizer in both rainfed farming and irrigated agriculture has been routine since the early sixties. According to estimates of the Cooperative Central Bank which supervises 95% of the fertilizer imports, cereal production annually consumes 17-20,000 t of 16-20-0 or 20-20-0 fertilizer at application rates of 150 kg/ha (Ashiotis, 1987). Also, 3,000 t of calcium nitrate (26-0-0), 1,700 t of ammonium nitrate (33/34-0-0), and 1,000 t of urea (46-0-0) are applied annually to cereals as top dressing at rates of 75 kg/ha of nitrate and 40 kg/ha of urea.

Wheat production in Districts I and III receives higher rates of nitrogen as top dressing. Food legumes such as faba beans receive basic phosphate fertilizer of 60 kg/P_2O_5/ha and an initial application of 25 kg/ha of nitrogen (N) as ammonium sulphate (Xenophontos, 1984). Forage crops receive slightly higher applications of nitrogen as top dressing.

Plant Protection

In the last 35 years, pest and weed control has developed into a routine cultural practice in all farming systems. Most of the imported chemicals are used on irrigated crops, rainfed vines, and olives. Few pesticides are applied on cereals or other field crops and none are applied on forage crops. The cereal leaf miner, once a serious pest, is now of limited occurrence (Serghiou, 1975; Meliphronides, 1972). Aphidicides are used for the control of aphids on faba beans.

Selective herbicides have been used successfully on faba beans to control weeds and orobanche (Americanos, 1986; Americanos, 1983). A standard practice for protection of wheat and barley is control of broad leaf weeds using liquid sprays of 2-4D esters. For the control of wild oats, if economical, pre- and post-emergence herbicides are applied for heavy infestations.

Yield Relationships

Grain yields for rainfed farming can vary from no yield during severe drought to 2-3 t/ha in an average year. Droughts resulted in no grain production in 1931-33 and 1972-73. Historical evidence demonstrates that Cyprus suffered from drought and famine at the end of the third century and the beginning of the fourth. The years of 1469, 1510, and 1873 are listed as years of severe droughts (Hadjioannou, 1987).

Apart from total crop failures, grain yields are known to fluctuate widely from year to year depending on the amount and distribution of rainfall during the growing season. This is clearly borne out in production records received by the Cooperative Association for Cereal Grain Concentration (Neophytou, 1987) on behalf of the Cyprus Grain Commission (Annual Report, 1986) during the period 1980-81 to 1985-86 (Table 19.18). Average grain yields over a 6 yr period from 1981-86 were:

I: 1.9 t/ha barley and 2.0 t/ha wheat;
II: 3.4 t/ha barley and 3.6 t/ha wheat; and,
III: 2.8 t/ha barley and 3.8 t/ha wheat.

Table 19.16. Cropping patterns.

	District I		District II		District III	
	barley	wheat	barley	wheat	barley	wheat
Land area (ha)	4,400	244	3,892	249	1,025	225
1981						
Yield (t)	8,004	458	11,726	927	2,777	839
Area (ha)	1.8	2.0	3.0	3.2	2.7	3.7
Rainfall (mm)	265		396		490	
1982						
Yield (t)	9,585	484	12,328	972	3,176	866
Area (ha)	2.2	2.2	3.2	3.9	3.1	3.8
Rainfall (mm)	228		318		395	
1983						
Yield (t)	4,183	354	10,759	865	2,820	1,291
Area (ha)	1.0	1.6	2.8	3.5	2.8	5.7
Rainfall (mm)	138		251		407	
1984						
Yield (t)	10,310	569	12,883	873	2,535	1,004
Area (ha)	2.3	2.5	3.3	3.5	2.5	4.5
Rainfall (mm)	240		317		315	
1985						
Yield (t)	13,937	502	19,131	1,177	3,212	702
Area (ha)	3.2	2.2	4.9	4.7	3.1	3.1
Rainfall (mm)	277		428		305	
1986						
Yield (t)	5,244	266	12,776	682	2,339	500
Area (ha)	1.2	1.2	3.3	2.7	2.3	2.2
Rainfall (mm)	145		205		248	

Sources: (Pashiardis, 1987; Dept. of Stat. and Res., 1986)

The soils for cereal production under rainfed conditions are inherently low in nitrogen (total soil nitrogen is 0.1%) and phosphorous but are adequately supplied with potassium. Thus, in addition to insufficient rainfall, nitrogen and phosphorous may be factors which limit grain yields.

Consistent responses to nitrogen by wheat were reported (Loizides, 1958). However, responses to phosphorous have been shown to be more pronounced in years of relative drought (Matar, 1977; Krentos and Orphanos, 1979) when rainfall was below 250 mm. Barley responded to phosphorous more than did wheat. Because rainfall is erratic, adequate rates of nitrogen and phosphorous, 20-30 kg/ha, must be applied at sowing together with an additional application of 15-25 kg/ha in order to make best use of whatever rainfall occurs. It was also shown that nitrogen not taken up because of drought is available to the following crop sown (Krentos and Orphanos, 1979). The combination of a high rate of nitrogen fertilization with supplemental irrigation can result in very high grain yields for the short straw varieties of wheat which withstand lodging.

Improvement of durum wheat varieties through selection from indigenous populations and new introductions from international institutions (CIMMYT, ICARDA, and FAO) has resulted in the release of 3 high yielding durum varieties: Aronas, Mesaoria, and Karpasia (Hadjichristodoulou, et al., 1977, 1982, and 1984). Their potential for increased yields is particularly pronounced under high rainfall or supplemental irrigation (Hadjichristodoulou, et al., 1984; Josephides, 1987).

Livestock and Forage Production

About 30% of the holdings are engaged in animal husbandry of dairy cattle, sheep, and goats. Table 19.19 provides values of production revenue and total head of livestock.

Table 19.19. Livestock (ruminant) population and gross output (1,000 Cyprus pounds).

	Sheep	Goats	Cattle	Revenue (1,000 Cyprus pounds)
District I				
Total livestock (head)	13,000	8,000	2,200	
Gross output				
(1,000 Cyprus pounds)	793	552	691	2,036
District II				
Total livestock (head)	40,000	8,200	3,200	
Gross output				
(1,000 Cyprus pounds)	2,440	566	1,005	4,011
District III				
Total livestock (head)	17,000	8,000	500	
Gross output				
(1,000 Cyprus pounds)	1,037	552	157	1,746

Ruminants (in particular sheep, goats, and cattle) require adequate supplies of roughage such as hay and straw to balance nutrition for sustainability of production. In large measure, three interdependent development projects on mixed farming, hay-making, and pasture introduced the practices of adding forage crops into fallow, promoting forage conservation, and improving pasture lands. These efforts have had an impact in all districts but especially in District II with the largest livestock population and the highest forage and straw production. The land area planted to forage in District II is three times larger than in Districts I or III. In addition, a favorable rainfall distribution ensures a more stabilized production of cereal grain and straw than is possible in District I under rainfed conditions. The impact of supplemental irrigation on forage production can be evaluated using subjective observation to support results of appreciable gains in Districts I and III when this technique is practiced (albeit not in significant numbers) in years of normal rainfall.

Table 19.20. Costs and returns for one hectare of cereals, forage as hay, and food legumes.

District	Cereal grain		Forage hay		Food legumes	
	wheat	barley	barley	mixed	faba	chickpea
I. Astro-Akaki						
grain (t/ha)	2.0	2.0	—	—	3.0	1.3
straw (t/ha)	2.0	2.0	—	—	3.0	—
hay (t/ha)	—	—	8.0	6.0	—	—
Gross revenue (Cyprus pounds)	282	290	536	444	990	520
Total var. costs (Cyprus pounds)	178	171	160	168	430	162
Gross margin	104	119	376	276	560	358
II. Larnaca Coastal						
grain (t/ha)	3.6	3.4	—	—	—	—
straw (t/ha)	3.6	3.4	—	—	—	—
hay (t/ha)	—	—	8.0	6.0	—	—
Gross revenue (Cyprus pounds)	508	493	536	444		
Total var. costs (Cyprus pounds)	178	171	160	168		
Gross margin	330	322	376	276		
III. Paphos Coastal					cowpeas	haricot beans
grain (t/ha)	3.8	2.8	—	—	2.2	2.7
straw (t/ha)	3.8	2.8	—	—	—	—
hay (t/ha)	—	—	8.0	6.0	—	—
Gross revenue (Cyprus pounds)	536	406	536	444	2420	2025
Total var. costs (Cyprus pounds)	178	171	160	168	374	857
Gross margin	356	235	376	276	2046	1168

Cost of Production

In the last 15 years, several agro-economic studies have evaluated production costs, the bulk of this work undertaken by the Agricultural Research Institute (Papachristodoulou, 1987). The costs and returns for wheat, barley, forages, food legumes, and other crops have been investigated on a national basis but the data for these particular field crops can be generalized to the 3 districts. Table 19.20 provides the cost and returns of field crops in a summary form.

Recent technical and economic surveys of herds with 100 head of sheep and goats found that gross revenue per head of sheep was 85 Cyprus pounds and per head of goat was 91 Cyprus pounds. Meat production contributes 48% of this revenue and milk or milk products contribute 46% (Agric. Res. Inst. Annual Rpt., 1987). Gross profit, subtracting the total variable costs, was 61 Cyprus pounds per head of sheep and 69 Cyprus pounds per head of goat. Family labor represents more than 75% of the fixed costs. Similar results were found for dairy cattle by surveys administered by the Agricultural Research Institute, showing a gross profit of 314 Cyprus pounds per head (Papachristodoulou, 1987).

Comparison of Supplemental Irrigation Resources and Farming Practices in 3 Districts

This section evaluates data collected by questionnaire and compares the results with reference to the resources and constraints of the 3 districts for water resource management.

Table 19.21. Demographic data.

Item	District I	District II	District III
Total population	14,000	14,000	11,000
No. of villages	8	17	18
No. of households	4,000	3,700	3,000
Members/household	3.53	3.65	3.55
Male/female ratio	1,016	1,017	1,029
Average age/household			
< 5 yr	0.33	0.40	0.36
5-9 yr	0.27	0.29	0.28
10-14 yr	0.28	0.30	0.30
15-65 yr	2.24	2.24	2.29
> 65 yr	0.40	0.42	0.31

Source: (Dept. of Stat. and Res., Nicosia)

Demographic Information

The 3 districts differ only slightly from one another in demographic character (Table 19.21). The 8 villages of District I have larger populations than do those of the other two districts which have 17 villages each.

Land Area and Land Use

Total holdings are 6,500 occupying a total land area of 26,000 ha. Table 19.22 is a compilation of data collected from comparable sources which show total area and the major categories of land use, number of farms, degree of fragmentation, size of farms, and ownership.

Table 19.22. Land area and land use.

	District		
Category of use	I	II	III
Total land (ha)	8,809	8,037	8,921
Non-arable	48	316	820
Total arable			
not in production	348	595	1,305
rainfed	4,235	6,500	4,680
fallow	1,438	560	1,145
supplemental irrigation	683	none	none
water harvesting	none	none	none
Total no. of farms	2,189	1,340	3,000
Fragmentation (%)			
rainfed	84	82	88
supplemental irrigation	n.a.	n.a.	n.a.
Ave. no. of fragments	10	8	5
Type of ownership (%)			
own	61	44	78
rent	14	38	15
own and rent	25	18	7

Source: (Dept. of Stat. and Res., 1987)

Soils

Soils are calcareous and low in organic matter, nitrogen, and phosphorous but rich in potassium. Nonetheless, they are productive with applications of fertilizer, adequate soil moisture, and proper management.

District I soils are red (Terra rossa), shallow, calcareous and classified as

Rhodo-chromic Luvisols usually having an argilluvic B horizon. They have developed from parent materials of surface lime crust or igneous pebbles and some may have a high clay content but in general they are loams to clay loams with very good physical properties and good tilth (Grivas, 1987).

District II soils are deeper and classified as Calcaro-chromic Cambisols, the predominant type. They are of medium texture occurring on soft limestones or on surface lime crusts. To a lesser extent, orthic Rendzinas are also found having developed on extremely calcareous colluvial material.

Soils in District III are deeper and of a higher clay content. Calcaro-chromic Cambisols predominate. The soil is fertile but the heavier soil must be carefully managed particularly with irrigation.

Infrastructure and Rural Support Services

Education in Cyprus is compulsory to age 15; therefore, elementary schools, grades 1-6, are in each village for both male and female. In addition, many villages have pre-elementary or nursery schools. Secondary education, grade 7-12, is available in regional gymnasiums or technical schools as in District I, or in towns with students from rural areas commuting; for example, Larnaca serves District II and Paphos serves District III (Papadopoulos, 1987). Literacy is high throughout the farming communities with only a 3% illiteracy rate.

In general, the infrastructure and rural support services (Table 19.23) are satisfactory. There are adequate credit centers, with at least 1 Cooperative Credit Society in each district. Also, each district has agricultural extension and veterinary care with medical care centers in each village (Dept. of Coop. Dev., 1982). All provide good services.

All villages are supplied with electric power from the national grid while a number of houses have roof solar heaters for heating water. Gas and electricity are universally used as household fuel and potable water supplies are on house-to-house connections.

The road network connecting villages and urban centers is adequate but year to year improvement occurs on rural roads. Markets for agricultural produce are accessible with cereal grain purchased by the Cyprus Grain Commission through the Cooperative Association for Cereal Grain Concentration (Cyprus Grain Comm. An. Rpt., 1986; Neophytou, 1987).

Water Resources and Management

Available surface water and groundwater resources rely entirely on rainfall. Direct dependence of field crops on rainfall results in yield instability and production fluctuation from year to year. Drought is a common experience in the farming community even in normal rainfall years.

The water source in District I is principally an aquifer enriched by winter river

Table 19.23. Infrastructure and rural support services.

	District		
Service	I	II	III
No. of schools	9	18	16
Level of school:			
Elementary (Grades 1–6)	8	14	12
Secondary (Grades 7–12)	1	4	4
Illiteracy, %	2	2	3
No. of credit societies	8	16	12
Extension centers	2	2	2
Veterinary centers	1	2	1
Medical care centers	1	2	1
Medical sub-centers	4	6	2
Number of Cooperatives	11	18	14
Credit	8	16	9
Agricultural inputs	3	2	5
Members enrolled	90	90	95
Electric power	8	17	18
Source of electricity	EAC	EAC	EAC
Household fuel	gas/electric	gas/electric	gas/electric
Potable water supply	8	17	18
Pressure system	8	17	18

Sources: Nicosia: Ministry of Education
              Dept. of Cooperative Development
              Dept. of Agriculture
              Dept. of Veterinary Services
              Ministry of Health
              Electric Authority of Cyprus (EAC)

flows, low rainfall (305 mm annual average), and direct river flows during the winter and until April or May. The supply is used for supplemental irrigation of both field crops and perennially irrigated permanent crops such as citrus and olives. Groundwater is exploited through many private and some communal boreholes either managed by an irrigation division or as domestic supplies. Demand for water leads to a continuous decline in the water table even though annual replenishment occurs from seepage through permeable river beds. Winter river flows have an important role in water use and management as sources of supplemental irrigation of winter field crops or vegetables and tree crops.

In District II, annual average rainfall of 358 mm is higher than District I and in most years this amount assures a good cereal grain yield. Irrigated agriculture is of lesser importance occupying less than 20% of the cultivable land area. Although rainfed crop production is widespread, irrigated crops of vegetables, flowers, and citrus are restricted to specific villages. All irrigation water is pumped from groundwater resources.

District III, apart from having the highest annual average rainfall (430 mm), is also served by major waterworks including a dam of 51 M m^3 capacity, a 12 km open concrete lined canal, a 20 km long main pipeline, and a comprehensive distribution network for on-demand sprinkler irrigation. In addition, private boreholes supplement individual irrigation needs. The cost of water supplied to the farmers is subsidized by the government for the irrigation of citrus, table grapes, bananas, vegetables, avocados, and summer crops. The management of the project as a whole is achieved through an expanded committee with members from the Department of Water Development, the district Agricultural Officer, the district Administrative Officer, and representatives of the farmers. From the inception of the Paphos Irrigation Project, the policy was to convert rainfed cereal production into perennially irrigated land for high cash export crops. Although cereals are not encouraged by this policy, there is a potential for using river overflows in the winter for supplemental irrigation of wheat grain production.

Irrigation Practices

Historically, grain production constituted a much more important activity than it does today. Supplemental irrigation was practiced on a relatively large scale for cereals grown in the central plain and in some coastal river fans.

Water resource studies of 20 years ago estimated that 10% of the total water utilization was by a method locally known as "spate" irrigation (Min. of Agric., 1984). However, spate irrigation was possible only when erratic weather events associated with rainfall of long duration and high intensity caused river overflows to reach distant areas in the central plain.

District I lies close to the source of high runoff, the northern reaches of the Troodos mountains, and in normal years of rainfall, supplemental irrigation is practiced by virtue of communal and individual *ab antiquo* water rights. Five of the 8 villages in the district have exclusive rights on river water, usually available from February to April and into May. In effect, the two tributaries serving these communities are intermittent streams and not perennially flowing rivers.

District III had supplemental irrigation but with construction and operation of major water works, this practice lost its significance for crop production.

Water Harvesting

No water harvesting is practiced for irrigating field crops. However, roof runoff from plastic sheet covered greenhouses is collected and stored in plastic lined ponds of 300-400 m^3 capacity for drip irrigation in winter and early spring. The quality of the water makes it particularly useful for situations where a salt build-up has occurred from saline well water and high fertigation (Kalimeras, 1987).

Larger ponds with plastic sheet linings and storage capacities of 50,000 m³ to 270,000 m³ have been constructed on off-stream locations in a hilly terrain zone at 500 m elevation as part of an integrated rural development project. Stored water is used to irrigate deciduous fruit trees and vegetables.

Soil conservation works do not qualify by definition as water harvesting methods although they can impede and slow excessive runoff and flooding as well as store water in the soil.

Cropping Practices and Crop Production

Cropping practices adapt to the specific local conditions, but in general, they do not differ significantly from district to district. Cereal crops usually follow a continuous rotation of cereal – forage or cereal – legume – vegetable. Only 10-15% of the cropping area is left fallow.

Tillage operations are usually two cultivations for both rainfed and supplemental irrigation, one by the end of October and one concurrently with sowing.

Barley is sown from early to late November and wheat is sown from mid-November to mid-December. In general, seeds are drilled with simultaneous application of fertilizers at depths of 4-5 cm and row spacings of 16-20 cm. Results of the Agricultural Research Institute's studies, show that the best date for sowing barley is the beginning of November and, for sowing wheat, the beginning of December using seed rates of 100-120 kg/ha (Photiades and Hadjichristodoulou, 1984).

In years of adequate rainfall (above 300 mm), cereals respond to nitrogen applied at rates of 30 kg/ha at sowing and another 20 kg/ha as top dressing at the boot stage. Phosphorous also has been shown to increase yields of grain, particularly barley in years of drought (Krentos and Orphanos, 1979).

Chemical weed control with 2,4-D esters in liquid sprays is used almost universally for broad leaf weeds such as mustard. Wild oats are controlled by pre- or post-emergence herbicides.

No insect control is used for cereals except against the leaf miner in isolated areas. Only for faba beans are chemicals used for the control of aphids.

The scarcity of water is a major constraint to crop production. Inadequate supply of water in the form of rainfall affects rainfed crops directly during the growing period from November to April and, indirectly, the replenishment of aquifers, the source of perennial irrigation.

Supplemental irrigation of cereals, forages, and food legumes increases yields, particularly in years of drought. This is convincingly borne out by the farmers' own experience and from experimental work as well as from large scale seed production data. Of particular relevance are the yields of 1987 (Hadjisavvas, 1987). In District I, a farmer reported the following grain yields:

Cereal	Grain Yield (t/ha)
Barley	
1 irrigation	3.4
2 irrigations	5.3
rainfed	none
Durum wheat	
3 irrigations	
Aronas variety	4.1
Mesaoria variety	5.4
rainfed	0.2 - 0.6 (average 0.5)
Faba beans	
3 irrigations	3.0
rainfed	none

In another village in District I, a farmer reported grain yields of irrigated barley as 4.9 t/ha but only 2.0 t/ha under rainfed conditions. In District III, a farmer who contracted with the Seed Production Center of the Department of Agriculture to produce Durum basic seed delivered 55 t of wheat grain from a 12 ha field (4.6 t/ha). He applied 50 mm irrigation in March and another in April. Baled straw production was 3.0 t/ha (Xenophontos and Mytillineos, 1987). Similarly, a contractor farmer in the Limasol coastal area delivered 46 t of durum wheat seed from a 9 ha field (5.1 t/ha).

Earlier work carried out by the ARI, Agronomy, in District III during the 1977-83 period showed that supplemental irrigation of 50-100 mm resulted in grain yields of selected durum wheat varieties of 8 t/ha, double production of rainfed conditions (Photiades and Hadjichristodoulou, 1984). Similar work in 1983-86 showed that in 1983-84 season, 100 mm of supplemental irrigation applied in February and March resulted in doubling grain yields of wheat from 4.0 t/ha-8.0 t/ha. Results for the seasons of 1984-85 and 1985-86 were similar. Table 19.24 presents the effects of supplemental irrigation in comparison to 3 regimes of rainfed production (Josephides, 1987). The precipitation at the respective experimental sites during the season, 1985-86, is given in Table 19.25.

Table 19.24. Yield performance at 4 sites of durum wheat cultivars in 3 rainfed and 1 supplemental irrigation cropping regime.

Durum variety	Grain yields (t/ha)			
	Rainfed			Suppl. irrigation
	Dromolaxia, I	Akhera, II	Akhelia, III	Akhelia, III
Kyperounda	1.44	1.38	2.72	4.43
Aronas	1.90	1.38	3.40	7.09
Mesaoria	2.23	1.34	3.81	7.69
Karpasia	2.08	1.59	3.66	8.32

Table 19.25. Precipitation, 1985-86.

Month	Rainfed			Suppl. irrigation
	Dromolaxia	Akhera (mm)	Akhelia	Akhelia (mm)
October	48	18	80	80
November	23	15	32	32
December	77	55	73	73
January	68	21	78	78
February	39	46	40	40
March	6	10	7	6 + 75
April	5	15	6	5 + 75
May	16	67	11	16
Total	282	247	327	477

These results indicate not only the striking effect of supplemental irrigation on grain yields but also the effect of amount and distribution of rainfall on grain yields in the 3 districts under rainfed conditions.

Livestock Production

In Districts I and II when water resources allow, alfalfa is grown for animal feed as green fodder or as cured hay. Roughage in years of drought can become a serious problem for livestock producers; therefore, cereal straw is baled mechanically and sold for feed. Similarly, forage hay can be baled but is usually produced by individual livestock owners. The high percentage of concentrates and the low contribution of natural pastures indicate the intensive system of feeding for dairy cattle, and to a lesser extent, for sheep and goats. Barley grain and other concentrates are purchased at subsidized prices from the Cyprus Grain Commission. The need for roughage as well as the conservation of crop residues and agro-industrial by-products were the topics of two separate studies (Economides, 1984 and 1985). The contribution of feed type to the total nutritional requirements of ruminants is presented in Table 19.26.

Table 19.26. Contribution of feed type to the total requirements of ruminants.

Feed type	District		
	I (%)	II (%)	III (%)
Cereal straw	20	26	18
Cereal legume hay	14	11	19
Crop residues	3	1	4
Concentrates	53	54	50
Natural pastures	7	6	6
By products	3	2	2

Socio-Economic Aspects

A characteristic feature of present day agriculture in Cyprus is the part-time farmer. According to a recent study (Ansell, et al., 1984), the incidence of part-time farming in the coastal areas (Districts II and III) and the dryland zone (District I) is shown in Table 19.27. Currently, the trend is shifting more towards farmers with full-time off-farm employment. This is considered the consequence of channeling the labor force to other sectors of the developing economy without necessarily affecting the productivity of the agricultural sector.

Table 19.27. Incidence of part-time farming in coastal and dryland zones.

Type of farmer	Zone	
	Coastal (%)	Dryland (%)
Full-time	61	68
Part-time off-farm work	10	7
Full-time off-farm work	29	25

Another feature, not entirely disassociated from part-time off-farm employment, is the decreasing number who are in agriculture and livestock production without affecting the overall productivity of the sector (15% of the total economically active population).

Employment in agriculture expressed in units of the full-time equivalent of individual persons is 80% landholders and family members and 20% hired labor in District I; 88% landholders and family members and 12% hired labor in District II; and, 84% landholders and family members and 16% hired labor in District III. Whereas hired labor is engaged primarily in irrigated tree and vegetable production, rainfed agriculture (including supplemental irrigation) uses no hired labor because most crop production is fully mechanized. Seasonal labor is hired from neighboring districts for citrus contract harvesting.

Temporary and permanent migration to urban centers is more common from predominantly rainfed areas in the pursuit of employment in non-agricultural sectors such as construction, manufacturing and services.

The most extensively cultivated field crops at a ratio of 1:10 are wheat and barley. Food grains no longer possess the central position in subsistence agriculture since crop production is market oriented and more than two thirds of the country's grain needs are imported. Grain marketing is exclusively undertaken by the Cyprus Grain Commission, importing grain at international market prices and purchasing locally produced wheat and barley at subsidized prices fixed every year by the Government. The costs of production are comparable nationwide because input costs and cultural practices are common throughout Cyprus.

Straw is purchased from cereal producers but forages and hay are grown by livestock producers either on their own land or on rented hectares. Food legumes such as faba beans play a minor role in field crop production but, only in District I, are they in the traditional cropping system. Even so, land sown to faba bean has declined substantially because of accumulated stocks due to low market demand.

In the 3 districts, livestock is integrally associated with the winter cropping system; but, rainfed farming, supplemental irrigation agriculture, perennially irrigated crops, and even livestock production are practiced in a mixed farming system with no clearcut boundaries.

The contribution from livestock production to total farm income is given in Table 19.28. The percentage of the farmer's total income received from the farm, off-farm agricultural labor, and non-agricultural employment is given in Table 19.29.

Table 19.28. Contribution of crop and and livestock production to total farm income.

Production type	District	Rainfed (%)			Supplemental irrigation (%)			Perennial irrigation (%)		
		I	II	III	I	II	III	I	II	III
Winter crops		11	19	7	8	—	2	—	—	—
Tree crops		—	4	8	—	—	—	17	6	44
Vegetables		—	—	—	—	—	—	25	12	14
Livestock		39	59	25	—	—	—	—	—	—

Table 19.29. Sources of farmer's income.

District	Own farm (%)	Off-farm Agriculture (%)	Non-agriculture (%)
I Astromeritis/Akaki	88	6	6
II Larnaca Coastal	50	5	45
III Paphos Coastal	75	7	18

Future Potential for Supplemental Irrigation

A comprehensive plan for the rational development of water resources constitutes the core of overall efforts since 1965. Since this date, a number of dams have been constructed or are presently under construction thus ensuring development of most of the surface water resources for agricultural, domestic, and industrial purposes.

Indicative of the effort for investment in water resources development for district or remote regions that suffer from water shortage for intensive crop production is the planned expenditure of 300 M Cyprus pounds by the end of 1988 for a total dam storage capacity of 300 M m^3 (Louca, 1986). The high cost of water imposes economic constraints and limits its use to irrigation of high cash crops such as vegetables and fruit in competition with low return field crops.

Unless there are pressing reasons such as the cultivation of cereals in rotation with seasonally irrigated vegetables for disease control, the potential for supplemental irrigation would be unprofitable. This situation is of particular relevance to the cropping system of District III where prior to the construction of a dam, canal, and distribution network, supplemental irrigation was practiced on a larger scale. The institutional and technological infrastructures are available for the application of supplemental irrigation to increase field crop production, providing that this technique could be proven economically viable.

Even in District I, where supplemental irrigation is practiced by virtue of river water rights, its use is not likely to increase because river flows are erratic and dependent on rainfall runoff from higher elevations. The existing infrastructure of river intakes and channels does seem to cope adequately with the needs of the communities and individuals with water rights. In District II, supplemental irrigation is not practiced but there is a future prospect of using reclaimed sewage water for the irrigation of cereal and forage crops predicated on water quality measures. In the southern coastal river-fan area of Limassol, supplemental irrigation of wheat continues on a small scale.

Summary and Conclusions

Geographical position imparts to the island a typical Mediterranean semi-arid climate characterized by hot, almost rainless summers and rainy, rather changeable winters. Average annual precipitation is 477 mm but can vary from 182-759 mm.

Intense solar radiation, a relatively dry atmosphere, and prevailing winds result in high evaporation rates especially during warmer months. Pan evaporation reaches average annual values of 1,800-2,000 mm, with the pan evaporation from November to April contributing one fourth of this amount with more than 200 mm during March and April.

The success of rainfed winter field crops such as cereals and food legumes depends on rainfall for adequate soil moisture throughout their time of active growth. In contrast to complete crop failures under rainfed farming during seasons of insufficient rainfall, water applied as supplemental irrigation at the critical growth stages ensures high grain yields. The efficiency of this method for cereal grain production under conditions of drought is confirmed by the 1987 harvest. In District I, barley irrigated twice produced 5.3 t/ha, of grain compared to no yield for barley cultivated under rainfed conditions. Wheat irrigated 3 times yielded 5.4 t/ha as compared to wheat grown under rainfed conditons which produced 0.4 t/ha. Grain yields of durum wheat in District III were of similar magnitude with 5 t/ha, with field experiments indicating that the yield potential of some durum wheat varieties is nearly 8 t/ha under optimal soil moisture conditions.

The existing infrastructure and the necessary technology are available to sustain higher yields; however, supplemental irrigation as a pragmatic means to achieve increased grain yields of cereals and food legumes has yet to be economically justified from the point of view of the national economy. One criterion is the competition for irrigation water between field crops and vegetables. Another criterion is that cereal grain is already heavily subsidized and it is unrealistic to use costly water, which is highly subsidized also, to increase grain yields. In the case of District III, where major resources have been developed for perennial irrigation of tree and vegetable crops, cereals could be grown in rotation for controlling soil-borne diseases.

In District I, there is already the infrastructure for supplemental irrigation of cereals and food legumes from river flows during winter but at peak demand in late March to April other remunerative irrigated crops such as potatoes and vegetables come into competition for water. Only in exceptional years of abundant flows are demands satisfied; in years of low flows, even basic demands covered by water rights cannot be met. Therefore, the development and use of supplemental irrigation in District I is more or less in equilibrium with the supply and demand of water as prescribed by the existing communal and individual water rights.

The possiblity of recycled sewage effluent for supplemental irrigation of cereals and forage crops in District II remains an attractive alternative for use of wastewater.

The present trend in crop production restricts the expansion of supplemental irrigation beyond existing practices of applying low cost water from intermittent river flow through established surface irrigation networks. Such areas can be identified as Districts I and III where traditionally supplemental irrigation has been practiced. In addition, the coastal plain of Limassol which shares close similarities with these two districts could use supplemental irrigation advantageously for winter crops.

Acknowledgements

The author, V. D. Krentos, wishes to express his sincere thanks to the professional staff of the Department of Agriculture, and in particular, to the District Agricultural Officers, the staff of the Agriculture Research Institute, the Director and staff of the Department of Statistics and Research and, in particular, Mr. P. Philippides, Head of Agricultural Statistics, and the Heads of other government departments, services, and organizations of Cyprus who have freely provided all the information needed for this survey. Due thanks go to village heads and farmers who have helped in many ways by providing answers to the questionnaires.

References

Agricultural Insurance Organization, 1986. Annual Report 1985. (in Greek); Nicosia.
Agricultural Research Institute, 1987. Annual report 1986. Nicosia.
Americanos, P. G., 1983. Control of orobanche in broadbeans; Technical Bulletin 50. Agric. Res. Inst., Nicosia.
Americanos, P. G., 1986. Herbicides for broadbeans; Technical Bulletin 76. Agric. Res. Inst., Nicosia.
Ansell, D. I., C. Bishop, and M. Upton, 1984. Part-time farming in Cyprus. Development Study No. 26. University of Reading, Dept. of Agric. Econ. and Mgmt.
Ashiotis, A., 1987. *Personal communication*. Cooperative Central Bank. Nicosia.
Cyprus Agricultural Research Institute, 1986. Annual Reports, 1980-86. Nicosia.
Cyprus Grain Commission, 1986. Annual Reports 1984-85. (in Greek); Nicosia.
Department of Agriculture, 1986. Annual Report, 1985. Nicosia. 197 pp
Department of Cooperative Development, 1982. Report for the years, 1978, 1979, 1980, and 1981. Nicosia.
Department of Statistics and Research, 1986. Agricultural Statistics, 1985, No. 17. Nicosia. 145pp.
Department of Statistics and Research, 1987. Census of Agriculture, 1985 (in press). Nicosia.
Economides, S., 1985. The roughage situation in Cyprus in relation to animal health and productivity; Technical Bulletin 18. Agric. Res. Inst., Nicosia.
Economides, S., 1984. Inventory of crop residues and agro-industrial by-products in Cyprus; Misc. Reports 15.
Grivas, G., 1987. *Personal communication*. Dept. of Agriculture, Nicosia.
Hadjichristodoulou, A., Athena Della, and C. Josephides, 1977. A new durum wheat variety, Aronas; Technical Bulletin 22. Agric. Res. Inst., Nicosia.
Hadjichristodoulou, A., C. Josephides, and A. Karl, 1982. Performance of a new durum wheat variety Mesaoria under rainfed conditions; Technical Bulletin 41. Agric. Res. Inst., Nicosia.
Hadjichristodoulou, A., and A. Karl, 1984. Karpasia, a high yielding new durum wheat variety with improved grain quality characteristics; Technical Bulletin 57. Agric. Res. Inst., Nicosia.
Hadjioannou, L., 1987. The climate of Cyprus: past and present.
Hadjioannou, L., 1987. Climate of Cyprus, Nicosai. 9 pp (unpublished).
Hadjisavvas, S., 1987. *Personal communication*. Farmer from Peristerona village, District I.
Josephides, C., 1987. *unpublished Experimental results*.
Kalimeras, P., 1987. *Personal communication*. Dept. of Agric., Nicosia.
Konteatis, C. A. C., 1967. The water resources of Cyprus, their conservation and development. Department of Water Development, Nicosia.
Krentos, V. D. and P. I. Orphanos, 1979. Nitrogen and phosphorous fertilizers for wheat and barley in a semi-arid region; Jour. of Agric. Science. Cambridge 93:711-717.
Land Consolidation Authority, 1980. Annual Reports, 1979-80. Nicosia.
Loizides, P. A., 1958. Fertilizer experiments in Cyprus; Vol. II, Cereals, Empire Jour. of Expl. Agric. 26:25-33.
Louca, A., 1986. Annual Report of the Ministry of Agriculture and Natural Resources for 1985. (in Greek). Nicosia.
Matar, A. E., 1977. Yields and response of cereal crops to phosphorus fertilization under changing rainfall conditions; Agron. Jour. 69:879-881.
Mavrommatis, A., 1987. *Personal communication*. Department of Cooperative Development, Nicosia.

Meliphronides, J., 1972. The cereal leaf miner. (in Greek). Dept. of Agric., Nicosia.
Meteorological Note Series No. 3; Meteorological Service, Nicosia.
Meteorological Service, 1985. Some studies on the amounts of precipitation on Cyprus; Met. Paper No. 1 (revised), Nicosai. 26 pp.
Meteorological Service, 1986. Basic meteorological features in Cyprus; Met. Paper Series No. 7, Nicosia. 41 pp.
Meteorological Service, 1986. Evaporation from Class A pan and potential evapotranspiration in Cyprus; Met. Paper Series No. 5 (third edition). Nicosia. 8 pp.
Ministry of Agriculture and Natural Resources, 1984. Agriculture in Cyprus: An analysis and evaluation. Nicosia. 84pp.
Neophytou, M., 1987. *Personal communication.* Secretary- Director, Cooperative Assn for Cereal Grain Concentration, Nicosia.
Panayiotou, A., 1983. Geology of Cyprus; *In* Cyprus Today, 11:2-10.
Papachristodoulou, S., 1987. Norm input-output data of the main crops of Cyprus. Revised edition, (in press). Agric. Res. Inst., Nicosia.
Papachristodoulou, S., 1987. *Personal communication.*
Papadopoulos, A., 1987. *Personal communication.* Ministry of Education, Nicosia.
Pashiardis, St., 1987. *Personal communication.* Meteorological Service, Nicosia.
Philippides, P. and Char. Papayiannis, 1983. Agricultural regions of Cyprus: a comparative statistical and techno- economic analysis; Agricultural Studies. Report No.1.
Planning Bureau, 1987. The fifth five-year plan, 1987-1991. Nicosia. (unpublished).
Photiades, I., and A. Hadjichristodoulou, 1984. Sowing date, sowing depth, seed rate, and row spacing of wheat and barley under dryland conditions; Field Crops Research 9:151-162.
Serghiou, C. S., 1975. *Syringopais temperatella* in Cyprus; Jour. Econ. Entomology, 68:489-494.
UNDP/FAO, 1969. Cyprus Water Planning Project.
Xenophontos, E., 1984. Food legumes. (in Greek) Dept. of Agric., Nicosia.
Xenophontos, E., and J. Mytillineos, 1987. *Personal communication.* Seed Production, Dept. of Agric., Nicosia.

Chapter 20

Potential of Supplemental Irrigation in Iran

HAMID SIADAT

Geographical Characteristics

The country is situated on a plateau between the Tigris and the Indus Rivers with a land area of 165 M ha. Iran is located within the North Latitudes of 25° to 40° and East Longitudes of 44° to 64°. The natural boundaries are the Caspian Sea in the north, the Persian Gulf and Sea of Oman in the south, and a few rivers along the northern, western, and eastern borders.

Physiography

Iran is a mountainous country with 80% of the land more than 500 m above sea level. From a physiographic point of view, 4 sections can be distinguished:

1. the triangular elevated central plateau bounded by the Alborz mountain range in the north and the Zagrus Range in the northwest to south and south-east with elevations varying from 500-2,500 m;
2. the lowland areas of the Caspian Seacoast which are actually the northern piedmont plains of the Alborz Mountains with some land areas below mean sea level;
3. the Khuzistan Plain and southern littorals that include mostly low hills and plains stretching from the south-west to the south-east of the country; and,
4. the valleys and mountainous areas on the western slopes of the Zagrus chain of mountains.

Climate

Although most parts can be classified as arid to semi-arid, Iran has a wide spectrum of climatic conditions. Both latitude and elevation have a major influence on climate in the various regions (Figure 20.1). Seasonal variations in precipitation are common. For dry farming areas of the country, Mirnezami (1972) reports that a probability level of 75% is a dependable value and he

E.R. Perrier and A.B. Salkini (eds), Supplemental Irrigation in the Near East and North Africa, 367-388.

presents the following linear relationship:

DP75A = 0.82 AMP-11.7,

where,

DP75A = dependable annual precipitation (mm) at 75% probability; and,
AMP = annual mean precipitation(mm).

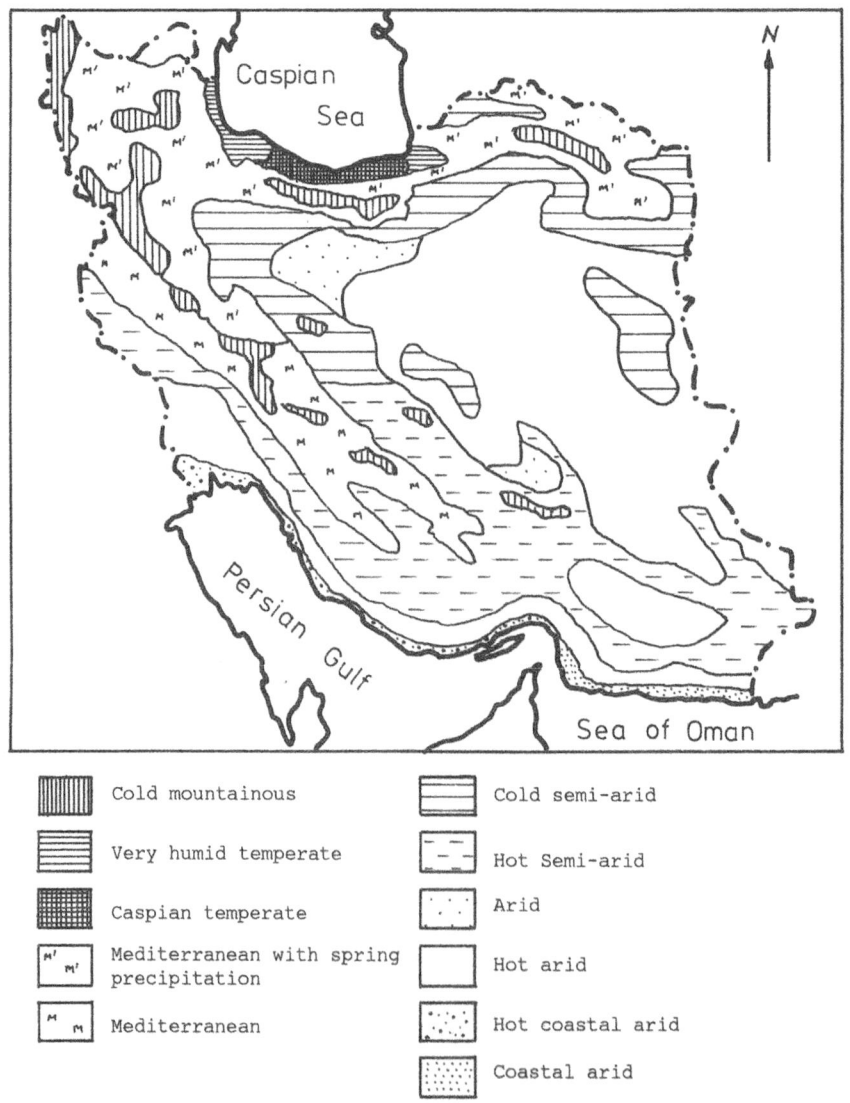

Figure 20.1. Climatic regions of Iran.

With the exception of the western and northern regions and certain locations in the central plateau, precipitation usually falls both as rain and snow. Freezing temperatures can occur as late as April. Summers in the south and south-east are very hot; maximum daily temperatures of 50°C have been recorded at some stations. Detailed information of climatic conditions in Iran has been presented by Ganji (1955), Wallen and Perrin (1962), Ganji, *et al.* (1965), Slabbers (1970), and Rafiq (1976).

Land and Water Resources

Total land area having agricultural value is estimated to be 50 M ha. Eventually, land which could be brought annually into cultivation (cropped and fallow) is believed to be only 30 M ha. However at the present time, in view of the limitations of topography and availability of water, the land cropped and fallowed annually is only 18 M ha. (Bur. of Ag. Infor. Stat., 1985). Of these 18 M ha, irrigated agriculture includes 5-6 M ha; dryland cropped areas (under rainfed conditions) are 6 M ha.; and, the remainder (5-6 M ha) is under annual fallow.

Major soil types, according to the FAO/UNESCO Soil Map of the World (Vol. VII, Sheet VII), have the characteristics of Xerosols, Arenosols, Regosols, Solonchaks, and Lithosols. The expansion of agriculture is not limited by soil resources although achievement of high yields on these soils requires suitable management techniques.

Water resources are considered the most important natural factor limiting expansion and development of agriculture. The annual volume of precipitation is estimated between 350-400 B m^3 and an additional 10 B m^3 from the border rivers' flow; but, only 115 B m^3 is potentially available for use (Masoumi, 1984). The difference is lost to evaporation from bare soil surfaces, evapotranspiration by natural vegetation, and outflow of the rivers into the Persian Gulf, the Caspian Sea, and neighboring countries. Presently, some 60 B m^3 of water is available for agricultural production and this volume comprises 95% of the total water used.

More detailed information on soil and water resources of Iran can be found in Dewan and Famouri (1963), various publications of SWRI, Banai (1977), Bookers, *et al.* (1975), and Masoumi (1984).

State-of-the-Art of Water Management in Iran

The basic ideas of supplemental irrigation and water harvesting are well known in Iran. In some parts of the country such as Baluchistan in the south-east and Khorasan in the east, farmers have used canal and dike systems for hundreds of years for spreading flood water over parcels of land in order to increase water penetration into the soil profile and conserve water for crop production.

However, the extent of the areas where these techniques are practiced is limited and the locations are sporadic.

Supplemental Irrigation

This method has been practiced where either surface or groundwater resources are available and easily accessible during spring and summer. The potential yield or price of produce is high enough to encourage some dryland farmers to accept the higher cost and increased management required for supplemental irrigation of their fields during dry spells.

Currently, farmers in the medium to high rainfall zones decide to use supplemental irrigation of cereals during the growing season but not in advance of the production season. The most important regions using this technique are parts of the Caspian Seacoast in the north and, in the north-west and west, particularly Bakhtaran Province. On the Caspian coast, a variety of crops such as tea, citrus, wheat, barley, cotton, and oil seed are grown as rainfed crops but some farmers do use supplemental irrigation. In the north-west and west, wheat, barley, chickpeas, and lentils are grown under rainfed conditions but cereals are irrigated once or twice in the latter part of the growing season during years when water is available and topography allows gravity irrigation.

In addition to farmers' use of supplemental irrigation, the Soil Water Research Institute (SWRI) has carried out research programs related to supplemental irrigation. Results obtained for tea, cotton, and soybeans in the Caspian coastal area are promising. Also, observation trials with wheat in the western and northwestern regions have shown encouraging results (see the section on Potential Yields in this chapter).

Water Harvesting

In contrast to supplemental irrigation, water harvesting is a traditional farming practice in areas where available water is limited. This method is usually practiced by accident. Some farmers in the central plateau and in the eastern and south-eastern areas grow their limited grain crops by harvesting flood water. In these areas, the annual precipitation falls in a few showers within a relatively short time, thereby leading to surface runoff and floods of various intensities. Different systems are employed to increase infiltration of this runoff for crop production.

In the first system, flood water from a plain upstream flows to a lower level through gaps separating two mountain ranges and is diverted into a wide channel constructed along the direction of flow. Secondary canals are connected at different angles to this main channel to convey water to land parcels enclosed by 0.5-1 m earthen walls. The flood water trapped in these high-walled basins infiltrates into the soil or evaporates into the atmosphere.

When no standing water is on the soil surface, farmers begin sowing operations.

In a second system, fields at the base of small hills are shaped into basins with walls or borders of 50 cm. Runoff water from the hills accumulates in these basins during intense storms and gradually infiltrates into the soil. Grains are sown in these basins when the soil is desaturated.

In yet a third method, the dry beds of old river channels with deposits of silt and clay are used for crop production. During flooding or increased flow in the current river channel, excess river flow is diverted through an inlet at the junction of the 2 channels or through a connecting canal to the dry river bed. Check dams of 1-2 m are constructed in the old river channel to store the diverted water for infiltration where seeds will be sown.

Whatever the method, basically, water harvesting supplies water for pre-sowing irrigations. Although these methods are helpful for seed emergence and early plant development, yields from these fields remain low because of inadequate soil moisture during the latter part of the growing season. Success of the agricultural activities using these harvesting methods is absolutely accidental and dependent on how many times the fields receive flood water. Besides traditional methods, in recent years efforts have been made to implement modern techniques of water harvesting. Results have been promising in a few places and further work by government agencies is forthcoming.

National Goals and Objectives

The important role of supplemental irrigation and water harvesting in agricultural production, so far, has not attracted the attention of policy makers and planners. This is, in part, because research institutes have yet to present concrete recommendations for these techniques. Pertinent research programs are scarce and only in recent years have a few projects been conducted. Thus national goals and objectives related specifically to supplemental irrigation and water harvesting have not been formulated. However, the following are recommendations in support of implementing these techniques to better achieve these national objectives.

1. Increase yields per unit area and stabilize level of agricultural outputs, particularly in areas of dryland farming with high production potential. These needs can support conversion of fully irrigated cereal fields into diversified production of other crops.
2. Increase national production of cereals to decrease imports from other countries.
3. Increase farmers' incomes in dryland farming zones thereby improving their living standards.
4. Optimize the use of Iran's water resources for agricultural production.
5. Decrease migration of farmers to urban areas.

Agricultural Zones

Considering the variability of climate and soil and water resources, it is of no surprise to see wide variations in agricultural crops and farming systems in different regions. Even for a specific crop, cultural practices differ depending on the regional environment. For instance, while most wheat in the north, north-west, and west is produced under rainfed conditions, in the remaining areas economic yields are obtained only with irrigation. Another example, the citrus plantations in the coastal area of the Caspian Sea are grown totally under rainfed farming; whereas in the south the plantations receive more than 20 irrigations in a year.

The growing season is not the same in different regions. In most areas, vegetables such as tomatoes and cucumbers are produced during spring and summer, but in the south these crops are grown in the fall and winter months. Iran can be divided into many agricultural zones with different potentials; however, 3 major agro-ecological zones which include large land areas can be described:
1. a *humid zone* with average annual precipitation of more than 700 mm but with rainfed farming and supplemental irrigation agriculture practiced (this zone is confined to the area of the Caspian Sea and has a variety of crops: rice, wheat, tea, citrus, cotton, and oil seed);
2. a *semi-arid zone* with average annual precipitation of 250-700 mm with dryland farming as well as supplemental irrigation prevailing (found in the north-west and west as well as in parts of the central plateau with major crops of wheat, barley, alfalfa, cotton, sugar beet, vegetables, and fruit); and,
3. an *arid zone* with average annual rainfall of less than 250 mm where crops are totally irrigated (includes most of the central plateau, southern coastal areas, and the east with localized agricultural zones with major crops of wheat, alfalfa, sugar beet, dates, sugar cane, citrus, pistachio, and vegetables).

More detailed classifications have been presented in other reports. For example, Bookers and Hunting (1975) divide the country into 10 agricultural zones coninciding with the provincial boundaries at the time of the study and they distinguish sub-zones based on the differences in soil characteristics, topography, and local climate in relation to crop ecology. Also, Rafiq (1976) presented a report on crop ecological zones of Iran in which he proposes 7 different zones on the basis of latitude, physiography, climate, and soils. Futher information in this regard also can be found in Wallen and Perrin (1962), Mahler and Van de Weg (1969), and Dehsara (1973).

The District Perspective

As stressed in previous sections, the most important dryland farming zones are located in the west and north (Caspian coast). The suitability of these areas for

rainfed farming and their potential for supplemental irrigation agriculture can be recognized by considering the following features prevailing in most parts of those regions.
1. Average annual precipitation is more than 300 mm but can exceed 600 mm in certain localities (see Figure 20.2).
2. Precipitation starts in fall and continues into spring.
3. Most annual evaporation occurs in late spring and summer.
4. Large areas of agricultural lands are flat or with gentle slopes and land grading is minimal for irrigation.
5. Soils are deep and relatively fertile.

In consideration of these 5 points, questionnaires were administered in selected districts. The following summaries of information are descriptions of each of these districts.

Miandorood District

This District is located in Mazandaran Province in the coastal area of the Caspian Sea and includes 52 villages. Rice is the major irrigated crop and wheat, cotton, and soybeans are grown under rainfed conditions in the typical agricultural community.

Total population is more than 30,000 with an average of 5-6 members/household. Female members outnumber males and actively take part in the field labor, particularly for rice production. The community has a high percentage of young people, with 65% of the population estimated to be under 20 years of age.

Most facilities and services are available in towns. There are schools in every village and 13 medical care units in the district. All villages have electrical power, 45 villages have pressurized domestic water, and 5 consumer cooperatives serve the total district. Asphalt and dirt roads connect villages and transportation is easily available. Use of agricultural machinery is well known to farmers with tractors used in all villages.

Total land area is 27,000 ha of which 18,500 ha is cultivated under rainfed conditions and 100 ha of cotton and soybeans are produced with supplemental irrigation. A large farm of 3,000 ha (presently operated by government agencies) is located in this district but data from this source is not reflected in the survey results. Different crops such as wheat, maize, and soybeans are grown on this large farm and the wheat crop receives 1-2 irrigations per year. For the rest of the district, supplemental irrigation is applied only on some cotton and soybean fields using sprinkler systems.

Soils have a medium texture, deeper than 1 m, and relatively fertile but with some salinity detectable. The major problem seems to be slow infiltration which can cause problems of surface drainage and surface crusting that reduce rates of germination and create difficulty for tillage operations.

The major water source is groundwater located at depths of 10 m and lower.

The quality is fair but there is some fear that installation of too many wells or over-exploitation from existing wells will eventually cause increased salinity because of salt water intrusion from the Caspian Sea.

Average annual precipitation is 600-700 mm which falls throughout the year although fall and winter rains account for more than 60% of the total. Figure 20.2 shows average monthly temperature and rainfall data recorded at Dasht Naz station located in Miandorood District.

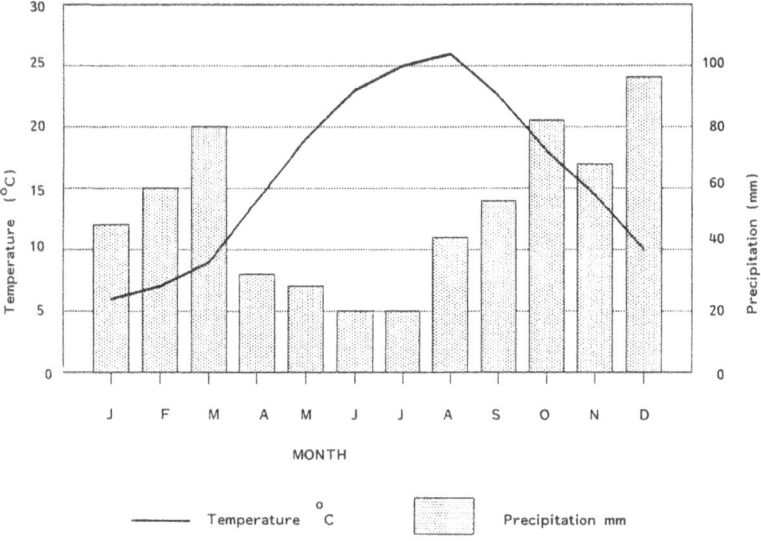

Figure 20.2. Average monthly temperature and precipitation for Miandorood District.

The major winter crops are wheat, barley, and some faba beans. On rainfed farms, the common rotation is wheat (or barley) – soybean (or cotton). However, some farmers follow a three-course rotation consisting of cotton – soybean – wheat. The estimated average yield of rainfed wheat is 2,500 kg/ha and for rainfed barley is 2,750 kg/ha. For wheat produced with supplemental irrigation, the estimated average yield is more than 3 t/ha reaching a maximum of 4 t/ha. These values clearly reflect the general suitability for expansion of crop production using supplemental irrigation.

It should be emphasized that there are constraints to expanding agricultural production using supplemental irrigation. First, a detailed study of water resources in the area should be carried out to determine the permissible annual volume of water which can be pumped for irrigation purposes. Second, training programs on principles of supplemental irrigation should be conducted to familiarize the farmers who presently practice dryland farming. Other possible constraints should also receive due attention such as economic shortcomings of farmers or lack of proper pumping and irrigation equipment.

Agh-Ghola District (south)

Agh-ghola District is situated in the Gorgan Governorate along the coast of the Caspian Sea and east of Mazandaran. Gorgan is a most important agricultural region where both irrigated and rainfed agriculture prevail. Irrigation is practiced only on the flat lands of the valley while rainfed farming exists both on the plain and in the mountains. Agh-ghola District is on the plain's southern section and has 30 villages. The total population is 50,000 with an average of 6 persons per household with a 1:1 ratio of male to female in a family unit.

Elementary schools exist in all villages but there are only 4 high schools. Farmers and their families use the medical facilities in Gorgan city which is 15 km away. There is one medical care unit operating in Agh-ghola's central village. All villages have electrical power and 27 of the 30 have pressurized potable water for domestic use as well as some public water stands. Both asphalt and gravel roads can be used throughout the year. Use of agricultural machinery is common with about half of the available equipment owned by farmers.

Total land area is 50,000 ha of which 29,000 ha are rainfed and 10,000 ha use supplemental irrigation. Major crops are wheat, barley, cotton, and watermelon which are grown both under rainfed conditions and with supplemental irrigation. However, wheat receives supplemental irrigation only in some years and only by some farmers.

Agricultural soils are deeper than 1 m with textures ranging between medium to heavy. These soils are generally fertile even though they have some salinity problems. Most salinity is a seasonal phenomenon which increases during fallow and/or dry periods. Due to the fine texture of the soil, there are problems with infiltration of water and surface crust formation. However, soil tilth is fair and cultivation practices are carried out without undue problems.

Groundwater is the main water supply although some seasonal surface streams flow close to a few villages. Artesian wells are common but the average pumping depth for irrigation is 150 m. Water quality is limited by salinity therefore yields of crops grown with supplemental irrigation are not as high as expected by farmers.

Wheat and barley are the most important winter crops and are sown in 30,000 ha. A small area of 50 ha is cultivated under faba beans in winter but these are mostly for family consumption. The major rotation is wheat – barley – cotton. Under rainfed conditions, two types of 3-year rotation are used. In the first type, wheat or barley is planted in the fall of the first year and harvested in the summer of the second year then the field is left fallow until spring of the third year when cotton is planted and harvested in the fall of the same year. In the second type, wheat or barley is planted for two successive years with cotton planted in the spring of the third year. For fields under supplemental irrigation, a 2-course rotation is used: wheat or barley is seeded in the fall and immediately after harvest in late spring, the field is sown to cotton.

The average yield for wheat under rainfed conditions is 1,500 kg/ha; for barley, the average yield is 1,100 kg/ha; and, for cotton, the yield is 1,500 kg/ha. When supplemental irrigation is used, average yields increase to 2,500 kg/ha for wheat and 2,000 kg/ha for barley and cotton. Considering that only one irrigation is applied for wheat and barley in the years with average or below normal rainfall, the benefit of using supplemental irrigation becomes clear. In the case of cotton, the number of irrigations for years with normal rainfall is 3;

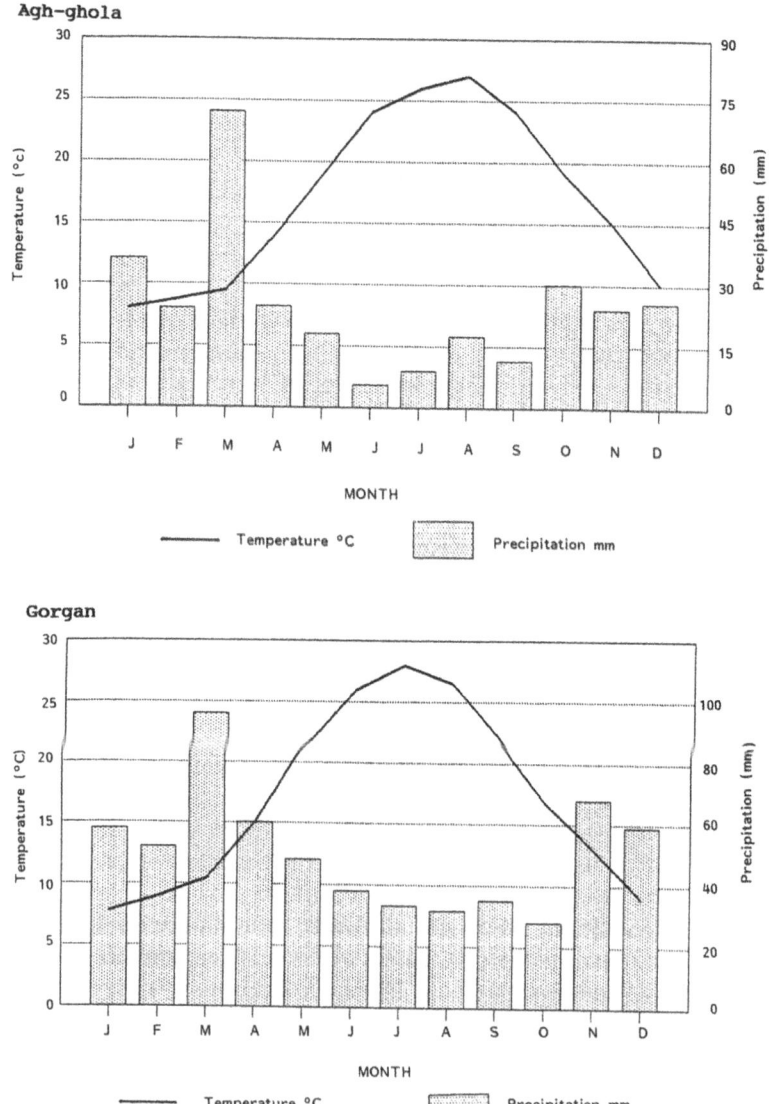

Figure 20.3. Average monthly temperature and precipitation at Agh-ghola and Gorgan

5 for years with below normal rainfall; and, only 2 in years of above normal rainfall.

Average annual precipitation is 300 mm whereas in Gorgan city which is 15 km distant the average annual rainfall is 650 mm. In both areas, precipitation falls throughout the year with the mimimum amount occurring during the summer. Figure 20.3 shows seasonal changes in rain and average temperature for Aga-ghola and for Gorgan.

Agh-ghola District as a whole is suitable for programs of supplemental irrigation agriculture. Detailed studies on existing water resources in the area would be of value in evaluating program feasibility. The most critical constraints to farmers for management of irrigated fields are the factors of modern farming knowledge and land grading. The need for the development of skills calls for training programs and agricultural extension projects. The constraint of land grading must be dealt with by supplying the district with proper equipment and well trained operators.

Bala-Darband (Sarab-Nilofar) District

Bala-Darband District is to the north-west of Bakhtaran city situated in the west. The Zagros chain of mountains is in the east of Bakhtaran Province; consequently, the moisture-laden atmosphere coming from the Mediterranean Sea deposits most of the moisture here. This condition has led to a very suitable environment for rainfed farming in this district as well as in the western provinces.

There are 56 villages in Bala-Darband with a total population of 10,000. The district has 1,700 households with an average of 6-7 members per household. Males outnumber females with a ratio of 55:45. Half the population is less than 15 years of age with 2% older than 65 years of age.

In 30 villages, there are elementary schools (grades 1-5) and one intermediate school (grades 6-8). Those youth who want to continue their studies can do so in Bakhtaran city, 20 km away.

For medical services, farmers go to the city but one medical care unit operates in Bala-Darband. Fifteen villages have electrical power and 40 villages have pressurized potable water for domestic use. Most roads have gravel surfaces in fair condition that can be used throughout the year for motor transportation. Tractors, combines, and seed drills have been used for the last 20 years with 95% of the available machinery farmer-owned.

The District has a total land area of 35,000 ha: 24,000 ha is cultivated annually for rainfed farming; 2,500 ha is left to fallow; 1,000 ha is used for supplemental irrigation of wheat and barley; and, 1,000 ha has irrigated corn, beans, chickpeas, and summer crops.

Soil texture in fields varies from clay loam to silt loam while, everywhere, soil is deeper than 1 m. Fields are generally level and stone-free. Fertility is high and there is no salinity problem. The infiltration rate is moderate for most soils although surface crust formation is common. The crust formation is not hard and soil tilth is generally good. However, after chickpea production, local experience supports that soil cultivation becomes difficult.

The main sources of water for supplemental irrigation are the springs in the area. Groundwater and intermittent streams are also used. Average depth to water table is 5 m and average pumping depth is 12 m; nevertheless, there are 18 wells that are deeper than 50 m. Quality of water from all sources is excellent with no salinity problem.

Wheat and barley, sown in September, are the major winter crops although chickpeas, sown in March, are more important than barley. The land area planted annually to wheat is 15,000 ha; to barley is 2,500 ha; and, to chickpeas is 7,500 ha. Another important crop is lentils which are sown in March on an area of 1,500 ha. The principal rotation used by farmers is a 4-course rotation of wheat – chickpea – wheat (barley) – fallow. This rotation is used both in rainfed farming and supplemental irrigation agriculture.

Average yields for wheat under rainfed conditions are 1,800 kg/ha and for barley are 2,000 kg/ha. The yields for both crops using supplemental irrigation increase to 4,000 kg/ha. These differences in yield show a dramatic 120% increase for wheat and a 100% increase for barley with only two supplemental irrigations. The irrigations are applied one in fall and a second in spring with the applications scheduled when rainfall is either normal or below normal. In years of above normal precipitation, no irrigations are applied.

Bala-Darband has an average annual rainfall of 450 mm as indicated by the data from the meteorological station in Bakhtaran city. Monthly averages of precipitation together with average monthly temperature data of this station are shown in Figure 20.4. The rainy season begins in fall and continues to the end

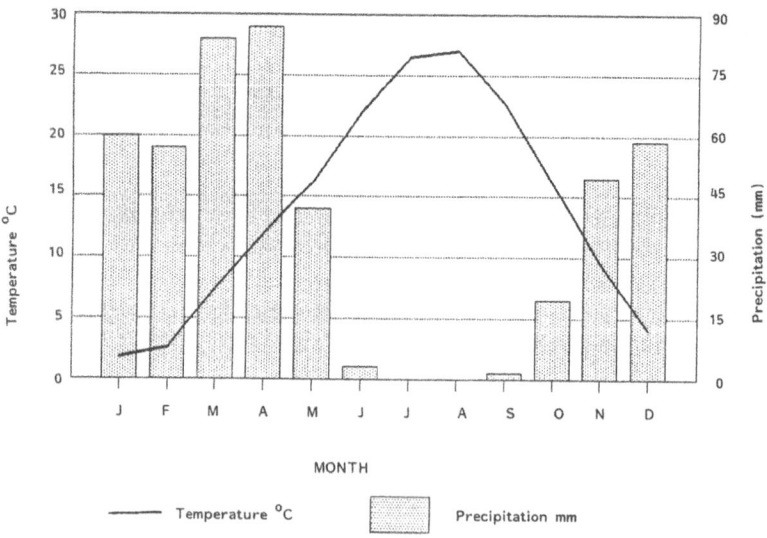

Figure 20.4. Average monthly temperature and precipitation at Bala-Darband.

of May. During the period between June through September, rainfall is nil while temperature is at its maximum. In June or July, one supplemental irrigation is applied to wheat and barley.

This district and others in Bakhtaran Province are suitable for supplemental irrigation. Water resources are ample but there is little use being made of these resources. This, together with favorable soil and climatic conditions, offers excellent opportunities for expansion of supplemental irrigation agriculture in the province. In spite of these positive conditions, farmers seem to lack adequate incentives for such a program. One reason could be that they obtain relatively high yields without irrigation. Also, illiteracy and lack of adequate knowledge of modern farming techniques are believed to be responsible for this attitude.

Expansion of literacy programs in Bala-Darband District and agricultural extension work in conjunction with skills training would certainly help and prepare the way for implementation of large scale projects. Further agricultural research should be conducted and is highly recommended.

Mahidasht District

Mahidasht District is located 10 km to the south-west of Bakhtaran city. There are 124 villages and a total population of 26,000 with 4,700 households having an average of 6-7 members each. The ratio of male to female in the population is 55:45 with 60% of the population less than 15 years of age.

There are elementary schools in 62 villages and 4 intermediate schools in the district. For medical services, farmers and their families can refer to 3 medical units operating in Mahidasht, although for serious illnesses they usually go to Bakhtaran city. Electrical power is supplied to some villages and pressurized water is available in 25 villages. Where pressurized water is lacking, water from shallow hand-dug wells is used for domestic purposes. Secondary roads are gravelly and in fair condition and accessible throughout the year.

Farmers are familiar with the use of agricultural machinery. Tractors, combines, seed drills, and sprayers have been used for the last 25 years. Sixty percent of available machinery belongs to farmers and the remaining 40% are owned by machine operators.

The total land area of Mahidasht is 60,000 ha. Some 1,000 ha is steppe and pasture and 38,000 ha is annually under rainfed cultivation of wheat, barley, chickpea, and lentil. There are 4,000 ha with supplemental irrigation of wheat and barley and 10,000 ha of fallow. The rest of the agricultural fields are used for production of fodder and summer crops as well as vegetables and some tree crops. Major crops are wheat, chickpea, barley, lentil, sugar beet, and alfalfa. In recent years, corn has been gaining in popularity but land allocated for its cultivation is still limited.

Soils are deeper than 1 m and have a slightly heavy to heavy texture. These are fertile soils with no salinity problem and a moderate infiltration rate. Slope and stoniness impose no limitation for agricultural activities. The single soil characteristic which causes some problem, particularly in dry periods, is crust formation which is an extensive phenomenon observed in all fields. Nonetheless, soil tilth is generally fair if proper equipment is used for cultivation.

Groundwater is the main source of water and is exploited by wells of varying depths. Although the average depth to water table is 5 m, the pumping depth of wells varies from 9-80 m and can yield an average flow of 50 l/sec. Other sources of water are galleries and intermittent streams. Quality of water is fair with salinity being no serious problem for agricultural production.

Major winter crops are wheat and barley. Chickpea planted in late winter is another important rainfed crop. The land area annually sown to wheat is 25,500 ha; to barley is 6,000 ha; and, to chickpea is 10,000 ha. Two principal rotations are used: a 4-course rotation of wheat (barley) – chickpea – wheat – fallow is practiced on rainfed fields by 80% of the farmers; whereas, on fields with supplemental irrigation, farmers use a rotation of wheat – sugar beet corn clover in alternate years. Some farmers plant barley instead of wheat in this rotation but the land allocated is a smaller area.

Under rainfed conditions, average yields obtained for wheat are 1,200 kg/ha and for barley are 1,800 kg/ha. Corresponding values using supplemental irrigation are 1,700 kg/ha for wheat and 2,500 kg/ha for barley. These increases generally are obtained with only one irrigation, although for years with below normal or delayed rainfall, farmers may irrigate wheat with two applications. Timing of supplemental irrigation depends on rainfall amount and seasonal distribution. In years with a total precipitation below normal, usually one irrigation is applied in late spring. However, when fall precipitation is delayed beyond October, farmers who have wells or other sources of water will irrigate their fields in the fall. This irrigation is before or after sowing, depending on soil moisture content. A pre-sowing irrigation is applied only if seedbeds were not prepared prior to sowing time.

Average annual precipitation is 397 mm as recorded at the SWRI station located in the district. Rainfall starts in October and continues through May with 45% of the total annual precipitation falling in winter, 20% in fall, and 35% in spring. Figure 20.5 shows the average monthly precipitation and air temperature.

Environmental conditions in Mahidasht are quite suitable for expansion of supplemental irrigation programs. However, the same constraints exist here as in Bala-Darband District: low levels of literacy and a lack of adequate farming knowledge particularly of irrigated agriculture. Some fields have problems with topography which lead to difficulties of water control during irrigation. Future agricultural development should focus on solutions to these problems and farmers will take care of the rest.

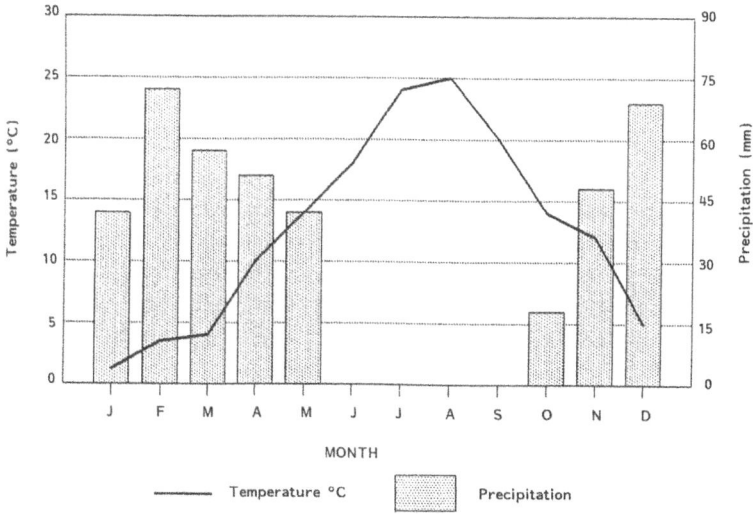

Figure 20.5. Average monthly temperature and precipitation at Mahidasht District.

Potential for Supplemental Irrigation in Iran

Potential Areas for Implementation

Supplemental irrigation has not been the focus of any large scale investigation in Iran and previous studies of general concern to agricultural planning for the country have ignored this system of crop production. Even Bookers, *et al.* (1975), who make brief and sporadic comments on the subject and present a general land capability map with suggested areas for implementing supplemental irrigation do not present their selection criteria or district level information.

In the selection process for the districts discussed in this chapter, the following criteria were considered in determining agro-climatic suitability and potential for implementing supplemental irrigation.
1. An average annual rainfall of at least 300 mm (for cereals, not to exceed 600 mm). Rainfall less than the lower limit is considered insufficient for profitable dryland agriculture (Mirnezami, 1972 and Bookers *et al.*, 1975); whereas, for cereal production, annual rainfall of more than 600 mm seems to decrease yields because of factors such as lodging, fertilizer leaching, and damage from certain plant pathogens (Hashemi, 1973).
2. Seasonal distribution of precipitation should be such that a greater portion of rainfall occurs during the active growth period which, for cereals, is during the spring season.
3. Soils should be permeable and have sufficient depth to store a high percentage of rainfall in the root zone. The soil should lack any serious salinity problem.

4. Topography should be such that grading requirements of the fields and hazards of water erosion are minimal for the purposes of irrigation.
5. Water resources with moderate to good quality should be available particularly after the rainy season.

Certain parts of Iran do meet the requirements for supplemental irrigation programs after considering the stated criteria and information presented in the previous sections of this chapter. Extensive areas along the coast of the Caspian Sea, in the western region, and some north-western parts of the country have good potential. In fact, an estimated 5-10% of the total fields under rainfed farming in these regions is presently receiving supplemental irrigation. Also, in the southern parts of the Zagros Mountains, there may be localized areas with similar conditions which need further investigation.

To state specific recommendations for expanding supplemental irrigation to fields in each district, the feasibility should be evaluated of using both surface water and groundwater. Contributions from surface water to such expansion may not be large since the main volume of runoff comes from melting snow. This runoff usually ends by June which coincides with the end of the rainy season and the beginning of high crop water requirements. Use of surface water for supplemental irrigation would be significant if storage facilities were constructed to save water for use in late spring and summer.

Groundwater is probably more important as a ready supply since timing for water use may be decided by the farmers. The potential for further development

Table 20.1. Groundwater resources.

Agricultural zone	Occurence of aquifers		Approx. annual abstraction ($m^3 \times 10^6$)	Additional potential
	Superficial deposits (alluvium)	Limestone		
Caspain Sea Coast	Continuous plain of 16,000 km². Large outwash fans along mountain front. Many deep, permeable aquifers.	Limited, mainly in areas of high relief	1,700	Considerable, 1,000 Mm³/yr can be developed in coast plain and fans.
North-western	Few intermountain basins, extensive alluvial outcrops at Lake Uromieh. Total area of alluvial outcrop, 21,000 km².	Limited	1,100	Considerable, 1,000 Mm³/yr can be abstracted from alluvium and limestone.
Central Zagros (including Bakhtaran)	Mainly small intermountain basins. Total area of alluvial outcrop, 13,000 km².	Extensive, highly permeable A Samari Limestone	1,800	Considerable 1,000 Mm³/yr can be abstracted from alluvium and limestone.

of groundwater is presented in Table 20.1. These values are taken from Bookers *et al.* (1975) and must be seen not as current data but rather as an indication of the possibility of using groundwater for the areas considered suitable for supplemental irrigation. A final evaluation of the extent to which the practice of supplemental irrigation can be expanded in each zone can only be made when an up-to-date inventory of water resources is available.

Potential Crops

Although wheat and barley are the main crops considered for supplemental irrigation in the present study, there are many other crops with economic significance that could receive similar attention. For example, on the Caspian Seacoast, thousands of hectares are planted to crops of tea, citrus, cotton, soybean, and sunflower which are predominantly rainfed but in a few places do receive supplemental irrigation. In the west, crops of chickpea, lentil, and sunflower are cultivated under similar conditions and have a high potential for production using supplemental irrigation systems. Any national program for expansion of supplemental irrigation systems should pay due attention to these crops.

Potential Yields

National agricultural statistics for 1985 released by the Ministry of Agriculture show that rainfed farming of wheat and barley represent 64% of the total land area sown to these crops while the yields obtained under these conditions are only 37% of the country's total production. According to statistics for rainfed cereal production, average yields of wheat are 643 kg/ha and of barley are 587 kg/ha. For irrigated fields, the yields for wheat are 1,852 kg/ha and for barley are 1,960 kg/ha. These yield data clearly indicate, on a national scale, a crop's response to environmental moisture conditions. Of course, part of this drastic difference is due to the fact that some unsuitable lands, particularly in recent years, have gone to dryland production of wheat and barley. Therefore, nationally, the average yields from traditionally dryland areas appear lower than that obtained in localities noted for medium to high potential under rainfed production. Notwithstanding, most of the difference observed nationally between the average yields from the two systems of production is due to the positive effect of more moisture available for plant growth with the use of supplemental irrigation.

Studies of the relationship between amount of precipitation (or soil moisture status) and yield of dryland wheat indicate similar conclusions. Hashemi (1973) obtained a correlation coefficient of 0.78 for the positive linear relationship between wheat yield and amount of precipitation in various parts of Iran except when rainfall exceeds 600 mm. In a study by Mirnezami (1972), the author concluded that precipitation is the most important limiting factor affecting yield of rainfed wheat. The linear relationships between yield of wheat in these areas and six moisture indices considered in Mirnezami's study give a correlation coefficient of 0.928-0.981. He also concludes that a satisfactory yield for a rainfed system is "difficult – if not impossible – to obtain" for areas with an annual precipitation of less than 295 mm.

An experiment carried out during 1983-87 at Tikmeh Dash Research Station located in the north-west shows that supplemental irrigation of rainfed wheat at heading and grain formation results in an average yield of 1,748 kg/ha. In the same experiment, one irrigation applied at either of these stages resulted in a yield of about 1,500 kg/ha (unpublished data of the regional office of SWRI, Tabriz). Comparing these data with the average yield of dryland crops which is below 1,000 kg/ha, shows the potential for increasing yields with supplemental irrigation in that region. Also, work in the west (unpublished data of Olfati, *et al.*) shows that 2 irrigations applied at the heading and milk growth stages resulted in a 3-year average of about 2,800 kg/ha when average yields of rainfed crops in the area were 1,200 kg/ha. (Data from Mahidasht Research Station, 1982-85).

Results of such studies can be interpreted as showing the potential of increasing yields and the benefits that could be expected if supplemental irrigation were applied. In this respect, information gathered by this author and presented in the section characterizing 4 districts in this chapter should also be mentioned. These data are summarized in Table 20.2 and demonstrate yield increases of 50% to more than 100% as reported by the regional officials who were interviewed in the survey. Although these values are not precise measures, they do provide indications of expected increases in yield with the application of supplemental irrigation.

Table 20.2. Yields obtained for wheat and barley grown under rainfed and supplemental irrigation systems.

District	Yield (kg/ha)			
	Wheat		Barley	
	Rainfed	Supp. Irrig.	Rainfed	Supp. Irrig.
Miandorood	2,500	3,500	2,750	3,500
Agh-ghola	1,500	2,500	1,100	2,000
Bala-darband	1,800	4,000	2,000	4,000
Mahidasht	1,200	1,700	1,800	2,500

Research investigating other crops under rainfed conditions also have shown significant results. Experimental data on tea production (usually cultivated under rainfed conditions) collected by the research staff of SWRI from water and fertilizer trials of 4-years duration show appreciable increases in yields when using supplemental irrigation. The harvest obtained in 1970 reflects an average increase of 2,179 kg/ha; and, in 1971, an increase of 1,593 kg/ha as the result of 4 supplemental irrigations (Nikmaram, 1984). Also, recent investigations of the effect of irrigation on rainfed cotton and soybeans on the Caspian Seacoast reveal a highly significant response to supplemental irrigation. For soybeans, 3 supplemental irrigations totaling 95 mm of water resulted in an 80% increase in bean production when compared to yields received under rainfed conditions. In the case of cotton, a 145% increase in yield was obtained upon application of 212 mm of water applied in 5 supplemental irrigations (unpublished data of Fallah and Maleki for the 1986 season).

It can be concluded that increases in yields can be obtained with supplemental irrigation but that these increases depend on factors such as seasonal precipitation, type of crop, and environmental conditions of the specific area. With the addition of supplemental irrigation, a 50% increase in yield is a reliable estimate for most cases.

Problems and Constraints

Potential in Iran for the development and expansion of supplemental irrigation systems cannot be fully realized unless a number of problems are dealt with properly by the authorities. A discussion of these 5 problems is needed:
- economic factors;
- research and training;
- fragmentation of land ownership;
- mechanization factors; and,
- agro-ecological restraints.

Economic Factors

Low economic returns and higher risks of production in the agricultural sector as a whole, and in dryland farming of cereals in particular, prevent significant private investment for development activities. Furthermore, the average farmer in rainfed areas, individually, cannot provide the capital necessary for implementing such a program even if encouraged by increases in yields as a result of using supplemental irrigation. This type of development program, at least in the early stages, should be subsidized by the government. With the provision of adequate credit and proper agricultural insurance protection, policies can encourage farmers to accept changes in the practices of the existing traditional farming system of production and accept new technology.

Research and Training

Information on soil resources in these areas is available; whereas, data is scarce that describes the extent of available water resources. Up-to-date information is unavailable to formulate a national program of expansion of supplemental irrigation. Research activities have been very limited in number and data reflect local conditions only. The ramification is the lack of training or extension work in this respect. Dryland farmers in suitable locations must become acquainted with the principles of water application, irrigation scheduling, pump maintenance, and equipment operation. They should be informed of the amount and timing of precipitation in due time for planning irrigations. Research topics should focus on rainfall probability analysis, evaluation of different irrigation methods, timing and amount of supplemental irrigation, salinity control, water-fertilizer interaction, suitable crop varieties, and control measures for weeds, pests, and diseases under irrigated conditions. A comprehensive study needs to be performed of the seasonal availability of water and changes in quality as well as collection methods of rainfall runoff and water storage.

Fragmentation of Land Ownership

Fragmentation of land ownership is a widespread problem in areas which have a potential for supplemental irrigation of cereals. Local experience would indicate that if proper educational programs and extension work were presented then these could be effective in circumventing this problem. Water conveyance systems and irrigation methods could be adapted to a situation when farmers insist on managing their fields separately. Under this condition, a system of underground pipelines or above ground tubing could be used together with temporary sprinkler equipment.

Mechanization Factors

In most areas suitable for supplemental irrigation, farmers have used tractors and related implements on their fields for the last two decades. However, problems are manifested in low operational standards, inadequacy of machinery, inappropriate types of equipment, and difficulties of repair and maintenance of spare parts' inventories. These problems need to be resolved before the advantages of mechanization can be realized. Once this is accomplished, increases in yields and decreases in production costs can result which would improve profitability of supplemental irrigation and further encourage investment in water management technology.

Agro-Ecological Constraints

The conditions cited in the section describing areas with a potential for implementing or expanding supplemental irrigation do not prevail everywhere in the zones under consideration. For example, in the coastal zone of the Caspian Sea, there are locations with serious soil and water salinity problems. These constraints could be removed in some cases with the guidance of research results from studies conducted to verify the extent of the problem. Further investigation of available resources in each zone should be carried out to determine the exact extent to which supplemental irrigation could be expanded.

Summary

Most of Iran can be classified as arid to semi-arid, although a spectrum of climatic conditions prevails. Supplemental irrigation and water harvesting are both practiced but their use is limited and locations are sporadic. Supplemental irrigation agriculture is practiced in areas with average annual precipitation exceeding 300 mm; whereas, water harvesting is traditionally used by farmers in drier and hotter semi-arid regions.

Potential zones for expansion of supplemental irrigation systems are on the Caspian Seacoast, and in the north-western and western parts of the country (Central Zagros Valleys). The important crops that presently receive supplemental irrigation are wheat, barley, cotton, soybeans, chickpeas, tea, and citrus and these crops deserve additional attention in the national program. From cereal production, a 50% increase in yield could be expected with the use of supplemental irrigation. This value is reflected in the information gathered in the 4 districts. Similar, or even extreme, yield increases might be obtained from other crops in regions compatible for their production.

The major constraints to further development of supplemental irrigation derive from economic factors, lack of adequate research and training, fragmentation of land ownership, problems related to agricultural machinery, and some local agro-ecological limitations such as salinity.

More extensive and elaborate research studies are required in order to determine the exact extent to which supplemental irrigation could be expanded in each area.

Acknowledgements

I sincerely express my deep appreciation to all who helped me in the different stages of preparation of this chapter. Comments and explanations by Dr. E. Perrier and Mr. A. Salkini of ICARDA were valuable in planning and conducting the study. I also appreciated the general consultations with Dr. M. H. Roozitalab and Engineer M. H. Banai of the Soil and Water Research

Institute of Iran (SWRI). For cooperation in selection of the districts and administration of the questionnaires, I extend my gratitude to my colleagues of the regional offices of SWRI in Mazandaran, Gorgan, and Bakhtaran, Engineers V. M. Fallah, K. Tavakkoli, and M. Olfati as well as Engineers Rouintan and Farzaneh of the Agricultural Service Centers in Bakhtaran. I also thank Ms Espandi who typed the original manuscript.

References

Banai, M. H., 1980. Soil moisture and temperature regime map of Iran. Soil Institute of Iran. Publ. No. 579.
Bookers Agric. and Tech. Serv. Ltd. and Hunting Tech. Serv. Ltd., 1975. Nat'l Cropping Plan. Ministry of Agric. and Natural Resources of Iran.
Bureau of Agric. Information and Statistics, 1986. Agricultural Statistics of 1984-1985. (In Persian). Ministry of Agric. Islamic Rep. of Iran.
Dehsara, M., 1973. An agro-climatological Map of Iran. Arch.Met. Geoph. Biokl. Ser. B, 21, pp393-402.
Dewan, M. L. and J. Famouri, 1964. The soils of Iran, FAO, Rome.
Fallah, V. M. and G. Maleki, 1987. (In Persian) Irrigation experiments on rainfed cotton and soybeans. SWRI of Iran. (Manuscript in preparation.)
Ganji, M. H., 1956. The climate of Iran. Bull. de la Soc. de Geo. d'Egypt, 1955-56. No.19, pp 196-299.
Ganji, M. H. and N. Najafabadi, 1965. Climatic atlas of Iran. Univ. of Tehran, Inst. of Geography.
Iran Center of Statistics, 1985. Iran in the Mirror of Statistics.
Mahler, P. J. and R. F. van de Weg, 1969. Outline map of agricultural land zones of Iran. Soil Inst. of Iran.
Masoumi, Alamooti, A., 1984. Potentials and limitations of Iran's water resources and the necessity of saving in water use. (In Persian). Presented at Conference on Save in Water Uses for Agriculture, Drinking, and Industrial Purposes, Ministry of Energy, Tehran, Iran.
Mirnezami, H., 1972. The relationship between the climate and dry farmed wheat in Iran. M.Sc. Thesis. Utah State Univ., Logan, Utah.
Nikmaram, S., 1984. Results of experiments on the combined and separate effects of fertilizer and water on tea production. (In Persian). Gilan Reg. Off. of SWRI.
Rafiq, M., 1976. Crop ecological zones of Iran. In "Crop Ecological Zones of Nine Countries of the Near East Region.", FAO, Rome.
Slabbers, P. J., 1970. Potential evapotranspiration calculation from meteorological and experimental data for Iran. (First Draft) SWRI, Iran Publ. No. 224.
Wallen, C. C. and G. A. Perrin de Brichambout, 1962. A study of agroclimatology in semi-arid and arid zones of the Near East. FAO/WMO/OMM/UNESCO.

Chapter 21

Supplemental Irrigation Systems of Iraq

SHIFA'A A. MAHMOOD

The project discussed in this chapter is located on the right bank of the Tigris River. The land area is situated on a flood plain of 75,000 ha surrounded by gently rising hills in the north, west, and south. Inhabitants have lived here since ancient times which is evident from the many tels (ancient settlement mounds) in the region of which the largest covers several hectares and reaches a height of 20-30 m. The soils in the project area are sandy loams to silty clays and are calcareous with 20-30% lime. They are slightly gypsiferous with 0.05-0.2% gypsum which is not harmful to agriculture.

The climate is characterized by a fairly hot, almost rainless summer from June to September with a rainy winter from December to February. Rains usually start in November and continue well into early May. Average annual rainfall is 300-350 mm.

Rainfed farming is practiced with wheat and barley in fallow rotation. Traditionally, crop cultivation is combined with livestock, sheep and some cattle. Cropping intensity is 70% with average grain yields of 900 kg/ha for barley and 700 kg/ha for wheat. The Tigris River is the principal water source for supplemental irrigation. Groundwater may contribute to a limited extent.

Irrigation Systems and Objectives

Many designs of irrigation systems have been developed in the last 50 years, all with different objectives. Three categories of planning activities are involved in designing and implementing an irrigation system:
1. determining optimal land units, alternative sowing dates, and irrigation schedules to make the most efficient and effective use of the water (allocated or otherwise available) in any time period;
2. allocating the water to varying time periods throughout the irrigation season from the principal source of supply; and,
3. determining the general land areas for development of irrigation using the principal water supply.

The first category determines the optimal productivity of various subunits of land as a function of the volume of water that could be available and the reliability of supply. The second category uses the analytical results of the first category to determine the volume of water (and reliability of supply) that could be delivered to the project's sub-units during any given time period. The third category then matches the potential of the project to the determined water supply; this includes consideration of providing storage to improve the time sequence of water availability and, more important, to improve the reliability of that supply.

In the first category of planning, the farmer is given the possibility of choosing which area is sown to any crop grown on the land. Most of the variables of production are deterministic at the start of the season but others are stochastic such as precipitation (and, in some instances, water supply) and are inputs for operation of the irrigation system. The farmer's objectives can be expected to maximize economic returns (or the equivalent in food, fuel, and fibre) and to avoid excessive risk.

Ismail and Saeed (1979) studied the effect of supplemental irrigation applied by sprinkler on wheat yields in Northern Iraq. The study was to determine whether or not supplemental irrigation would be beneficial. Several plots were sown under rainfed and irrigated conditions, using identical cultural practices for both methods, and comparing the yields obtained from each, rainfed and supplemental irrigation. Average yields increased considerably: 58% for rainfed and 100% for supplemental irrigation. These data show that adding supplemental irrigation to dryland farming is extremely beneficial and results in higher economic returns. This study concluded that higher yields can be obtained with supplemental irrigation as compared to rainfed conditions.

Collection and Assessment of Data

Data that were collected are presented below. It should be noted that supplemental irrigation is not yet in practice because the project is under construction.

Two major crops are considered, wheat and barley. Using irrigation, wheat production can be expected to increase substantially from the present average of 700 kg/ha. With good management, fertilizer application, and timely irrigation (especially at the critical growth stages of flowering and grain setting), the yields of Mexipak and other improved varieties could increase 5-fold or more than 4,000 kg/ha.

Experiments were conducted in northern Iraq by the Agricultural Mechanization Division, College of Agriculture and Forestry Irrigation, (Mosul University). The studies examined the effect on wheat of supplemental irrigation using sprinklers under the prevailing climate. These results showed that supplemental irrigation helps prevent crop failures and poor yields in years of low rainfall. In 1975-76, a year of late rains, 2,283 kg/ha were obtained from crops with supplemental irrigation (a 58% increase in wheat grain yield). For this same season, 1,442 kg/ha were obtained from rainfed production of wheat. In 1976-77, rainfall started on time but the annual precipitation was abnormally low, yields from plots with supplemental irrigation were 1,556 kg/ha, 110% higher yields than the 742 kg/ha obtained from rainfed fields. The experiments were conducted on reddish brown, calcareous clays with a zone of carbonate accumulation at the 30-50 cm depth. Fertilizer application (superphosphate), land preparation, and sowing methods were the same for trials on fields under rainfed conditions and those with supplemental irrigation.

Table 21.1 shows that with supplemental irrigation target yields of the crops described are expected to increase gradually. An assumption was made that, during the first 5 years, yields would increase rather steeply; whereas thereafter, from years 6-10, an increase of 1.5% per year would take place.

Table 21.1. Target yields for crops grown under rainfed conditions with supplemental irrigation.

Crop	Year						
	0	1	2	3 (kg/ha)	4	5	10
Wheat and Barley	700 900	1,500	2,000	2,500	3,000	3,500	4,000

The rapid increase in yields for this first 10 years of the project development is based on an assumption that all constraints to direct cultivation will have been removed. Gradual increases after the tenth year are expected on the basis of increasing experience in management and organization of agricultural research as well as the introduction of new technology and other aspects of production. Yet for evaluation and determination of costs and benefits for project operations, constant yield levels were assumed after the tenth year. Wheat and barley are assumed to have the potential for equal grain target yields of 4,000 kg/ha although barley will have a lower straw output. Sugar beet production of 25 t/ha after the tenth year seems to be justified, considering current production values for Mosul.

Projections for crop inputs conform to the expectations of crop yields. Fertilizer requirements are shown in Table 21.2. During the first 5 years of project development, after the completion of the infrastructure, a gradual increase in fertilizer need will occur. The type of fertilizer applied will depend on soil characteristics. High pH requires the use of sulphate of ammonia instead of urea and sulphate of potash should be used instead of muriate of potash. Pesticide and herbicide applications usually involve a few kilograms or litres of active ingredients per hectare and current use is not expected to change considerably over the developmental period. The cost per hectare is estimated to be 90 US dollars based on the rates of application and cost of pesticides. The chemicals planned for use are:

| Dimethoate | Malathion | Carbary 1 85% | Metasystox | Pyramin |
| MCPA 25% | Dipterex | Dithane | Zineb | TCA |

Table 21.2. Fertilizer requirements.

Crop	Type of fertilizer	Year of input						
		0	1	2	3	4	5	10
					(kg/ha)			
Wheat	N	—	50	60	70	80	90	110
	P_2O_5	—	20	25	30	35	40	45
	K_2O	—	15	20	25	30	35	35
Barley	N	—	50	60	70	80	90	110
	P_2O_5	—	20	25	30	35	40	40
	K_2O	—	15	20	25	30	35	35

Establishment of field units are planned for farmer operation by farmers who will receive all mechanical assistance from the Machinery Rental Stations or who will own and operate their own equipment. Wheat can be grown continuously on these farms if soil fertility is maintained by applications of adequate amounts of fertilizer. Attention must be given to the diseases of rust and smut therefore preventive spraying may be performed.

Present Land and Water Use

Rainfed Farming

Existing agriculture in the project region is a rainfed farming system of winter cereals, wheat and barley, in conjunction with migratory flocks of sheep which graze stubble between time of harvest and mid-February when flocks are moved south in search of seasonal natural pasture.

Average annual rainfall is low, 200-350 mm, and correspondingly, the yields are poor. Crop production is extensive and fully mechanized with farm machinery owned both privately and by joint cooperatives. The State Organization for Agricultural Mechanization in Mosul, supported by the General Establishment for Operation and Maintenance, might be requested to provide additional equipment in years of good harvests. Fertilizers and sprays are not used in crop production.

Survey data show that 82% of the land is cultivated each year with the remaining 18% left fallow. The percentage of fallow is marginally greater in the northern and eastern sectors where up to 25% of the land can be in fallow at any given time.

Land Ownership and Farming Practices

Farm type is governed by land ownership in accordance with the Agrarian Reform Law of 1970. Most land is privately owned under the law and farmed cooperatively, but some cultivated land is farmed privately. Land farmed non-cooperatively is frequently cultivated by share cropping with a machinery contractor providing seeds and equipment, and receiving an agreed percentage of the yield at harvest. There are no state farms within the project area.

Irrigation

No irrigation is now practiced within the project area although there are small areas of irrigated farming on the northwestern boundary of the Jezira Plains at the foothills of Jebel Sinjar.

Crops

The present agronomy of the area is basic and dependent upon natural rainfall. The growing season is long for 3 reasons:
1. winter temperatures are low, retarding the growth and development of the crops (ground frost can be expected at least 30 nights/year);
2. rainfall is low and intermittent, with a duration of 6 months or longer; and,
3. for cereal crops, in particular, the length of the growing season is extended as a result of livestock grazing of the crop immediately after plant emergence because of insufficient fodder at this time of year.

Wheat and barley are the only crops grown in the area. Sowing should start in October and finish in December but usually commences in November and has even been observed still in progress in late January. The soil is sometimes plowed or disced but is frequently drilled with a disc coulter drill. Seeding rates are low in order to reduce consumptive water use of the crop because of the limited soil moisture. Plant populations are planned for much lower densities to minimize competition for moisture and therefore sustain growth. From soil pit observations, roots have been observed to penetrate to 80 cm.

Crop rotations vary with respect to the quantity of land in fallow. In some areas, wheat is planted consecutively for up to 4 years to 100% of the land before fallow is introduced. In others, wheat and barley are alternated 4-6 years before fallowing. In yet another type of rotation, half the land is sown to wheat or barley and the other half is left in fallow; i.e. only 50% land utilization. In the 1979-80 season, data from field observations and aerial photographs show that 82% of the land area was cropped.

Harvest begins in late May and continues until July with wheat and barley harvested with combines. Some combines are tanker models and transfer the grain in bulk to trucks for transport to local cooperatives while others are bagger models. The latter are used more by the private farmer than by those farmers who work cooperatively. Yields can vary from zero yield in a bad year to 800 kg/ha in a good year. Late rainfall in April and May is the most important factor contributing to good yields.

Irrigation Methods

Several irrigation systems and methods are available in Iraq, ranging from the traditional method of surface irrigation to sophisticated equipment systems of sprinkler and drip irrigation. A comparison of the advantages and disadvantages of surface irrigation with large scale sprinkler irrigation systems is presented qualitatively in Table 21.3. For purposes of this chapter, the discussion will be confined to 2 types of system:
1. surface irrigation with long furrows and borders to be fed by siphons; and,
2. large scale sprinkler irrigation systems cooperatively operated and maintained.

Table 21.3. Comparison of surface and large scale sprinkler irrigation systems

Factors	Surface	Sprinkler
Labor requirement	high	low
Cadastral changes	considerable	limited
Social adaptation	considerable	limited
Water economy	low	high
Night storage	yes	no
Drainage requirement	complete system	limited
Land leveling	yes	no
Tertiary canals/structures	yes	no

In general, surface irrigation does not involve complicated machinery which may break down as a result of insufficient maintenance. However, there are considerable drawbacks to this method. Surface irrigation is labor intensive such that each farmer could not farm more than 7-10 ha. Consequently, to farm the planned project's land area effectively, many families would be required to cultivate the total area or a portion of the land would need to remain in fallow. Introduction of surface irrigation to a region that is thinly populated but with established farming practices of rainfed farming would place considerable constraints on project operations and could negatively affect the economic feasibility of the project.

Surface irrigation also has technical drawbacks: the irrigation efficiency is less than that of sprinklers. Considering that water is the limiting factor and the volume pumped expensive, water economy is a factor in selection of any irrigation method or system. If surface irrigation were selected, night reservoirs would be needed; with sprinkler irrigation they are not. An even more critical consideration is the water table which might be expected to rise with the use of surface irrigation, necessitating the early construction of a sub-surface drainage system. With sprinkler irrigation, surface drainage can be of simple construction and sub-surface drainage could be postponed for several years and might not be needed at all.

Sprinkler irrigation requires more knowledge for operation and maintenance; however, sprinkler systems offer many advantages over surface irrigation if the correct system is installed. Sprinkler systems for project purposes could be designed so each farmer could manage an independent delivery system but each farmer would have to operate and maintain this equipment, a labor intensive and expensive choice. Alternatively, large scale sprinkler systems could be installed and operated within a cooperative organization. These large systems would irrigate land units of 100 ha or more and could be operated and maintained by skilled technicians employed by each cooperative.

Use of cooperatives and skilled technicians would minimize social adaptation of farmers to new farming practices. This approach should not necessitate changing cadastral boundaries and most farmers could continue cultivating the same land as before project implementation. Irrigation would be managed by cooperatives and cropping patterns could then be changed from traditional practices of rainfed farming to supplemental irrigation agriculture to maximize returns. Local farmers would be allowed to gradually adapt new practices within existing social environments without the immigration of large farming populations for additional labor.

Large scale sprinkler systems would be designed as follows. A main canal would convey water to laterals which in turn would divert water to distributaries. Large sprinkler equipment would pump directly from the distributaries to irrigate land units of 100 ha. Two sprinkler systems will be considered for project use: center pivot and linear move systems. The linear move system is considered best suited for conditions in the project area.

The technical advantages of large scale sprinkler systems would be: no night storage reservoirs, no land grading, and fewer open canals. These advantages support greater accessibility to cultivated fields with less drainage construction and achieve a higher water economy.

Conclusions

Average annual rainfall is low, between 200-350 mm. In the north-east, the average annual rainfall is 400-450 mm and yields can reach 1,500-1,600 kg/ha. Except for these areas in the north-west, grain yields are correspondingly poor amounting to 800 kg/ha for wheat and barley.

The main wheat varieties grown are Sabirbed, Agiba, and Mexipak. Mexipak is more successful in high rainfall areas with fertilization. Barley is grown, but the varieties of the Black and White types are unknown.

Cropping patterns include a fallow rotation with a cropping intensity of not more than 70% and 50% land utilization.

Rainfall is not well distributed, especially during critical stages of plant growth. Germination is in October and November with grain formation in March and April. Moisture requirements are not assured during these two growth stages. Few supplemental irrigations would be required to alter this situation: one application of 30-50 mm during germination and 1-2 applications during the grain formation. One or two supplemental irrigations could increase yields 100% with increased applications of fertilizer.

Table 21.1 shows expected yields (kg/ha) for wheat and barley for the production year. These target yields are estimated using supplemental irrigation and are calculated to progressively increase from 700 kg/ha to 4,000 kg/ha from year 0 to year 10.

No quantified data is available for evaluation of national resources for agricultural development with the integration of supplemental irrigation into traditional rainfed farming practices. Observations of yield increases when additional rainfall is available for crop production are a positive indication of the need for an inventory of national resources for future application.

References

Al-Fakhri, Abdullah Kasim, 1982. Agriculture in the Arab land, Books House for Printing and Publishing, University of Mosul, Iraq.

Al-Fakhri, Abdullah Kasim, 1982. Rainfed agriculture in Iraq, (in Arabic), Books House for Prtg. and Publ., Mosul, Iraq.

Al-Fakhry, M. K., M. Asrar, and L. K. Ismail, 1978. Effect of different seeding practices on the yield of wheat crop under Hammam Al-Alil conditions; Mesopotamia Jour. of Agric., Vol. 13:1, Univ. of Mosul, Iraq.

Al-Mauof, Mahmood Ahmed, 1982. A key to beans agriculture in Iraq. Books House for Prtg. and Publ., Univ. of Mosul, Iraq.

Al-Younis, Abdul Hamid Ahmed, and Abdul Sattar Abdullah Al- Girghui, 1981. Agriculture of industrial crops in Iraq, Books House for Prtg. and Publ., Univ. of Mosul, Iraq.

Arab Organization for Agricultural Development, 1984. Technical and economical feasibility study for crop agriculture in dry regions in Iraq, AOAD.

Central Statistical Office, 1971. Agricultural census, Gov't. of Iraq, Baghdad, Iraq.

Central Statistical Office, 1976. Annual abstracts of statistics, Gov't. of Iraq, Baghdad, Iraq.

Chakvavarty, H. L., Plant weal of Iraq (A Dictionary of Economic Plants), Botany Dept., Ministry of Agric., Baghdad.

College of Agriculture and Forestry, 1979. Rainfed agriculture study in north Iraq, study of agricultural production references and the scientific measures for its improvement, Univ. of Mosul, Mosul, Iraq.

Council of Scientific Research, Current research in Iraq, Scientific Documentation Center, Baghdad, Iraq.

Council of Scientific Research, Research abstracts in Iraq, Scientific Documentation Center, Baghdad, Iraq.

El-Sherif, S. M., A. Y. Al-Sayegh, and M. Al-Samerrai. 1976. Interaction between rate of water application on yield of cotton plant, Mesopotamia Jour. of Agric., Vol. 11:1, Mosul, Iraq.

1977. Studies on effect of irrigation water and fertilizers on field crops (response of cotton),

Mesopotamia Jour. of Agric., Vol. 12:1, Mosul, Iraq.

First Scientific Conference on Applied Agricultural Research, 1984. Different research on wheat and barley in the Governorates of Sulaymania & Ninevah, Baghdad, Iraq.

Gov't of Iraq, 1942-1978. Rawi statistical abstracts for Iraq, Meteorological Service, Baghdad.

Harza Engineering Co. and Binnie and Parterns, 1963. Hydrological survey of Iraq – Final Report.

Harza Engineering Co. and Binnie, Deacon, and Gourley, 1959. Summary of monthly precipitation of stations in Iraq, 1887- 1958.

Institute of Applied Research on Natural Resources, 1971. An integrated reconnaissance survey of the natural resources in the environs of Tel-Afar, Tech. Bulletin No. 31, Baghdad, Iraq.

Ismail, L. K., and Mohammed Saleh, 1979. Effect of supplemental irrigation to sprinkler on wheat yields in northern Iraq; Agric. Mechanization in Asia, Vol. X:1 (winter), The Farm Industrial Res. Corp., Tokyo, Japan.

Journal of Research for Agriculture and Water Resources. Agric. and Water Resources Res. Cntr, Baghdad, Iraq.

Mesopotamia Journal of Agriculture. College of Agriculture and Forestry, Univ. of Mosul, Iraq.

Ministry of Agriculture, Annual report for the implementation of the annual agricultural conference recommendations, planning, and follow-up office. Baghdad, Iraq.

Ministry of Agriculture, Iraqi Agric. Magazine, Baghdad, Iraq.

Ministry of Planning, 1973-1978. Annual abstracts of statistics. Central Bureau of Statistics, Baghdad.

Nedeco, Ministry of Irrigation, 1982. Final planning report, Shimal Irrigation Project, Vol.4, Baghdad.

Swiss Consultant, 1984. Draft planning report, South Jazira Irrigation Project; Gen'l. Establishment of Jazira Irri. Proj., Mosul, Iraq.

Tecniberia, Planning report, East Jazira Irrigation Project, Vol. 5, Gen'l Establishment of Jazira Irri. Proj., Mosul.

Teaching Staff, College of Agric. and Forestry, 1982. Research Abstracts, Vol. I; 1966-1982, Univ. of Mosul, Iraq.

Teaching Staff, College of Agric. and Forestry, 1985. Research Abstracts, Vol. II; 1983-1985, Univ. of Mosul, Iraq.

Iraqi Journal of Agricultural Science (Zanco). Univ. of Salahalddin; Arbil, Iraq.

VIO Selkhozpromexport, Ministry of Irrigation, 1979. Distribution of irrigation land in Iraq up to the year 2000, Gen'l. Establishment of Studies and Design, Gen'l. Scheme of Water Resources and Land Development in Iraq, Stage II, Gov't of Iraq, Baghdad, Iraq. (in Arabic), Books House for Prtg. and Publ., Mosul, Iraq.

Chapter 22

The Farming Systems in Jordan: Rainfed, Water Harvesting, and Supplemental Irrigation

A. A. JARADAT

Geographical Characteristics

Jordan is located in the eastern Mediterranean region between latitudes 29° 30' N and 32° 31' N with a land area of 96,000 sq km. The country's terrain is marked by stark contrasts, the most striking feature being the Dead Sea, the lowest point on earth at 395 m below sea level. The Jordan Valley is north of the Dead Sea and is rapidly being developed for agricultural purposes by the construction of a sophisticated irrigation network. The rift valley is bordered by precipitous hills that within a short distance drop from an average of 1,067 m above sea level to the Jordan Valley and Dead Sea. Beyond these hills are the central highlands which stretch from the north of Ma'an to the Syrian border widening slightly like a wedge towards the north. The terrain varies from extensive, slightly sloping plains to abrupt hills too steep for cultivation. To the east lies the steppe and steppe desert, only occasionally broken by oases or catchments where cultivation is possible (Behairi, 1973 and Gubser, 1983).

Climate

The climate is predominately Mediterranean characterized by mild rainy winters and hot dry summers; however, it displays a few variants caused by the effect of latitude, altitude, and the shadow effect of the mountain range. These factors bring about marked differences in temperature and the amount and distribution of rainfall (Abandah, 1978). Climatologically, Jordan is divided into 3 distinct regions: the Jordan Valley, the highlands, and the steppe and steppe desert (El-Kawasma, 1983). The rainy season extends from October to March or April with an average annual rainfall related to topography, with the mountain region receiving the highest rainfall and the steppe desert, the lowest (Figure 22.1). Over 50% of the total land area receives less than 100 mm of average annual rainfall; however, in the major dryland agricultural areas, the average annual rainfall is 250-600 mm. Rainfall is erratic, irregular, and highly variable with occasional storms of high intensity especially in the highlands.

E.R. Perrier and A.B. Salkini (eds), Supplemental Irrigation in the Near East and North Africa, 399-423.
© *1991 ICARDA.*

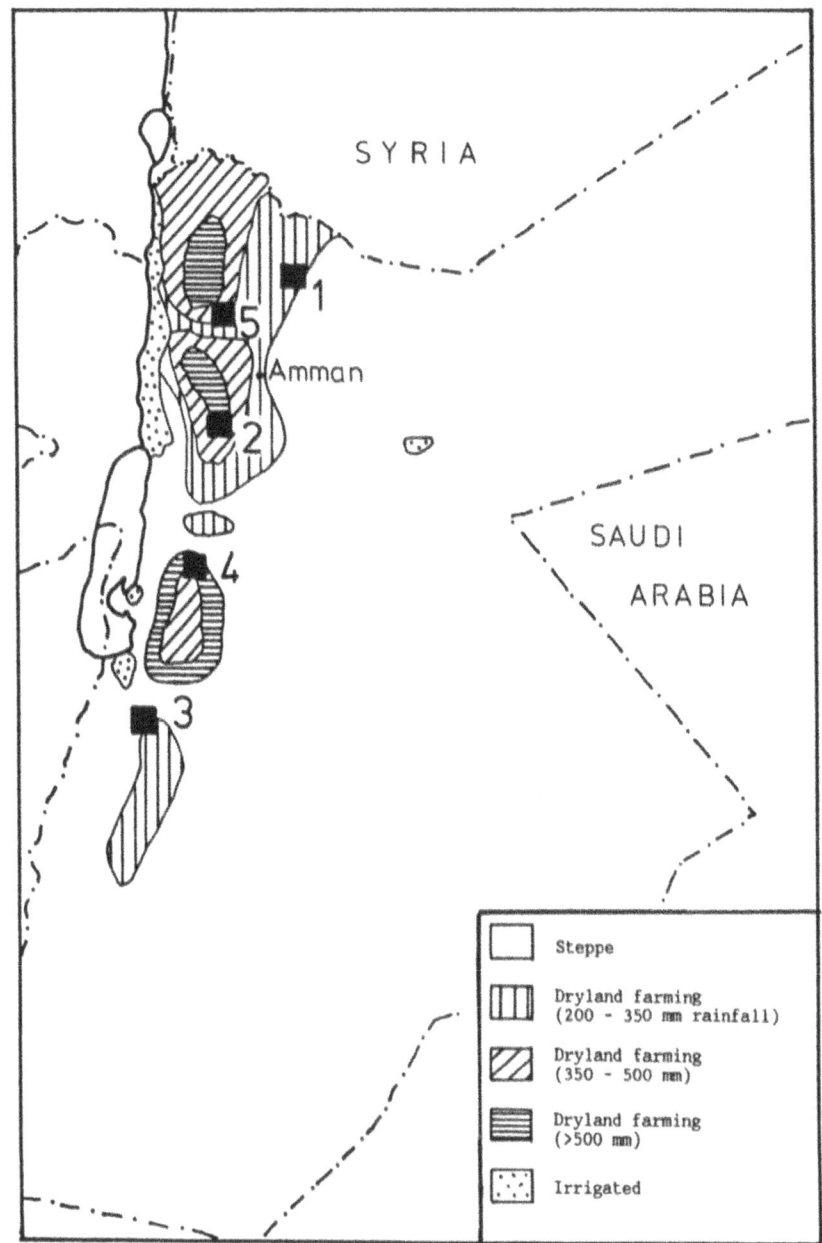

Figure 22.1. Agro-climatological zones and locations of surveyed districts in Jordan.

Land Resources

Soils have developed under 2 major regimes: the xeric and the torric (aridic) moisture regimes. Xerochrepts and Chromoxererts developed under the first regime and Aridisols and Entisols developed under the second (Qudah and Jaradat, 1987). Several authors (Draz, 1979 and Kasapligil, 1956) discuss Jordan's land resources and their distribution according to average annual rainfall. The 5-Year Plan of 1976-80 (NPC, 1975) partitioned the total land area into 4 regions (Ministry of Agriculture, 1974). Potentially, over 8% can be considered arable whereas the majority is classified as arid desert receiving less than 200 mm of average annual rainfall.

Land use in the rainfed agricultural sector is inefficient in part because of a lack of information on capability of the soils for crop production. There are large areas of underutilized soils in this sector and these are undergoing a rapid depletion of fertility under prevailing land use practices (Qudah, 1983). Opportunities for improved use of Jordanian soils for crop production do exist and are highlighted in many reports (Mitchell and Howard, 1980) and feasibility studies (Ministry of Agric., 1974 and Natural Resources Auth., 1977). The rainfed lands are essentially a fixed resource and any future additions to cropped lands must come either from conversion of natural grazing lands or from reclamation of lands that are not being used for agriculture (Ministry of Agric., 1974). The conversion of grazing lands to cropped lands in the low rainfall zone, less than 250 mm, has resulted in soil erosion and expansion of the desert during the last 2-3 decades (World Bank, 1981).

The Agricultural Context

Agricultural Zoning

The drylands have been divided into 4 agro-ecological zones (Ministry of Agric., 1974) (Table 22.1). This division is based on the factors of rainfall, topography, temperature, altitude, natural vegetation, and soil groups. Over 90% of the area is in the arid zone where cultivation is prohibited by law; however, cultivation occurs during years of above normal rainfall and, as a consequence, the natural vegetation has been destroyed (Tadros, 1984). The

Table 22.1. Profiles of agro-ecological zones in Jordan.

Agro-ecological zone	Ave. annual rainfall (mm)	Land area (ha)	Total population (%)
Arid	0-200	8,456,900	31
Marginal	200-300	563,400	41
Semi-arid	350-500	135,900	19
Semi-humid	500-800	98,900	9

arid zone, traditionally, serves as natural pasture land for sheep, goats, and camels. The marginal, semi-arid, and semi-humid zones, together, comprise less than 10% of the country's area. Nonetheless, 41% of the population live in the marginal zone and a slightly smaller percentage (31%) lives in the arid zone. The remaining 28% live in the semiarid and semi-humid zones (Dept. of Statistics, 1981).

The Agrarian Sector

Heterogeneity is a major characteristic of the agrarian structure (Lanzendorfer, 1985) and results from the varying production potential in different regions of the country. Dryland agriculture is dominated by the cereals, wheat and barley. In areas of moderate to high rainfall (350-500 mm), other crops are grown, e.g. lentils, chickpeas, summer vegetables, and tobacco. In some parts, drought-tolerant tree crops are grown such as olives, grapes, and almonds.

Agriculture accounts for 12% of the gross domestic product (GDP), with employment in the agricultural sector constantly and steadily declining, reaching a low level of 15% in the early 1980's (El-Hurani, 1987). The public and private sectors invested heavily in irrigated agriculture during the 1970s when development of rainfed and arid lands received lower priority. During the 1980s, the government has directed major efforts towards developing rainfed farming to maximize local production and reduce rising food imports (Ministry of Planning, 1986).

Production Environment of the Dryland Regions

Two major factors which play a role in the development of rainfed regions are small farm size and fragmentation of land holdings and availability of improved inputs.

Small farm size and fragmentation of land holdings

These factors adversely affect the technical and economical feasibility of using modern inputs and services in rainfed farming and are believed to be a principal cause of low productivity and a major impediment to the development of the rainfed lands. Available data indicate that the average size of a landholding in the dry areas is 8 ha with this single landholding divided into 2.5 fragments. This reduces the farm size in one location to 3.3 ha. This small farm size hinders the use of improved inputs and services such as modern agricultural machinery, fertilizer, and irrigation.

The Farming Systems in Jordan

Availability of improved inputs

A fundamental requirement for improving dryland agriculture is the availability of inputs and agricultural services in the rural areas where farmers can easily obtain them. This includes improved seeds, herbicides, fertilizers, and proper machinery. Unfortunately, private agribusiness has not been as interested in investing in dryland areas, even for supplemental irrigation, as has been the case with irrigated projects.

Supplemental Irrigation

Existing supplemental irrigation agriculture is limited to fruit trees and some vegetables. Tobacco, as a summer cash crop, receives 1-2 supplemental irrigations early in the growing season (Dept. of Statistics, 1982). Farmers, in rainfed areas, are hesitant to use limited water resources for cereal or forage production because the gross margin of fruit trees is 12 times higher per hectare and 4 times higher per day than that of cereal production (FAO, 1974). However, if production potential is considered from a technical perspective then forage production using supplemental irrigation where water resources are available would double land productivity and triple labor productivity.

The final report of a feasibility study in the Shoubak area concludes that supplemental irrigation is highly profitable but even sub-optimal rainfed fruit production is still preferable economically to the farmers than is growing ordinary field crops (Dietz, 1983).

Goals and Objectives of Agriculture

In the 5-Year Plan for Economic and Social Development (Ministry of Planning, 1986), there is no clear cut reference to supplemental irrigation but one stated goal in the water and irrigation section is to increase irrigable lands to the maximum extent that available water resources permit. In the agriculture section, no mention is made of supplemental irrigation. However, a report of an FAO Programming Mission (FAO, 1980) stated "Better understanding of soil moisture characteristics can lead naturally to other refinements such as the use of water harvesting to impound water during the wet season when rain falls on saturated soils. This water can be used for supplemental irrigation to make up deficits during dry spells later in the season or for bringing the soil moisture storage up to a sufficient level at the end of the season to enable the growing of a summer crop. Such systems are not difficult to incorporate in certain kinds of soil conservation works."

Rainfed Farming Practices

Farmers in rainfed farming broadcast and plow in the seed in one operation, except where land is fallowed (Tamini, 1981). In a crop-fallow system, an initial tillage operation for weed control is accomplished anytime during summer or early fall. Farmers implement 4-6 tillage operations during summer and fall to ensure that wheat and barley are sown in clean fallow (Bab, 1976).

A new package of practices was introduced into dryland farming regions (Goetze, 1976) whereby a first tillage operation is done in mid-April at a medium depth of 7-13 cm. A second tillage might be needed to completely eradicate weeds and reduce evaporation. A summer tillage is performed at a shallow 5-8 cm depth and only when necessary to control weeds. These practices usually limit moisture loss from the tilled soil.

Sowing Operations

Methods: Two methods are common for sowing small grains: broadcasting and drilling. Broadcasting is done by hand and the seed covered by a tillage implement. Most barley and wheat is seeded by broadcasting (Tamini, 1981), a technique which can be used in soils too wet for a drill (JCCIP, 1984). Availability of seed drills is increasing and drilling does produce a more uniform stand with less weed problem and less lodging than does broadcasting. Wheat and barley are generally drilled to a depth of 5-8 cm at a row width of 25-30 cm.

Date: Dryland farmers learned through trial and error that sowing small grains before enough rain is risky (Tamini, 1981). With light, early showers, seeds may germinate in the absence of sufficient moisture but can then dry out. Delaying sowing until enough moisture is stored in the soil profile to ensure germination and seedling growth, leads to yield reductions (Jaradat, 1979). Recommendations are to place the seed at a depth of 6-8 cm to ensure that heavy rains will germinate the seed (JCCIP, 1984) because premature sowing could result in a poor stand establishment if early showers are not enough to sustain proper germination and early seedling growth.

Experiments of date of sowing show that mid-November is the best time for small grains. Lentils and chickpeas are sown during spring (Haddad, 1983). Farmers plant the wheat or barley crop first then the lentils or chickpeas, usually after the first few rains and the germination of weed seed so they can ensure a weed-free seedbed.

Haddad and Arabiat (1985) surveyed cultural practices for sowing lentils in Jordan. They found that 80% of the farmers plant lentils during a period from December to February. Sowing is done manually with seeding rates from 50-200 kg/ha with the higher rates associated with higher rainfall. In a detailed study, Haddad (1986a) recommended from late October to mid-December as sowing dates for lentils. His results showed that the earlier sowing gave 100% more grain and straw yield than sowing in January.

Farmers usually sow chickpeas in the spring to escape the Ascochyta blight which occurs with winter seeding. With the introduction of resistant varieties, sowing has been pushed to an earlier date in January with yield increases up to 160% over later planting (Soub, 1984).

Seeding rate: Appropriate seeding rates vary according to variety, seeding method, and sowing date and farmers' practices reflect these varying rates. Seeding rates of 100-140 kg/ha of wheat and barley are commonly used, especially in higher rainfall areas and when sowing with seed drills. Farmers tend to use a rate of 80-100 kg/ha when broadcasting seed (JCCIP, 1984).

In a field survey of 50 villages in 4 regions, 109 farmers were interviewed about cultural practices they follow in growing lentils (Haddad and Arabiat, 1985). Results show that sowing rates ranged from 50-200 kg/ha, with an average of 115 kg/ha. Higher rates are associated with higher rainfall. The regression equation expressing the relationship between sowing rate and yield is calculated as:

$$Y = 555 + 16 X$$

where Y is the expected yield; and,
 X is the sowing rate in kg/ha.

Increasing the sowing rate within the range used by farmers could lead to an increased yield, especially with increased rainfall.

Seed depth: Seed depth has a marked effect on seedling emergence and stand establishment, especially under dryland conditions. In low rainfall zones, moisture at sowing time can be low in the top soil layer, adversely affecting stand establishment and reducing yield potential (Hadjichristodoulou, 1979).

Broadcasting small grain seeds with a later tillage operation to cover them, results in varying seed depths across the field. As a result, seed germination, seedling emergence, and stand establishment will not be uniform or optimal. With modern seed drills, sowing depth can be controlled to ensure proper stand establishment by placing seed in moist soil.

Generally, local wheat and barley landraces and cultivars can emerge from a depth of 10-15 cm because of their longer coleoptiles. Modern semi-dwarf varieties with short coleoptiles fail to emerge if sown deeper than 5-6 cm (Jaradat, 1979).

Fertilizer Use

Fertilizer consumption per hectare of arable land in Jordan is only 6 kg/ha of nitrogen and 11 kg/ha phosphorous which is the lowest among countries in the Middle East (Aresvik, 1976). Obviously, fertilizers are not used at optimal rates (Arabiat, *et al.*, 1983; Oglah, 1984).

The Jordan Cooperative Cereals Improvement Project (JCCIP, 1984) recommends that 30 kg of N/ha be applied at sowing of wheat with an additional 30 kg of N/ha added, if rainfall amounts are higher than average, as top dressing at tillering. In areas of low rainfall (250-350 mm), the recommended rates are 30 kg of N/ha at sowing with an additional 30 kg of N/ha at tillering only if timely rainfall occurs.

Barley is traditionally grown on marginal lands and in the low rainfall zone (200-300 mm). Farmers usually do not fertilize barley because of the limited response with low rainfall (Jaradat, 1987).

Traditionally, lentils and chickpeas are not fertilized because farmers feel that nitrogen would not improve the low yields. Farmers have noticed that wheat yields are lower when grown after an unfertilized crop of lentils or chickpeas than are yields of wheat grown after fallow (Haddad & Arabiat, 1985). Nitrogen requirements of lentils and chickpeas in arid and semi-arid regions are estimated to be 20 kg of N/ha, especially when both crops are planted early in the winter season (Haddad, 1986a and 1986b). This practice leaves an acceptable level of residual nitrogen for wheat or barley in the rotation.

Limited soil tests and phosphorous response data during the early 1970s suggests that phosphorous deficiency is widespread in the rainfed wheat growing regions, particularly in the red soils of the higher rainfall zone of more than 450 mm (Moore, 1975). Researchers recommend 30 kg P_2O_5/ha for wheat and barley throughout different agro-ecological zones (JCCIP, 1984). A general recommended rate, in arid and semi-arid regions, is 40 kg/ha of P_2O_5 for lentils and 80 kg/ha for chickpeas (Abu-Aien, 1986).

Time of application: Ammonium sulfate, a commonly used source of nitrogen for cereal production, is either drilled with the seed at sowing, or broadcast and incorporated (Jaradat, 1987). When a split application is practiced, one half of the nitrogen is broadcast at tillering (JCCIP, 1984). Phosphorous fertilizer is broadcast and incorporated before sowing.

Tillage

Until recently, tillage in rainfed areas was characterized by:
1. many tillage operations: 4-6 were common during the fallow period of 16-18 months;
2. use of unsuitable implements for tillage such as heavy discs and moldboard plows;
3. plowing up and down the slopes, a practice which aggravates the problem of soil erosion; and,
4. during the fallow year, some farmers tend to delay spring tillage for weed control and leave the weeds for grazing by the animals until early summer.

Primary tillage in rainfed areas is performed from June-November. Farmers practice fall tillage to break open the soil surface to increase infiltration and store more moisture in the soil profile (Tamini, 1981). During a fallow year, spring tillage is performed from March to May to control any weed growth and reduce moisture loss from evapotranspiration. Secondary tillage is done to further pulverize the soil and prepare a fine seedbed. The disc is the most popular secondary tillage implement; but, more recently, field cultivators and harrows have been introduced.

Weed Control

Weed control in traditional dryland farming systems has been achieved through tillage, crop rotations, and hand weeding. Hand weeding and tillage operations were the only alternatives for weed control in Jordan until 20 years ago (Duwayri & Saghir, 1983). However, these weed controls which supplemented tillage operations in wheat fields, are no longer economic (Haldorson, 1976; Arabiat, et al., 1983).

Problem weeds are numerous and compete with major field crops for limited resources of soil moisture and plant nutrients. Yield losses of major crops, a result of weed competition, can range from 30-80% (Abu-Irmaileh, 1982). Chemical weed control under rainfed conditions is recent (Goetze, 1976). Most farmers use cultural practices and, on a limited scale, hand weeding for control (Tamini, 1981).

Some farmers still practice delayed tillage during the fallow year to provide grazing material for their animals (Saghir, 1977). The use of improper tillage implements has resulted in poor weed control (Lanzendorfer, 1985). Both practices contribute to the loss of soil moisture and allow weeds to drop their seeds further increasing future weed populations.

The use of herbicides as substitutes for tillage or hand weeding is limited with only an estimated 16% of the wheat being sprayed with herbicides, mainly 2,4-D (Qasim, 1982). Wild oats (*Avena spp.*) is the most grassy weed and is a serious problem in wheat fields (Anonymous, 1982). Several herbicides have been tested for their effectiveness in the control of wild oats but Triallate (Avadex BW) is the most promising (Duwayri and Saghir, 1983). The major weed species that grow in Jordan are *A. sterilis, A. longiglumes, A. hirtula, A. weistii, A. barbata, and A. clauda* (Kanan and Jaradat, 1987). Chemical weed control of broadleaf weeds dates back to the early 1970s with 2,4-D being the first herbicide introduced into the rainfed areas (Goetze, 1976).

Supplemental Irrigation and Water Harvesting Practices

Characterization of 5 Districts

District I: Balama, Mafraq

Demography: Balama District constitutes the western part of Mafraq Governorate (see Figure 22.1). During the past 20-30 years, the population has been in transition from half Bedouin to sedentary farming. Population density in the district is below the national average.

Land area: The total land area is 15,000 ha. The eastern fringes are in the arid zone with an average annual rainfall of less than 200 mm and are usually limited to natural grazing grounds. Land use in the cultivated areas is divided between annual crops of wheat, barley, lentils, and fallow. A small area is under supplemental irrigation using groundwater and water harvested through soil conservation measures. This land is planted mainly to fruit trees and some summer vegetables. Fragmentation of holdings, multiple ownership, and small farm size are major problems for farmers.

Soil characteristics: Clay loams and loams are the predominant soils. The land in the wadis and on the hillsides which is under supplemental irrigation has clay soils. Shallow soils predominate in rainfed farming; whereas, the soils used for supplemental irrigation are heavy, deep clays.

Infrastructure: Thirteen villages and small settlements are scattered all over the district, especially in the wetter western part. Although public roads are adequate, some agricultural roads cannot be accessed during rainy winter months. Farmers consider that transportation for agricultural inputs and outputs is expensive. There are public schools in each village, students of both sexes attend schools until the secondary level. Electricity is available already in 8 villages and 10 villages have potable water systems. The Agricultural Credit Corporation (ACC) is the major source of credit although farmers consider that the amount and type of credit provided by the ACC is inadequate.

Water supply: Balama District lies within the upper Zerqa River Basin and is part of a basalt system, the Wadi Dhulil groundwater aquifer. The baseflow of the basin is 46.5 M m^3. There are 10 wells in the district; 7 wells are owned by individual farmers and 3 are owned by the Water Authority. The average depth to the water table is 120 m with an average yield of 150 m^3/hr.

Supplemental irrigation: Five hundred hectares in the district utilize supplemental irrigation. Surface irrigation is the major method but there is drip irrigation in a small area. Supplemental irrigation is used exclusively for fruit trees and vines which include (in descending order of importance) olives, peaches, and grapes.

Water harvesting: Water harvesting was a secondary objective when soil and water conservation measures of bench terraces, contour ridges, and check dams were implemented in the district. The collected rainfall is used mainly for fruit trees; but, because of little or no ongoing maintenance, the water harvesting facilities are endangered by erosion and, consequently, the amount of water collected is inadequate. Farmers face major field problems of runoff from the poorly maintained facilities and deep percolation in the heavy clays caused by cracks in the soil surface.

Crop production: Major field crops which are produced under rainfed conditions in the district are barley, wheat, and chickpeas. Farmers sow field crops after the first few rains to ensure moist seedbeds for seed germination and seedling growth. The major sowing dates for wheat and barley are October and November and lentils and chickpeas are sown in January. Sowing is hand broadcast with covering of the seed in a single operation. No seed drills have been used in the district.

First and second tillages are done with a duckfoot harrow, the first operation is performed during August and the second is done during November to cover the seed. Sowing depth cannot be controlled with the current tillage practices therefore the seed depth can vary from 7-10 cm. The seeding rate of wheat and barley is 100-120 kg/ha and the rate for lentils and chickpeas is 50-70 kg/ha. No fertilizer is used for field crops; however, fruit trees are fertilized at a rate of 100 kg of NPK/ha.

The major crop rotation is 2 years with wheat or barley production followed by a fallow period of up to 18 months. Some farmers practice a 3-year crop rotation including a legume crop in the cycle. The average yields obtained under rainfed conditions are low with 1 t/ha for wheat, 1.5 t/ha for barley, and 0.8 t/ha for lentils.

Mechanization of farm operations is limited to tillage. A small percentage of farmers (15%) are using combine harvesters but the majority rely on hand labor and draft animals.

Animal husbandry (sheep and goats) contributes 40% of the total farm income in the district. Farmers depend, to a large extent, on natural pastures and grazing land for livestock feed.

District II: Umalbasateen, Madaba

Demography: District II is a typical farming community located to the southwest of Amman in Central Jordan. Population density is high because the district is near the nation's capital. Eighteen thousand people live in 15 villages (not including Madaba) on a total land area of 13,000 ha.

Land area: Rainfed farming land is about 13,000 ha. Forty percent is non-arable with the remaining land under production. Land holdings are generally small (0.3-0.5 ha); however, a few are as large as 300 ha. Fragmentation of holdings, small farm size, and access to land in winter are the major problems faced by farmers under rainfed conditions. Five hundred hectares have supplemental irrigation and are planted mostly to fruit trees but with some vegetables. Tobacco, grown as a summer cash crop, occupies 25% of the 500 ha.

Soil characteristics: Heavy clay to slightly heavy clay loam soils are predominant. Rainfed farming and supplemental irrigation agriculture are practiced on both soil types.

Infrastructure: There are 15 villages and a population of 18,000. Public roads, schools, and health centers are available and adequate. All villages have electrical power and potable water systems. Although credit is available through the ACC, farmers rely mostly on family income as a first priority for support of agricultural projects.

Water supply: The district is a part of the Zarqa-Mai'n aquifer, the baseflow is about 20 M m^3. Five wells owned by individual farmers operate in the district. The average pumping depth is 300 m and average yield is 60 m^3/hr.

Supplemental irrigation: The land area under supplemental irrigation is small as compared to the total rainfed area. Only fruit trees, winter vegetables, and tobacco are grown with supplemental irrigation. Surface flow is the traditional method of irrigation, more recently, the use of drip irrigation systems is increasing. The major problem for fruit tree producers and tobacco growers is the limited amount of water available for supplemental irrigation.

Crop production: The district is a part of the central plains. Major field crops grown are wheat, barley, lentils, and chickpeas. New crops are being introduced such as field peas, tobacco, cucumber, and cabbage; and these crops, except for field peas, are grown with supplemental irrigation.

A common practice has been to sow field crops in dry soils, however, there has been a shift towards sowing after the first few rains to achieve better germination and seedling emergence. Although November is the major sowing month for wheat, barley, field peas, and lentils, chickpeas are sown during late February to early March. Seed drills are used for wheat and barley but other crops are manually broadcast. Tobacco is transplanted during March and is irrigated 1-2 times early in the season.

Seeding rates used by farmers are lower than recommended for the region. The rate for wheat and lentils is 100 kg/ha; for barley and chickpeas, 70-80 kg/ha; and, for field peas, 50 kg/ha. Fertilizer is commonly used in the district; the application rate used by farmers is 100 kg of NPK/ha.

The Farming Systems in Jordan

Weeds are a major problem and 50% of the farmers use chemical herbicides for control. This is done through the Jordan Cooperative Organization (JCO) which has a machinery station in the district.

Three tillage operations are performed for rainfed crops: a summer tillage during July-September using a moldboard plow; a fall tillage during October using a chisel plow; and, a final disc harrow before drilling the seed in November.

A 2-year or 3-year crop rotation is practiced. Wheat (or barley) – fallow (or tobacco) is an example of a 2-year rotation and tomato (or watermelon) – beans – cabbage is an example of the 3-year rotation.

Average crop yield for wheat is 2.0 t/ha, for barley is 1.5 t/ha, and for lentils is 2.0 t/ha. Tobacco produces 0.9-1.0 t/ha of dry leaves.

Mechanization of farming operations is well advanced because of the establishment of a JCO machinery station. Specialized implements are available for reduced tillage as well as seeders and combine harvesters. Legume crops are still manually broadcast and harvested.

Livestock is not a major source of income in the district. Sheep and goats provide 25% of the family income. Feed for livestock largely depends on natural grazing grounds during late winter and spring, and on wheat and barley stubble during summer.

District III: Shoubak, Ma'an

Demography: Shoubak is a sparsely populated region with limited natural resources in the north-western part of the Ma'an Governorate. Settlements have been established only where soil and water resources are sufficient for permanent use of the land. A population of 14,000 lives in 16 small villages and settlements.

Land area: Of the total area of 20,000 ha in Shoubak District, 50% is either non-arable or not in crop production. Only 300 ha have supplemental irrigation even though groundwater resources are available in the region. Small farm size, fragmentation of land holdings, and access to land in winter are the major problems of rainfed farming. Regulations on water supply imposed by the Water Authority, small size of farms, and high production costs are the constraints for farmers using supplemental irrigation.

Soil characteristics: Clay and clay loam soils are predominant in the rainfed as well as in the supplemental irrigation lands. These soils are confined to the western part of the district; eastwards, lighter soils of loam, silt loam, and sandy loam predominate. Soils are generally shallow, brown in color, and slightly to moderately sloping.

Infrastructure: Sixteen small villages and settlements, scattered on a north-south axis in valleys and on hill tops, make up the Shoubak District. Population density is lower than the average. Schools and public roads are adequate and there is an agricultural school and a community college. All villages have electrical power and potable water for domestic use. Medical care is available in all villages but veterinary services are available only in Nijil, the major town.

Water supply: Groundwater is available and 22 wells are being drilled. About 1.5 M m^3 are extracted from 7 wells with tested yields of 476 m^3/hr. Five more wells have been tested recently with a total yield of 782 m^3/hr and an average yield of 120 m^3/hr.

Supplemental irrigation: Fruit trees (mostly apple) are the only crops grown with supplemental irrigation. Seventy percent of the 300 ha is irrigated by drip irrigation with the remaining area irrigated by surface flow.

Crop production: Major winter crops are wheat, barley, and lentils. In the last 2-3 decades, several government and commercial apple orchards have been planted.

Under existing farming practices, sowing of winter crops places seed in dry soil before the rains but some farmers delay sowing until enough moisture is stored in the soil profile. Sowing is done by broadcasting the seed and covering it with one pass of a disc harrow. No fertilizers are used for wheat, barley, or lentils but fruit trees are fertilized at a rate of 100 kg of NPK/ha. The major crop rotation is wheat (or barley) – fallow (or lentils).

Average yields of wheat are 1.25 t/ha, of barley are 1.0 t/ha, and of lentils are 0.9 t/ha. Wheat and barley are sown either early in the season during October before rainfall, or later during December or January, after the first few rains. Lentils are sown in February. Seeding rates for wheat and barley range from 70-80 kg/ha and for lentils range from 60-70 kg/ha.

Mechanization has been limited to tillage; however recently, combine harvesters have been introduced.

Animal husbandry is a major occupation for the Bedouin who became settlers in the last 20-30 years and own a large share of the land. Livestock, mainly sheep and goats, generates about 30% of the family income. Most herds depend on natural grazing grounds in the steppe to the east of the district and on wheat and barley stubble during the harvest season for livestock feed.

District VI: Alqaser, Karak

Demography: Alqaser District constitutes a major part of Karak Governorate in southern Jordan. The farming community is the oldest in Jordan with a population density above the average. A total of 20,000 people inhabit an area of 8,000 ha.

Land area: The total land area is 8,000 ha, of which, 1,200 ha are non-arable. Two thousand hectares are rainfed farming with established measures of soil and water conservation. Fragmentation of land holdings, small farm sizes, and low rainfall are problems for rainfed farming. Farmers practicing water harvesting face the first two problems. Also, there is difficulty in reaching these lands during the winter months.

Soil characteristics: Clays and clay loams are the major soils in areas under rainfed conditions and water harvesting. Shallow soils of less than 50 cm are predominant in some rainfed areas, especially in the eastern parts. Most are red Mediterranean soils, slightly sloping, with no salinity problems, and moderate infiltration and crusting.

Infrastructure: Twenty thousand people live in 16 villages. Schools are adequate with an agricultural school at Rabbah. All villages are connected to the national power supply and all have potable water for domestic use. Health services are available in health centers or clinics.

Water harvesting: The district constitutes a part of the Mujib basin which has a baseflow of 41 M m^3. A limited amount of this water is utilized for agricultural purposes locally.

Soil conservation measures were built on 2,000 ha and, secondarily, have been utilized for water harvesting. The major methods used for this purpose are conservation bench terraces, tied ridges, and contour ditches. Field crops and, to a lesser extent, fruit trees are grown on these lands. Major constraints are maintenance of water harvesting facilities and soil erosion.

Crop production: The District is one of the best for wheat and lentil production. Both crops are produced under rainfed conditions and water harvesting. Because of the availability of seed drills and special tillage equipment, farmers are able to sow seed in dry soil before the rain. The major sowing time for wheat and barley is October, at a 5-7 cm depth and a seeding rate of 70-80 kg/ha. Lentils are sown in February by broadcasting seed at a rate of 90 kg/ha. Wheat is fertilized at sowing time with 80 kg of N/ha. A new package of tillage practices was introduced during the 1970s performing one chisel plowing during October and another before seeding in November.

The major crop rotation is 2-years, with wheat (or barley) – fallow. A minor 3-year crop rotation is practiced in the wetter areas where lentils, chickpeas, or vetches are included in the rotation. Vetch was introduced in the region as a forage crop and now 20% of the land with legumes grows vetch. Most farmers practice weed control: 50% use herbicides and 20% do hand weeding.

Mechanization of farm operations is well developed in comparison with other districts. Tractors have been used for the last 30 years with the introduction 15 years ago of combines and seed drills.

Livestock generates 40% of a family's income in rainfed areas and only 10%

of the family income comes from water harvesting areas. This is a major district for sheep and goat husbandry. Flocks rely on natural grazing grounds in the steppe for feed.

District V: Jarash WMU-II

Demography: Jarash District is one of 5 watershed management units in the Zerqa River Project. The district is mountainous and has the largest evergreen forest in Jordan. It is located in the north-western part of Jarash Province with a high population density because of the presence of 2 refugee camps. A total population of 41,000 lives in 18 villages.

Land area: Although the total land area is 43,000 ha, only 11,000 ha is utilized for agricultural production. Soil conservation measures have been built on a land area of 2,120 ha; however, most of the land will be brought eventually under the Zerqa River Project and soil conservation measures will be expanded to most of the land area.

Most farmers have problems of fragmentation of land holdings, small farm sizes, and access to land during winter months. Most holdings are less than 1.0 ha.

Soil characteristics: Slightly heavy soils predominate in both rainfed and water harvesting areas. Soils in lower hillsides and valley floors are heavy clays. Soils are of the red Mediterranean type, moderately to slightly sloping, especially in areas with soil conservation measures.

Infrastructure: Infrastructure is well developed. All villages and refugee camps are supplied with electrical power and potable water systems for domestic use. Schooling is available at all levels in all villages. A veterinary center and health clinics are available. Public roads are adequate except for some mountainous roads which become inaccessible during winter.

Water supply and management: Average rainfall is 400 mm with storms of high intensity. The maximum amount of rainfall recorded in 24 hr is 158 mm. Although the Zerqa River runs at the southern end, little water remains to irrigate citrus trees and vegetables because most water is channeled into the Jordan Valley.

Water harvesting: Water harvesting is practiced in an area of little over 2,000 ha. The major methods used for soil conservation and water harvesting are stone walls, gully control structures, conservation bench terraces, and contour cultivation. A larger land area will be conserved in the future through the Zerqa Valley Project by implementing additional soil conservation measures.

Crop production: The district is one of a few which has a diverse cropping system. The major field crops produced are wheat, barley, lentils, chickpeas, and vetches. Summer vegetables are grown with field crops in a 3-year crop rotation or are planted between rows of young fruit trees. Apple and stone fruit trees are grown extensively. A sizeable area is under natural or planted forests. Upon the completion of the Zerqa River Project for soil conservation, a shift is expected from cereal production to more fruit trees and fodder crops.

Wheat and barley are sown during October and November by broadcasting manually at a seeding rate of 100 kg/ha then plowed in with a disc plow. Lentils and chickpeas are broadcast at a seeding rate of 60-70 kg/ha during February and March. Vetches are sown during December at a rate of 100 kg/ha.

Both 2-year and 3-year crop rotations are followed in this district. Wheat or barley are followed by a fallow year in the 2-year rotation and a legume crop follows the fallow year in the 3-year rotation. Weed control is limited to manual weeding because of the hilly topography of the region and the difficulty of using tractors.

In spite of high rainfall, poor management results in low crop yields. Wheat and barley grain yields range from 700-1,100 kg/ha although green vetch cut for feed can produce as high as 10 t/ha.

Livestock generates a small portion of family income, not exceeding 15%. Sheep (21,000) and goats (15,000) are the major farm animals. Dairy cows (1,000) are increasing with the availability of green fodder. The district is a net importer of animal feed although it is expected to become a net exporter when the Zerqa Project is completed.

Water Resources

Jordan is an arid country with limited surface and groundwater resources (Arar, 1978). Water is important for agricultural development, this is especially so because of an increasing demand from other competing uses such as municipal and industrial purposes (Natur, 1985). The database for water resources is limited and sometimes inconsistent. Irrigation, including supplemental irrigation, was probably practiced around springs on a limited scale before recorded history (Gubser, 1983). Public irrigation projects of modern times began in the Jordan Valley in the late 1940s and early 1950s, when the Government started building small irrigation schemes on side wadis. The land areas irrigated by these schemes were usually larger than could be irrigated by the baseflows and partial irrigation was the rule (NWMP, 1977 and Natl. Resources Auth., 1977).

Irrigation projects in the highlands were established within a program of groundwater exploration to test the aquifers under long term pumping, investigate potential irrigation of crops and production technologies, and, to plan for future expansion (Joudeh and Abu Taha, 1978). The planned projects, some of which will involve supplemental irrigation, are listed in Table 22.2 with the area to be irrigated (Ministry of Planning, 1986).

Table 22.2. Planned irrigation projects in the highlands of Jordan.

Project name	Area (ha)
Shoubak	450
Abu Makhtub	210
Udruh	150
Qurein	270
Adir	200

Surface Water

Surface water resources of rivers and wadis flowing into the Jordan Valley are summarized in Table 22.3. Total water resources in an average rainfall year are estimated in Table 22.4. Sixty percent of the catchment area is not gauged (NWMP, 1977). Runoff measurements are generally insufficient; only annual average flows are available, and seasonal variations are not recorded (Natur, 1985).

Table 22.3. Surface water resources of the Jordan Valley.

River or Wadi	Annual average volume (Mm^3)
Yarmouk River	408.6
Wadi Al-Arab	28.8
Wadi Ziglab	9.5
Wadi Jurm	11.2
Wadi Yabis	3.3
Wadi Kufranji	6.1
Wadi Rajib	7.1
Zerqa River	67.3
Wadi Shuieb	7.9
Wadi Kafrein	14.3
Wadi Hisban	5.0

Table 22.4. Water resources of Jordan in an average rainfall year.

Resource	Volume of water (M m^3)
Total rainfall	8,065
Rainfall related to Jordanian territory	6,000
Total recharge	580
Available groundwater	220
Stored underground water	12,000

Several dams to recharge groundwater were planned and built on some of the wadis (see Table 22.3); whereas, others were built to collect flood water for use in irrigation. A preliminary study was prepared for 14 dams in different locations where flood water might be collected and stored (Ministry of Planning, 1986).

Groundwater

Groundwater replenishment is estimated at 580 M m^3/yr, of which 380 M m^3/yr is surface runoff during the dry season. This leaves 200 M m^3/yr as the net available water resource (NWMP, 1977). Four main aquifer systems are recognized:
1. the Disi group, a sandstone system exploited in southern Jordan for municipal, industrial, and agricultural purposes, is the deepest and oldest aquifer;
2. the Kurnub-Zerqa group, a sandstone system, discharges as springs and baseflow in the escarpment of the rift valley;
3. the carbonate rock aquifer system is the most important and the Amman-Wadi Sir is a part of this system; and,
4. the shallow aquifer system includes the basalt aquifer of Azraq-Wadi Dhuleil.

The groundwater system is divided into 11 basins as shown in Table 22.5.

Table 22.5. Groundwater resources.

		Groundwater resources (M m^3)			
No.	Basin	Springs	Wells	Baseflow	Total
1	Yarmouk		3.5	218.0	221.2
2	N. Escarpment, Jordan Valley		2.7	91.1	123.5
3	Jordan Valley Floor	15.4	8.1	56.0	91.2
4	Zerqa River	37.8	66.4	41.8	90.4
5	Central Escarpment, Dead Sea	79.6	8.0	137.5	157.2
6	Escarpment, W. Araba		0.2	14.7	35.7
7	Red Sea Basin		1.7	0.6	9.1
8	Jafer Basin	2.4	5.3	2.1	67.8
9	Azraq Basin	15.0	4.7	15.1	67.8
10	Sirhan Basin			35.5	39.5
11	Wadi Hammad		0.5	5.0	15.0

Problems and Constraints Limiting Supplemental Irrigation

Supplemental irrigation is practiced on a limited land area and on crops of economically high gross margins. This is due mainly to the scarcity and remoteness of adequate water supplies. Several sectors are competing for the limited water resources of the country. Municipal and industrial sectors are the major users with the agricultural sector receiving the smallest share except in the Jordan Valley.

The cost associated with irrigation and supplemental irrigation is high. Insufficient data are available of the costs of on-farm irrigation systems. Binnie and Jouzy (1979) give price estimates of capital costs for sprinkler irrigation at about US$ 3,000/ha and a total variable cost for operating the system at US$ 1.30/hour. Natur (1985), in a more recent study estimated the capital cost for a solid set sprinkler system at US$ 4,000/ha. This high cost, together with other reasons, explains why Jordanian farmers do not invest in supplemental irrigation of cereal and forage crops.

Theoretically, water resources, under a sound management system, could meet supplemental irrigation needs in parts of the country. However, a better understanding of the database of water resources is needed to achieve this purpose (Natur, 1985). Experts in different governmental agencies dealing with water resources agreed on the following as the major problems facing the water resource management sector:
1. the database of water resources is incomplete;
2. flood flows in many areas of the country are not measured;
3. hydrological data are lacking on intermittent streams;
4. data on underground water are inadequate;
5. the groundwater monitoring network needs to be expanded; and,
6. underground aquifer characteristics need to be reevaluated.

Planners at the Water Authority (Ministry of Planning, 1986) concluded that to develop the agricultural sector, especially in the highlands, water resources must be maximized. This can be achieved through the storage of flood water, exploration of underground water in these areas, and the collection, treatment, and reuse of sewage for supplemental irrigation.

Governmental Policies in the Dryland Regions

The Government recognizes the important role of rainfed agriculture to the Jordanian economy since 93% of all lands are considered rainfed agricultural resources. The Five-Year Development Plan (1976-1980) has set national goals to improve production conditions and to increase food production:
1. to increase production of the principal dryland crops by 1980: wheat to increase by 36%, barley by 25%, lentils by 25%, grapes by 110%, and vegetables (partially produced in rainfed areas) by 50%;
2. to increase fodder and forage production;
3. to develop improved agricultural methods and support extension services; and,
4. to change the pattern of landuse for rainfed farming by relating the crops planted to the agro-climatic conditions; to reduce the grain-growing land

area from 340,000 ha to 240,000 ha, shift about 90,000 ha from wheat to barley production, and increase planting of fruit trees.

This first 5-year plan did not allocate enough capital resources to improve the production environment in the rainfed regions. The largest proportion of investment was spent on irrigated projects.

The second 5-year plan (Ministry of Planning, 1986) essentially has restated the goals of the first plan for the rainfed regions. The most important goals were to limit the production to those areas most suitable for field crops and to plant 24,500 ha to fruit trees, giving priority to lands with a greater than 9% slope.

Total production of field crops has decreased during the last 10 years. To better understand the reasons behind this result, it is necessary to review the current dryland agricultural policy. There are 3 significant policies: price subsidy for agricultural inputs; local price support policy for wheat production; and, price subsidies for bread and wheat consumption.

Price subsidy for agricultural inputs

The Government provides some general subsidies to the agricultural sector as duty-free importation of agricultural inputs and machinery. However, all agricultural inputs are sold by private firms who import directly from international markets. Therefore, the prices of these inputs reflect international supply and demand (El-Hurani, 1987).

Some projects, subsidized by the government, are aimed at increasing the land area planted to fruit trees, especially in hilly regions with steep slopes. The FAO/UNDP Program provides consultants, olive tree seedlings. and pays food subsidies to farmers planting these trees and incorporating conservation measures on their lands (Jaradat, 1987).

Local price support policy for wheat production

Dryland farmers are encouraged to produce wheat in rainfall zones of 300 mm and above and in areas with slopes from 0-8%. The goal of this policy is to increase domestic wheat production to reduce the amount of imported wheat. For many years the government has supported prices by offering to buy domestically produced wheat at a higher price than that of the international market. Although this price support program has been practiced for many years, it has never succeeded in reducing imports because of inefficiency and poor timing of program implementation.

Price subsidies for bread and wheat consumption

Government subsidizes the wheat consumed by Jordanians, a policy which has been in effect for many years. The purpose of this market intervention is to provide bread at low prices to the urban population. The government buys most of the wheat on the world market, sells to mills at a reduced price to control the price of flour, and consequently, stabilizes the price of bread (El-Hurani, 1987).

A study of the estimates of future food requirements stresses the need to work for long-term solutions to reach a higher percentage of food self-sufficiency for Jordan. A promising approach is to maximize crop production in areas where the technologies of water harvesting and supplemental irrigation can be employed economically.

Future Plans

Crop yields in the rainfed agricultural sector vary from year to year, largely because of differences in rainfall (Jaradat, 1987). Farmers in the rainfed areas, who have access, use water for supplemental irrigation of fruit trees and cash crops. Economically, supplemental irrigation of these crops is more profitable than field crop production. However, where resources are available for supplemental irrigation of field crops, this technique should be implemented to stabilize and increase yields. The costs associated with supplemental irrigation can be high; therefore, careful consideration must be taken of how much water to apply and at which growth stages irrigation should occur to achieve maximum economic benefits.

A computer simulation study is being undertaken at Jordan University of Science and Technology (JUST) to:
1. assess the impact of supplemental irrigation on wheat yield under varying rainfall and supplemental irrigation levels;
2. determine the proper amounts of water added as supplemental irrigation for maximum economic yields;
3. determine the growth stage(s) for supplemental irrigation to achieve maximum water use efficiency; and,
4. plan field experiments to verify the findings in simulated studies.

Preliminary simulated results indicate that the net income from wheat can be increased 7-10 times with supplemental irrigation. A field experiment at the JUST farm is planned for 1987-88 to verify this work.

Summary and Conclusions

Jordan is an arid country with limited surface water and groundwater resources. Over 50% of the land area receives less than 100 mm of average annual rainfall. In the major dryland agricultural areas, the average annual rainfall ranges from 250-600 mm. Rainfall is erratic, irregular, and highly variable with occasional storms of high intensities, especially in the highlands.

Farming practices in the rainfed highlands have been developed under the assumption that inadequate soil moisture is the major factor limiting crop growth and production. Nevertheless, little to improve land productivity and crop production has been accomplished through soil conservation, water harvesting, or supplemental irrigation.

The Government provides some subsidies to the rainfed sector. These are: price subsidy for agricultural inputs; local price support policy for wheat production; price subsidy for bread and wheat consumption; and, subsidies for fruit tree planting and soil conservation. These subsidies have failed in motivating the private sector to invest in supplemental irrigation in the rainfed regions of the country.

The results of a survey of 5 agro-ecological zones found that supplemental irrigation and water harvesting are practiced on a limited land area and on crops of high gross margins. This is due mainly to scarcity and remoteness of water resources. Moreover, other sectors, e.g. municipal and industrial, compete with agriculture for the limited water.

Supplemental irrigation has the potential to stabilize and increase the low crop yields in rainfed agriculture. Preliminary results of a computerized simulation study indicate that the net income from wheat can be increased 7-10 times using supplemental irrigation. Costs associated with supplemental irrigation can be high and careful consideration must be taken of how much water to apply and at which growth stage(s) to schedule the application(s) to achieve maximum economic benefit.

References

Abandah, A., 1978. Long-range forecasting seasonal rainfall in Jordan. Dept. of Meteorology Publ. No. JNWS/ 78/11. 17 p

Abu-Aien, A., 1986. Effect of legumes in crop rotations on wheat and barley production under semi-arid conditions in Jordan; M Sc thesis, Faculty of Agriculture, Univ. of Jordan, Amman.

Abu-Irmaileh, B., 1982. Weeds of Jordan. Univ. of Jordan, Amman. 433 pp

Anonymous, 1982. Rainfed agriculture in Jordan. (in Arabic). Proceedings of Workshop, 17-20 May 1978, Amman. 60 pp

Arabiat, S., S. Nygaard, and K. Somel, 1983. Factors affecting wheat production in Jordan. JCCIP Report.

Arar, A.A., 1978. Some considerations for increasing the supply of and reducing the demand for usable water in Jordan. In Proceedings of the National Water Symposium, 19-22 March 1978, Amman. Natural Resources Auth.

Aresvik, O, 1976. The agricultural development of Jordan. Praeger Publishers, New York.

Bab, E.N., 1976. Integrated agricultural development of the rainfed areas. Draft project proposal, Amman.

Behairi, S., 1973. Geography of Jordan. (in Arabic) Univ. of Jordan. 256 pp

Binnie, H., and D.Jouzy, 1979. Mujib and southern Ghors irrigation project; a feasibility study submitted to JVA. Amman.

Department of Statistics, 1981. Yearbook of statistics. Amman, Jordan.

Department of Statistics, 1982. Survey of agricultural land irrigated by artesian wells .

Dietz, M., 1983, Integrated rural development in Shoubak area. Pre-feasibility study. 100 pp

Draz, O., 1979. Rangeland development and stabilization of nomadic sheep husbandry. FAO Report on Jordan.

Duwayri, M., and A.R.Saghir, 1983. Effect of herbicides on weeds and winter cereals in Jordan. Dirasat 10:115-128.
El-Hurani, M.H., 1987. Governmental policies in the dryland regions of Jordan. In A.A.Jaradat, (ed.) An Assessment of Research Needs and Priorities for Rainfed Agriculture in Jordan.
El-Kawasma, Y., 1983. Climatic water balance in Jordan: characteristics and applications; PhD. Thesis, Univ. of Kent., England. 278 pp
FAO, 1974. Development and use of ground water resources in East Jordan. Rome. 114 pp
FAO, 1980. Jordan. Report of FAO Programming Mission, Rome. 80 pp
Gubser, P., 1983. Jordan: Crossroads of Middle Eastern Events. Westview Press. 139 pp
Goetze, N., 1976. Jordan wheat research and production. Final Report, USAID/OSU. 206 pp
Haddad, N., 1983. Effect of date of planting and plant population on yield and other agronomic characters of lentils. Dirasat 10:153-167.
Haddad, N., 1986a. Recommendations for growing lentils in Jordan. (in Arabic) Ag. Ext. Bull. No. 3. 15 pp
Haddad, N., 1986b. Recommendations for growing chickpeas in Jordan. (in Arabic) Ag. Ext. Bull. No.4. 20 pp
Haddad, N.and S.Arabiat, 1985. Methods and problems of planting lentils in Jordan. (in Arabic) Dirasat 12:31-74.
Hadjichristodoulou, A., 1979. Preliminary agronomic studies on wheat and barley in 1977-1978 seasons. ICARDA Report, Syria.
Haldorson, D., 1976. Chemical herbicides application. In Goetze, N., (ed.) Jordan Wheat Research and Production; Final Report; USAID/OSU. 206 pp
Humphreys, H., 1978. Water use strategy, North Jordan. Summary Report submitted to Jordan Valley Authority.
Jaradat, A.A., 1979. Comparison among several wheat varieties in relation to yield, yield components, and root systems. M Sc. Thesis, Univ. of Jordan.
Jaradat, A.A., 1987. An assesment of research needs and priorities for rainfed agriculture in Jordan. Publ. by JUST and USAID.
(JCCIP) Jordan Cooperative Cereals Improvement Project, 1984. Report on the Jordan cooperative cereals improvement project. ICARDA, Syria. 94 pp
Joudeh, O.M.and M.Abu Taha, 1978. Present and needed information on water resources in Jordan. In Proceedings of the National Water Symposium, 19-22 March 1978. NRA, Amman.
Kanan, G.and A.A.Jaradat, 1987. Genetic diversity in Avena species in Jordan (in preparation).
Kasapligil, B., 1956. The vegetation profiles of the forest and grazing lands in Jordan. FAO Report, Rome.
Lanzendorfer, M., 1985. Agricultural mechanization in Jordan. Edition Herodot; GMBH, Gotengen. 320 pp
Ministry of Agriculture, 1974. Agroecological zones over Jordan; Working paper. 48 pp
Ministry of Planning, 1986. Five year plan for economic and social development, 1986-1990. Amman. 574 pp
Mitchell, C.W. and J.A.Howard, 1980. Land system classification; A Case History: Jordan. FAO Publ., Rome.
Moore, F., 1959. The soils of Jordan. FAO Report No. 1132, Rome.
National Planning Council (NPC), 1975. Five Year Development Plan, 1976-1980. Amman. 325 pp
National Water Master Plan (NWMP), 1977. Vol I.; NRA and GTZ.
Natural Resources Authority, 1977. Rainfall in Jordan. Technical paper No. 49. 130 pp
Natur, F.S., 1985. Water supply for the agricultural sector. In A.B.Zahlan, The Agricultural Sector of Jordan: Policy & Systems Studies. A. H. Shoman Foundation, Amman. 411 pp
Oglah, M., 1984. Infrastructural constraints to technology change in wheat production in Jordan. JCCIP Report.
Qasim, J.R., 1982. The competitive and allelopathic effects and chemical control of Syrian sage in wheat. M Sc. Thesis submitted to the University of Jordan.
Qudah, B., 1983. Land use in Jordan. Publ. by Ministry of Agric.; (with maps and figures). 27 pp

Qudah, B.and A.A.Jaradat, 1987. Soil resources of Jordan. *In* A.A.Jaradat, (ed.), An Assesment of Research Needs and Priorities for Rainfed Agriculture in Jordan.

Saghir, A.R., 1977. Weed control in wheat and barley in the Middle East. PANS 23:282-285.

Soub, H., 1984. Effect of planting date and population on nodulation, yield and other agronomic characteristics of two chickpea genotypes. M Sc Thesis, submitted to the Univ. of Jordan.

Tadros, K.I., 1984. Effect of grazing intensity by sheep on the production of *Atriplex nummularia* in Jordan. Research Proposal, June, 1984.

Tamimi, S., 1981. A brief review of research on rainfed agriculture. FAO Publ., Baghdad.

World Bank, 1981. Kingdom of Jordan rainfed agriculture subsector. Report No. 3285-50. Washington D.C., USA.

Chapter 23

Supplemental Irrigation and Water Harvesting Systems in Libya

SAAD AHMED AL GHARIANI

Geographical Characteristics

Libya is situated in the center of the North African coastline between latitudes 24° and 32.5° north occupying a land area of 1.75 M km^2 . In spite of a distinctive and separate geographical identity, Libya can be considered, for discussion purposes, as a transitional zone of Africa, bearing features of the east, west, north, and even, of the southern portions of the continent. The physical and geographical structure of the country clearly indicates this: central and southern regions are distinctly integral to the large African platform; whereas, the northern region reflects the disturbances characteristic of the central and northern Mediterranean area.

Topographically, Libya possesses a slightly buckled and deformed edge of the African gneissic and granitic plateau with the complex basement of this plateau, dating back to the Archaean Age, underlying most of the country. In the extreme south, the basement surfaces to form the Tibesti Mountains. Above the basement, sedimentary strata of varying thickness and of a later age occur in horizontal layers of marine and aeolian origin. These layers increase in thickness towards the north where faults of tectonic origin run in the east-west direction. These faults define most of the coastline east of Benghazi (see Figure 23.1). In the northwest, the fault line begins west of Misurata at the coast and continues due west inland to form the highlands that represent the hills of the western mountains known as Jabal Al-Gharbi. The lowlands between the coast and these hills are the Jafara Plain, triangular in shape, containing most of the agriculturally productive soils of the western region. South of the western mountains lies the Hamada, a stony plateau composed mainly of red sandstone frequently overlain by basaltic outflows to form highland ridges.

Further south, the land surface ends in small basins covered by desert sand dunes. Artesian groundwater may occur at the bottom of these basins and support small settlements at oases scattered over the southern desert. In the north-east, the Jabal Al-Akhdar Mountain is formed by an anticline of limestone from the Eocene Age. High altitude and a seaward aspect of this

E.R. Perrier and A.B. Salkini (eds), Supplemental Irrigation in the Near East and North Africa,
425-447.
© 1991 *ICARDA.*

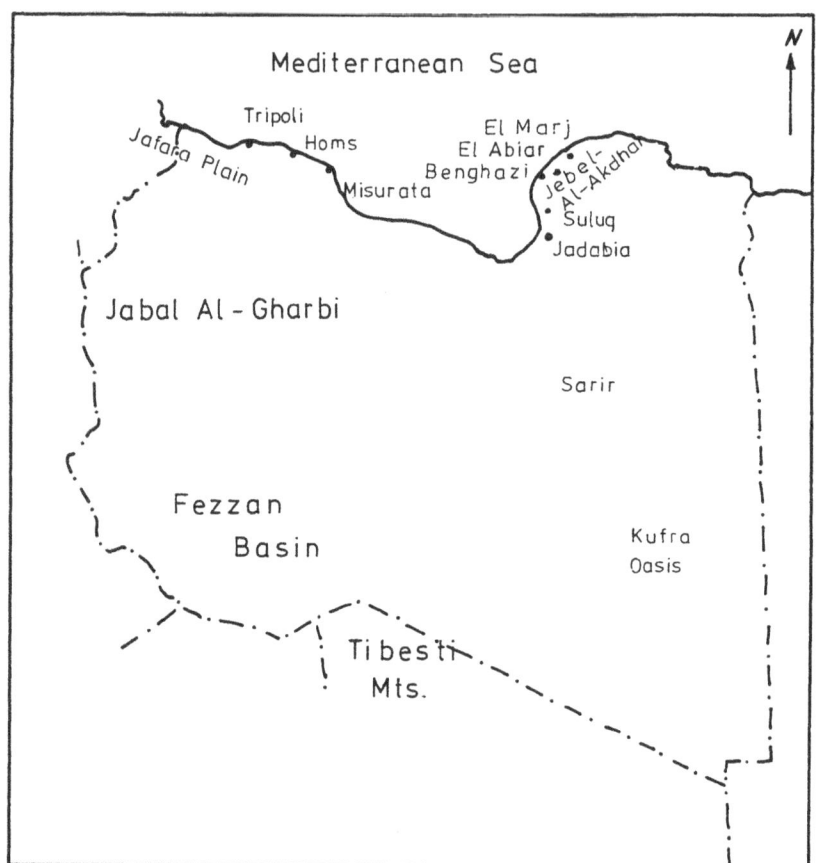

Figure 23.1. Map of Libya.

eastern mountain beyond the Gulf of Sirte provides the highest rainfall in Libya where average annual precipitation rates can exceed 500 mm. South of Jabal Al-Akhdar, sand dunes occur, forming great mobile seas of sand.

Libya is a country of extremes possessing an interaction of maritime and continental desert climates. The hot desert affects the entire area inland beyond 100 km south of the coastline. Generally speaking, climate is influenced by the arid environment which prevails with the exception of a narrow coastal strip and the mountainous areas to the east and west.

Rainfall is capricious in its formation, intensity, distribution, and duration with local flooding and significant amounts of runoff. The storms are convective and orographic, characterised by isolated thunderstorms of high intensity, short duration, and limited distribution. Cyclonic storms of migratory systems which extend over larger areas occur less frequently. This frontal rainfall is of longer duration and lower intensity and falls on highly permeable soils, producing little runoff and is not conducive to water harvesting as is the case with the orographic and convective storms.

Annual precipitation is variable both in quantity and distribution during the rainy season from October to April. Winter crops including cereals and forages can be grown successfully without irrigation or supplemental irrigation along most coastal areas, in the western and eastern mountains, and on the Jafara and Benghazi plains. The coastal strip receives an average annual precipitation of 150-400 mm whereas more than 500 mm are common in mountain locations. The western and northern slopes of these mountains receive more water than the southern and eastern slopes where rainfall decreases drastically.

The differences in geography and topography as well as the direction of the coastline is reflected in the temperatures. The narrow coastal strip enjoys a moderate climate and a high relative humidity due to significant evaporation and cool north-eastern and north-western winds blowing across the sea during most of the year. The maximum average temperature in coastal areas occurs during July and August but rarely exceeds 30°C while the minimum average temperature, rarely below 13°C, occurs in January. In the desert interior, maximum averages of 40°C are not uncommon. A local hot dry wind, the Ghibli, originates in the interior loaded with dust and blows northward during late spring and early summer bringing devastation to northern agriculture.

The high intensity of rainfall and rocky nature of the western and eastern mountains produce high volumes of sediment loaded runoff each year. This runoff forms ephemeral streams which flow in gullies and wadis during the rainy season. The streams discharge this fresh water and fertile sediment on the plains of Jafara, Marj, and Benghazi; or, the runoff may reach the coastline and discharge directly into the Mediterranean Sea. This runoff causes severe soil erosion and loss of valuable water resources. A comprehensive plan is critically needed to integrate water harvesting techniques with afforestation, soil conservation, dam construction, and modern farming practives. Al-Ghariani (1987) estimates that, annually, more than 2.9×10^{10} m^3 of water falls as precipitation over the land area of Libya but only a negligible amount (less than 1%) is conserved for a supply of water. Research programs whose results could maximize utilization of this resource at the expense of evaporation and runoff should be implemented as soon as possible to meet increasing demands for fresh water and to augment dwindling groundwater.

The absence of perennial rivers and fresh water bodies requires that all economic activities in the country depend on groundwater. This major source supplies 95% of the total amount of water used. Existing sources of groundwater may be classified into 5 major hydrogeological zones:
1. the western mountains and the Jafara Plain;
2. the central zone;
3. the Jabal Al-Akhdar area;
4. the basins of Fezzan; and,
5. the Kufra-Sarir zone.

A detailed description of quantity, quality, and occurrence of groundwater is elaborated in greater detail in the sections concerned with supplemental irrigation in this chapter.

The descriptions of geography, topography, climate, and water resources give an overview of the vastness and variability of the country and expose the potential for developing natural resources under constraints of environment, human resources, and capital investment. Archeological studies have revealed that under Roman rule Libyan agriculture achieved tremendous success but experienced a disastrous decline because of civil disorder, vandalism, and foreign invasions. Until recently, the chances for modern development were quite limited; but, with the availability of huge oil reserves and a high income from foreign exchange, the situation has changed radically. Millions of dollars are being spent on resource development for agriculture and several plans have been implemented making promising circumstances. Nevertheless, a sound research program is indispensible in achieving a delicate equilibrium between commercial exploitation and a fragile natural habitat of marginal productivity.

Extreme aridity precludes sustainable agricultural productivity without supplemental irrigation especially deep in the interior where rainfall is negligible. Therefore, new projects using automated sprinkler systems (mostly center pivots) have been erected wherever groundwater supplies permit adequate pumping rates. Average annual rainfall in the coastal strip can support successful plant growth of some grain crops and forages without irrigation. This is especially true in the Western Jafara Plain and in the vicinity of Jabal Al-Akhdar where rainfall exceeds the 200 mm isohyet. Nonetheless, reliance on rainfall alone could be disastrous in years of low precipitation or of poor distribution of storms during the season.

The water holding capacity of most soils is low and prolonged stress to plants during a critical growth stage is enough to cause crop failure. Crop yields are well below average and do not exceed 2.5 t/ha for wheat and barley as compared to yields of 5 t/ha of these grains under full irrigation. To stabilize crop yields and secure adequate returns on investment, most farmers within the 200 mm isohyet tend to meet extra crop water requirements with supplemental irrigation when water supply, required equipment, and labor are available.

Land cultivated with supplemental irrigation in the Jafara Plain is estimated to be 220,000 ha of which 20% is winter grain crops with the remaining land devoted to green forages, vegetable crops, and citrus. When compared to dry farming areas with greater than 500,000 ha and the severe constraints to rainfed farming of climate and limited water, supplemental irrigation is well established throughout the region. The same could be said about the Jabal Al-Akhdar and the surrounding plains. Supplemental irrigation, however, is still erratic and spontaneously applied depending on the temperature and mood of the farmers.

With available water, better conservation practices, higher irrigation efficiencies, and comprehensive plans of action and organization, the potential exists for extending supplemental irrigation to new areas or for increasing crop productivity on the existing land. These objectives have been incorporated into the national plans of development and are considered to be among the most important national goals for achievement in the near future. The conveyance of fossil waters from the huge basins of the Fezzan to the coastal zones is one step in this direction. Agricultural research centers and institutions are becoming more involved nowadays in water harvesting, resource conservation, and arid zone research than ever before. In the next section, the author will probe the possibility for success of these efforts based on evaluation of freshly gathered information from comprehensive questionnaires provided by ICARDA.

Agro-Ecological Classification

If a uniform agro-ecological zone is defined as homogeneous, having a non-varying form of agricultural science, then Libya can be classified into 4 major zones (see Figure 23.1):
1. Zone I includes the Jabal Al-Akhdar and a surrounding plain with an average annual rainfall of greater than 400 mm but not less 300 mm in 2 out of 3 years;
2. Zone II includes most of the Jafara Plain and parts of the western mountains with an average annual rainfall of 200-400 mm but not less than 200 mm in 2 out of 3 years;
3. Zone III includes the central coastal zone between Misurata and Jadabia and extending 100 km inland to the south with an average annual rainfall of 50-200 mm but not less than 100 mm in 2 out of 3 years; and,
4. Zone IV includes the remaining southern parts with an average annual rainfall of less than 100 mm and, in most areas, 0 mm.

Rainfed agriculture is limited to average annual precipitation above 200 mm; consequently, Zones III and IV are potentially unfit for sustainable dryland farming or supplemental irrigation without reliable water supplies. Even the marginal lands in Zone III cannot be developed with existing groundwater for supplemental irrigation. Pipeline transport of fossil water from the Tazerbo-Sarir Basin to this zone is under construction and would bring these marginal lands into sustainable production with supplemental irrigation or conventional irrigated agriculture.

Existing lands appropriate for introduction and expansion of supplemental irrigation and water harvesting practices are located in Zone I and II. Only these zones will be considered in detail in this chapter. Continuous changes and modifications of local political boundaries within each of these zones makes discussion more comprehensive and less confusing if these areas are described as independent geographical units rather than single administrative districts.

Jabal Al-Akhdar and Surounding Plains

This geographical unit, with an average length of 600 km, is situated in the north-east, bounded by the Mediterranean Sea at latitude 32° north; the town of Burdia at latitude 30° south; the town of Sulug at longitude 25° east; and, west at longitude 20°.

The climate is a sub-humid Mediterranean type dominating the slopes of the mountain proper, semi-arid in the Benghazi Plain, and arid around Sulug. Northwesterly winds occur 45% of time and are moist, rain-bearing winds during winter. Altitude and these winds give the area its climate. This zone experiences the hot, dry scorching Ghibli several times during the year. The dust from this wind is destructive to grain crops, vegetables, and fruit trees. Average annual rainfall varies from the highest, more than 500 mm on the northern slopes around the town of Shaat, to 350-400 mm in the Marj Plain, and to 250-300 mm on the Benghazi Plain.

In spite of a recently established network of weather stations, the weather station at Shaat provides the only reliable continuous records representative of Jabal Al-Akhdar. Figure 23.2 presents these descriptive profiles of rainfall, relative humidity, average temperature, minimum and maximum temperatures, and sunshine hours of the day.

All values are monthly averages. Average temperature varies from 10°C in December to 23°C in July and August. Maximum temperature during these months rarely exceeds 28°C while minimum temperature may drop to 5°C during January and December. Rainfall ranges between a maximum of 132 mm during January down to negligible traces during the summer months. Proximity of the region to the coast and dominant north winds give the highest relative humidity in the country: 80% in winter and 70% during the rest of the year. Fifty percent humidity can occur in late spring and early summer when the hot, dry Ghibli blows from the south.

High intensity rainfall and the topography of mountain slopes produce high amounts of runoff which seep through fissures and cracks in the limestone layers to be stored in aquifers of varying thickness and depth. Water tables are shallow close to the coast but become deeper (300-500 meters) in the mountains. Water quality is good and unthreatened by sea water intrusion because of high values of slope in the piezometric level. Potential groundwater resources in the plains of Benghazi, Al-Marj, Al-Abiar, can be summarized as follows.

1. *Benghazi plain*: The main reservoir occurs in the fissures and cracks of Miocene limestone structures. Aquifer thickness is 50-60 m and can supply up to 11 M m³/year of good quality water.
2. *Al-Marj plain*: Groundwater is present in three aquifers. The first is of Quaternary silt and clay deposits of the late Pleistocene Age and is shallow with low yields of highly saline water. The second is formed of Tertiary mixed deposits of clay, sand, and gravel of the Pliocene Era with a depth to water table of 25-90 m. Storage capacity is limited and water quality is poor. The third is located at a depth of 150-500 m with its major water-bearing strata composed of Eocene limestone which has reasonable yields of good quality water.
3. *Al-Abiar plain*: Groundwater is present in two separate strata: the

Supplemental Irrigation and Water Harvesting Systems in Libya 431

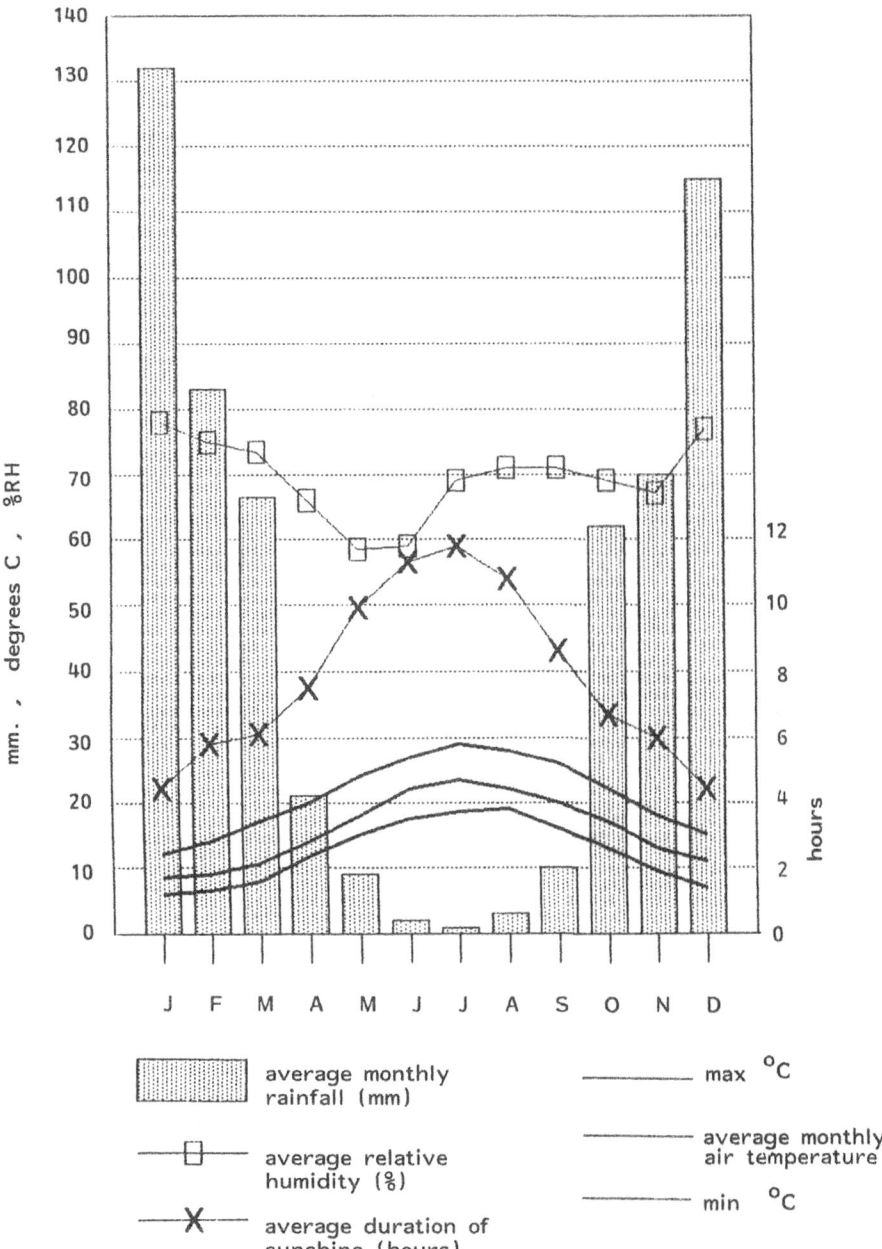

Figure 23.2. Weather profile for the city of Shaat in the Jabal Al-Akhdar Zone.

Oligocene and the Miocene deposits. Depth to water table is 80-120 m in the Oligocene strata and 200-300 m in the Miocene strata.

Total rainfall is estimated to amount to more than 1.4×10^{10} m^3 of water. Apart from deep percolation and water use by the vegetative cover, most of this is runoff to the sea and into the far depressions of the south. To minimize these losses, water harvesting has been practiced in recent years by construction of dams and dikes to reduce flood damage and soil erosion and to combine collected surface water with groundwater supplies for supplemental irrigation. This effort, in a short period of time, has resulted in bringing more than 15,000 ha under cultivation with supplemental irrigation. Compared to 1.2 M ha of rainfed farming, the areas with supplemental irrigation are negligible. The transport of Tozerbo-Sarir fossil water to the Benghazi Plain at a rate of 1.0 M m^3/day for supplemental irrigation is expected to increase cultivation by 20,000 ha.

The soils of this zone (210,766 ha) have been studied in detail. These have been classified into 4 major capability classes.

1. Productive and fertile suited to all agricultural crops and fruit trees without restrictions to potential yields. This class includes only 1,625 ha of flat, alluvial soils of less than 20% gravel, less than 15% calcium carbonate, deep profiles, and average rates of infiltration.
2. This class has 51,688 ha with good productivity for several agricultural crops and fruit trees with shallow roots. The soils are deep loamy sands or sandy loams containing variable percentages of gravel, low to average salinity, a low percentage of gypsum, and a calcium carbonate content not exceeding 40%.
3. Total area is 40,000 ha of average productivity for cereal crops such as wheat and barley and a few fruit trees. The soils have variable depths, are shallow, stony, and have clay pans with some salinity and gypsum. Gravel is less than 5% and calcium carbonates are less than 40%.
4. There have been 97,780 ha studied and found to be of low productivity feasible for limited cultivation of forest trees and pastures. These soils are shallow and rocky with steep slopes and are susceptible to severe erosion.

Production practices are determined by soil conditions and available water supplies. Rainfed farming systems support grain crops in the plains, animal grazing in natural pastures or on mountain slopes, and fruit trees on terraced mountain slopes with collection basins. Rainfed wheat and oat production are confined to the 250 mm isohyet. Barley, on the other hand, can grow successfully with less than 250 mm rainfall. Average yields for wheat are 1.2-1.5 t/ha; for barley, are 2.0 t/ha; and, for oat hay, 5.0 t/ha.

Vegetable production is limited to irrigated areas and greenhouses only. Most local vegetable consumption is produced within the area. Four thousand hectares are devoted to vegetables and produce 140,000 t during the summer and 43,162 t during the remainder of the year. Greenhouses are limited to 219 ha, contributing more than 49,000 t of fresh vegetables each season.

Annual livestock production is estimated to yield 6,019 t of meat from sheep, 894 t of beef, 8,400 t of poultry, 190 million eggs, and 37,800 t of milk. Planning for future production levels in this geographical unit estimates a capability of supplying more than 30% of the average per capita consumption of animal proteins.

To achieve this level of productivity will be difficult if measures to conserve water are not taken rapidly to conserve water. This achievement is required to support expansion of irrigated hectares to increase grain and forage crop yields and to improve pasture quality and other associated animal feeds. Rainfed grain production is low when compared to yields of 5 t/ha achievable with supplemental irrigation.

Constraints of limited water supplies and high costs of pumping do not encourage expansion of land for supplemental irrigation. For now, increased productivity through better management and improved cultural practices seems a more profitable approach. Training programs and technical help are needed critically by farmers (mostly of Bedouin origin) in techniques of fertility, mechanization, pest and weed control, and soil and water conservation. Prospects for contour terracing and water harvesting systems must be tapped to a greater extent in mountainous areas. Also, there are indications that fruit trees such as apples, pears, and stone fruits could be planted successfully on the gentle slopes if soil and water conservation methods were implemented.

The Jafara Plain and Surrounding Western Mountains

In spite of a comparatively small land area of 18,000 km^2, the Jafara Plain represents the most important economic region for agriculture in Libya because of the resources available for development and a high population density of two million, 50% of the total population. Socio-economically, it is well endowed with the heaviest investments in agriculture, industry, and societal infrastructures realized during the 3 decades since independence.

The Jafara Plain forms a triangle in the north-west. The western side is 100 km wide, extending more than 150 km long with the apex stretching to the south-west and, west of Homs, touching the coast. The Plain is bordered by Tunisia in the west, the western mountains continuing along the southern boundary, and to the north, the Mediterranean Sea. The western mountains meet the seacoast at Homs, encompassing the agriculturally productive heights of Misallata and Tarhuna, emerging as a natural barrier to isolate the Jafara Plain from the desert of the central zone. The Plain and the surrounding terrain, therefore, become a unique geographical entity dominated by an arid and unstable climate as a direct result of the opposing influences of the Sahara Desert and the Mediterranean Sea. This zone has several established farming systems which include the plantations and small garden oases of the fertile coastal belt, the shifting rainfed cultivation of the steppe, and, in the sur-

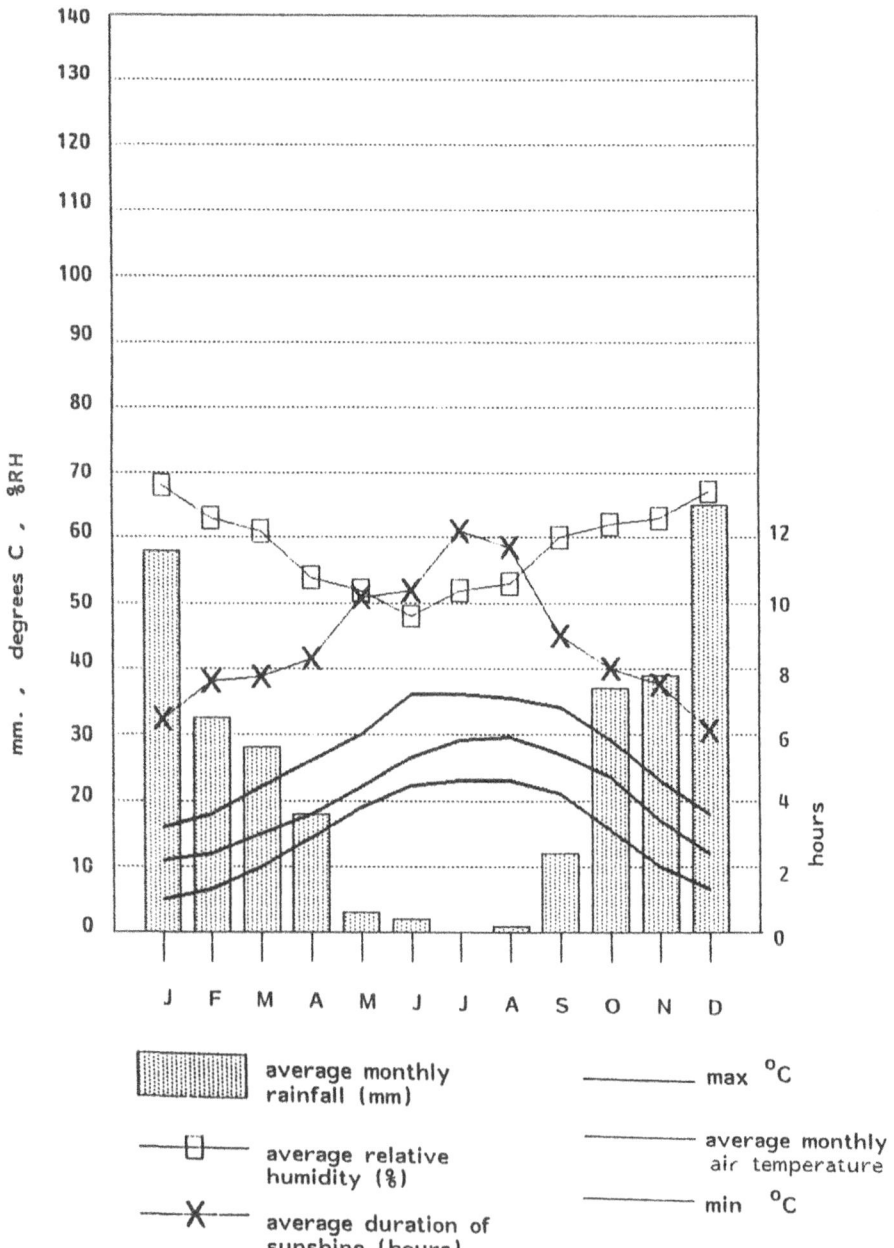

Figure 23.3. Weather profile for the city of Tripoli in the Jafara coastal belt.

Supplemental Irrigation and Water Harvesting Systems in Libya 435

rounding mountains, the specialized dryland farming using water harvesting techniques. For reasons of clarity, this geographical unit can be subdivided into three sub-zones:
1. a fertile coastal belt;
2. the interior; and,
3. the piedmonts and escarpments of the mountains.

Fertile coastal belt

This belt includes a narrow cultivated strip from the city of Misurata in the east to the town of Zoara in the west. The depth varies from less than 1 km in some locations to more than 20 km in others. In general, a maritime type of climate exists with winter rainfall and the influence of the Mediterranean Sea. Rainfall occurs in high intensity storms of short duration with the yearly total falling from October to March as indicated in Tripoli's weather profile presented in Figure 23.3.

Weather data of Tripoli is considered representative of the three sub-zones and is used because there are no continuous records from other weather stations in the Jafara Plain. Maximum rainfall occurs during December and January. Temperatures range from a mean minimum of 8°C in January to a mean maximum of 30°C during June to September. Relative humidity is high and dew is appreciable in late summer and can contribute to available moisture for crops and vegetation.

The cultivated area under conventional irrigation is 200,000 ha divided into 202,168 holdings which can be classified by farm size into 5 groups as indicated in Table 23.1. Most land is under irrigation with more than 60,000 wells as the only water source. Twenty eight million fruit trees are planted: olives, citrus, dates, stone fruits, and figs. Livestock production is estimated at 45,600 head of cattle, 838,000 head of sheep, 97,000 head of goats, 1,000 head of camels, and 2.3 million poultry.

Table 23.1. Distribution of land holdings by size of farm.

Group	Farm size (ha)	Number of farms	Total holdings (%)
1	1	106,200	53
2	1-3	43,658	22
3	4-6	26,467	13
4	7-15	19,382	9
5	>15	5,634	3

Source: Ministry of Agriculture.

Groundwater is pumped at an annual rate of 636 M m^3; whereas the recommended safe yield of the aquifers is calculated at an annual recharge of 199 M m^3. This over-pumping of the aquifers has resulted in an annual deficit of 437 M m³ of water which has led to a continuous decline in the piezometric levels of groundwater and has led to contamination by sea water intrusion at several locations. This destructive exploitation is of great alarm to both officials and the public.

Several alternatives to increasing water supply have been formulated: water conveyance from the Fezzan Basin; sea water desalination; or, modification of the farming systems. Two alternatives, water conveyance from the Fezzan Basin and desalination of seawater, are very expensive using current technology and might be acceptable for municipal purposes but are not feasible economically for agriculture. The third alternative, modification of existing farming systems, is a more profitable means. The intention now is to reduce, through calculated steps, the conventionally irrigated 125,000 ha to 32,000 ha by:

1. shifting 34,700 ha of irrigated wheat and barley to rainfed farming or supplemental irrigation. Research findings (Al-Ghariani, 1986) would indicate that supplying only a part of crop water requirement through 1-2 supplemental irrigations stabilizes crop yield and higher crop water use efficiency.
2. reducing to 9,600 ha the vegetable production with open field irrigation.
3. reducing fruit production from 43,120 ha to 21,630 ha by removal of ageing and low yielding trees.
4. relying on treated wastewater for irrigating 27,280 ha of forage crops.

These steps, if executed properly, should reduce groundwater consumption by irrigation from 636 M m³ to the recommended safe yield rate of 198.8 M m³ and sustain annual recharge of aquifers. The average utility of the underground aquifers in the coastal belt could then be prolonged an estimated 125 years instead of 25 years at the present rate of pumping.

This discussion indicates unlimited prospects for research and development of supplemental irrigation and increased water use efficiency at several levels of sophistication and modes of operation. Examples are introduction of deficit irrigation of field crops by sprinklers with localized systems for orchards, water stress tolerance, short growing seasons, evaporation suppressants, etc. Unfortunately, research along these lines is lagging and still in initial stages. International cooperation and experience is needed in this direction.

Interior sub-zone

The interior includes all remaining areas of the Jafara Plain south of the coastal belt and has a hotter, arid climate. Average annual rainfall decreases to 200 mm with increasing annual and seasonal fluctuations. Once every 4-5 years, farmers, under rainfed farming, can experience serious droughts leading to crop failures. These interior lands, far removed from the moderating effects of the sea, experience a greater contrast between winter and summer temperatures than does the coastal belt. The mean maximum temperature in August is 40°C; whereas, the mean minimum in January is 3-4°C. Relative humidity is usually higher in winter than in summer.

Until recently, grazing of rainfed crops especially of winter grain has been the rule. High costs of well drilling and groundwater pumping prevented the introduction of irrigation except in limited cases. Erratic rainfall and the scorching effects of the Ghibli winds make crop production and living conditions unstable and uncertain when dependent on dryland farming. Several settlements have been constructed and irrigation on a limited scale has been introduced successfully. The soils are productive but the major constraint to implementation is insufficient water supplies and the Bedouin inhabitants' lack of experience in water management for supplemental irrigation.

Conventional irrigation or supplemental irrigation is practiced on 22,491 ha and rainfed farming on 471,125 ha. With water transported from the Fezzan Basin and development of local water resources, land cultivated with supplemental irrigation could be increased to 50,000 ha. Grain yields under rainfed conditions rarely exceed 1.5 t/ha as compared to 5 t/ha under supplemental irrigation. Investments in supplemental irrigation could be profitable but no comparative studies have been made in Libya to provide documentation for this conclusion. Low productivity under dryland farming cannot be ascribed to water shortages alone but could be lack of management techniques and the efficient cultural practices of crop varieties, mechanization, fertilization, pest and weed control, and soil and water conservation. This region is, however, potentially suitable for supplemental irrigation if major constraints were alleviated.

Surrounding mountain areas

Abrupt alterations in topography, ranging between 400-700 m, are associated with concurrent changes in climate. Rainfall is highest in the central and eastern parts (300-350 mm) and decreases in the westerly direction. The altitude intercepts cloud-bearing north-west winds to form orographic and convectional rainstorms of high intensity but short duration. This rainfall combined with low infiltration rates produces severe soil erosion on the rocky slopes and unloads torrential sediment-laden runoff into gullies then discharges this flood water across the Jafara interior. Some gullies or large ephemeral streams from wadis reach the coastal belt and discharge into the Mediterranean Sea.

The conservation practice of water harvesting by small cultivated collection basins or by contour terraces or even constructed dams in gullies and wadis were used, in ancient times, to sustain production of olives, figs, grapes, almonds, and grain crops in excess of local consumption. From Tarhuna and Misallata, high quality yields were exported via Leptis Magna to Rome and the Roman Empire. Poor management and inferior maintenance lead to destruction and negligence of the systems with the eventuality of soil erosion and low productivity. Renewed interest has been generated for reconstruction and development of these ancient microcatchments for agricultural production which gives assurance for the future.

Mean minimum temperatures are similar to the desert but, with the influence of altitude, the mean maximum temperatures more closely reflect those of the coastal belt. Relative humidity is low, reaching 25% in summer. Snowstorms are rare but not unknown. The worst storm recorded occurred in 1949 when 1 m of snow accumulated on mountain slopes during 3 days of precipitation. Hail is more frequent and can cause severe crop damage. Table 23.2 gives a summary of weather data from different parts of the Jafara Plain and the western mountains.

Table 23.2. Climatic means for selected stations in Jafara and the Jebel.

Site	Jan	Feb	Mar	Apr	May	Jun	Jul	Aug	Sep	Oct	Nov	Dec	Year
Tripoli													
°C	12.0	13.3	15.2	18.2	20.6	23.9	25.6	26.3	25.6	22.7	18.1	13.6	19.6
RF mm	76.9	42.9	24.6	9.3	5.3	1.3	0.5	0.8	10.2	36.5	65.9	93.8	368.0
Rel H	62	61	60	59	62	63	65	65	63	60	61	62	62
Zuara													
°C	11.4	13.2	15.3	18.0	20.2	23.5	25.9	26.5	25.4	22.6	17.9	13.2	19.4
RF mm	42.8	27.1	19.0	9.4	6.4	1.2	0.0	0.1	5.7	24.8	42.9	37.4	216.8
Rel H	69	68	65	64	70	72	73	71	69	66	66	65	68
Homs													
°C	13.2	14.0	15.6	18.5	20.5	24.3	26.2	26.9	25.8	23.5	19.0	14.7	20.2
RF mm	57.1	39.2	19.6	9.3	5.7	1.3	0.0	0.8	7.3	29.9	48.9	52.3	271.4
Rel H	66	65	64	62	65	64	65	67	67	65	66	66	65
El Azizia													
°C	11.3	13.0	15.7	19.6	23.1	27.0	28.7	28.8	27.5	23.6	17.7	13.0	20.8
RF mm	47.5	33.2	21.1	9.6	1.2	4.6	0.1	0.1	6.6	16.0	27.1	50.5	217.6
Rel H	62	58	54	46	42	37	41	45	49	50	57	60	50
Ben Gascir													
°C	11.7	13.1	15.9	19.0	22.0	26.1	27.6	28.0	26.6	23.4	17.9	12.8	20.3
RF mm	62.4	48.0	23.3	8.2	3.9	1.2	0.8	0.7	8.9	23.9	41.3	66.2	288.8
Rel H	57	55	51	43	43	40	41	45	47	50	57	60	49
Garian													
°C	8.3	10.0	12.7	17.0	20.7	24.7	26.5	26.8	24.1	20.4	15.3	9.9	18.0
RF mm	74.1	51.5	39.9	16.0	12.7	2.1	0.5	0.7	11.4	23.8	39.5	54.0	326.2
Rel H	57	54	48	40	36	28	30	31	39	43	53	58	43
Nalut													
°C	7.8	9.3	13.5	17.2	21.3	25.9	27.9	27.3	25.1	20.6	15.2	9.3	18.3
RF mm	21.1	25.4	21.2	12.9	6.4	1.3	0.3	0.5	4.4	10.1	17.3	16.8	137.7
Rel H	61	62	52	41	41	36	36	36	44	51	56	52	47

°C = Temperature, Degrees Centigrade
RF = Rainfall
Rel H = Relative Humidity

Existing agriculture in the high hills is limited to olive and almond plantations left from the time of Italian colonialists and a few collection basin gardens of fruit trees planted by Libyans. Cultivation of rainfed cereal crops and forage occurs in developed soils of depressions or as inter-row cropping in olive plantations. Yields are comparable to those obtained in the interior plain described earlier. No irrigated agriculture on a significant scale has been

Figure 23.4. A photographic comparison between (1) an open large rainfed plantation and (2) a small collection garden. [Photos taken in Jafara region].

introduced and depth of groundwater (1,000 m) puts a severe constraint to implementing this technology in the foreseeable future. Possible agricultural development, now, should be achieved through establishment of water harvesting systems and associated agricultural sciences. Figure 23.4 shows a comparison between garden collection basin cultivation constructed by local farmers and open large rainfed plantations based on the Italian style of farming.

Evaluation of Collected Information

Precoded questionnaires from ICARDA were administered: one in each of the two selected agro-ecological zones. The degree of sophistication and the wide spectrum of the questions asked, and the lack of accurate statistics in the central government agencies and at the district level extension offices, made the job of data collection a very difficult one. The answers obtained are approximate for many questions while several others were answered using out-dated information. Both policy and technique of Libyan agriculture are rapidly changing without comtemporary updating of statistical information. Nevertheless, no effort was spared in responding to the questionnaires to comply the best available data from existing (published and unpublished) information. With these cautious remarks in mind, a brief assessment of the information is presented.

The Jafara Plain

Population numbers 2,000,000 people, or 50% of Libya's population, and this is considered the most populous region. Five major cities are located here including the capital, Tripoli. Most population is concentrated in the cities and the irrigated coastal belt with the majority involved in several economic activities. This diversity of economic interests increases the difficulty of knowing how many depend on farming as income.

This zone has the most developed infrastructure for delivery of public services. There are 818 primary schools, 351 preparatory schools, and 125 secondary schools attended by both male and female. Literacy rates among farmers are 40% with highest literacy in the irrigated coastal belt. There are 13 credit centers, 13 agricultural extension offices, 16 veterinary care centers, 41 medical centers, and 92 cooperatives. The major credit source is the agricultural bank but cooperatives provide farmers with agricultural inputs and equipment services. Survey results indicate that all farmers enjoy a modern system of new roads, high tension lines for electric power, and pressurized potable water. Transportation and marketing facilities are easily accessible.

Arable land is 1.3 M ha of a total 2.2 M ha. Only 0.75 M ha of the arable land is cultivated with 471,000 ha under rainfed farming, 223,000 ha with supplemental irrigation, and 53,000 ha using water harvesting. These farming systems are divided into 220,000 farms, 60% using irrigation involving 60% of the farmers. Management problems encountered by the majority of farmers are land fragmentation and multiple ownership of land.

Dominant soils for supplemental irrigation agriculture and rainfed farming are deep, yellow or brown, and of moderate slope. They are stone-free, of low salinity, and have moderate infiltration rates, but with good tilth and no surface crusting problems. Soils used for water harvesting, however, have shallow stony soils with steep slopes and low infiltration rates.

Average annual rainfall is 300 mm/yr of high intensity and short duration orographic storms with high volumes of runoff and severe soil erosion. Groundwater is the major water supply for supplemental irrigation. There are 70,000 productive wells with an average flow of 40 m^3/hr and an acceptable water quality. Pumping depth averages 75 m but is increasing because of excessive exploitation. These wells provide water for irrigation of 200,000 ha as well as domestic and livestock requirements. Well construction and pumping equipment are expensive with costs of US$ 20,000 to 50,000.

Sprinklers are the principal equipment used for supplemental irrigation. Limited areas have been converted to drip irrigation in recent years but still are considered experimental. These are expected eventually, at least, to replace sprinkler irrigation for tree crops. For winter crops, especially cereals, farmers apply an average of 2-3 irrigations per season. Available groundwater supplies are limited and do not allow for expansion of production using supplemental irrigation. The exception is development of the deep aquifers in the Jafara interior subzone.

Water harvesting using small collection basins has been practical to a limited extent since ancient times but terraced slopes have been newly added and this technique, while promising, suffers from poor management and lack of maintenance and escalates soil erosion.

Barley, wheat, and olives are the major crops cultivated under rainfed farming; wheat, alfalfa, and citrus are grown using supplemental irrigation; and, olives, fruit trees, and barley are cultivated using water harvesting farming. Average grain yield of wheat is 1.5 t/ha under rainfed conditions, 3.5 t/ha under supplemental irrigation, and 2.0 t/ha with water harvesting. Average grain yield for barley is 1.4 t/ha for rainfed farming, 3.6 t/ha with supplemental irrigation, and 1.6 t/ha with water harvesting.

Both wheat and barley are sown from late October to late December depending on rainfall. Most farmers, especially those using supplemental irrigation, employ current cultural practices of mechanization, fertilizer application, weeding, and control of pests and diseases.

Survey results indicate that most farmers own animals. Table 23.3 gives the distribution of livestock in different farming systems. Livestock usually depend, notably in irrigated areas, on forages and animal feed concentrates (either produced or bought from the market) or on grazing of crop residues and natural pastures. Natural pastures are estimated to cover 200,000 ha but are not managed to their full potential.

Table 23.3. Distribution of livestock by type of farming system.

Livestock	Farming system		
	Rainfed	Suppl. irrig.	Water harvesting
Sheep	250,000	838,000	200,000
Goats	30,000	97,000	25,000
Cattle	1,000	45,600	150
Camels	500	1,000	200
Horses	300	1,000	1,200
Donkeys	1,500	500	2,800

Socio-economic data indicate that benefits could be augmented with expansion of supplemental irrigation and water harvesting. But, as mentioned in previous sections, existing water resources and technical farming experience place severe constraints on the implementaion of these technologies. These constraints should be alleviated.

The Jabal Al-Akhdar Zone

There are 860,000 inhabitants who predominantly live in the cities of Benghazi, Baida, Marj, and Derna. Farmers are of Bedouin origin and reside in large tribal communities of strong social ties. The area is less advanced than the Jafara Plain but has a potential for future development. Communication systems, medical care, educational facilities, credit centers, and other public services are considered sufficient to enhance any economic development.

Average annual rainfall in this region is the highest, greater than 500 mm, particularly in the mountains. Other climatic parameters are suitable for cultivation of many crops. Total land area is 2.35 M ha, with only 1.23 M ha of arable land, with the remainder either rocky mountains and valleys with difficult access or steppe and forests of trees and shrubs only suitable for livestock grazing. Presently, 230,000 ha are cultivated: 200,000 ha are rainfed farming, 15,000 ha have supplemental irrigation, and 15,000 ha are cultivated with water harvesting. Cultivated lands are divided into 70,000 farms varying in size from 0.5 ha, the smallest holdings with supplemental irrigation, to 250 ha, the largest holdings under rainfed farming.

The dominant soils are of average depth, slightly stony, and of red color. On irrigated and rainfed plains, the soils are level, with moderate infiltration, low salinity, and no crusting problems. On the slopes used for water harvesting, the soils are of moderate slope and of high productivity.

Although surface runoff in the mountains is heavy, control has not been sufficient for maximal use of this water. Groundwater is the major supply for supplemental irrigation but is comparatively deep and of low productivity. The 20,000 wells have an average pumping depth of 120 m and a flow rate of 25 m³/hr of reasonable quality water. Most well water is for municipal and industrial uses in the urban areas. Construction and operation of these wells vary in cost from US$ 38,000 – 75,000. Irrigation systems are almost exclusively sprinklers with only 1,000 ha using drip irrigation. There is a tremendous potential for supplemental irrigation if the groundwater resources were exploited to capacity.

Water harvesting is restricted to 15,000 ha and is used mainly for fruit trees, especially apples. With higher technical skills and concentrated efforts, this area could be increased to 100,000 ha in a short period of time. The most compatible water harvesting technique is level contour terracing with stone walls to prevent soil erosion.

The major crop cultivated on the plains is wheat which produces an average of 1.6 t/ha under rainfed conditions and 4 t/ha with supplemental irrigation. Barley is produced with lower rainfall and average yields are 2 t/ha under rainfed conditions and 3.5 t/ha with supplemental irrigation. Forages and vegetables are minor crops produced with irrigation. Table 23.4 shows the number of livestock supported by natural pastures distributed in the farming systems. Livestock depend on green forages in the irrigated areas and for natural pastures and crop residues in rainfed areas. Concentrates, both imported and locally manufactured, are available in the local markets. There is an impressive potential for intensifying livestock production with pasture management and by increasing grain yields with supplemental irrigation.

Table 23.4. Number of livestock able to graze natural pastures.

Livestock	Farming system		
	Rainfed	Suppl. irrig.	Water harvesting
Sheep	1,120,000	90,000	— —
Goats	40,000	8,000	— —
Cattle	1,000	5,000	— —
Camels	150	50	— —
Horses	200	700	— —
Donkeys	1,000	600	— —

Although pumping costs are high, results of the survey indicate that with good management and improved skills, farmers could show profits if supplemental irrigation were introduced into existing farming practices. More successful farmers with the addition of supplemental irrigation could earn US$ 15,000 income per year. Planned transfer of water from southern basins at a rate of 1 M m³/day is expected to increase available water supplies and therefore, with supplemental irrigation, increase cultivated land by at least 20,000 ha.

Potential for Development

The preceding discussion of major agro-ecological zones for implementing supplemental irrigation reveals two important facts about the present state-of-the-art and possible ways of development. First, in view of the severe constraints imposed by limited groundwater in the coastal belt of the Jafara Plain and the plains of Jabal Al-Akhdar, any further expansion of irrigated land cannot occur without increasing water resources for this purpose. Water transport projects from the Fezzan Basin to the eastern and western parts hold uncertain promise for the future; their impact on agricultural development cannot be evaluated in the present. Second, high annual precipitation of these zones amounts to a rate of 2.9×10^{10} M m^3/yr but this has not been managed for maximum distribution nor with sound practices of water conservation. These facts clearly demonstrate that prospects for present development lie in the following possibilities:

1. increasing the irrigated lands in the Jafara interior subzone by further exploitation of local groundwater supplies and accepting the risk of depletion over a shorter duration;
2. formulation, design, and construction of comprehensive water harvesting systems to achieve maximal production from the precipitation, favorable climate, and topography of the eastern and western mountains; or,
3. increase land productivity and achieve crop water use efficiency on the existing irrigated lands with improved cultural practices and intensive applied research.

Development of the Jafara Interior

Soils of 100,000 ha are generally productive and suitable for agriculture with supplemental irrigation. These include:
1. undeveloped sandy soils with poor structure and susceptibility to wind erosion. These soils overlay most of the plain's interior could be put under irrigation;
2. the aeolian deposits covering most of the flood areas of the wadis such as Wadi al Mejenin and Wadi Al Ramel;
3. the alluvial soils in the eastern parts of the interior transported through wadis and gullies from the mountains and deposited in the flood basins on the plain. These soils are composed of weathered particles of shale, sandstone, and calcium carbonates; and,
4. the clayey sandy soils existing in certain places such as Wadi Alatal and Wadi Alhira.

Surface water occurs as runoff from mountain slopes flowing in several wadis which have been diked and dammed for storage and groundwater recharge. Major dams have been constructed on Wadi Al Mejenin, Wadi Ghan, and Wadi Kaam to store an annual volume of 30 M m^3. Groundwater from 4

major reservoirs is the major supply for agricultural, industrial, and municipal purposes.
1. A Quaternary aquifer, at a depth of between 40-120 m, consists of various limestones. This aquifer is the most exploited in the plain to furnish a reasonable volume of high quality water (total dissolved solids of < 1 g/l).
2. A Tertiary aquifer extending to a depth of 250 m, composed of limestone; high yielding but with poor water quality (> 1.5 g/l).
3. The Kikla sandstone formation extending to a depth of 500 m which yields good quality water in appreciable amounts.
4. The Al Azizia limestone formation extending to a depth of 700 m and containing large amounts of water of an acceptable quality.

These 4 sources could be combined to expand supplemental irrigation in the fertile soils. The infrastructure and public services exist to support development. An intricate network of modern asphalt roads, high voltage power lines, and sophisticated communication systems have been constructed during recent years. Public schools, hospitals, gasoline stations, housing facilities, and markets have been established already. Irrigation equipment and associated components are easily accessible. The critical constraint to sustain this technology is a stable water supply.

Comprehensive Water Harvesting Systems

Water harvesting systems seem to be the most promising prospect for development. This technology could capture high amounts of runoff and decrease evaporation through increased soil infiltration and storage for crop use while preventing erosion and increasing fertility and, therefore, crop productivity. Although traditional water harvesting has been practiced on the slopes of the western and eastern mountains since ancient times, modern techniques have not been exercised in a systematic manner. More than 2 M ha could be cultivated with modern techniques of water harvesting farming.

Interest in historical water harvesting systems has been regenerated which has led to experimental trials to construct terraces on 53,000 ha around Tarhuna, Misallata, Urban, Assabaa, and other locations. Another 1,500 ha were terraced in the Jabal Al-Akhdar zone for cultivation of apple and cherry trees. The traditional stone walls and small collection basins in the western mountains have been improved and expanded to support 5 M apple, almond, and fig trees. Experience gained from these trials has encouraged efforts of afforestation and fruit tree production in general. The only existing constraint to expansion is the availability of skilled farmers to occupy and manage these newly established water harvesting farms.

Collection basins and cisterns could be constructed throughout the open pastures for purposes of livestock water. Livestock production is more often limited by availability of stock water than by a lack of natural vegetation for animal feed. Other methods of water harvesting by land alteration, chemical application, and micro-catchment modification should be tried to verify the feasibility of alternative designs for site specificity. Local research is lacking and a sustained research program must start immediately to determine the methods of water harvesting adapted to local conditions.

Crop Productivity and Water-Use Efficiency

Generally speaking, improved cultural practices that increase production also improve water-use efficiency in both rainfed farming and supplemental irrigation agriculture. In the eastern and western agro-ecological zones, crop yields are not limited by water supply alone but by other production factors such as fertility, pest and disease control, cultivation practices, sowing dates, and weed control. In fact, these factors may be the major causes of low productivity and, hence, reduced water-use efficiency.

Research programs are needed to determine optimal density of plant population to maximixe water-use efficiency. Pasture and range management should be designed to maximize infiltration and minimize runoff. Overgrazing results in wind and water erosion and in decline of productivity. Contour row cropping, mulching with crop residues to reduce losses to evaporation, determining optimal sowing date, plant breeding for stress resistance, and other techniques should be investigated and adapted to specific areas and specific crops.

Systems chosen for supplemental irrigation should improve water-use efficiency through methods of application. Most existing production is under sprinkler irrigation; and, since this method results in high spray evaporation losses (up to 25% in some locations during hot windy days) shifting to a more efficient localized system would save water which might be used to expand irrigated production or for other agricultural purposes. Since limited land area is not a factor to expansion, existing water resources could be more efficiently used if only a part of the crop water requirement were provided for crops sown to a larger land area rather than meeting the full water requirements of a smaller cropped area. There are indications that this might be successful, at least, for alfalfa, barley, oats, and grain sorghum (Al-Ghariani, 1986).

Summary and Conclusions

The geographical location of Libya between the southern coast of the Mediterranean Sea and the heart of the great Sahara Desert, combined with a large land area accounts for variable climate, hydrogeology, and topography. Annual rainfall divides the country into four major agro-ecological zones, each having similar systems of production. The Jafara Plain and the surrounding western mountains and the Jabal Al Akhdar and the surrounding plains have an average annual rainfall of more than 200 mm/yr which is considered adequate for rainfed farming and supplemental irrigation agriculture. The slopes of the mountains within these major zones offer an ideal opportunity for the development and expansion of rainwater harvesting and runoff agriculture. Rainfall in the central zone and the rest of the country ranges from less than 200 mm to 0 mm precipitation; therefore, these zones have not been considered for rainfed agriculture and were excluded from discussion in this chapter. The central zone, however, does offer a limited opportunity for water harvesting and runoff agriculture.

Research of rainfed agriculture and supplemental irrigation is unavailable for Libya, but the information collected during the survey demonstrates that significant benefits could be made, bringing potential land under cultivation with supplemental irrigation. Two major constraints, however, are limited water resources and low technical skills of local farmers. The first constraint can be alleviated by developing new water supplies and the second constraint can be removed by extensive applied research and intensive training of farmers.

Rainwater harvesting and runoff agriculture offers a better prospect for development through efficient use of the 3.0×10^{10} m^3 of rainfall received from the mountain altitudes. Although water harvesting farming has been well known and actually applied, to a limited extent, for fruit tree production in small collection basins and stone wall contour terraces, further expansion and improvement are required. Potential areas for expansion are virtually unlimited and recent field trials of terraces have proved successful. The major constraint to expansion is the availability of farmers who are willing to settle and manage the area developed through these trials. Increasing population pressures on the fertile irrigated farms will eventually force many farmers to migrate and settle on these farms in the future.

References

Al Ghariani, S. A., 1987. Development of agricultural resources of Libya: a plan for action. Report of a seminar; Agricultural Research Center, Tripoli, Libya.

Al Ghariani, S. A., 1986. Water requirements of alfalfa in the western parts of the coastal belt. A technical report submitted to the National Authority for Scientific Research. Agric. Res. Center, Tripoli, Libya.

Government of Libya, Ministry of Agriculture, 1980. Handbook of Agriculture in Libyan Jamahiriya: Facts and Figures. Tripoli, Libya.

Government of Libya, Secretariat of Planning, Surveying Dept., 1978. The National Atlas of Libya. Tripoli, Libya.

Chapter 24

Supplemental Irrigation Systems in Morocco

AMBRI ABDEL ILAH

Geographical Characteristics

Morocco is located in north-west Africa and is encircled, to the north, by the Mediterranean Sea; to the west, by the Atlantic Ocean; to the south-southeast, by the Sahara Desert; and, to the east, by Algeria. The total population in 1982 was 20.409 M living on a land area of 69 M ha; 57% are rural dwellers.

Climate

A Mediterranean climate is characterized by a rainy period during the coldest months from autumn to spring with a hot, dry summer. The weather reflects a semi-arid environment, presenting variations according to latitude, altitude, and ocean effects (Table 24.1).

Table 24.1. Climatic data for Morocco.

	Months of year											
	S	O	N	D	J	F	M	A	M	J	J	A
Berrechid Station, total annual rainfall, 400.4 mm												
Rainfall (mm)	4.0	80.2	55.0	75.1	50.0	48.2	58.7	40.0	25.0	4.2	0	0
T, min (°C)	13.1	11.8	10.0	8.1	7.6	8.6	9.2	9.2	10.0	12.1	14.2	14.8
T, max (°C)	28.0	25.0	21.0	18.3	17.7	18.3	19.7	20.2	23.3	26.6	30.0	30.1
Ain Taoujdate Station, total annual rainfall, 556.7 mm												
Rainfall (mm)	6.1	40.2	82.0	65.3	80.4	92.0	70.0	72.3	29.1	17.0	1.1	1.2
T, min (°C)	14.0	11.4	7.6	4.5	4.2	5.4	7.9	7.9	10.6	13.6	17.3	16.6
T, max (°C)	29.6	24.4	19.7	15.8	15.8	17.1	19.8	19.8	23.4	29.7	34.4	33.6
P. Evap (mm)	18.4	16.4	8.2	7.0	57	83	87	98	124	150	225	183
Taourirt Station, total annual rainfall, 275 mm												
Rainfall (mm)	18.4	27.7	24.5	25.8	26.1	26.1	40.4	46.1	28.6	10.0	3.5	2.7
T, min (°C)	15.5	11.7	7.5	4.9	3.6	4.2	6.2	8.3	10.6	14.2	16.9	17.8
T, maxc (°C)	30.7	25.0	19.4	15.7	15.2	16.7	19.3	22.5	25.1	30.1	34.5	34.6

E.R. Perrier and A.B. Salkini (eds), *Supplemental Irrigation in the Near East and North Africa*, 449-461.
© 1991 *ICARDA*.

Precipitation diminishes from north to south: the region of Tanger receives more than 700 mm; Gharb, 500-700 mm; Chaouia, 400 mm; and, Souss, 200-400 mm. Precipitation, also, decreases from west to east: the Oujda region records 200-400 mm; whereas, in the Larache region, at the same latitude, more than 600 mm is recorded. Altitude has an effect on distribution spatially as is exemplified by the mountainous chains of the Rif, Middle, and High Atlas regions which receive about 900-1,200 mm. Average annual rainfall varies proportionally, 1:4, demonstrating an inter-annual as well as an intra-annual variability.

Mean annual temperatures are 18°C except in the mountains, Sahara Desert, and pre-Saharan regions. Average minimum temperatures are below 10°C but can fall below 0°C in the mountains; nonetheless, frost is rare in the Atlantic Plains. Average maximum temperatures are 30°C excluding the land along the oceanic coastlines. Evaporation is high with values of 800 mm at Tanger, 2,200 mm at Marrakech, and 4,800 mm at Ouarzazate.

Land Use

Cultivable land is 7.72 M ha, equivalent to 11.2% of the total land area: 800,000 ha has irrigation and 6.92 M ha is rainfed. Thirty percent of the cultivated land is left fallow for a cropping intensity of 70%. Wheat and barley are grown on 60% of the land; food legumes on 6%; and, industrial crops on 1.2%. Cereals, sugar beet, and fodder crops do not receive irrigation except in the northern areas of the Gharb and Loukkos irrigation schemes.

Water Resources

One hundred fifty billion m^3 of water comes from atmospheric precipitation, only 30 B m^3 (20%) is effective rainfall. The balance, 120 B m^3, is water lost to agricultural production through evaporation and transpiration. The annual available water for production purposes is 30 B m^3 which can be partitioned into 22.5 B m^3 of surface flow and 7.5 B m^3 for groundwater recharge.

Currently, only 10 B m^3 are utilized: 7.5 B m^3 of surface water and 2.5 B m^3 of groundwater. In the long term, 21 B m^3 could be utilized for production: 16 B m^3 from surface flow and 5 B m^3 extracted from groundwater with the balance of 9 B m^3 not exploited because of economical and technical constraints.

By the year 2000, 14.5 B m^3 of water are planned to be in use in Morocco, of which, 12 B m^3 (83%) will be allocated for irrigation to augment existing irrigated production for a planned cultivated area of 1.22 M ha.

Existing Areas with Supplemental Irrigation

Techniques of supplemental irrigation are encountered in different agro-ecological zones. The average annual rainfall of these zones varies from 250 mm to more than 600 mm. In view of rainfall irregularities recorded in recent years, farmers have started to practice supplemental irrigation, seeking to stabilize crop production and improve yields. Besides the large scale irrigation schemes of Tadia and Doukala where winter crops are produced with supplemental irrigation, 3 agro-ecological zones of rainfed farming can be distinguished where supplemental irrigation is practiced.

Plains and Valleys: Middle and High Atlas Mountains

The principal water supply for irrigation is surface flow from precipitation runoff as flood water and snow melt, although groundwater from wells is used, also. Ain Taoujdate District in Meknes Province and Ait Ourir District in Marrakech Province are representative of these plains and valleys. Irrigated crops are vegetables, forages, cereals, and fruit trees. The yields of field crops under rainfed cultivation are doubled or tripled with supplemental irrigation. Two to four irrigations are applied depending on seasonal rainfall (Table 24.2).

Table 24.2. Yield comparisons of field crops under rainfed farming and supplemental irrigation.

District/crop	Yield (t/ha) Rainfed	Yield (t/ha) Suppl.irrig.	No. of irrigations	Ave. rainfall (mm)
Ain Taoujdate				500
(Meknes Province)				
Wheat	1.5-2.0	3.0- 4.5	2-3	
Barley	1.5-2.0	3.0- 4.5	3-4	
Dry Forage	3.0-2.0	9.0-12.0	1-2	
Ait Ourir				450
(Marrakech Province)				
Wheat	1.0-1.2	3.0- 4.0	2-4	
Barley	1.0-1.4	3.0- 4.0	3-4	
Berrechid				400
(Settat Province)				
Wheat	1.0-2.0	3.5- 4.5	2-3	
Dry Forage	3.0-3.6	9.0-12.0	1-2	
Maize	0.8-1.0	3.0- 4.0	4-5	
Had Kourt				500
(Kenitra Province)				
Sugar Beet	2.0-4.0	4.0- 8.0	2-4	
Taourirt				275
(Oujda Province)				
Wheat	0.4-1.0	2.0- 2.5	5-6	

Atlantic Plains and Plateaus of the Eastern Areas

Berrechid District in Settat Province and Taourirt District in Oujda Province are representative of this zone. Average annual rainfall is 400 mm in Berrechid District and 250 mm in Taourirt District. Most irrigation water is pumped from groundwater with individual farmers using their own pumps to irrigate vegetables, cereals, and forages in the Berrechid District, and forages, cereals and olives in the Taourirt District. For field crops with supplemental irrigation, yields are doubled in Berrechid District and quadrupled in Taourirt District. Near Berrechid city on the route to Settat, a new irrigation scheme has been constructed with center pivot sprinkler systems to irrigate 1,000 ha. Average yields for 1985-86, with supplemental irrigation, were 4.0 t/ha for wheat and 6.0 t/ha for maize. Under rainfed conditions, average yields were 1.0 t/ha for wheat and 0.6 t/ha for maize (Table 24.2).

Had Kourt District, Kenitra Province

Farmers usually cultivate sugar beet under rainfed conditions with an average yield of 2.0 t/ha, but during the last dry years some farmers started to irrigate. Last season, 430 ha of sugar beets were cultivated with supplemental irrigation, resulting in average yields of 5.0 t/ha. Water is pumped from the Sebou and Rdate Wadis and applied using 1-4 irrigations.

Research Findings

Oulad Gnaou Experimental Station

Since 1975, experimental work on supplemental irrigation has been performed at Oulad Gnaou Experimental Station in Beni-Mellal Province. Average annual rainfall at this site is 420 mm. Experiments have been conducted to find what effect water deficits have on yields of different crops. These studies were performed in conjunction with experimental work of crop water requirements.

The results show that supplemental irrigations of 100 mm have a high impact on grain yield of wheat ranging from 3.0-4.2 t/ha. In the 1975-76 season, with 311 mm of rainfall, a 98 mm supplemental irrigation was applied at germination which resulted in 4.2 t/ha. A sugar beet yield of 8.3 t/ha was obtained the same season with 96 mm of supplemental irrigation applied at germination. In 1979-80, with 224 mm of rainfall, a 79 mm supplemental irrigation was applied to wheat at germination and produced a grain yield of 3.5 t/ha.

In another series of experiments on the timing of supplemental irrigation at different growth stages and the relative effects on the yields of wheat, results showed that:
1. the early irrigation at germination of 70-100 mm ensured a good yield;

2. treatments irrigated both at stages of germination and flowering produced the highest yield; and,
3. increases beyond certain levels (150 mm) in water applied had no significant effect on yields.

Settat Experimental Station

Experimental studies have investigated the impact of supplemental irrigation on yields of field crops at Settat Experimental Station in conjunction with the "Dryland Farming Project" supervised by the National Agricultural Research Institute (INRA) of Morocco in coordination with the Mid-America International Agricultural Consortium (MIAC) of USAID. The objectives of these studies were:
1. to compare the yield response of different cereal genotypes grown under different water deficits; and,
2. to establish a simulation model for wheat growth under different water deficits and cropping techniques.

The first experiment on barley was conducted during the 1982-83 season. The 1982-83 season was dry with 140 mm of rainfall. Different amounts of irrigation were applied up to 160 mm. Although rainfed plots produced 1.5 t/ha of dry matter, they produced no grain. The plots receiving 160 mm of supplemental irrigation produced 60 t/ha dry matter and 1.4 t/ha grain.

During the 1985-86 season, two interesting studies on the effects of supplemental irrigation on the yield of field crops were carried out at the Settat Experimental Station. The first experiment compared the yield response of different wheat and barley genotypes grown under different water deficits and seeding rates. Three species were compared: 3 varieties of bread wheat, 2 varieties of durum wheat, and 2 varieties of barley. Two seeding rates, 50 kg/ha and 100 kg/ha, were used for each variety. Total rainfall was 276 mm; this is below average annual rainfall, but the rains were well distributed. Experimental data are presented in Table 24.3.

First, highest yields for both bread and durum wheat were obtained for the supplemental irrigation (SI) treatment of 185 mm; whereas, the highest yield for barley was obtained with the SI treatment of 105 mm. Second, yields were doubled for wheats and 1.6 times greater when comparing the moist treatments to the dry. Third, increasing the seeding rate to 100 kg/ha had a negative effect on the yields of 3 species with the supplemental irrigation treatments.

The second experiment was to verify a simulation model (SIMTAG) developed by ICARDA for wheat grown under semi-arid conditions. In this experiment, the wheat was grown under different cropping techniques and received different amounts of supplemental irrigation. Two varieties of bread wheat and two sowing dates were tested with 3 treatments of supplemental irrigation. Yields more than doubled with 150 mm of supplemental irrigation for both varieties and both sowing dates. The results are summarized in Table 24.4.

Table 24.3. Yields of cereal crops and seeding rates under different amounts of irrigation water, 1985-86.

Seed rate (kg/ha)	Species	Grain yield (kg/ha) Supplemental Irrigation Treatments		
		moist 185 mm	intermediate 105 mm	dry 0 mm
50	Tender wheat	41.3	35.7	22.2
	Durum wheat	37.8	30.8	19.0
	Barley	49.8	51.8	30.8
100	Tender wheat	39.5	35.1	17.6
	Durum wheat	27.8	25.1	17.9
	Barley	39.2	46.6	24.2

Table 24.4. Yield of wheat under different amounts of irrigation water and planting dates, 1985-86.

Planting date	Grain yield (kg/ha) Supplemental Irrigation Treatments*		
	moist 150 mm	intermediate 75 mm	dry 0 mm
Semi-early, 12 Nov 1985			
NESMA	45.3	31.7	22.0
POTAM	51.6	26.7	22.3
Semi-late, 26 Dec 1985			
NESMA	36.9	24.6	14.0
POTAM	48.9	39.2	25.2

* Rainfall = 276 mm

Supplemental Irrigation in 3 Districts

Berrechid District, Settat Province

Berrechid District is in Settat Province 40 km south of Casablanca within the important agricultural zone of Chaouia which is known nationally for cereal production. INRA of Morocco in cooperation with MIAC is carrying out an important research program of "dryland farming" within this zone.

Total population is 163,000 with 288 villages. Total cultivable land is 203,000 ha: 193,000 ha are cultivated under rainfed conditions and 10,000 ha have supplemental irrigation.

Average annual rainfall recorded for the last 30 yr at this site is 400 mm. The minimum annual rainfall of 180 mm was recorded during the 1980-81 season and the maximum annual rainfall of 610 mm was recorded in 1970-71. Average minimum temperature is 10°C and average maximum temperature is 23°C.

Groundwater is the main supply for irrigation and is pumped from 3,000 wells. In some locations such as Soualem, a seasonal decline in pumping capacity occurs during the summer. Depth to water table is 15-40 m.

Most land is privately owned and the size of holdings is 10-50 ha with rainfed farming and more than 10 ha with supplemental irrigation. Soils are of medium depth to deep and heavy textured (45% vertisols). Existing land use under rainfed conditions and with supplemental irrigation is shown in Table 24.5.

Table 24.5. Land use, Berrechid District.

	Land use (%)			
Crop	Rainfed		Supplemental irrigation	
Cereal	(wheat, barley, maize)	65-70	(wheat, maize)	30
Legume	(peas, faba beans)	10-12		
Forage	(barley, Dats, V.A.)	8-10	(alfalfa, berseem)	12
Vegetable				88
Fallow		4- 5		
Cropping intensity		95		130

The principal crop rotation practiced by 45% of the farmers in rainfed farming is a 2-course rotation of cereals – legumes; 25% use a 3-course rotation of fallow – wheat – barley; and, 20% use a 4-course rotation, fallow – wheat – forage – barley. With supplemental irrigation, a 2-course rotation is used of cereals – vegetables. High cropping intensity with supplemental irrigation increases agricultural inputs. Although most farmers use some fertilizers and pesticides with supplemental irrigation, a more intensive study should be done of phosphorus and nitrogen as well as pesticides and herbicides to isolate criteria for more judicious use. Under rainfed conditions, only 26% of the farmers use fertilizers, pesticides and herbicides, and improved seeds.

Farmers who cultivate under rainfed conditions obtain average yields of 1.0 t/ha for wheat, 1.4 t/ha for barley, and 0.6 t/ha for maize. Farmers with supplemental irrigation systems obtain yields of 3.5-4.5 t/ha for both wheat and maize and 12.0 t/ha for dry forages.

A farmer's net income increases substantially using supplemental irrigation. Making the assumption that the farmer would obtain the average yields shown in Table 24.6, a cost benefit analysis shows that the farmer's net income could be quadrupled using supplemental irrigation. Income from livestock production of dairy and meat by-products has also increased using supplemental irrigation.

New work opportunities have been created with the introduction of supplemental irrigation. These increased needs for labor contribute towards

Table 24.6. Cost benefit analysis of different farming systems.

Farming systems	Crop	Yield (kg/ha)	Price (DH/kg)	Gross income (DH/ha)	Total inputs* (DH/ha)	Net income (DH/ha)
Berrechid District						
Rainfed	Wheat	1,000	1.8	1,800	1,200	1,600
	Barley	1,400	1.4	1,960	800	1,160
	Maize	600	1.7	1,020	600	420
Total net income						3,180
Supp. irri.	Wheat	3,500	1.8	6,300	2,580	3,600
	Maize	4,000	1.7	6,800	2,540	4,260
	Forage	8,750	0.6	5,250	1,800	3,450
Total net income						11,330
Ain Taoujdate District						
Rainfed	Wheat	1,500	2.0	3,000	1,400	1,600
	Forage	3,000	0.75	2,250	1,400	850
	Faba Bean	1,200	3.0	3,600	1,430	2,170
Total net income						4,620
Supp. irri.	Wheat	3,000	2.0	6,000	2,300	3,700
	Forage	8,000	0.75	6,000	2,300	3,700
	Potato	18,500	1.60	29,600	10,200	19,400
Total net income						26,800
Taourirt District						
Rainfed	Wheat	400	2.0	800	810	(−) 10
Supp. irri	Wheat	2,000	2.0	4,000	3,500	(+) 500

* Cost of land rent is not included in the total inputs.

diminishing migration of rural population to the urban area of Casablanca, 40 km distant.

Ain Taoujdate District

Ain Taoujdate District is in Meknes Province situated 150 km from Rabat. The total land area is 62,000 ha and is considered a favorable environment for rainfed farming of most field crops. Average annual rainfall recorded during the last 20 yr is 550 mm. Average minimum temperature is 9.5°C and the average maximum temperature is 23°C with an average annual potential evaporation of 1,500 mm (see Table 24.1). Population is 65,000 living in 78 villages scattered throughout the district.

Of the total cultivable land of 42,000 ha, 38,000 ha are rainfed with the balance under irrigation. Irrigation water is mainly surface flow supplied from important sources (Bittit, Sebaa Aioum, etc.) where the supply is year around. Depth to water table is 20-50 m but with extension of irrigated land, increasing cropping intensity, and farmers digging private wells to satisfy crop water requirements, a seasonal decline in pumping capacity occurs in August. Surface irrigation is the major method of irrigation: farmers use surface flooding for cereals and forages, furrows for vegetables, and basins for tree crops.

Land is privately owned by 78% of the farmers. The size of holdings is 10 ha for rainfed farming and 5-8 ha with supplemental irrigation systems. Soils are 50-100 cm deep and heavy to moderately light textured. Large programs of removing stones from fields have been carried out and 4,000 ha have been treated in this manner. Existing land use under rainfed and with supplemental irrigation systems is presented in Table 24.7.

Table 24.7. Land use, Ain Taoujdate district.

Crop	Land use (%)	
	Rainfed	Suppl. Irri.
Cereals (wheat, barley)	70-75	0.5
Dry forage crop	10-15	
Forage crop		20
Legumes (faba beans, lentil, chickpea)	8-10	
Vegetables (potato, onion, etc.)		50
Tree crops (olive, apple)		50
Fallow	5	
Cropping intensity	95%	120%

The principal crop rotation practiced in the rainfed areas is a 2-course rotation of cereals – forages (legumes). Under supplemental irrigation, a 3-course rotation is used of cereals – legumes – onions (potatoes). In general, cereal sown in irrigated areas is not irrigated; but last season, 20 ha received 2-3 irrigations. Farmers hesitate to irrigate cereals even when water is available because this crop can be affected negatively when intense storms follow immediately after irrigation. The problem of timing supplemental irrigation to coincide with growth stage was stated as a concern by farmers because they believed water had no effect at flowering but when applied during the milk-stage, yields of wheat could be doubled. Research data to support recommendations of scheduling of supplemental irrigations are needed as timing of application relates to status of soil-water and crop growth stage.

Most farmers in both farming systems use the agricultural inputs of fertilizers, improved seeds, pesticides and herbicides but still do not utilize these inputs efficiently. More extension activity should focus on which agronomic factors will increase cereal production. Instruction is needed regarding the impact and alternative practices of soil tillage, sowing date, crop variety, fertilizer (types and rates), and controlled use of pesticides and herbicides.

Introduction of supplemental irrigation into the farming systems of this district has increased the farmer's net income. Cost benefit analysis for major crops grown with rainfed farming and with supplemental irrigation shows that net income could be increased even more, up to 6 times (see Table 24.6)

Taourirt District

The Taourirt District is in Oujda Province in the east with a population of 90,200, 122 villages, and a total land area of 662,000 ha. Only 14% of this land is cultivable, the remaining 86% is desert shrub, steppe, and natural pasture. The main agricultural activity depends on an extensive farming system of livestock production (mainly sheep). The principal source of animal feed is grazing vegetation of the steppe and natural pastures. Average annual rainfall is 200-275 mm but wide inter-annual disparities are observed (a scant 74 mm in 1978-79 and 530 mm in 1972-73, double the average annual rainfall). Average annual minimum temperature is 10°C and the average annual maximum temperature is 24°C; but, higher temperatures can occur during summer with average maximum temperatures in July and August of 35°C. Potentially high rates of irrigation might be expected to occur in this area.

Of the 62,000 ha of cultivable land, 4,000 ha of vegetables, olives, and forages, the major crops, have conventional irrigation. Water for irrigation is surface flow supplied from wadis. Supplemental irrigation is used on 250 ha and is farmed by individual farmers who obtained loans and equipped their wells with pumping stations. To ensure good yields, they produce olives, forages, and cereals with supplemental irrigation. Development of this farming system type would be meaningful if additional financial aid and technical advice were available to farmers. In fact, these farmers are new irrigators and have experienced problems with their irrigation techniques and systems. Energy consumption is high for pumping because the equipment has not been adapted to the flow rates of local wells. This, in turn, has contributed to rising costs (to 0.15 DH/m^3) of irrigation water.

The percentage of land cropped annually depends firmly upon the amount of rainfall before sowing time. During 1986-87, 3,500 ha were cultivated to cereals; 60% to barley and 30% to wheat. No specific crop rotation is practiced. Average yields are low with 0.4 t/ha of grain; when supplemental irrigation is practiced, grain yields increase to 2.0-2.5 t/ha. Low yields with supplemental irrigation are associated with poor soil fertility.

Cost benefit analysis for wheat shows that a farmer's net income might be negative with rainfed farming; whereas, with supplemental irrigation, 500 DH/ha is the average economic gain. The reason for this low gain is the high cost of irrigation water in this district (see Table 24.6).

Potential for Development

Berrechid

Rechargeable groundwater resources in the Plains of Chaouia, the location of Berrechid District, have been estimated to equal 150 M m^3 (1980). Ten thousand hectares are irrigated in the district, of these, 3,000 ha are sown to cereals. If smaller amounts of supplemental irrigation (100-150 mm) were used for cereals then increased productivity would be possible on existing farms where vegetables are grown now. However, an evaluation of groundwater resources must be done before further extending irrigation production.

Ain Taoujdate

As mentioned previously, water resources are principally surface flow supplied from various sources. Although water is available for supplemental irrigation during winter and spring, only 430 ha of cereals and forages have been cultivated using this technique. Most farmers do not use supplemental irrigation because traditional practices designate cultivation of cereals under rainfed farming. Scheduling of supplemental irrigation to correspond to plant growth stages was stated as critical by farmers. Supplemental irrigation can be implemented in this area if farmers could be convinced that constraints of scheduling of supplemental irrigation and volume of applications would be resolved with instruction.

Taourirt

Cost of water is high and constitutes a serious constraint to supplemental irrigation agriculture. Extension staff are advising farmers of alternative choices of irrigation equipment; but in Taourirt, most farmers are new irrigators and more comprehensive instruction from extension is required describing irrigation management, soil fertilization, and pest control. Besides increasing existing cereal yields, introduction of supplemental irrigation would enhance cultivation of forage and, therefore, livestock production which is the major agricultural activity.

Had Kourt

Had Kourt is situated in a favorable rainfall zone with average annual rainfall of 450-500 mm. Total cultivable land is 100,000 ha, of which, 98,000 ha is now rainfed farming with 65% sown to cereals; 12% to legumes; and, 8% to various crops of sunflower, tobacco, and sugar beet. Cultivation during 1986-87 using supplemental irrigation included 2,000 ha divided between 1,500 ha of citrus trees and 500 ha of sugar beets. Water resources are surface flow and stored water pumped from the Wadis Sebou, Wargha, and Rdate. The possibility here of increased use of supplemental irrigation is substantial since only a small amount of water is needed for irrigation of winter crops and a supply is available from the wadis the year around.

Ait Ourir Region

Many valleys similar to Ait Ourir Valley are located along the foothills of the Middle and High Atlas Mountains. These are suitable for implementation of supplemental irrigation with a water supply provided from precipitation runoff as floods and snowmelt. Traditional networks of irrigation systems exist but rehabilitation is needed to achieve higher irrigation efficiency.

Conclusions

Many valleys at the foot of the Atlas Mountains, large land areas in the Atlantic Plains (Berrechid) and in the Gharb Region (Had Kourt), and the Plateaus of Eastern Morocco (Taourirt) bear considerable capacity for developing supplemental irrigation systems in Morocco.

Sources of water for supplemental irrigation are surface flow, in the case of the valleys, or groundwater in the case of the Atlantic Plains and the Eastern Plateaus. However, the cost of pumped water for irrigation might be considered a constraint in some of these areas.

Supplemental irrigation for cereals and forages of 75-200 mm, depending on seasonal variation and distribution of precipitation, can guarantee a yield of 3-4.5 t/ha with average annual rainfalls of 350 mm or higher.

Economic analysis of supplemental irrigation has shown that this technology sustains a favorable net outcome. Net income per hectare after implementation could be increased 4-6 times that of income gained from rainfed farming.

References

Arar, Abdullah, 1984. Report on consultancy to ICARDA on supplemental irrigation. Land and Water Division, FAO, Rome.
Rapport d'activite de programme aridoculture, 1985-86. INRA du Maroc et MIAC des U.S.A.
Rapport du seminaire sur les "Ressources en eau au Maroc", 1980. *organise* par ANAFID et AIPC, June.
Statistiques agricoles, 1983-84. DPAG, Ministere de l'Agriculture et de la Reforme Agraire.

Chapter 25

Supplemental Irrigation in Pakistan

M. RAFIQ

Geographical Characterisics

Pakistan has a population of 86 M living in an area of 834,000 km² and is situated between longitudes 60°-76° east and latitudes 24°-37° north. There are 3 physiographic regions: mountains, Indus River Plain, and Sind Plain. The mountains are in the west and the north at altitudes of 1,500-3,000 m intersected by large valleys with steep to very steep, rocky slopes. In the north, the slopes become steeper with narrower valleys. The Indus Plain is 1,600 km long and 150-300 km wide and has a level surface with minor local relief. The region of the Sind Plain extends eastwards over an extensive area as well as the interfluve of the Indus and Jhelum Rivers. The surface is undulating with sand ridges.

Climate

The plains have hot subtropical continental weather; whereas, the mountains benefit from influences of the highland's cooler sub-tropical environment. These principal characteristics can be categorized (on the basis of precipitation) into arid, semi-arid, sub-humid, and humid climates. Precipitation occurs in 2 distinct seasons, one extending from July to September and the other from January to April. Summer rains, caused by monsoon currents, are of high intensities (100-160 mm/hr) whereas, winter rainfall, the result of Mediterranean disturbances, has a lower intensity of 10-20 mm/hr. Only the northern regions of the plains and the mountains have sub-humid or humid climates which are acceptable for rainfed farming.

Geology and Hydrogeology

The mountains, north of Peshawar, are part of the Himalayan and Karakoram Ranges which are the highest in the world with peaks rising above 7,000 m. Uplifted during the Tertiary and Pleistocene periods, these ranges are formed

E.R. Perrier and A.B. Salkini (eds), Supplemental Irrigation in the Near East and North Africa,
463-496.
© 1991 *ICARDA.*

of granites, micaceous or siliceous schists, gneisses, shales, and slates. Valleys are narrow and rainfall runoff as surface flow from these is the major water supply for irrigation.

Western mountains are principally limestone with some sandstones and shales. The gravelly alluvium of the intermontane valleys contains groundwater for irrigation even though the water occurs in small pockets and in limited quantities. Arable land is not limited but water is; consequently, water scarcity is a serious problem in the western region.

The Indus Plain is a basin filled with Pleistocene and Recent alluvium to a thickness of more than 300 m, underlain by rocks of mostly Pre-Cambrian Age. The alluvium is of silty, loamy, or clayey materials to depths of a few meters, forming an unconfined, extensive aquifer composed of sandy materials. Quality of the groundwater varies and is suitable for irrigation in only 40% of the area.

In the Pothwar area (Rawalpindi, Jhelum, Chakwal, and Attock), there is a thin covering of loess material overlaying tilted beds of shale, sandstone, and some limestone. Groundwater occurs in small pockets in the floodplains of rivers and streams.

Land Resources

Out of the total land area of 194.64 M acres, land use statistics are available for 143.61 M acres (Table 25.1).

Table 25.1. Land utilization statistics (1985).

Category	M acres
Geographical area	196.64
Total area reported	143.64
Forest	7.81
Not available for cultivation	57.43
Cultivable waste (arable)	27.64
Cultivated area	50.73
Fallow land	12.65
Net area sown to crops	38.09
Area sown more than once	10.79
Total cropped area	48.88

Source: Agric. Statistics of Pakistan, 1985.

These statistics show that of 50.73 M acres in cultivation only 38.09 M acres are planted annually to crops and 12.65 M acres remain in fallow primarily because of insufficient water for production. On 10.79 M acres, more than one crop is produced during a year for a cropping intensity of 97% although climate warrants production throughout the year. Even with irrigation, cropping intensity is low, 100-120%, due to water shortages. Water scarcity is clearly evident.

Water Resources of the Indus Plain

Important water resources data are presented in Table 25.2. In the Indus Plain, 96 M acre ft of water, both surface flow and groundwater, are available at the farm gate and about 47 M acre ft could be added to this volume if development potential were realized. Currently, irrigated land is 38.5 M acres but 12.3 M acres are cultivated without irrigation. Three million of these 12.3 M acres could be furnished with facilities for irrigation and with optimal development of available water resources, the cropping intensity could be increased from 90-110% to 135-140%.

Table 25.2. Water resources in the Indus Plain.

Source of supply	Volume of resource (M acre feet, Maf)
Existing resources	
Total available surface flow	
(Indus, Jhelum, Chenab rivers)	137.27
Water diverted into irrigation canals	101.00
Groundwater pumped through tubewells	40.8
Canal water at farm gate	58.94
Tubewell water at farm gate	37.04
Potential for development of resources	
Within the Indus Plain	
Surface water at farm gate	12.0
Groundwater at farm gate	19.0
Through water course improvements	12.0
Through minor canal lining	4.0
Outside of the Indus plain	
Groundwater potential	1.4

Environment and Cropping Systems

Existing Supplemental Irrigation

Supplemental irrigation is practiced in 3 agro-ecological zones, with groundwater and, in a few instances, surface water. Methods of supplemental irrigation can be classified into 3 systems.
1. In the Gujrat zone, major crops of rice and wheat, but also some maize, clover, and sorghum, are irrigated with groundwater which is pumped from depths of 10-30 m by tubewells. With supplemental irrigation, cropping intensity increases to 150-200% and a 50% yield increase occurs. Under rainfed farming with a cropping intensity of 100%, major crops are wheat and sorghum or millet grown in rotation with 1 year fallow.
2. In the Rawalpindi and Talagang zones, the major crops, wheat and millet,

are grown under rainfed conditions but with supplemental irrigation maize replaces millet. North of Attock, tobacco is an important crop. Using supplemental irrigation, cropping intensity increases to 150-200% from the 100% of rainfed farming and crop yields increase by 50% and stabilize.
3. In the Rawalpindi zone, surface water is diverted from the Dor and Haro Rivers for supplemental irrigation. With supplemental irrigation, the main crops are maize, wheat, and tobacco with wheat and sorghum with rainfed cultivation.

In all 3 systems, supplemental irrigation generates changes in cropping patterns in favor of cultivating, in summer, the more remunerative crops of rice, maize, and tobacco to replace millet and sorghum of rainfed farming. This increases cropping intensities, improves yields, and the gross and net production per unit of land multiplies many fold. Inputs such as fertilizer, seed, and improved crop varieties proliferate and the trend is towards mechanization of agricultural operations.

In the Gujrat zone which possesses an unconfined aquifer containing good quality water, there is still ample opportunity for improvement in productivity. In other zones, conditions are not so favorable because groundwater is either unavailable or is at a depth considered uneconomical for exploitation. Developing mini-dams or farm ponds is a more feasible means of storing rainfall runoff for later application with supplemental irrigation.

In the cultivated areas of rainfed farming, expansion of supplemental irrigation systems, where feasible, is among the priorities of the national agricultural policy. The government provides 24,000 Pakistani Rupees as subsidy for installation of a tubewell at feasible sites. Mini-dams or farm ponds, also, receive a subsidy of 60-70% of the cost of construction. A mini-dam usually can supply water to irrigate 1-2 ha of land. Small dams are constructed at government cost to provide water for supplemental irrigation with farmers asked to pay a nominal charge only for water.

Potential Zones for Supplemental Irrigation

According to an FAO sponsored study, Pakistan can be divided into 17 agro-ecological zones (Rafiq, 1976). Three zones (9, 10, 11) have characteristics that meet the criteria for selection of areas feasible for implementation of supplemental irrigation: Gujrat, Rawalpindi, and Talagang (Figure 25.1).
9. *Gujrat zone* has a sub-humid subtropical continental climate with average annual rainfall of 500-900 mm. Crops are produced both by supplemental irrigation agriculture and rainfed farming systems.
10. *Rawalpindi zone* has similarities to zone 9, Gujrat, except that zone 10 has a higher proportion of severely eroded lands. Crop production is rainfed.
11. *Talagang zone*, has an abundance of eroded land, but a climate which is semi-arid subtropical with continental effect. Production depends predominantly on rainfall (annually 300-500 mm) for crop water requirements.

Supplemental Irrigation in Pakistan 467

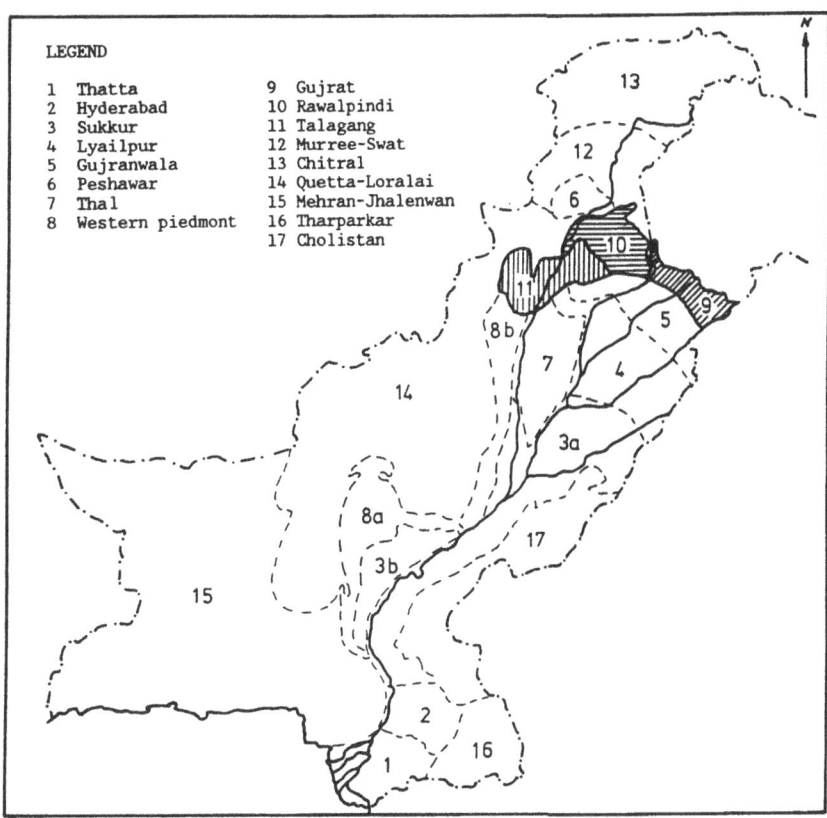

Figure 25.1. Map of Pakistan showing 3 agro-ecological zones where supplemental irrigation is feasible.

Gujrat zone: With management, climate and rainfall is congenial to crop production. Mean maximum temperature in June is 40.6°C and the mean minimum is 26.8°C. In January, the mean maximum temperature is 18.7°C and the mean minimum is 5.7°C. Mean annual potential evapotranspiration is 2,110 mm. Climatic data are given in Table 25.3. Soils are predominantly brown silt loams, loams and clay loams, with calcareous or non-calcareous soils of weak structure and brown, calcareous loamy soils without structure. All soils are deep: more than 100 cm (Eutric Cambisols, Chromic Vertisols, and Calcaric Fluvisols), (Table 25.4) The natural vegetation is open forests with shrubs and grasses. Indicative species are: *Dalbergia sissoo, Acacia nilotica, Acacia modesta, Zizyphus mauritiana, Cymbopogon jawarancusa*, and *Eleusine compressa*. Groundwater is usable throughout the area (Punjab P & D, 1987); but, in 50% of the zone, it is deeper than 20 m therefore requires turbine pumps for exploitation. The cost of water is not profitable for production with existing technology. The water table in the remaining 50% of the zone is shallow enough to use centrifugal pumps. Water quality in some areas is affected by problems of sodicity because of high RSC (residual sodium carbonate) or high SAR (sodium adsorption ratio).

Table 25.3. Climatic data for Gujrat zone: Sialkot and Gujranwala.

Location/ month	Mean temperature (°C)				Mean no. of rain days	Rainfall (mm)	
	Daily		Extreme			Heaviest in 24 hr	Mean monthly
	Max	Min	Max	Min			
Sialkot[a]							
Jan	18.7	5.7	26.1	−2.2	2.9	114	50
Feb	20.8	7.5	31.1	−2.8	3.0	60	42
Mar	26.7	12.3	40.0	1.7	3.2	45	38
Apr	33.6	18.2	45.0	8.3	2.1	77	25
May	39.0	23.5	46.7	12.8	2.0	64	25
Jun	40.6	26.8	48.3	18.3	3.5	163	61
Jul	36.3	26.4	48.3	18.9	9.0	199	212
Aug	34.6	25.7	43.3	18.9	9.0	254	236
Sep	23.7	26.0	42.2	15.0	3.8	169	87
Oct	33.0	16.7	40.0	7.2	0.8	82	9
Nov	27.0	9.5	35.0	1.1	0.4	52	4
Dec	20.8	5.7	27.8	−1.7	1.3	70	18
Annual	30.5	16.8	48.3	−2.8	41.0	254	
Total annual rainfall							808

Gujranwala[b]		
Month	Mean monthly rainfall (mm)	Evaporation (mm)
Jan	33	66
Feb	31	77
Mar	29	135
Apr	18	205
May	20	334
Jun	46	375
Jul	155	308
Aug	166	241
Sep	65	228
Oct	8	188
Nov	5	124
Dec	15	82
Total	589	2,110

[a] Temperature, rainfall, and ave. no. rainy days (days with more than 0.1 mm of rain) data from 1896-1940; heaviest rain in 24 hr, 1896-1964.
[b] Rainfall data, 1941-60; evaporation from a 48 inch diameter pan, data 1962-64, WASID, WAPDA.

Source: Reconnaissance Soil Survey Report, Gujranwala Area, Soil Survey of Pakistan, Lahore, 1965.

Table 25.4. Extent of different soils in selected agro-ecological zones.

Zone soil type		1,000 acres
Gujrat		
Structured,	level to nearly level,	
	calcareous, loamy	205.0
	calcareous, moderately fine texture	143.3
	non-calcareous, loamy	318.6
	non-calcareous, moderately fine texture	251.9
	non-calcareous, very clayey	187.7
Structureless, level to nearly level,		
	calcareous, loamy	392.7
Other minor soils		427.4
Total		1,926.6
Rawalpindi		
Structured,	level to nearly level	
	calcareous, moderately fine texture	84.0
	calcareous, fine texture	79.0
	non-calcareous, moderately fine texture	135.9
Structured,	level to nearly level, coarse surface	
	calcareous, moderately fine texture	74.1
Structureless, gently sloping, calcareous, loamy		368.0
Other minor soils		565.6
Total		1,306.6
Talagang		
Structured,	level to nearly level,	
	calcareous, loamy	298.9
	calcareous, moderately fine texture	155.6
	calcareous, fine texture	316.2
	calcareous, moderately coarse texture	130.9
Structureless, level to nearly level,		
	calcareous, coarse texture	182.8
Structureless, gently sloping,		
	calcareous, coarse texture	177.8
Structureless, shallow, gently sloping,		
	calcareous, coarse texture	98.8
Other minor soils		758.3
Total		2,119.3

Rawalpindi zone: Climate and rainfall could support crop production. In June, the mean maximum temperature is 39.7°C and the mean minimum is 24.4°C. In January, the mean maximum temperature is 16.3°C and the mean minimum is 2.7°C. Mean annual potential evapotranspiration is 1,528 mm. The climatic data are presented in Table 25.5. Soils are of varying types but, principally, they are brown/dark brown, non-calcareous, silt loams, loams, silty clay loams, and clays with weak structures; brown, calcareous silt loams and silty clay loams with structure; and brown structureless, calcareous loams and silt loams. All soils are deep, more than 100 cm (Eutric Cambisols, Chromic Vertisols, and Calcaric Regosols, see Table 25.4). Open forests with tall grasses is the natural vegetation. The indicative species are: *Dalbergia sissoo, Acacia milotica, Acacia modesta, Olea cuspidata, Zizyphus nummularia, Dodones viscosa, Cymbopogon montanus, Heteropogon contortus*, and *Digitaria decumbens*. Only 10% of the zone has groundwater which could be used for supplemental irrigation but only half of this available reserve is in sufficient quantities and at shallow enough depths for easy exploitation (Punjab P & D, 1987). Available resources could deliver water for supplemental irrigation to only 3% of the

Table 25.5. Climatic data for Rawalpindi zone: Jhelum and Haripur.

Location/ month	Mean temperature (°C)				No. of rain days	Rainfall (mm)	
	Mean Daily		Extreme			Heaviest in 24 hr	Total monthly
	Max	Min	Max	Min			
Haripur							
Jan	16.3	3.9	23.3	0.6	4	41	66
Feb	19.4	6.4	28.3	1.7	3	28	38
Mar	24.4	10.4	33.3	2.2	6	46	65
Apr	29.2	14.9	39.4	8.3	4	28	41
May	35.7	19.4	44.4	8.9	1	56	23
Jun	40.7	24.7	46.7	16.1	1	20	6
Jul	37.2	25.0	46.1	20.0	7	175	164
Aug	34.6	24.0	40.6	17.8	8	84	187
Sep	34.1	21.7	38.9	16.1	4	58	106
Oct	30.7	15.8	36.7	10.6	1	46	25
Nov	24.9	9.4	32.2	1.7	1	43	17
Dec	19.8	5.8	25.6	−1.1	2	203	34
Total annual rainfall							771
Jhelum							
Jan	18.6	5.2	25.0	−1.7	4	58	54
Feb	22.0	7.3	33.9	1.1	3	76	34
Mar	26.8	12.6	35.6	3.9	4	79	66
Apr	33.9	17.4	43.3	8.9	2	23	20
May	39.2	22.7	47.2	14.4	2	48	19
Jun	41.1	26.1	49.4	17.8	3	53	39
Jul	36.4	26.2	45.6	21.1	9	168	248
Aug	34.4	25.2	41.7	19.4	10	114	241
Sep	35.2	22.9	40.6	16.7	4	91	106
Oct	32.9	16.4	38.9	8.9	1	107	25
Nov	27.3	8.6	33.9	0.6	1	66	14
Dec	21.1	5.1	27.2	−2.8	2	46	24
Total annual rainfall							891

* A rain day is one with more than 0.1 mm of rain.

Source: Reconnaissance Soil Survey Report, Rawalpindi Area. Soil Survey of Pakistan, Lahore. 1967.

cultivated land. Twenty small dams and 172 mini-dams were constructed to store water for supplemental irrigation of 21,300 acres. Sites for 32 small dams await funding and 170 eligible sites are practical for small dams (Punjab P & D, 1987). Three percent of the cultivated land could use supplemental irrigation facilities in conjunction with small dams. Mini-dams or farm ponds could be constructed on innumerable sites to store rainwater which then could be used for supplemental irrigation of an additional 2-5% of the total cultivated land (Punjab Government, 1986).

Talagang zone: Climate and rainfall can support production with resource conservation and management. Mean maximum temperature in June is 40.8°C with a mean minimum of 17.0°C; and, in January, a mean maximum is 18.6°C and the mean minimum is 5.1°C. The mean annual evapotranspiration is 1,345 mm. Climate data are presented in Table 25.6. The soils are mainly brown, calcareous sandy loams, silt loams, loams, silty clay loams and silty clays with weak structures; and also, some gently sloping, brown, calcareous, very sandy soils both deep and shallow. All soils are more than 100 cm deep (Haplic Yermosols and Calcaric Regosols, see Table 25.4). There are scattered trees and bushes with open stands of grass. The indicative species are: *Acacia modesta, Acacia nilotica, Prosopis cineraria, Zizyphus mauritiana, Zizyphus nummularia, Capparis decidua, Rhazya stricta, Cymbopogon jawarancusa,* and *Chrysopogon aucheri.* Three percent of the land has available groundwater resources occurring along the Soan River which could be allocated for supplemental irrigation. In the Bannu Basin, some land is served by perennial canals.

Existing Rainfed Farming Systems

Cropping patterns

Gujrat zone: Blocks of fields are sown to crops while other blocks are left fallow for a year. The rotation practiced produces a winter crop which is followed immediately by a summer crop, then, fallow for 1 yr. During winter, 60-70% of the cropped land is sown to wheat; 20-25% to a wheat and chickpea intercropping; 10% is to lentil; 10% to fodder; and 3-5% is sown to mustard. In summer, pearl millet is the major crop grown on 70% of the cropped land whereas mung bean is sown on 15% and fodder on 7%. Some maize and cotton are also grown. Of the total cultivated land, 55-60% is cropped in winter and 40-45% is cropped in summer giving a cropping intensity of 95-105%. Often mustard is intercropped with wheat and pulled out green for fodder. Crop residues of wheat straw and millet stalks are used as dry matter for animal feed. Also, mustard and turnip can be cropped separately as fodder crops.

Wheat is sown after 1 yr of fallow. Fields are cultivated after each rain during

Table 25.6. Climatic data for Talagang zone: Bannu and Mianwal.

Month	Temperature (daily)			Rainfall (mm)	Mean rainy days	% Relative humidity		
	Max (°C)	Min (°C)	Mean (°C)			5AM	8AM	5PM*
Bannu								
Jan	18.6	5.1	11.8	22	2.1	73	64	45
Feb	20.7	8.2	14.4	30	2.9	67	63	37
Mar	25.2	12.8	19.1	46	4.0	70	59	37
Apr	31.4	18.0	24.7	31	2.7	59	50	29
May	38.1	23.4	30.7	14	1.8	50	39	25
Jun	40.8	27.0	33.9	21	1.2	50	43	24
Jul	38.8	27.9	33.4	69	3.7	73	62	43
Aug	37.3	27.2	32.3	55	3.4	77	67	47
Sep	36.4	24.1	30.3	18	1.1	73	62	41
Oct	32.4	17.5	24.9	3	0.3	65	53	35
Nov	26.6	10.6	18.6	3	0.4	68	54	41
Dec	21.0	6.2	13.6	11	1.0	75	62	46
Total Annual	30.6	17.3	23.9	327	24.6	67	57	37

Month	Rainfall (mm)	Temperature		V.P. (mb)	Wind (km/day)	Sunshine (hr)	Energy (cal/cm²/day)	PET (mm/day)
		Min (°C)	Max (°C)					
Mianwali								
Jan	16	4.6	18.7	8.5	31	6.4	263	1.1
Feb	20	6.7	21.3	9.7	39	5.9	342	1.9
Mar	53	12.2	26.1	12.7	47	5.3	420	3.0
Apr	33	17.2	32.1	14.9	50	5.2	525	4.8
May	22	22.3	37.7	15.2	63	4.9	586	6.6
Jun	20	26.4	41.4	19.9	61	4.5	538	6.6
Jul	83	27.5	38.8	28.7	65	4.3	501	5.8
Aug	118	26.5	36.7	29.4	57	5.1	490	5.1
Sep	50	23.4	35.1	24.1	42	5.7	446	4.3
Oct	10	16.5	32.8	16.3	34	6.9	398	3.3
Nov	5	9.6	26.0	11.5	26	7.8	317	1.7
Dec	8	5.1	20.8	9.0	25	7.6	261	1.1
Total 437 Annual Average		16.5	30.6	16.7	45	5.8	424	3.8

* West Pakistan Standard Time

Source: Robertson, G.W., 1985. Rainfall probabilities in Mianwali Area. Pakistan Agric. Res. Council, Islamabad.

the summer preceding the season of production in order to conserve moisture and control weeds. Tillage operations are performed 5-8 times during July through September. Wheat is sown from the end of October to the end of November with a 1-row drill drawn by bullock or a seed drill worked with a tractor. For the summer crop, land is cultivated after the first rain in July and the crop sown immediately.

Farmyard manure is applied to some fields near the village which are used for fodder, maize, or wheat. However, if farmers, in addition to rainfed fields, own land with irrigation, they will apply the farmyard manure to these fields. Under rainfed farming, chemical fertilizers are becoming common with wheat receiving 18 kg of nitrogen and 12 kg of phosphorous per acre. A few farmers (33%) use nitrogen only.

Rawalpindi zone: The crop rotation begins with a wheat crop sown after a year of fallow. The fallow land is tilled after every rain storm during the summer of fallow to control weeds and conserve moisture. A total of 5-10 cultivations are done to prepare the land for sowing of wheat beginning the end of October and continuing until the end of November. A tine-cultivator drawn by a tractor or a single-row cultivator worked with bullocks is used for these operations. Seeding, at a rate of 30-40 kg/acre, is with a multi-row drill worked with a tractor or with a single-row drill drawn by a pair of bullocks. New dwarf varieties are produced by 50% of the farmers while the others cultivate the tall cultivars.

Wheat is harvested in May with the crop stubble grazed. In July after the first rain, land is cultivated and sown to pearl millet or sorghum which is harvested in October. Twelve months of fallow succeed these two crops. The purpose of fallowing land is to restore soil fertility and conserve summer rainfall in the soil profile for the wheat's germination and emergence with reliance on the winter rains for the crop's maturation.

Lentil and mustard are minor crops in winter and groundnut, maize, and mungbean are minor crops in summer. The growing of lentils during fallow year as a cash crop is gaining in importance and the land area sown to this crop has increased dramatically in recent years.

Fallow fields are communal lands grazed by the livestock of the total village. This practice of communal grazing of fallow explains the layout of the village's fields into 2 blocks: one is in fallow and a second is sown to crops.

Wheat is commonly intercropped with mustard which is pulled green to feed animals; in fact, rarely is fodder mustard grown in a pure stand. During summer, pearl millet or sorghum is sown in thick stands and thinning is done as the crop grows to provide additional green fodder.

Forty to fifty percent of the land is non-arable but is common grazing for village livestock. Grass grows abundantly on these gullied or rocky lands during the spring months and the monsoon season of July to September. This grass is freely grazed, decreasing fodder requirements from farm fields at this time.

Talagang zone: Wheat is the major crop sown in the same fields year after year. Systematic crop rotation is not practiced: usually, land remains fallow during the preceding summer or for 18 months preceding the wheat crop depending upon the volume of summer rainfall. Chickpea is an important crop in some parts, especially on sandy loam soils. Groundnut and mustard are minor crops.

Mustard is intercropped with wheat or chickpea and is pulled green for use as green fodder. Land sown to groundnut is increasing.

Preparation of land for wheat begins in July after the first summer rain. During July, August, and September, fields are tilled after each rain with a cultivator to control weeds and conserve moisture. A wooden plank is then dragged over the land after each cultivation to break clods, compact the soil, and smooth the surface. Crops are sown with seed drills of either a multi-row drawn by tractors or a single-row pulled by bullocks.

Cropping intensity is 90-100%. Wheat and chickpeas are sown on 50-70% of the cultivated land during winter (percentage seeded depends on amount of rainfall); and, pearl millet is produced in summer on 25-30% of the area. Little fertilizer is applied on either wheat or millet. The switch from wheat – chickpea to pearl millet in a field is made after a year of fallow. Wheat crops can fail once in 4-5 yr because of irregular or insufficient rainfall. Millet often fails to produce grain and is planted as animal fodder more than for a grain crop.

Fields in fallow and uncultivated land (40% of the land) are available for common grazing of village livestock. Cultivated land is divided into blocks: each complete block is either sown to a crop or left in fallow for livestock grazing. Crop stubbles are grazed and the dry stalks of pearl millet and wheat straw are used as dry fodder. During winter, mustard intercropped with wheat is pulled for green animal feed. In summer, pearl millet, seeded in thick stands, is thinned and these plants are also used for green fodder. Through the monsoon season of August and September and in spring, grass grows well in the uncultivated land and furnishes grazing. Supplemental feeding of livestock in stalls is rarely required.

Manure and fertilizer use

Rawalpindi zone: Traditionally, only the land in the environs of the village compound receives farmyard manure and is intensively cropped. Each farmer has a plot of such land and grows fodder or maize. The fields away from the compound seldom receive any manure. However, the situation has changed dramatically during the last 10-15 yr since chemical fertilizer was introduced and adopted by the farmers. At present, 70-80% of the farmers use chemical fertilizers for wheat, applying 18 kg of nitrogen and 12 kg phosphorous per acre. However, one third of these farmers use only nitrogen. Little fertilizer is used for pearl millet or sorghum because they are of secondary concern to wheat.

Talagang zone: Only a small percentage of land adjacent to the village compound receives farmyard manure to grow wheat and fodder. Most crops receive no manure or fertilizer. However, land is fallow for 1-3 seasons to restore fertility to the soil as well as to conserve moisture. Chemical fertilizers are not economic due to the low and uncertain rainfall.

Farm power

Gujrat zone: Tillage operations are accomplished either with animal traction (bullocks) or tractors. Sixty percent of the farmers do tillage with tractors. Most farmers sow wheat, chickpeas, and millet with bullocks with only 30-40% of the farmers using seed drills drawn behind a tractor. Harvesting is done manually but the threshing of wheat is with mechanical threshers operated with tractors (Table 25.7).

Table 25.7. Farm power.

District/ zone	Type of power	No. of farms	No. of fragmented parcels		
			< 1	2.5–5	25–50 (acres)
Sialkot/Gujrat	Bullocks	41	15	36	29
	Tractors	36	74	45	36
	Bullock-Tractor	22	7	18	34
Gujrat/Gujrat	Bullocks	50	19	45	37
	Tractors	31	64	39	28
	Bullock-Tractor	18	10	15	35
Rawalpindi/Rawalpindi	Bullocks	66	74	67	53
	Tractors	16	18	18	13
	Bullock-Tractor	16	4	14	32
Jhelum/Talagang	Bullocks	56	29	53	58
	Tractors	14	37	20	8
	Bullock-Tractor	25	2	20	31
Attock/Talagang	Bullocks	48	21	43	49
	Tractors	21	58	31	5
	Bullock-Tractor	31	20	25	46

Source: Pakistan Census of Agric., 1980. Agric. Census Org., Gov't. of Pakistan, Lahore.

Rawalpindi zone: Tractors are becoming more common with 40% of the farmers having operated tractors, and many farmers hire equipment from their neighbors. Most tillage operations are mechanized but some farmers still use bullocks for tillage and other farm operations. Sowing of crops on 40-50% of the cropped land is done with tractors; on the remainder, sowing is with bullock-drawn single-row drills. With inappreciable exception, wheat harvesting is done with manual labor. Threshing of wheat is totally mechanized using threshers powered by tractors. Manual labor is used for threshing pearl millet, sorghum, maize, and mungbean. Groundnut is dug, manually, by family members (see Table 25.7).

Talagang zone: Tractors have become common during the last two decades. In 1980, 44% of the farmers reported having used tractors; by now, this value would be 55-60%. Farmers who cultivate 5.0 acres or less more often use tractors than farmers with more land; but, most hire the equipment. Half of these farmers who use tractors depend solely on this equipment, whereas others use bullocks also. Most tillage operations are mechanized and sowing is done with tractors on 50% of the cropped land but harvesting is done manually. Threshing depends on threshers powered by tractors, except for pearl millet which is threshed using manual labor. Groundnuts are dug, manually, by family members (see Table 25.7).

Table 25.8. Livestock production.

Livestock farm type	No. of farms with livestock				
	Gujrat 128,279	Sialkot 141,762	Rawalpindi 131,972	Jhelum 81,704	Attock 101,347
Cattle					
% farms	72	51	78	84	87
Ave. no./farm	2.8	2.6	3.4	4.1	4.7
Ave. no./non-farm	2.1	2.0	2.1	2.3	2.4
Buffalo					
% farms	87	88	40	38	23
Ave. no./farm	3.4	3.2	1.9	2.2	2.0
Ave. no./non-farm	2.1	2.0	2.2	2.0	2.0
Goats					
% farms	45	23	56	62	67
Ave. no./farm	3.0	2.5	3.6	4.8	4.8
Ave. no./non-farm	3.3	2.7	3.0	4.3	3.6
Sheep					
% farms	33	7	5	12	27
Ave. no./farm	8.5	3.6	7.5	9.1	7.4
Ave. no./non-farm	7.5	5.0	3.7	11.1	4.7
Donkeys					
% farms	35	8	23	47	46
Ave. no./farm	1.1	1.1	1.1	1.1	1.2
Ave. no./non-farm	1.4	2.4	1.3	1.4	1.5
Chicken					
% farms	60	63	75	54	70
Ave. no./farm	5.0	5.0	8.2	5.5	6.5
Ave. no./non-farm	4.4	4.0	13.9	5.3	6.3

Source: Pakistan Census of Agric., 1980. Agric. Census Org., Gov't. of Pakistan, Lahore.

Livestock

Gujrat zone: Seventy percent of the farmers have an average 2.8 cattle, ordinarily bullocks. Most farmers (90%) keep an average of 3.3 buffalos for milk; 25-50% have goats (2.5-3.0 head); and, 10-20% own sheep (3.5-8.5 head). Ten to thirty percent of the farmers own an average of 1-2 donkeys and 50-70% of the households have 4-5 chickens/household.

Households without farmland (non-farm) provide casual labor for agricultural operations but livestock is the households' major source of income. Each of these maintain an average of 2.2 head of cattle, 2 buffalos, 2.8 goats, 6.5 sheep, 1.8 donkeys, and 4.6 chickens. Their livestock graze in the fallow fields and uncultivated land of the village where grass grows abundantly during the monsoon season (July-September) and spring (Table 25.8).

Rawalpindi zone: Eighty percent of the farmers have an average 3.4 cattle/farm; 40% keep 1.9 buffaloes; goats (3.6 head/farm) are owned by 56%. Only 5% keep an average of 7.5 head of sheep. Donkeys (1.1 head) are owned by 23% of the farmers and 75% of the farmers own 8.2 chickens/farm.

Non-farm households do not cultivate land; however, they hire out as casual farm labor for other farmers but retain livestock for essential revenue. These non-farm households each maintain 2.1 cattle, 2.2 buffalo, 3.0 goats, 3.7 sheep, 1.3 donkeys, and 13.9 chickens (see Table 25.8).

Talagang zone: Unlike the other two zones, animals gain in significance because crop production is not reliable. Eighty five percent of the farmers keep 4.4 cattle/farm; 30% hold 2.1 buffalo; 64% average 4.8 goats; 20% have 8.2 sheep; 46% own 1.1 donkeys; and, 62% keep 6.0 chickens.

Family members of the non-farm households (35% of the total households) provide casual labor for other's farming operations and keep livestock as a source of livelihood. Each household has an average of 2.3 cattle, 2.0 buffalo, 4.2 goats, 7.9 sheep, 1.4 donkeys, and 10.1 chicken. (see Table 25.8).

Farm size, land tenure, and farm fragmentation

Gujrat zone: Average farm size is 8.0 acres; 40% of the farms are 5.0 acres or less (or 12% of the cropping land) and 1.5% are more than 50 acres (or 7% of the cultivated land). Table 25.9 presents data on farm size. Owners operate 58% of the farms; owner-cum-tenants operate 31%; and, tenants operate 10% (Table 25.10). Most farms (67%) are fragmented: 68% of the owner-operated farms; 87% of the farms with owner-cum-tenants; and, 47% of the tenant operated farms (Table 25.11).

Rawalpindi zone: Average farm size is 6.1 acres, although farms can vary from less than 1 to more than 150 acres. Fifty four percent of the total farm land is

Table 25.9. Number and size of farms.

Admin. district	No. of farms	Size of farm (acres)	Farms (%)	% farm/ total area	Average farm size (acres)
Gujrat	128,275				8.4
Gujrat zone		< 1.0	5	< 0.5	0.5
		1.0- 2.5	14	3	1.6
		2.5- 5.0	21	9	3.5
		5.0- 7.5	18	12	5.9
		7.5- 12.5	24	27	9.6
		12.5- 25	15	28	16.2
		25 - 50	3	13	31.9
		50 -150	1	7	64.8
		> 150	< 0.5	1	217.8
Sialkot	141,766				7.6
Gujrat zone		< 1.0	6	< 0.5	0.5
		1.0- 2.5	14	3	1.6
		2.5- 5.0	21	10	3.5
		5.0- 7.5	20	15	6.0
		7.5- 12.5	24	30	9.6
		12.5- 25	12	26	16.2
		25 - 50	2	10	32.0
		50 -150	1	5	65.8
		> 150	< 0.5	1	227.3
Rawalpindi	131,972				6.1
Rawalpindi zone		< 1.0	13	1	0.5
		1.0- 2.5	27	7	1.5
		2.5- 5.0	21	11	3.3
		5.0- 7.5	15	14	5.8
		7.5- 12.5	13	21	9.7
		12.5- 25	8	21	16.5
		25 - 50	3	14	31.2
		50 -150	1	8	69.5
		> 150	< 0.5	2	211.2
Jhelum	81,702				11.4
Talagang zone		< 1.0	3	< 0.5	0.5
		1.0- 2.5	11	2	1.7
		2.5- 5.0	17	5	3.5
		5.0- 7.5	19	10	5.9
		7.5- 12.5	24	24	9.5
		12.5- 25	16	24	16.8
		25 - 50	7	19	31.7
		50 -150	2	14	67.4
		> 150	< 0.5	6	239.8
Attack	101,347				8.4
Talagang zone		< 1.0	5	< 0.1	0.5
		1.0- 2.5	13	2	1.6
		2.5- 5.0	15	7	5.9

Table 25.9. (Continued).

Admin. district	No. of farms	Size of farm (acres)	Farms (%)	% farm/ total area	Average farm size (acres)
		5.0- 7..5	15	7	5.9
		7.5- 12.5	21	16	9.5
		12.5- 25	20	28	16.5
		25 - 50	8	22	31.7
		50 -150	2	14	69.5
		> 150	< 0.1	7	173.8
Pakistan					
National averages		< 0.5	34	8	
		5.0- 12.5	40	30	
		12.5- 25	17	26	
		25 - 50	3	17	
		50 -150	3	13	
		> 150	< 0.1	6	

Source: Pakistan Census of Agric., 1980. Agric. Census Org., Govt. of Pakistan, Lahore

Table 25.10. Type of land tenure.

District	Owner (%)	Owner-cum-tenant (%)	Tenant (%)
Gujrat			
Gujrat	64	27	15
Sialkot	52	36	12
Rawalpindi	81	14	5
Talagang			
Jhelum	68	27	5
Attock	58	27	9

Source: Master Plan for Barani (Rainfed) Area Development Project, Punjab Planning and Development Dept., Lahore. (1987).

Table 25.11. Fragmentation of farms.

District	No. of farms	Owner (%)	Owner/ tenant (%)	Tenant (%)	Average no. of fragment farm
Gujrat	80	81	93	43	5.8
Sialkot	75	66	92	61	3.9
Rawalpindi	80	81	97	54	6.8
Jhelum	90	88	98	81	6.0
Attock	77	71	93	65	3.7

Source: Master Plan for Barani (Rainfed) Area Development Project, Punjab Planning and Development Dept., Lahore, (1987).

partitioned into 88% of the farms: 19% of the land comprises farm units of 5 acres or smaller but which encompass 60% of all farms; whereas, 35% of the farm land is divided into farms of 5-12.5 acres or 28% of total farms (see Table 25.9). Eighty one percent of the farms are operated by the landowners, 14% by owners-cum-tenants, and 5% by tenants (see Table 25.10). Most farms (80%) are fragmented into an average of 6.8 fragments/farm. Eighty one percent of the owner-operated farms are fragmented; 97% of the farms operated by owner-cum-tenants; and, 54% of the tenant operated farms (see Table 25.11).

Talagang zone: Average farm size is 11.7 acres but size can vary from less than 1 acre to more than 200 acres. Seventy five percent of the farms are less than 12.5 acres and comprise 42.5% of the total farm land. Only 2.5% of the farms are larger than 50 acres but they comprise 20% of the total cropping land (see Table 25.9). Owners operate 63% of the farms; owner-cum-tenants operate 27%; and, tenants operate 7% (see Table 25.10). Most farms (83%) are fragmented into an average of 4.8 parcels/farm: 79% of the owner-operated farms; 95% of the farms with owners-cum-tenants; and, 73% of the tenant farms (see Table 25.11).

Farm income

Data on farm income is for rainfed farming areas of the Punjab Province but for all 3 zones the story is similar (Table 25.12). An average income for an owner farmer is P. Rs. 7,090/annum of which only 10% is received

Table 25.12. Farm income per household.

Farming system	District	Land tenure	Net farm income (P. Rs)	Source of revenue % total income
Rainfed	Attock	Owner	7,090	10% farm; 90% other
	Rawalpindi Jhelum	Owner/tenant	−3,320	54% loss from farm; non-farm pays loss.
		Tenant	−1,360	56% loss from farm; non-farm pays loss.
Rainfed/Irrigated	Gujrat	Owner	11,820	66% farm; 33% other
	Sialkot	Owner/tenant	13,540	90% farm; 10% other
	Jhang	Tenant	5,900	80% farm; 20% other
Irrigated	All districts	Owner	14,920	88% farm; 12% other
		Owner/tenant	16,940	90% farm; 10% other
		Tenant	8,940	78% farm; 22% other

Source: Master Plan for Barani (Rainfed) Area Development Project. Punjab Planning and Development Dept., Lahore.

directly from the farm. An owner-cum-tenant allocates 54% of non-farm income to assuage farm related losses of P. Rs. 3,320/annum. Tenant farmer allots 56% of non-farm income to satisfy farming losses of P. Rs. 1,360/annum.

Supplemental Irrigation Farming Systems

Gujrat zone: Supplemental irrigation facilities on 40% of the land depend on groundwater exploited from tubewells. Persian wheels on open wells are still used but are being replaced rapidly by tubewells. In 50% of the area, an unconfined aquifer is available with the water table at a depth of 10-20 m which can be pumped with centrifugal pumps. For remaining sites, water is deeper and the initial costs and operation of a tubewell could be expensive because exploitation would require turbine pumps. Major crops with supplemental irrigation in winter are wheat and berseem (Egyptian clover) with rice produced in summer. Minor crops include maize, sugarcane, pearl millet, and vegetables but some fruit orchards use irrigation also.

With a stable water supply adequate for supplemental irrigation, crop yields are stablized and much higher than those for rainfed farming. Most farmers use chemical fertilizers as well as using higher rates of application. For example, the yield of wheat with supplemental irrigation is 50% higher than those yields received with dryland farming. Cropping intensity is also higher: 150-180% compared to 100% of rainfed production. More fodder is produced and farmers maintain increased numbers of livestock. In addition, the landless rural households (non-farm) raise more livestock (buffaloes, goats, and sheep) by feeding fodder purchased from local farmers.

Rawalpindi zone: A small proportion (2%) of the farmland occurring in blocks or strips has facilities for supplemental irrigation. One block is north near the town of Haripur; another is north-west on the Indus River floodplain, and another strip is located along the Soan River. Water resources for supplemental irrigation are available to these sites either from shallow groundwater or diverted flow from the rivers. Major crops using supplemental irrigation are wheat in winter and maize and tobacco in summer with minor crops of vegetables and fruit trees. Cropping intensity is 150-200% with supplemental irrigation (depending upon the available water supply) in contrast to cropping intensity of 100% under rainfed practices. Yields are stabilized and improved (increased by 50%) with supplemental irrigation.

Talagang zone: Only 1% of the farmland has supplemental irrigation; this land occurs in narrow strips on the floodplains of the Soan River around Pindigheb and Talagang. With supplemental irrigation, wheat is grown in winter and maize in summer. Also under irrigation are tobacco, vegetables, and fruit orchards. Farming practices resemble those of the Rawalpindi zone. The cropping intensity increases to 150-200% with supplemental irrigation but remains at 90% for rainfed farming. Crop production using supplemental irrigation is stabilized and sustained and yields are doubled (100% higher than rainfed production).

Net production per acre

Gujrat zone: As Table 25.13 shows, the annual net profit is P. Rs. 4,000 per land unit from crop production with supplemental irrigation as compared to an annual net profit of P. Rs. 425 per land unit under rainfed conditions (1,000% difference). The net profit is accelerated with biannual production of 1 crop of wheat plus 1 crop of rice using supplemental irrigation contrasted to production every 2 yr of 1 crop of wheat plus 1 crop of pearl millet.

Table 25.13. Economics of production of wheat in Gujrat and Rawalpindi.

Item	Farming system (P.Rs.)	
	Rainfed	Irrigated
1) Seed: 40 @ 3/kg	120	120
2) Fertilizer: 18 kg N	108	
12 kg P_2O_5	54	
32 kg N		192
18 kg P_2O_5		81
3) Pesticide (seed treatment)	10	10
4) Herbicide	0	0
5) Seedbed preparation:		
5 tractor cultivations @ 40 each	200	
2 tractor cultivations @ 40 each		80
6) Sowing	80	80
7) Weeding (manual)	60	60
8) Irrigation: 4 @ 50 each	0	200
9) Harvesting @ 1/10 of produce	128	240
10) Treshing @ 1/10 of produce	128	240
Total cost of production	888	1,303
Gross production		
Grain yield : @ 2/kg	1,280	2,400
Rainfed, 640 kg		
Irrigated, 1,200 kg		
Straw yield : @ 0.1/kg	96	180
Rainfed, 960 kg		
Irrigated, 1,800 kg		
Total value of gross produce	1,376	2,580
Net profit	488	1,277

Rawalpindi zone: Net value is augmented 600% with added production of summer crops. The average annual net value of production per rainfed land unit is P. Rs. 428. For the same time period, the value is P. Rs. 2,430 per land unit with supplemental irrigation. The cropping intensity is 100% for rainfed farming and 180% for supplemental irrigation production (see Table 25.13). Costs are P. Rs. 60,000 for sinking a tubewell but this price is considered profitable to the farmer who cultivates a single parcel of 12.5 acres or larger.

Talagang zone: The difference in net production is 1,000% per acre between supplemental irrigation agriculture and rainfed farming. Assuming 180% cropping intensity, net income per acre of irrigated land is P. Rs. 2,470; whereas, with 100% cropping intensity (70% wheat; 30% pearl millet), net income for rainfed production is P. Rs. 245 (Table 25.14). Installation of a tubewell costs P. Rs. 60,000 but is economic for 12.5 acres or larger in one parcel.

Table 25.14. Economics of production of wheat in Talagang.

Item		Farming system (P.Rs.)	
		Rainfed	Irrigated
1)	Seed: 30 kg, rainfed		
	40 kg, irrigated	90	120
2)	Fertilizer: none	0	
	32 kg N		192
	18 kg P_2O_5		81
3)	Pesticide (seed treatment)	10	10
4)	Herbicide	0	0
5)	Seedbed preparation: @ 40 each		
	5 tractor cultivations,	200	
	2 tractor cultivations,		80
6)	Sowing	80	80
7)	Weeding (manual)	60	60
8)	Irrigations: 5 @ 50 each	0	250
9)	Harvesting @ 1/10 of produce	80	240
10)	Treshing @ 1/10 of produce	80	240
	Total cost of production	600	1,353
Gross production			
Grain yield : @ 2/kg		800	2,400
Rainfed, 640 kg			
Irrigated, 1,200 kg			
Straw yield : @ 0.1/kg		60	180
Rainfed, 600 kg			
Irrigated, 1,800 kg			
	Total value of gross produce	860	2,580
	Net profit	260	1,277

Scope for expansion

Gujrat zone: With an unconfined aquifer, a high rate of groundwater recharge and good water quality, expansion is favorable for supplemental irrigation. Water in half the zone is at a depth impossible to pump with centrifugal pumps; turbine pumps which are more expensive would be needed. Since this would increase the eventual water cost, only farms of 50 acres or larger with advanced levels of management and modern techniques of crop production could profitably install turbine pumps on tubewells. Consequently, conversion of land to supplemental irrigation when exploitation would be expensive is proceeding at a slow rate but is progressing rapidly when groundwater is shallow enough for economic development.

The main constraints to expansion in the Gujrat zone are:
1. initial capital investment: a tubewell 30 m deep with a centrifugal pump costs P. Rs. 50,000-60,000 and a deeper well with a turbine pump would cost P. Rs. 100,000-200,000 (A government subsidy of Rs. 24,000 for a tubewell covers 40-60% of the cost of drilling);
2. small farm size: a farmer with a parcel of 10 acres could economically sink a tubewell and install a centrifugal pump; but, only a farmer owning 25-50 acres could afford to sink a tubewell and install a turbine pump; and,
3. local farmers who are not proficient in modern technology and adequate management skills: most local farmers are experienced in traditional practices. Intensive efforts by extension staff would be necessary for instruction of modern irrigation technology and for transitional support to supplemental irrigation.

Rawalpindi zone: Only 10% of the land has groundwater which could be exploited for supplemental irrigation. In half the area, groundwater is too deep for economic pumping leaving only 5% of the land for development of supplemental irrigation. However, numerous sites are appropriate for small dams but paucity of funds delays construction. Mini-dams or farm ponds could be built at innumerable farms for storage of excess rainfall for domestic purposes as well as for supplemental irrigation. If developed to full capacity, these facilities could supply supplemental irrigation to 2-5% of the cultivated land. Farm ponds controlled by individual farmers or small groups of like-minded farmers operate more effectively. Management of farm ponds could improve if portable pumps were available for hire to farmers on demand.

Experience with small dams is not encouraging because the cost per unit of water has been high and supply continues to be erratic and unreliable and farmers are reluctant to pay the cost. According to an evaluation of small dams, the economic ratio of cost benefit is 0.32 with the cost per irrigated acre calculated as P. Rs. 2,320 plus 560 for construction and maintenance costs of ditches and land preparation (Punjab P & D, 1987). The government subsidizes 40-60% (P. Rs. 20,000) of the costs of installing a tubewell as well as 60-70%

of the costs of constructing a farm pond. This policy is designed to encourage farmers to increase productivity of cultivated land.

Talagang zone: Only 3% of the land has access to groundwater (Punjab P & D, 1987). In some places (1%), this is too deep for feasible and economic exploitation. However, construction of small dams and farm ponds for storing excess rainfall offers an alternative water supply for domestic and livestock purposes as well as supplemental irrigation for 1-3% of the farmland.

Constraints to installing tubewells and to building farm ponds are the absence of investment capital and size of farm units. These constraints together with insufficient groundwater inhibit development of facilities for supplemental irrigation.

Characterization of Locations

Demography

Sialkot tehsil (Gujrat zone): Total population is 802,537 residing in 694 villages of 113,179 households having an average of 7.2 members with most members older than 15 yr (4.2 members). Males slightly outnumber females in the population; the ratio of male to female is 100.5:100.

Haripur tehsil (Rawalpindi zone): Total population is 490,104 dispersed into 82,630 households of 375 villages. The male to female sex ratio in the population is 101.8:100. Households average 5.9 members with the age distribution split between young and old (3.2 members are 15 years of age or older).

Chakwal tehsil (Talagang zone): Total population is 367,772, living in 343 villages, with 67,544 households, and an average of 5.4 members/household. Males outnumber females in the population (100:96.5) and households have a majority of older members (3.4 members are older than 15).

Soils

Sialkot tehsil: Soils are principally of 2 types: brown, silt loams, loams and clay loams, calcareous and non-calcareous with weak subangular blocky structure and brown, calcareous loams and silt loams without structure. All soils are more than 100 cm deep. Table 25.15 describes the land use for these soils.

Haripur tehsil: Soils are of 3 types: brown, non-calcareous, silt loams, loams, silty clay loams, and clays with weak structure; brown, calcareous, silt loams

Table 25.15. Land use.

Sialkot tehsil	(acres)
Total land area	249,695
Non-arable, trees/shrubs	12,692
Non-arable, steppe and pasture, rocky	41,422
Arable, not in crop production	2,624
Arable, rainfed farming	23,527
Arable, fallow	2,361
Arable, supplemental irrigation	167,069
Arable, water harvesting	Nil
Haripur tehsil	
Total land area	439,190
Non-arable, trees and shrubs (forest)	17,705
Non-arable, steppe and pasture, rocky	260,029
Arable, rainfed farming	138,049
Arable, supplemental irrigation	23,407
Arable, water harvesting	Nil
Chakwal tehsil	
Total land area	642,242
Non-arable, trees and shrubs	28,610
Non-arable, steppe and pasture	34,719
Non-arable, rocky and gullied	241,152
Arable, rainfed cultivation	332,702
Arable, fallow	32,735
Arable, supplemental irrigation	5,055
Arable, water harvesting	Nil

and silty clay loams with structure; and, brown, calcareous loams and silt loams, structureness. All soils are more than 100 cm deep. (see Table 25.15)

Chakwal tehsil: Soils are generally level, brown, calcareous, sandy loams, silt loams, loams, silty clay loams, or silty clays with weak structure: all are more than 100 cm deep. Some soils can be gently sloping, brown, calcareous, very sandy of varying depths from shallow to 100 cm. (see Table 25.15)

Infrastructure and Rural Support Services

Sialkot tehsil: Illiteracy rates are decreasing but 76% of the farmers are still illiterate (Table 25.16). Both males and females attend school but males remain for more years. There are 782 schools: 654 schools are grades 1-5 and the remaining 128 are dispersed between grades 6-8 (70), 9-10 (53), and 11-12 (5).

Adequate services are available in a number of centers: 29 agricultural extension; 30 veterinary services; 46 for medical care; and 117 credit cooperatives. Agricultural cooperatives number 429 societies but only 31% of the farmers are members.

Supplemental Irrigation in Pakistan

Table 25.16. Education and literacy by gender.

Zone/Tehsil	% Literate		% Age 10-14 Primary pass		% Matriculated	
	M	F	M	F	M	F
Gujrat						
Sialkot	49	32	24	20	7.3	3.6
Shakargarh	37	12	19	8	4.1	0.9
Gujrat	47	25	27	20	7.3	2.6
Khariah	50	20	28	18	7.4	1.6
Rawalpindi						
Rawalpindi	62	38	32	26	11.5	5.4
Gujar Khan	55	21	30	17	7.0	1.9
Jhelum	52	29	36	25	7.8	2.5
Pind Dadan Khan	47	17	29	15	6.9	1.1
Talagang						
Chakwal	55	26	40	25	7.3	2.2
Attock	39	13	22	10	5.9	1.3
Fateh Jang	29	7	22	5	3.3	0.6
Pindi Gheb	35	10	23	7	4.4	0.9

Source: Master Plan for Barani (Rainfed) Area Development Project. Punjab Planning and Development, Lahore (1987).

Most farmers do not want to pay interest on loans and prefer family members or friends for credit purposes. Notwithstanding, institutional credit facilities have increased manifold during the last few years, and yet are not capable of fulfilling current demands.

Of the 694 villages, electrical power is in 243 and electricity will extend to 90% of the villages by the year 1990. All villages, now, have potable water with hand pumps but only a few have public systems of supply. Wood or shrubs are the primary source of fuel for heating and cooking.

Agricultural inputs and harvested produce is bought and sold in market towns an average of 20 km distance from the village. Most secondary roads have dirt surfaces but the number of asphalt surfaced roads is increasing rapidly.

Haripur tehsil: Although literacy is increasing, 86% of farmers are illiterate. Both males and females receive some formal education but males attend school more years than do females. There are 432 schools with most for grades 1-5 (347) and the others divided between grades 6-8 (35) and 9-10 (50). There are no schools for grades 11-12 (see Table 25.16).

Centers delivering rural services are not adequate for the population and land area: 3 agricultural extension, 23 veterinary services, and 43 medical care centers.

Only 7.4% of the farmers are members of the agricultural cooperatives: 121 credit and 124 service societies. Most farmers take loans from family members or friends. In spite of this, institutional credit is not adequate for local demands.

Of 375 villages, 234 have electricity and an estimated 90% will have electrical power by 1990. Wood and shrubs are used for cooking and heating. A few villages (75) have potable water available from hand pumps but only 6 have pressure systems.

Marketing of agricultural produce and purchases of agricultural inputs are done at towns an average of 35 km from the village. While used the year around and in fair condition, secondary roads are primarily dirt with a few having asphalt surfaces. More asphalt surfaced roads are being built to join villages to market towns.

Chakwal tehsil: While illiteracy rates in the population are decreasing, 82% of the farmers remain illiterate. Both males and females receive formal education but males attend school for more years than do females. There are 539 schools of which 440 are for grades 1-5; the remaining 99 are partitioned into grades 6-8 (52), 9-10 (34), and 11-12 (13) (see Table 25.16).

Albeit there are 28 centers for agricultural extension, 37 centers for veterinary services, and 52 medical care centers, these numbers can barely satisfy demands of the rural population for public services.

Twenty nine percent of the farmers are enrolled as members in 358 credit cooperatives. Farmers prefer loans owed to family members or friends because they do not like paying interest for credit. Even so, institutional credit is not of sufficient quantity to meet demands.

One hundred and fifty seven of 343 villages have electricity, and, 90% of all villages will have electric power by 1990. Wood and shrubs are the primary fuel for cooking and heating. Few villages (14) have potable water or water under pressure systems. Most villages obtain drinking water from public open wells which are not hygenic.

Sale of harvested produce and purchase of agricultural inputs are performed in towns an average of 25 km distance from the villages. Secondary roads are primarily dirt surfaces which are used the year around even though in poor condition. Some roads do have asphalt surfaces, linking villages to the market towns, with more under construction.

Water Resources for Supplemental Irrigation

Sialkot tehsil: Groundwater can be pumped from tubewells with centrifugal pumps. Average annual rainfall of 680 mm occurs in 2 seasons: summer and winter. Forty percent of the land of the Gujrat zone has facilities for supplemental irrigation. In Sialkot Tehsil, these facilities are available to only 5% of the arable land.

Haripur tehsil: Surface flow diverted from the Dor and Haro Rivers into canals without regulators is available for supplemental irrigation. Average annual rainfall is 770 mm in 2 seasons (summer and winter). In the Rawalpindi zone, only 1% of the cultivated land has facilities for supplemental irrigation. In Haripur tehsil, 1.5% of the cultivated land could be irrigated.

Chakwal tehsil: Groundwater is available to 5% of the land. Water is exploited with tubewells fitted with Persian wheels. Average annual rainfall is 300-500 mm received in summer and winter. Less than 1% of the total land of Talagang zone has facilities for supplemental irrigation; whereas 1.5% of the arable land in Chakwal tehsil has supplemental irrigation. Some farms have water harvesting systems but the extent is not known. Rough estimates conclude that 1-2% of the land cultivated would benefit from this technology.

Crop Production

Sialkot tehsil: Major crops under rainfed farming are wheat and pearl millet. With supplemental irrigation, rice, wheat, and berseem (Egyptian clover) are the major crops produced. Average yields (kg/acre) are:

Crop	Rainfed	Supplemental irrigation
wheat	560	1,000
pearl millet	400	
lentils	240	
mustard	400	
rice		1,000
berseem (green fodder)		40,000

Haripur tehsil: The major crops in rainfed farming are wheat and pearl millet. Crops produced with supplemental irrigation are wheat and maize with tobacco important in some locations. Average yields (kg/acre) obtained with each farming system are:

Crop	Rainfed	Supplemental irrigation
wheat	400	1,000
pearl millet	160	
maize		1,200

Chakwal tehsil: In rainfed farming, wheat and pearl millet are the major crops. With supplemental irrigation, wheat and maize are the major crops. Average yields (kg/acre) from the 2 farming systems are:

Crop	Rainfed	Supplemental irrigation
wheat	600	1,200
pearl millet	200	
maize		1,600

Livestock Production

The average head of livestock per household is given in Table 25.17.

Table 25.17. Average head of livestock per household.

Livestock	Sialkot	Haripur	Chakwal
Cattle	2.6	4.0	4.0
Buffalo	3.2	2.2	2.2
Sheep	3.5		
Goats	3.0	4.8	4.8
Donkeys		1.0	1.0
Chickens		5.0	5.0

Socio-Economic Factors

Sialkot tehsil: Under rainfed cultivation, farmers derive 90% of their income from the farm and 10% from non-farm sources. Using supplemental irrigation, farmers receive 100% of their income from farm production. Owners operate 58% of the farms; owners-cum-tenants operate 31%; and, tenants operate 10%. Revenue sharing for farm operations of owner-tenant, under rainfed farming, is 1/3 of the produce for owner and 2/3 for tenant. With supplemental irrigation, these shares are reversed, i.e. 2/3 produce for owner and 1/3 for tenant. Family members provide 2/3 of the labor required; 20% is from animal power, 10% is mechanized, and 5% is hired labor (Table 25.18). Tractors have been used for 10 years; seed drills (drawn by tractors), wheat threshers, and pesticide sprayers have been used for 6 years. Under rainfed farming, 60% of the farmers own tractors and 70% of the farmers who use supplemental irrigation own tractors. Some migration of labor to cities and to other countries occurs due to the lack of employment locally (Table 25.19).

Table 25.18. Labor force, rural (R) and urban (U).

District	% population, age ≥ 10, in work categories									
	Agric.		Tech. & admin.		Sales services		Transport		Total	
	R	U	R	U	R	U	R	U	R	U
Sialkot	13.5	2.2	1.0	1.9	3.3	7.3	9.4	10.4	27.4	23.0
Gujrat	14.1	1.6	0.8	2.3	2.7	7.5	7.2	10.7	26.0	22.1
Rawalpindi	11.7	0.8	1.2	1.8	3.3	8.6	5.0	10.4	22.9	22.6
Jhelum	11.7	1.1	1.1	1.8	3.2	8.0	5.9	11.6	22.7	23.1
Attock	14.5	1.9	1.2	2.4	3.3	7.8	5.4	9.1	25.6	22.9

Source: Master Plan for Barani (Rainfed) Area Development Project. Punjab Planning and Development Dept., Lahore (1987).

Table 25.19. Migration patterns.

Out migration % population	District				
	Sialkot	Gujrat	Jhelum	Attock	Rawalpindi
Migrant (M, F)	9, 16	6, 19	4, 17	1, 13	5, 25
% of total population	13	8	14	10	6
% of rural population	15	9	16	11	9
% of females	13	8	12	8	5
% of illiterates	18	10	16	11	7
% with degrees	67	47	88	45	23
% migrated, last 10 yr					
rural population	0.8	1.9	2.6	1.1	1.1
urban population	2.5	1.6	3.2	1.0	1.0

Source: Master Plan for Barani (Rainfed) Area Development Project. Punjab Planning and Development, Lahore (1987).

Haripur tehsil: Farmers practising rainfed cultivation earn 80% of their income from the farm while those with supplemental irrigation derive 90% of their income from farm production. Both groups of farmers receive the remainder of their income from non-farm sources. For a farmer practising rainfed farming, 70% of the revenue is from winter crops and 30% from livestock production. For a farmer using supplemental irrigation, 50% of income is from winter crops, 40% is from livestock, and 10% is from vegetables. The average size of farm is 6.1 acres with 80% of the farm units fragmented into multiple parcels. Owners operate 80% of the farms; owner-cum-tenants operate 14%; and, tenants operate 5%. Family members contribute 70% of the labor; 10% is mechanized; and 20% is from animal power (bullocks) (see Table 25.18). Some labor migration occurs to cities and to other countries (see Table 25.19). Forty percent of the farmers use tractors but the majority of these tractors are hired. Threshers are used for wheat but other crops are still harvested with manual labor.

Chakwal tehsil: Under rainfed cultivation, farmers derive 20% of their total income from non-farm sources and 80% from farming (70% is from winter crops and 30% is from livestock). Farmers with supplemental irrigation receive 10% of their total income from non-farm sources and 90% from farm production (60% from winter crops, 30% from livestock, and 10% from vegetables). Owners operate 63% of the farms; owner-cum-tenants operate 27%; and, tenants operate 7%. The harvest from rainfed production is shared by the owner and tenant in 2:1 ratio. When crop production is with supplemental irrigation, the ratio is reversed (1:2). Family members contribute 70% of the labor for rainfed farming, with 10% mechanized, and 20% from animal power. With production using supplemental irrigation, family members still perform 70% of the labor; 10% percent is mechanized, 10% is animal traction, and 10% is hired labor (see Table 25.18). Tractors are becoming common, in use by 44% of the farmers. Tractors have been used for 10 yr; whereas, seed drills and wheat threshers have been used for only 6 yr. Most farmers use mechanical threshers for wheat; but, harvesting is done by hand labor. Some labor migration (10-14%) takes place to cities and to other countries because there is no work locally (see Table 25.19).

Potential for Supplemental Irrigation

Zones with the most potential for supplemental irrigation are Gujrat, Rawalpindi, and Talagang. Facilities could be extended to the riverine areas occurring along the Indus and Jhelum Rivers. The land in the riverine area is flooded every year during the high waters of summer. Winter crops, wheat and chickpea, are produced using residual moisture from flooding. Since construction of the Mangla and Tarbela dams, the extent of these flooded lands has been reduced; but, fortunately, there is abundant groundwater of good quality at shallow depths, less than 20 m. Implementation of supplemental irrigation systems would be simple and economic.

Water Resources

Gujrat zone: The total area is underlain by an unconfined aquifer, but under half the area, the water table is too deep for feasible exploitation; moreover, 20% of the area has water with a sodicity problem due to high SAR or high RSC. Where depth to the water table is less than 30 m, tubewells have already been installed by farmers. Alternatively, with an average annual rainfall of 500-900 mm, excess rainwater could be stored in small dams or farm ponds for later use with supplemental irrigation.

Rawalpindi zone: Groundwater is available to only 10% of the area and 1/3 of this has water at depths too great to be economical. Some land is gullied and rocky (40%) and collection of rainfall runoff in small dams and farm ponds is feasible. Rainfall runoff collected and stored behind small dams and farm ponds currently provides supplemental irrigation for 20,000 acres. The possibility exists to extend these facilities for collection and storage of rainfall runoff for supplemental irrigation.

Talagang zone: Groundwater is available to only 5% of the land; and, in places, the water table is too deep to be economically developed. Some water harvesting is practiced; and, with improvements, these systems could divert excess rainfall to farm ponds for later application to fields by supplemental irrigation.

Riverine areas: Abundant groundwater of very good quality occurs at shallow depths in the entire area.

Component Needs

Gujrat zone: Facilities for supplemental irrigation already exist in 40% of the zone. Before extending these facilities to the remaining land, certain action must be taken.
1. A detailed survey and investigation of groundwater reserves to assess depth, quality and volume.
2. The subsidy (P. Rs. 24,000) for installing tubewells should be increased when depth to water table is more than 30 m.
3. Assure increased credit at low interest rates for tubewell constuction and a simplified procedure for obtaining loans.
4. Availability of credit for agricultural inputs should be extended.

Rawalpindi zone: Only 1% of the cultivated land has a water supply from tubewells or open wells and from canal water diverted from the two rivers. To implement supplemental irrigation into this area requires action.
1. Increase the subsidy of P. Rs. 24,000 for sinking a tubewell when depth to water table is more than 30 m.
2. Funds allocated for the expenditure of subsidy payments for tubewells and farm ponds have to be multiplied.
3. Credit reserves for costs of installing tubewells need to be amplified and the interest on loans reduced.
4. Funds have to be strengthened to allow credit to purchase agricultural input.

Talagang zone: Less than 1% of the area has facilities for supplemental irrigation. To extend these facilities, the component needs for this zone are the 4 points listed for Rawalpindi zone in the preceding paragraph.

Riverine areas: For extending supplemental irrigation facilities in these areas, the following actions must be taken.

1. Subsidies and funds for payments for sinking tubewells need to be established.
2. Electric power lines need to be installed to these areas.
3. Increase funds to loan for agricultural input purchases.

Problems and Constraints

Major constraints to supplemental irrigation development in the riverine areas and the agro-ecological zones, Gujrat, Rawalpindi, and Talagang include the following:
1. Farm size: Most farms are less than 12.5 acres (70-80%) a size which is considered sufficient only for subsistence farming.
2. Fragmentation of farms: Data of the sample sites in the 3 zones show that 80% of the farms are fragmented, posing problems for production using the modern technology of supplemental irrigation.
3. Land tenure: Farm operation of 30-40% of the farms is by tenants or tenants-cum-owners. Both the owner and tenant are reluctant to make capital investments for development of supplemental irrigation.
4. Illiteracy: Most farmers are illiterate and require intensive efforts by extension staff using specialized instruction methods and expanded services would be a prerequisite for successful adaptation by farmers of supplemental irrigation production.
5. Deficit funds for capital investment and credit: This constraint is especially critical in areas where the water table is more than 30 m deep.

Actions Needed

Constructive action is needed to encourage expansion of the land area with supplemental irrigation.
1. The subsidy (P. Rs. 24,000) for installing tubewells should be enlarged when depth to water table is more than 30 m.
2. Funds allocated for the expenditure of subsidy payments for tubewells and farm ponds have to be multiplied.
3. Credit reserves for costs of installing tubewells need to be amplified and the interest on loans reduced.
4. Funds have to be strengthened to allow credit to purchase agricultural inputs.
5. A detailed survey is needed to investigate groundwater reserves to assess depth, quality, and volume.
6. Electric power lines installed to service tubewells in contiguous lands with groundwater of good quality (for example, in the riverine area).
7. Semi-detailed soil survey of riverine areas to provide information to farmers on characteristics of the land.

Summary and Conclusions

Supplemental irrigation in contrast to rainfed cultivation, creates changes in cropping patterns, increasing cropping intensity and increasing net income per acre. Irrigated rice replaces rainfed pearl millet in the Gujrat zone and irrigated maize replaces pearl millet/sorghum in the Rawalpindi and Talagang zones. The cropping intensity of 100% under rainfed farming increases to 180% with supplemental irrigation. Annual net income/acre with supplemental irrigation increases dramatically to P. Rs. 2,400-4,000 (from P. Rs. 400 under rainfed farming).

In the 3 selected agro-ecological zones, the proportion of cultivated land having supplemental irrigation facilities is as follows.

Zone	Cultivated Land (M acres)	% Supplemental Irrigation
Gujrat	1.9	40
Rawalpindi	1.3	2
Talagang	2.1	1

A 100% increase in cropped land with supplemental irrigation is possible in all 3 zones, and especially in Rawalpindi and Talagang zones. If deep tubewell technology is applied and subsidies raised for construction where water tables are below 30 m, this permits installation of tubewells by the average farmer.

In addition to the 3 agro-ecological zones, supplemental irrigation development is feasible in 2.6 M acres of riverine land which were previously flood-watered for winter cropping. These lands no longer flood because of construction of the Mangla and Tarbela dams.

Supplemental irrigation in the Gujrat zone is used for rice and wheat production. Berseem (Egyptian clover), an important crop, is grown for animal fodder. Cropping intensity is 150-200% and the crop rotation is rice – wheat – berseem. Fertilizer use is common.

Supplemental irrigation in Rawalpindi and Talagang zones, is for production of wheat and maize. Tobacco is important in some locations. Crop rotation is wheat – maize (tobacco) and the cropping intensity is 150-200%. Fertilizers are commonly applied.

On the riverine land, chickpea, mustard, and wheat are grown using residual moisture stored in the soil from the flood waters of summer. No fertilizer is used and land remains fallow in summer. With supplemental irrigation, wheat, chickpeas, and cotton are grown with a crop rotation of wheat (chickpea) – cotton. Cropping intensity is 150-200% and fertilizer use is common.

References

Agriculture Census Organization-1, 1980. Pakistan census of agriculture 1980; Province Report Punjab. Agric. Census Org., Govt of Pakistan, Lahore.

Agriculture Census Organization-2, 1981. Pakistan census of agriculture 1980; Province Report N.W.F.P. Agric. Census Org., Govt of Pakistan, Islamabad.

Ali, M. Ashraf, 1971. Climate (of Pakistan), Chap. 3; *In* Soil Resources in West Pakistan and Their Development Possibilities. Soil Survey Project; AGL:SF/PAK-6. Technical Report 1, FAO/UN, Rome. pp 37-47.

Fraser, I. S., 1958. Report on reconnaissance survey of the landforms, soils and present land use of the Indus Plains, West Pakistan. Publ. for the Govt. of Pakistan by the Govt. of Canada.

Government of Pakistan, 1983. 1981 district census report of Abbottabad. Population Census Org., Govt of Pakistan, Islamabad.

Government of Pakistan, 1984-1. 1981 district census report of Rawalpindi. Population Census Org., Govt of Pakistan, Islamabad.

Government of Pakistan, 1984-2. 1981 district census report of Jhelum. Population Census Org., Govt of Pakistan, Islamabad.

Government of Pakistan, 1984-3. 1981 district census report of Sialkot. Population Census Org., Govt. of Pakistan, Islamabad.

Punjab Government, 1986. Soil conservation activities in Barani Tract. Directorate of Soil Conservation, Punjab, Rawalpindi.

Punjab Government, 1987. Veterinary institutes in Punjab. Directorate of Livestock and Dairy Development, Punjab, Lahore.

Punjab P & D, 1987. Master plan for Barani area development project; Interim Report Vol. II, Inventory and Analysis. Punjab Planning and Development Dept., Lahore.

Rafiq, M., 1976. Crop ecological zones of Pakistan; Crop Ecological Zones of Nine Countries of the Near East Region. FAO /UN, Rome. M-15.ISBN-92-5-100126-X.

Reconnaissance Soil survey reports of Gujranwala, Rawalpindi, Cambellpur, and Bannu Areas. Soil Survey of Pakistan, Ministry of Food and Agriculture, Govt. of Pakistan, Lahore.

Robertson, G. W., 1985-1. Rainfall probabilities in Rawalpindi – Islamabad area. Pakistan Agric. Res. Council, Islamabad.

Robertson, G. W., 1985-2. Rainfall probabilities in Haripur area. Pakistan Agric. Res. Council, Islamabad.

Robertson, G. W., 1985-3. Rainfall probabilities in Bannu area. Pakistan Agric. Res. Council, Islamabad.

Robertson, G. W., 1985-4. Rainfall probabilities in Mianwali area. Pakistan Agric. Res. Council, Islamabad.

Supple, K. R., A. Razzaq, Ibram Saeed, and A. D. Sheikh, 1985. Barani farming systems of the Punjab. National Agricultural Research Center; Pakistan Agric. Res. Council, Islamabad.

Chapter 26

Supplemental Irrigation Systems of the Syrian Arab Republic (SAR)

GEORGE SOUMI

Agro-Climatic Classification

The total land of Syria has been classified into 5 zones according to average annual rainfall regardless of other climatic and ecological factors and soil characteristics. These zones are shown in Figure 26.1.

Zone 1: Average annual rainfall is more than 350 mm but not less than 300 mm in any 2 yr out of 3 yr. This zone can be divided into 2 sub-zones: a) rainfall is more than 500 mm; and, b) rainfall of 350-600 mm.

Zone 2: Annual average rainfall is 250-350 mm but not less than 250 mm for 2 yr out of 3 yr.

Zone 3: Average annual rainfall is more than 250 mm but not less than 250 mm for 2 yr out of 4 yr.

Zone 4: Average annual rainfall is 200-250 mm but not less than 200 mm for 2 yr out of 4 yr.

Zone 5: Average annual rainfall is less than 200 mm. Fifty five percent (55%) of the land area of Syria is located in this zone.

Comparing potential evapotranspiration data (using the Blaney-Criddle Equation), these 5 zones, show no relationship between ET_o and zone classification. The ET_o values for the 5 zones are:

Zone 1a   :   1315 – 1511 mm/yr,
Zone 1b   :   1748 – 2002 mm/yr,
Zone 2   :   1883 – 2023 mm/yr,
Zone 3   :   more than 2023 mm/yr,
Zones 4 & 5: 2002 – 2319 mm/yr.

Water Resources and Their Use

Rainfall

Rainfall in 1979-83 was 38,684 – 51,073 M m^3. From analysis of rainfall patterns and volume of water for that period, the following observations can be made.

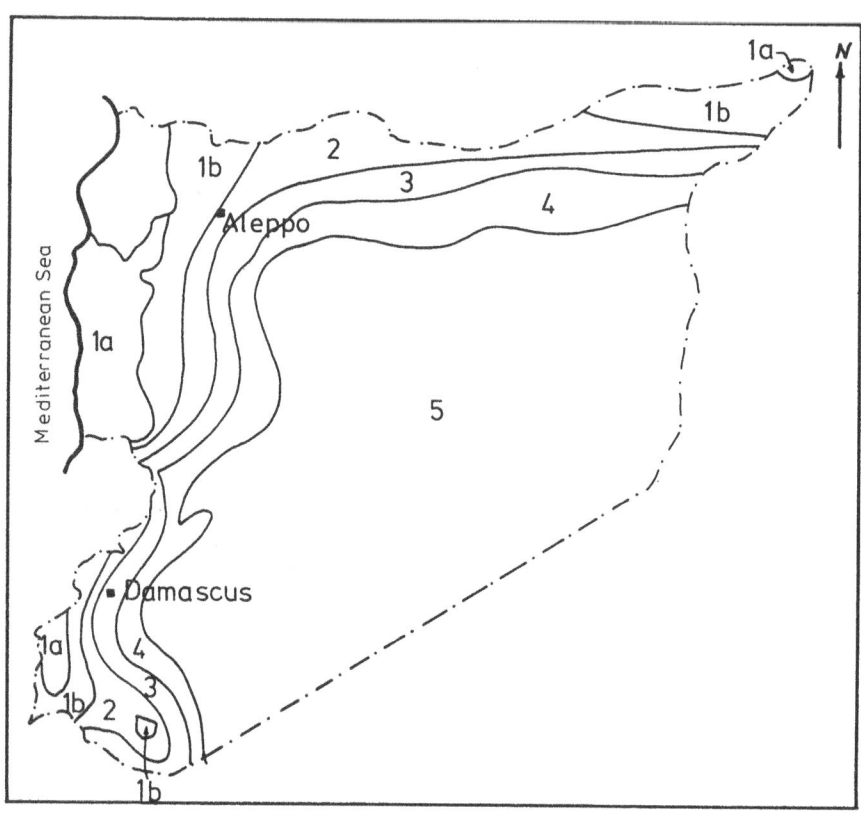

Figure 26.1. Map of Syria showing rainfall zones.

1. Where average annual rainfall is more than 600 mm, the production stability of the land area (4.4% of total Syrian area) can be inferred from the natural vegetation of the mountains and the forests. Eighteen percent of the total rainfall supports agricultural production of citrus trees and vegetables.
2. Twenty seven percent of the country's land area receives less than 200 mm of average annual rainfall; but in 1980, this average encompassed 40% of the area and, in 1983, 64%. Usually, 50% of total land area receives this average rainfall; therefore, rainfed farming is not viable (except in land depressions which receive runoff from neighboring catchments). The region is used mainly as natural pastures for grazing sheep.
3. The amount of land receiving 200-600 mm varies from one year to another. In 1983,
   31% was affected; in 1982, 64%; however, during 1979-1983, the average was 45.2% of the total area. The rainfall in this area may contribute as little as 19% of Syria's total rainfall in dry years and as much as 45% in good years.

Most agricultural land (85%) relies on rainfall for production; yields are generally low and fluctuate from year to year due to highly unpredictable and erratic annual rainfall.

Aquifers

Total water resources available for agriculture, industry, and human consumption are estimated as 23 billion m³/year, including water lost to evaporation and deep percolation. A hydrogeological and hydrological investigation of 8 aquifers shows a restriction of water resources rather than arable land resources. Of the total water resources, surface flow supplies 74% while the aquifers contribute only 8.4% and springs, 17.5%.

Water Use in Agriculture

The SAR seeks self sufficiency in agriculture in both production systems, crops and livestock, and interest has evolved in irrigation and land reclamation. Irrigated lands in 1985-86 were 651,903 ha, partitioned into 3 types of supply: rivers (flow), 129,255 ha; rivers (pumping), 204,342 ha; and, wells (pumping), 318,306 ha. Exploitation of water resources by pumping conveys the supply to 80.2% of the total irrigated area. Wells provide water for 49% of the total irrigated area. Sixty one percent of this area is owned privately or cooperatively.

Land in government projects (220,000 ha) is equipped with networks of irrigation and drainage and pumping stations. Project farms are cultivated by farmers who pay a small tax for project services. System maintenance and scheduling of water delivery is performed centrally by the governmental authorities.

Data from studies performed for the water plan of the 1986-87 season show that consumptive use for agriculture differs (0-100%) from one water basin to another. The water requirement for irrigation is estimated, for an average annual cropping intensity of 130%, at 7.1 B m³ water, incorporating irrigated land in crop rotation of winter production and a rainfall probability of 75%.

Water requirements for irrigation of the 229,000 ha of irrigated wheat production are estimated as 650-700 M m³/season, net, or 1.0-1.2 B m³/season total (water losses included). The 229,000 ha includes 25% of the irrigated area as well as 43% of the land used in crop rotations. With a rainfall probability of 75%, this estimate necessitates using 30% of total water reserves for annual production. Deficits in water requirement for wheat production, to be met by irrigation, are 0-500 mm/yr. If rainfall is deficient, irrigations applied for production (probability 75%) must come from water reserves in the aquifers: annual rainfall in zone 1a furnishes total water requirement of wheat; whereas, in zones 4 and 5, the deficit is 400-500 mm.

Volume of water used is actually no more than 50-60% of the crop water requirement. The problem defined for purposes of planned exploitation of aquifers is that farmers do no measure the exact water requirement of wheat. Recommendations for volume of water to be applied are over-estimated; this happens for two reasons:

1. some water resources are saved during the winter months for summer time irrigation; and,
2. some farmers do not like to irrigate wheat especially when the water supply is pumped.

The limited water resources are not well allocated. Since the supply of water is in excess of crop water requirement during winter and spring, administrators of planning and development should consider new strategies for constucting irrigation projects, improving water efficiency of water resources in irrigation projects, and interconnecting agricultural production with the goal of realizing food security especially for the strategic crops of wheat and forages. This can be achieved by assessing which projects could deploy excess rainfall during winter to irrigate crops (particularly wheat) that would establish yield improvement and stability then self sufficiency of production.

Land Use and Cropping Patterns

Land Use

Syria's total land area is 18.518 M ha which, according to 1985 estimates, can be placed in 4 categories:
1. arable land, 6.127 M ha or 33% of total land;
    a. cultivated land is 5.633 M ha, 92% of the arable land, and represents 30% of the total land area; and,
    b. non-cultivated land is 0.504 M ha, 8% of the arable land.
2. non-arable land, 3.547 M ha or 19% of total land;
3. natural pastures and range lands, 8.328 M ha or 45% of total land; and,
4. forest land, 0.546 M ha or 3% of total land.

Cultivated lands are further grouped into 3 categories: 1.663 M ha in fallow (27% of the arable land or 29% of cultivated lands); 3.318 M ha of rainfed production (54% of the arable lands; 59% of cultivated land); and, 0.652 M ha irrigated production (11% of the arable lands; 12% of the cultivated land).

The actual cropped land (21.4% of the total land) according to 1985 data is 3.970 M ha, 65% of the arable land or 70.6% of the cultivated land. Of this land in production, 16.4% has supplemental irrigation; the remaining 83.6% are cultivated under rainfed conditions. The cultivated land under rainfed farming is located throughout the agro-ecological zones as shown in Table 26.1. The cultivated land with supplemental irrigation is distributed between the agro-ecological zones as shown in Table 26.2.

Crop Rotations

According to 1985 data, different crop rotations are used on cultivated land depending on the farming systems prevalent in each agro-climatic zone.

Table 26.1. Distribution of land under rainfed production.

Zone	Land area (1,000 ha)	Rainfed (%)
1	1,153	35.3
2	1,017	30.3
3	446	13.4
4	484	14.6
5	218	6.4
Total	3,318	100.0

Table 26.2. Distribution of land under supplemental irrigation.

Zone	Land area (1,000 ha)	Suppl. irrig. (%)
1	245	37.6
2	160	24.6
3	43	6.6
4	43	6.6
5	161	24.6
Total	652	100.0

Crop rotations for irrigated lands

Table 26.3 lists winter and summer crops. Percentages are given for those crops grown in rotation with other crops. The crops rotated with cotton and their incidence of use (%) are: corn (8%), barley for forage (1.6%), tomato (1.1%), sesame (2.4%), fall potato (2.0%), and groundnut (1.6%). Part of the time (25%), other vegetables and field crops (7.3%) are used in rotation.

Table 26.3. Crops grown with supplemental irrigation.

	Rotation by season	
Crop	Winter (%)	Summer (%)
Wheat (local, high yield)	43	
Autumn beet	2	
Faba bean (seeds)	1	
Spring potato	1.6	
Lentil/chickpea	0.4	
Vegetables	3.2	2.4
Forage crop/other vegetables	5.8	8.1
Cotton		33.5

Crop rotations and cropping patterns on rainfed lands

Different cultural and farming practices are used contingent on the crop and rotation course chosen. Table 26.4 gives the general cropping pattern and average allocation of crops grown in rotation on rainfed areas as a whole. However, cropping intensity corresponds to the rainfall zone and reaches 98% in zone 1 using cropping patterns and crop rotation as shown in Table 26.5.

Table 26.4. Crops grown under rainfed conditions.

Crop	Rotation by season	
	Winter (%)	Summer (%)
Wheat (local, high yield)	21.6	
Barley	24.5	
Lentil	1.8	
Chickpea	2.0	
Vetch (seed)	1.2	
Watermelon/other melons		2.1
Legumes		1.4
Other crops	2.2	2.8

Table 26.5. Crops produced in rotation and relative cropping intensities.

Crop	Rotation by zone (%)			
	1	2	3	4
Wheat	46.5	27.7	10.2	
Barley	3.0	22.0	39.0	34.4
Lentil	6.0	1.5		
Chickpea	4.6	2.6		
Forage vetch (seed)	8.2	1.0		
Other crops	4.7	4.2	1.8	
Summer crops/vegetables	25.0	4.0		
Total cropping intensity (%)	98.0	63.0	51.0	34.4

Food Security

As in other developing countries, the gap between domestic food production and consumption has been growing. Reasons for this growth are: poor rainfall that prevailed in several years during the 1980s; partial failure to accomplish all agricultural development plans of the government, especially those of irrigation and land reclamation; increased levels of consumption; and, population growth. To indicate the food situation, data are presented in Table 26.6 for comparison of major food commodities' rates towards achieving self-sufficiency between the 1970s and the 1980s.

Table 26.6. Rates of self-sufficiency in the major food commodities in the 1970s and the 1980s.

Commodity	Self-sufficiency (%)	
	1970-72	1983-85
Wheat	72	59
Barley	107	84
Maize	82	19
Red meat	103	99
White meat and eggs	100	100

The situation of food security is expected to worsen during 1986-90 unless greater results are achieved from development plans and projects result in rapid and high returns. When the Sixth 5-year Plan for Social and Economic Development was set, projections of domestic production and consumption were made.

Domestic Demand

Domestic demand for major food commodities was estimated in consideration of human requirements for basic food components (calories, proteins, vitamins, amino acids, etc.) based on major sources of availability (plant and animal). Current deficits of some food components and potential for improvement were contemplated and population growth regarded as well. Predicted demands are shown in Table 26.7.

Table 26.7. Projections of demand for basic foods, 1986-90.

Food product	Projections of demand (1,000 t)				
	1986	1987	1988	1989	1990
Wheat	2,274	2,345	2,431	2,522	2,618
Barley	822	925	1,050	1,181	1,228
Legumes	92	97	102	107	112
Maize	276	291	308	325	344
Potatoes	325	357	393	430	472
Red meat	143	148	153	158	163
White meat	84	90	95	101	107
Milk	1,741	1,846	1,958	2,074	2,198
Eggs (M)	1,481	1,370	1,664	1,764	1,870

Production

Production goals are based on vertical expansion of agricultural production, and expected impact of agricultural development projects which will be accomplished during the planning period (1986-90). Table 26.8 gives production levels for basic food commodities during 1985, the base year, with projections for 1990.

Table 26.8. Levels of production for 1985 and projected levels for 1990 for basic food commodities.

Food commodity	Production levels (1,000 t)	
	1985 base year	1990 projections
Wheat	1,637	2,234
Barley	776	864
Legumes	149	166
Maize	115	231
Read meat	101	137
White meat	80	109
Milk	958	1,506
Eggs (M)	1,398	1,500

Projected Food Deficit, 1990

By comparing projected production and domestic demand, in 1990, the expected deficit (in units of 1,000 tons) for major food commodities would be: wheat, 386; barley, 464; sugar, 477; maize, 113; red meat, 26; fish, 4.5; and milk, 692. Three major factors could account for food deficit in 1990:
1. unsuitable utilization of natural resources, especially water resources;
2. inadequate policies for production and investment finance; and,
3. inadequate pricing policies.

Supplemental Irrigation

Productivity of Existing Supplemental Irrigation

The contribution from the agricultural sector to national income is low in contrast to other sectors of the economy. Data analyzed for the 1981-85 Plan show that agriculture's contribution was 15.4% (in market prices) and 13.6% (in static prices). Even though, the contribution from land with irrigation has not been evaluated as yet, this revenue will have a major role in the agrarian economy. Statistical data for supplemental irrigation for 1981-85 show these results.

1. The average contribution of winter irrigated lands to the total production of winter crops is 30.6% and becomes larger (48%) in below average rainfall years. This contribution from winter vegetables accounted for 90% of total winter production of vegetables. The contribution of supplemental irrigation to total production of winter crops and vegetables is disproportionally high since they are sown to only 10% of total area planted to these products.
2. For wheat production with yields of 2.3-2.8 t/ha, the contribution from land with supplemental irrigation was 33-40% although the irrigated area is only 15% of the total wheat land. During the same period, yields were 0.693-1.4 t/ha for rainfed production and yield differences become even greater during dry years. In 1984, total yield was 636,867 t from 1,922,114 ha of rainfed production; whereas, the total yield was 428,735 t (or 40% of total wheat production) from 184,841 ha of production with supplemental irrigation.
3. The relatively low yields (2.3-2.8 t/ha) from irrigated land in 1981-85 demonstrates that insufficient water was applied for irrigation.
4. Yields of high yielding varieties using supplemental irrigation are: zone 1, 3.12 t/ha; zone 2, 2.43 t/ha; and, zone 3, 2.40 t/ha. Yields in each of zone 2 and 3 equal 77% of the yield obtained in zone 1 which confirms that, in all zones, insufficient amounts of water were applied to the crop.
5. The yield from rainfed production for high yielding varieties was: zone 1, 1.78 t/ha; zone 2, 0.82 t/ha; and, zone 3, 0.70 t/ha.
6. The yield for lentils grown on irrigated lands was: zone 1, 1.17 t/ha; zone 2, 0.64 t/ha; and, zone 3, 0.54 t/ha. In comparison, yield for lentils with rainfed production was: zone 1, 0.87 t/ha; zone 2, 0.64 t/ha; and, zone 3, 0.40 t/ha. The rainfed yields showed 74% of the irrigated production in zone 1, 100% of the irrigated yields in zone 2, and 62% of the irrigated yields in zone 3.
7. The yields of chickpea grown under rainfed farming practices were compared to supplemental irrigation production and the proportion from rainfed crops revealed: zone 1, 0.86 t/ha, only 64% of the yields from supplemental irrigation production; zone 2, 0.55 t/ha or 44% of the supplemental irrigation yield; and, zone 3, 0.54 t/ha, 31% of the irrigated yields.

Potential for Further Development

Supplemental irrigation, using the minimum number of irrigations, could double the production of strategic crops such as wheat. By 1990, the food deficit is anticipated to be greater because of incremental increases in demand. Considering the claim for wheat alone, a 2.618 M t gross domestic demand is expected and 1.328 M t barley will be needed. The exigency indicates completion of irrigation projects to put more land into cultivation and furthermore to consider supplemental irrigation as the principal method to improve yields to 4-5 t/ha. The ability to increase yields using supplemental irrigation has been established by the results of a joint project between MAAR and ICARDA. These data demonstrate that yields of 5-6 t/ha are possible using scientific methods of supplemental irrigation with recommended application rates of fertilizers.

In a primary study of natural resources, results showed that a potential exists which has not yet been utilized. Efficient use of existing water resources could sustain agricultural productivity and contribute to food security. Nowadays, irrigation water for the agricultural sector is annually 30-35% of available water resources which demonstrates the presence of uncommitted reserves.

A means of improving the efficient use of water resources is to seriously contemplate new strategies for conservation by planning project implementation in locations of winter production with adequate resources to allow a 2-course rotation of wheat and forages. These locations would furnish the major proportion of agricultural production, and provide food security. The existing conditions in these locations and the logic of selection can be summarized:

1. available water resources are most accessible during winter and spring;
2. winter production has low crop water requirements (especially in zones 1 and 2) which is 25-35% of that used by summer crop production;
3. a lower percentage of water lost to evaporation and percolation;
4. possibility of fully mechanized production of cereals and forages;
5. possibility to enlarge cultivated land by extending irrigation networks; and,
6. possibility of excess water for summer crop production.

These conditions relate to all underground and surface water resources including the Khabour and Tigris Rivers in the northeastern region. Pricing policies would have an impact on these projects because these policies proffer an incentive, to small farmers pumping from rivers and wells, to irrigate strategic winter crops (especially wheat) in competition for the water supply between vegetables and other crops.

MAAR/ICARDA Research

Research studies for supplemental irrigation of wheat were carried out jointly by MAAR and ICARDA.

Objectives of the research

1. To determine the economic water requirement and irrigation regime for wheat according to several irrigation and fertilizer treatments.
2. To find relationships between water requirement and the application rate of fertilizers.

3. To determine the crop coefficient using Blaney and Criddle, Class A Pan, and Penman's Equation.

Statistical Design of the Field Studies

The design was a fully randomized split plot with 4 irrigation treatments as the main plot (Table 26.9) and 4 fertilizer treatments, i.e. 0, 50, 100, and 150 kg/ha, as the subplot. Treatments were replicated 4 times, phosphorous was added at a rate of 80 kg/ha for all treatments. Before seeding, 80 kg/ha of P_2O_5 was added with half the total amount of nitrogen. The seed variety was Sham I. The plot size was 5 m × 6 m with 2 m borders between the main plots or a total area of 30 m². The field trials were performed at the ICARDA Center at Tel Hadya (Aleppo, Syria).

Table 26.9. Supplemental irrigation treatment for field studies.

Irrigation treatment	% Field capacity	
	Minimum	Maximum
A	60	100
B	70	100
C	80	100
D	Control	Control

Statistical Results

A. Eighty percent of field capacity (FC) treatment with fertilizer treatment-2 had higher yields (6.82 t/ha) with an average of 0.79 m³ of water/1 kg wheat, followed by the 70% FC treatment with fertilizer treatment 2 (6.64 t/ha) on the average of 0.83 m³ water/1 kg wheat, followed by the fertilizer 3 (6.52 t/ha) on the average of 0.84 m³ water/1 kg wheat (Table 26.10)

B. For the irrigation treatments:
  1. 70% of FC treatment = 5.77 t/ha for 0.95 m³ water/1 kg wheat.
  2. 80% of FC treatment = 5.62 t/ha for 0.97 m³ water/1 kg wheat.

C. The best equation to determine the water requirement is Penman's Equation then Blaney & Criddle then the Class A Pan. The average values for the crop coefficient, K_c, are:
   0.71 using Penman's Equation,
   0.91 using the Blaney and Criddle method, and
   0.77 using the Class A Pan.

D. It is possible to use the monthly K_c value to determine the consumptive use of water.

Table 26.10. Yield and water use for supplemental irrigation studies.

SI treatment Yield (t/ha)	N treatment water use (m³/ha)	Total water (m³/ha)	Total SI (m³/ha)	Rainfall (m³/ha)	Fertilizer treatment
60% treatment					
4.368	4,679	4,480	2,217	2,263	N_1
	4,856				N_2
	4,359				N_3
	3,584				N_4
70% treatment					
5.771	5,587	5,489	3,226	2,263	N_1
	6,641				N_2
	6,521				N_3
	4,334				N_4
80% treatment					
5.616	5,755	4,763	2,500	2,263	N_1
	6,829				N_2
	6,350				N_3
	3,529				N_4
Rainfed					
3.688	4,160	2,563	300	2,263	N_1
	3,641				N_2
	3,902				N_3
	3,047				N_4

Technical and Economic Feasibility of Supplemental Irrigation of Wheat

The total land area under production for wheat and barley is 2.571 M ha. Wheat is given priority over barley and other strategic crops. Average annual land sown to wheat for the period of 1984-86 was 2.134 M ha for irrigated wheat and 0.950 M ha for rainfed wheat. Average annual wheat production is 1.493 M t/ha. Barley production with irrigation is only 0.146 M t/ha, but for rainfed, is 1.393 M t/ha with total barley production of 1.539 M t. The third major crop is cotton, sown to 0.165 M ha with a production of 0.452 M t.

Rainfed wheat is grown principally in agro-climatic zones 1 and 2 and competes with barley in zone 2 for the more fertile soils; therefore, barley is usually sown to the less fertile soils. Wheat grown under rainfed conditions does not receive regular water requirements during most of the years because of erratic and fluctuating patterns of rainfall and irregular distribution of rainfall in the growing season. Annual rainfall is greater than 600 mm in the wetter parts of zone 1; however, these wetter parts which have mostly woody vegetation and which are mountainous do not have major wheat production. Average annual rainfall in the drier parts of zone 1 is 350-600 mm; and, in zone 2, is 250-350 mm.

Crop water requirement for winter wheat production is 4,500-6,000 m³/ha consumed during the season November through May. Based on research conducted to estimate water requirements for supplemental irrigation regimes, wheat requires irrigation as shown in Table 26.11.

Table 26.11. Irrigation requirements for wheat.

Agro-climatic zones	Irrigation requirements (m³/ha)				
	Feb	Mar	Apr	May	June
1	—	400	600	800	1,800
2	500	700	825	1,200	3,225

The source of irrigation water for wheat and other agricultural production and the volume supplied are variable: 38% of the total supply is from groundwater pumping; 37% is from surface water pumping; and, 25% is from gravity flow without pumping. Water costs are variable as shown in Table 26.12 for a pumping unit of 60-70 m³/hr capacity which is the most common in Syria. These estimates are for 1 irrigation with a volume of 800-1,000 m³.

Table 26.12. Cost of irrigation water according to source and delivery system.

Irrigation source	Construction depreciation (SL)	Equipment depreciation (SL)	Fuel (SL)	Total (SL)
Gravity, rivers	29.4	—	—	29.4
Pumping, rivers	7.4	34	66	107.4
Pumping, shallow wells	4.4	38	110	192.0
Pumping, deep wells	10.0	80	220	310.0

Rainfall contributes 66% of the annual water requirement for wheat in zone 1 and 48% of the annual water requirement in zone 2 (see Table 26.11). The costs of irrigation in these zones, therefore, would be less than the estimates (see Table 26.12). Nevertheless, water costs of supplemental irrigation correspond directly with annual rainfall, i.e. irrigations deliver the difference in quantity between crop water requirement and actual annual rainfall. The total costs of supplemental irrigation per hectare are presented in Table 26.13.

Supplemental irrigation could achieve substantial improvement in yields but, would incorporate new and additional costs to production in the form of land shaping to facilitate irrigation, labor, and agricultural inputs e.g. seed and fertilizer. The technical and economic feasibility of supplemental irrigation of wheat can be ascertained by comparing the costs and returns of each water source and contrasting these results to rainfed wheat production.

Table 26.13. Cost of supplemental irrigation.

Type of water delivery	Water costs for supplemental irrigation	
	Zone 1 (SL)	Zone 2 (SL)
Gravity irrigation, river	90	105
Pump irrigation, river	330	390
Pump irrigation, shallow wells	591	699
Pump irrigation, deep wells	955	1,128

Costs and returns are calculated using official prices prevailing in 1987 and the added value per hectare of wheat is as follows:

Zone 1: 9,683 – 9,046 SL/ha (varies by source of water) for supplemental irrigation contrasted to 2,062 SL/ha for rainfed production of wheat;

Zone 2: 9,683 – 8,931 SL/ha for supplemental irrigation contrasted to 1,693 SL/ha for rainfed production.

However, net income estimates for wheat production were:

Zone 1: 9,831 – 7,339 SL/ha for supplemental irrigation and 1,643 SL/ha for rainfed;

Zone 2: 8,189 – 7,066 SL/ha for supplemental irrigation and 1,202 SL/ha for rainfed.

Profitability (net income) of wheat production contingent on source of water is listed in descending order:

1. gravity irrigation in zone 1,
2. gravity irrigation in zone 2,
3. pumping irrigation, rivers, zone 1,
4. pumping irrigation, rivers, zone 2,
5. pumping irrigation, shallow wells, zone 1,
6. pumping irrigation, shallow wells, zone 2,
7. pumping irrigation, deep wells, zone 1,
8. pumping irrigation, deep wells, zone 2,
9. rainfed production, zone 1, and
10. rainfed production, zone 2.

Conclusions

In 1985, the cultivated land was low, 3.970 M ha; this represents only 21.4% of the total area and 65% of the arable land.

To arrive at scientifically balanced patterns of land use requires classification maps of an acceptable scale depicting basic soil data, chemical and physical characteristics. Without such maps, it is difficult to draft effective agricultural policies for planning, land use, specifying cropping patterns, and establishing criteria to select and confirm priorities for transforming production systems.

Irrigated land in zones 1 and 2 is an estimated 62.1% of the total in the 5 agro-ecological zones, and offers the most opportunity for improvement with supplemental irrigation. The challenge is to make the opportunity available economically since annual rainfall satisfies a big proportion of the crop water requirement, particularly for wheat. Efforts to increase grain yields in other zones, notably 4 and 5, must contemplate exploiting precious water reserves because supplemental irrigation would supply most of crop water requirements in as much as rainfall is extremely low.

A major factor restricting development of supplemental irrigation is that most lands are irrigated by pumping water. Fifty one percent of the total area is owned by farmers who, because of costs, irrigate wheat or other cereals only when plants are under stress. Revenue from cereal production, including wheat, allots a trivial income for a farmer compared to possible gains from other crops such as winter vegetables.

Wheat production with supplemental irrigation constitutes 43% of the total irrigated land and 75.4% of the land irrigated in winter. This is the maximum amount which can be taken from rotation cropping. The other field crops, lentils, faba beans, and chickpeas, are sown on 3% of the total irrigated land. Barley is not included in the plan for irrigated lands.

Most wheat lands (46.5% of the cultivated land in Syria) are concentrated in zone 1 together with lentils, faba beans, and chickpeas which constitute 11% of the cultivated land in Syria. The potential for extending supplemental irrigation to all crops, but particularly to wheat production, prevails in zone 1 in view of the favorable circumstances of available land and sufficient rainfall for production. This production needs only a few supplemental irrigations to benefit total yields. Zone 2 also has a potential to develop supplemental irrigation.

Chapter 27

Cereal Cropping and Supplemental Irrigation in Tunisia

A. BOUZAIDI

The cereal sector is of great importance to Tunisian agriculture: it covers large areas, assures the necessary food production, and creates jobs either directly or indirectly. The area sown to cereals is 30-40% of the total arable land, i.e. an average 1.6 M ha.

Over the last two decades, cereal production has shown an annual average increase of 2.7% with improved cultivation practices and increasing use of quality seeds, fertilizers, and chemical herbicides. Despite the progress that has been made, cereal production is still insufficient to meet national needs and satisfies only 65% of the human consumption, topping the list of food imports. Besides, cereal yields remain below potential production which is estimated to be 17 M q in an average year and 24 M q in a good year. Production is closely linked to rainfall levels and frequencies; therefore, yields can double or even triple from one year to another, or can diminish drastically, e.g. 9.2 M q in 1983, 20.6 M q in 1985, and just 6.5 M q in 1986.

In order to reduce these fluctuations in production and to secure higher, stable yields, sufficient soil water reserves should first be assured for all growth stages throughout the season. The implementation of techniques which are most suited to rainfall is still inadequate to ensure a balanced water supply for cereal crops. This can only be achieved by supplemental irrigation. Once the water needs of a crop are guaranteed, the yield depends on other production factors: crop rotations and land preparation, seeding density, fertilizers, and weed control.

This chapter is divided into 2 main parts: the first part presents some characteristics of rainfed cereal cultivation and the cereal development program while the second part focuses on supplemental irrigation as experienced in Tunisia.

Characteristics of Rainfed Cereal Cultivation and the Development Program

Importance of Cereal Cultivation to the National Economy

The cereal sector is important in terms of food security and balance of trade.

E.R. Perrier and A.B. Salkini (eds), Supplemental Irrigation in the Near East and North Africa,
513-527.
© 1991 ICARDA.

Social Development Plan (1982-86), average cereal production reached 11.7 M q in contrast to 9.94 M q during the Fifth Plan (1977-81), establishing 17.7% increased growth in production. Despite this growth, cereal production remains in deficit. During the Sixth Plan, cereal products amounted to 42% of the total food imports of which 31% were destined for human consumption, i.e. an average of 88 M dinars/yr. The country's objective is to reach the highest level of self-sufficiency possible by 1991.

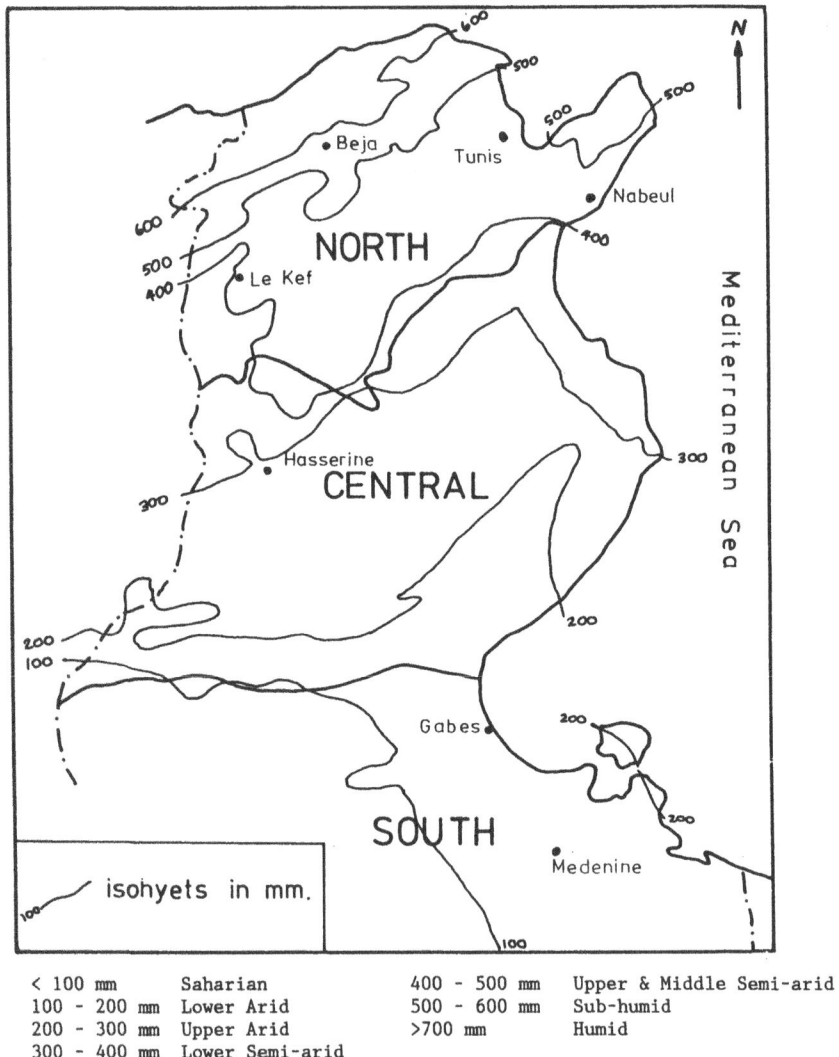

< 100 mm	Saharian	400 - 500 mm	Upper & Middle Semi-arid
100 - 200 mm	Lower Arid	500 - 600 mm	Sub-humid
200 - 300 mm	Upper Arid	>700 mm	Humid
300 - 400 mm	Lower Semi-arid		

Figure 27.1. Tunisia, showing climatic regions.

Areas of Cereal Cultivation

Cereals are cultivated in the north and the center of the country (Figure 27.1).

The northern area

The northern area is located towards the northern edge of the central mountain range. This land area includes several agro-climatic zones from humid to semi-arid. Average land area annually sown to cereals is 800,000 ha, mostly in the sub-humid zone where average annual rainfall is 500-600 mm but also in the upper and middle portions of the semi-arid zone with average annual rainfalls of 400-500 mm. Rainfed farming of cereals is practiced to produce 83% of the national cereal grain production (88% of the wheat and 67% of the barley). Two thirds of these crops (wheat and barley) are grown in the north-west where half the area receives average annual rainfall of more than 400 mm. Cereals are grown on every farm regardless of size, soil depth, or topographic detail of plain or slope because cereal grain constitutes the basic diet of the Tunisian population.

The central area

The central area is the steppe region situated between the central mountain range which extends from the Tebessa mountains to Cap Bon and along a line running from west to east at the latitude of Gafsa. This region is dominated by several agro-climatic zones: the lower semi-arid with average annual rainfalls of 300-400 mm; the upper arid zone with an average annual rainfall of 200-300 mm; and, the lower arid zone with an average annual rainfall of 100-200 mm.

The climate is characterized by insufficient and very irregular rainfall. Extensive dryland farming systems and varied sowing dates prevail depending on autumn weather conditions. Areas sown to cereals vary in size: over the last 20 years, the total fluctuation has ranged from 172,000 to 1,090,000 hectares. With rainfed cultivation, cereal crops are usually low yielding and yields are uncertain. Average yields are 3.59 q/ha for durum wheat, 4.59 q/ha for bread wheat, and 3.59 q/ha for barley. In the arid zones, rainfall is insufficient and cereal cultivation depends on structures for retaining runoff and distributing flood water for spate irrigation.

Farming Systems

The total number of farms in Tunisia is estimated to be 355,000, either state owned or privately owned. State owned farms include 800,000 ha and are managed by the Service of Public Lands (National Land Registry), by agricultural production cooperatives, or by agricultural development companies. In this sector 80,000 ha are allotted to cereal production each year. The structure of the private farms is somewhat complicated which makes it difficult to apply advanced agronomic techniques, especially supplemental irrigation of cereals.

The intensification of cereal crops by irrigation depends on several factors such as farm size, field patterns, land tenure, soil characteristics, production systems, and water supplies. In this complex situation, the size of farm is a clear indicator of the potential for intensifying cereal production. In Tunisia, there are 5 types of farms.

1. Small farms of less than 5 ha (44% of all farms) where cereals are grown for household consumption. Resources are limited; equipment is lacking; and, the labor is done by the farmer's family. In rainfed farming, durum wheat and barley are the main crops. When there is sufficient water for irrigation, the farm produces vegetables. Cereals, for household consumption, continue to be grown under dryland conditions.
2. Farms of 5-10 ha are more favorable for production than are smaller farms; but, cereals are still produced primarily for household consumption.
3. Farms of 10-20 ha with cereal production, particularly wheat.
4. Medium sized farms of 20-50 ha represent 11% of the land. They are planted predominantly to cereals and comprise 20% of the large scale farming with crop rotation. These farmers often rely on seasonal loans and have difficulties with repayments. Mechanization exists though it is limited. Farmers in this category are more receptive to cultivation technology and development processes and look forward to increasing their incomes by limiting the risks related to varying climatic conditions of the region. These farms are favorable for cereal crop intensification by means of supplemental irrigation.
5. Large farms of more than 50 ha are located mainly in Kef, Beja, Zaghouan, Bizerte, and Siliana. There are about 5,000 large farms in northern Tunisia of which 2,000 are more than 100 ha. Increasing the value of the land is considered important. Mechanization is generalized. Farming systems in this category operate to attain an acceptable income with the minimal risk. Despite different investment possibilities and good conditions for intensification, the farmers' attitudes are not yet favorable to development actions.

In farm categories 2 and 3 with farm sizes of 5-20 ha, individually owned equipment is limited; but, often, mechanical equipment is hired. Despite their limited resources, these two categories of farms have a potential for intensified farming especially if water is available for supplemental irrigation. However, maximization of production by supplemental irrigation is not a main objective shared by all farmers.

Crop Rotation

A 2-year wheat – fallow rotation prevails, especially in the north-west. To the north-east and a part of the sub-humid zone, 3- or 4-year rotations are modified to realize a proportion of 50% cereals. Vetch and oat fodder are often grown which constitutes a bad precedent for cereal production because these crops are grass not legumes.

Major Constraints

Different development plans and specific studies have identified constraints which could hinder improvement of the cereal sector. These constraints are diverse and of concern at many levels of cereal production.
1. Despite the adoption of new policies of subsidies, credits, and pricing programs to encourage production, the cereal sector is not yet competitive with other crops such as fruit trees or vegetables.
2. Implementation (into the existing farming systems) of congruent technical processes is difficult because of the diverse ecological conditions.
3. Accustomed to rainfed cereal cultivation, farmers are not always disposed to accept the new investments inherent with innovation.
4. Cereal cultivation is affected by several land tenure problems: fragmented fields, short term farmland leasing, high rental fees, and small size farms due to family inheritance.
5. Introduction of intensified cereal production requires specific solutions for each farm category: large farms are characterized by absentee ownership and extensive production; medium farms tend towards diversified cultivation; and, small farms have few resources and produce for household consumption.

Development Program for the Cereal Sector

The agricultural sector was given national priority in the Sixth Plan (1982-86) and in the Seventh Plan (1986-87) this priority was consolidated into a national objective of self-sufficiency of cereal production by the year 1991: total sufficiency of durum wheat and barley and 27% of bread wheat.

To reach this objective, rational efforts have focused on the mobilization and exploitation of the production potential of available land, water, and technology and implementation of policies to support the cereal producer. With this in mind, a development program was expounded for the cereal sector to define the actions for removal of constraints hindering development in all farm categories.

The Land Directorate has established a soil map for cereal cultivation. This map is of great interest to northern Tunisia where cereal cultivation prevails. This map establishes a hierarchy of the soils appropriate for cereal rotation and considers soil characteristics and the agro-climatic restraints. Soil analysis laboratories have been established to acquaint farmers with their soils for improved management.

A national project aiming at the development of supplemental irrigation of cereals was implemented in 1986-87 to improve production in irrigated areas and in those areas with a potential for irrigation.

Supplemental Irrigation of Cereals

Several experiments are investigating the irrigation of cereals. In this chapter, a summary of major results will be given with fundamental characteristics of the national supplemental irrigation project of cereals and the results of the 1986-87 season.

Major Findings of Experimental Results

Water requirement and critical growth stages of wheat

Since 1962, several experiments have been performed on wheat to determine crop water requirements and the critical stages of growth. The results have shown that the quantity of water needed can vary from 480-570 mm according to the cultivation cycle. The distribution of water throughout 5 different stages of growth in a cultivation cycle is shown in Table 27.1.

Table 27.1. Water requirements of wheat.

Time period of water use Month	20/11-10/12 Dec	10/12-20/2 Jan	20/2-20/3 Feb	20/3-20/4 Mar	20/4- Apr	Total
Water (mm)	80	120	100	100	80	480
Consumption (mm/day)	2	2	3	4	5	

Other experiments have determined the average daily consumption of water by wheat during the cultivation period. Total consumption is 480 mm from December to April. Data for a 10 yr period show that yields for durum wheat can vary between 50-60 q/ha with total water consumption of 400-500 mm. Analysts agree that wheat is sensitive to water deficiency during the growth stages of stem elongation, ear emergence, flowering, and grain development. These results agree with those in the FAO Irrigation and Drainage Bulletin No. 33 on water yields.

Influence of supplemental irrigation on wheat yields

The impact of supplemental irrigation on wheat yields has been tested at 2 experimental stations, Cherchef (upper semi-arid) and Hendi Zitoun (lower semi-arid, a "cutoff" zone). In both cases, irrigation used runoff water conveyed to previously leveled basin plots. Fertilizer rates averaged 150-200 kg of 45% super phosphate, 100-150 kg of potassium sulphate, and 250-300 kg of 33.5% ammonium sulphate. Cultivation practices used were similar to mechanized methods. The results in absolute terms were dependent upon soil management and successful chemical weed control.

Cherchef experimental trials: Cherchef Station is located in the upper semi-arid zone and has an average annual rainfall of 450 mm. Cereals frequently suffer from water deficits because of irregular rainfall distribution, particularly in April. Several trials have been completed on durum wheat at this station to evaluate the impact of supplemental irrigation on yields. Table 27.2 presents the results of 1982-83 research on the Karim variety. Note that the yield (20.6 q/ha) from rainfed cultivation doubled (43.4 q/ha) with 2 supplemental irrigations totaling 176 mm applied in February and March.

Table 27.2. Durum wheat, Karim variety, 1983 yields.

Treatment	Irrigation applied				Total water applied (mm)	Rainfall Dec-May (mm)	Yield (q/ha)
	17/2 (mm)	23/2	14/4	03/5			
T_0	0	0	0	0	0	197	20.6
T_1	78	0	0	0	78		22.3
T_2	78	98	0	0	176		43.4
T_3	78	98	65	66	307		44.2

In 1984-85, recorded rainfall was 610 mm (Table 27.3) and yields were 47.2 q/ha for rainfed farming. These yields increased to 54.9 q/ha by using rainfall runoff applied during the growing season as supplemental irrigations (81 mm). Yields were 0.95 kg of durum wheat grains for every 1 m^3 of irrigation.

Table 27.3. Monthly rainfall.

Season	Monthly rainfall (mm)								Annual rainfall (mm)
	Sep	Oct	Nov	Dec	Jan	Feb	Mar	Apr	May	
1982-83			143	15	68	0	19	0	15	260
1983-84			13	6	1	16	39	14	41	130
1985-86			3	0	3	19	6	22	7	160
1986-87	11	16	56	20	17	10	22	17	6	175

During 1986-87, a favorable season despite the deficit rainfall in April (24.6 mm) and May (16.4 mm), yield for the same variety (Karim) under rainfed conditions was 52.39 q/ha. Two irrigations, one in March (40 mm) and one in April (60 mm), achieved an increase from 52.3-62.6 q/ha, a productivity of 1.03 kg of grain for every 1 m^3 of irrigation.

These results for both seasons (1984-85 and 1986-87) indicate that even in a favorable season, supplemental irrigation can improve yields considerably, by supplying water to fulfill possible deficits. Temporary water deficits are frequent in both the upper and middle semi-arid zones and sometimes even in the sub-humid zones. In other words, water deficits frequently occur in the north, the largest region of cereal production.

These data, in addition to other results obtained in the High Valley of Mejezdah in the north-west, demonstrate that a supplemental irrigation of 100-150 mm is sufficient to compensate for weather dependent deficits in the northern areas.

Hendi Zitoun experimental trials: Hendi Zitoun Station is located in the lower semi-arid zone where the average annual rainfall of 338 mm is often unevenly distributed throughout the season and not adequate for economic crop production. At this station and in the central region, rainfed production of cereals, especially wheat, is uncertain; nevertheless, farmers still grow cereals whenever autumn rainfall permits. Yields are never guaranteed because drought regularly occurs in these areas.

Traditionally, farmers with shallow wells grow barley with irrigation, graze for green fodder, and then leave the crop for grain production. Farmers in areas affected by flood water distribution do not hesitate to combine the necessary structures for water spreading and cereal production.

Despite these traditions, irrigation of cereals is not yet exhaustively practiced using surface wells and recently developed irrigation schemes. For this reason, experiments have been carried out at Hendi Zitoun Station to evaluate the effect of supplemental irrigation on yields. Table 27.4 shows the results for 5 seasons.

When wheat seedling emergence takes place in a moist soil (T_0) or when the soil is irrigated during the first growth stage (T_1), yields vary between 13-21 q/ha dependent upon other cropping conditions. The same treatments also show that at Hendi Zitoun Station, on alluvial clayey soils with a retention capacity of 35% and 150 mm of water available to a depth of 80 cm, the yields are 15 q/ha with one irrigation of 50-75 mm and increases to 21 q/ha when 100 mm of water is applied.

When germination and emergence is assured either by rainfall or irrigation, yields depend upon conditions of water supply for plant growth during February, March, and April. When one irrigation is applied in February, the yield is 36 q/ha (T_2); but, when two irrigations are applied, one in February and one in March, the yields increase to 46 q/ha.

Yields could exceed 45 q/ha or even 60 q/ha if, during dry years, irrigations were applied regularly in February, March, and April (see Table 27.4). Supplemental irrigation of cereals is necessary to insure satisfactory yields

Table 27.4. Durum wheat yields, Karim variety, at Hendi Zitoun Station.

Treatment	Irrigation date, and amount (mm)								Total water applied (mm)	Yield (q/ha)
	Nov	Dec	Jan	Feb	Mar	Apr	May	Total		
1982-83					3	4	5			
T₀					0	0	0	0	260	15
T₁					90	0	0	90	350	26
T₂					90	65	0	155	415	29
T₃					90	65	155	277	530	33
1983-84		1		4		7				
T₀		0		0		0		0	130	0
T₁		108		0		0		108	238	21
T₂		108		102		0		210	340	37
T₃		108		102		100		310	440	60
1984-85		20		20	25	10 25	3			
T₀		0		0	0	0 0	3	0	130	0
T₁		48		0	0	0 0	0	48	178	13
T₂		48		0	60	0 0	0	108	238	36
T₃		48	80	0	70 70	0	268	298	46	
T₄		48		0	80	70 0	70	268	398	39
*1985-86**		1	11 28	24	1 14	7				
T₀		0	0 0	0	0 0	0	0	160	6 2	
T₁		60	50 50	0	50 0	50	0	420	55 63	
T₂		60	60 0	60	0 60	0	60	460	52 55	
T₃		60	0 70	0	0 60	0	70	420	48 51	
1986-87			31	2	3	6				
T₀			0	0	0	0		0	175	2
T₁			75	0	0	0		75	250	15
T₂			80	0	0	75		155	330	47
T₃			75	0	0	54		129	304	25

* Ben Bechir variety

Supplemental irrigation of cereals is necessary to insure satisfactory yields because the central region consistently experiences a shortage of water. To sustain increased yields, an adequate water supply has to be available during February, March, and April to secure germination and seedling emergence and crop growth needs. Depending on the annual rainfall, the critical volume of supplemental irrigation varies between 200-250 mm.

Large Scale Application of Supplemental Irrigation

Large scale supplemental irrigation has given encouraging results. In 1986, the average wheat yield on a land surface of 15,000 ha using supplemental irrigation was 20 q/ha in contrast to a yield of 4 q/ha for production under rainfed cropping. In Jendouba region, the organized sector achieved an average yield of 41 q/ha using supplemental irrigation and 6 q/ha with rainfed farming. In Lower Medjerdah Valley, the results were 34 q/ha with supplemental irrigation but only 12-15 q/ha under rainfed cultivation. In the Kairouannais, average yields were 18 q/ha with supplemental irrigation and 0-2 q/ha under rainfed farming. Supplemental irrigation of durum wheat varieties (Karim and Ben Bechir) and of bread wheat varieties (Salambo and Carthage) produced yields of 25 q/ha for durum and 26 q/ha for bread.

In Sidi Bouzid region, an ancient farming practice of flood water spreading on cereals survives, as well as modern irrigation techniques using public water works (PPI) and surface wells. Land cultivated for cereal production with irrigation reached 5,415 ha in 1986, an increase since 1983 from 3,100 ha. Average yields from irrigated agriculture were 25 q/ha for barley, 20 q/ha for durum wheat, and 30 q/ha for high yielding wheat varieties. Estimated average yields obtained using the flood water spreading method were 10 q/ha in contrast to yields of 2 q/ha for rainfed farming. Yields of 60 q/ha were recorded on experimental plots with scheduling of supplemental irrigation.

National Project for Supplemental Irrigation of Cereals

From the outset, justification of supplemental irrigation has been in two quite different contexts: the first is related to the extension service's sphere of influence with farmers who have water supplies in winter but who continue to rely on rainfed farming practices; the second (not directly related to irrigation proper) concerns the farmers who practice traditional flood water spreading (spate irrigation) for cereal production in the central and south-central regions. The concern was to convince farmers to develop water resources for supplemental irrigation of cereals to improve yields by investing in management techniques and irrigation equipment. These traditional practices needed encouragement and consolidation.

These two contexts for justification were outdated quickly as experimental data verified that supplemental irrigation could secure high yields and there was sufficient water in reservoirs in the northern area for supplemental irrigation of cereals. Water storage in the reservoirs was reviewed with due consideration to the average absolute minimum volume and capacity of the wadis as well as depth to water table from 15 January to 15 April.

Identification of water resources specified certain zones where supplemental irrigation systems could be implemented either within or around irrigation schemes, with exploitation of groundwater, or diversion of rainfall runoff from wadis.

A project establishing supplemental irrigation of cereal production was implemented on 63,400 ha: 36,400 ha in the north and 27,000 ha in the central region. This was to ensure economic profitability for the community and

individual farmers. Areas designated for supplemental irrigation are divided according to water resources:
1. 20,200 ha within state-owned irrigation schemes (PPI) (already equipped);
2. 4,200 ha in the PPI to be developed for irrigation;
3. 21,500 ha with rainfall runoff from wadis; and,
4. 17,500 ha using groundwater.

Implementation of this project was planned in 2 phases: to begin in 1986-87, on 38,500 ha (20,200 ha, already equipped, on PPI and 18,300 ha, to be developed) and, to begin in 1987-88, on 25,000 ha irrigated with runoff in wadis and groundwater.

Funds for project implementation were estimated at 32.23 M Tunisian dinars. Sustainable production is expected to total 2.85 M q which was based on average yields of 45 q/ha. Additional production of 1.9 M q/yr is estimated based on an additional average yields of 30 q/ha (average yields of 75 q/ha or a total yield of 3.75 M q/yr).

Several incentives have been offered to encourage farmers to use supplemental irrigation for cereal production; particularly as regards investing in equipment. The scheme for financing equipment purchases includes self-financing (10%), subsidies (25%), and loans (65%). At the present stage of project implementation, no fundamental constraint exists to water resources or credit availability for acquisition of necessary equipment for supplemental irrigation systems.

Where gravity irrigation is used (Lower Medjerdah Valley), programs for land leveling are planned. In the central region, farmers are encouraged to level lands around their wells. For farmers who have not yet acquired the necessary equipment, sprinkler systems are being recommended for supplemental irrigation of cereals to insure economical applications of water.

Results of the 1986-87 Season

The 1986-87 season is considered in the framework of the national project of supplemental irrigation of cereals. Irrigated cereal production, which is estimated to occupy 23,501 ha of which 20.701 ha or 88% of the total land area, is located in the center of the country and suffers from severe drought.

Table 27.5 shows maximum and minimum yields of irrigated cereals in 5 governorates. The data indicate that yields resulting from supplemental irrigations can be particularly high in the north and the center of Tunisia. For durum wheat, the highest yields amount to 43.99 q/ha (Kasserine) and 103.6 q/ha (Jendouba). For bread wheat, the highest yields are between 22.7 q/ha and 78.3 q/ha; and, for barley, the highest yields are between 26.7 q/ha and 88 q/ha.

The previous values show that cultivated varieties have a relatively high potential that can somehow be easily achieved in favorable conditions when there is an adequate supply of water for production.

Table 27.5. Maximum and minimum yields (q/ha) for various cereals with supplemental irrigation.

Location Governorate	Durum wheat		Bread wheat		Barley	
	Min.	Max.	Min.	Max.	Min.	Max.
Jendouba	19.8	103.6	17.6	65.9	—	—
Beja	52.2	72.5	54.4	78.3	—	—
Kairouan	13.5	60.9	11.8	61.1	8.0	88.0
Kasserine	16.1	43.9	—	—	25.3	33.3
Sidi Bouzid	5.2	58.7	13.3	22.7	9.3	26.7

Table 27.6 shows the average yields evaluated for irrigated and rainfed production of wheat and barley and the differences. Information given in this table demonstrates that, in the north, the average yields of wheat and barley have increased from 24.9 q/ha under rainfed conditions to 42.4 q/ha with supplemental irrigation, i.e. a difference of 17.5 q/ha or an increase in yields of 70%.

Table 27.6. Comparison of yields (q/ha) for various cereals under rainfed conditions (R) and with supplemental irrigation. (SI)

Location	Durum wheat			Bread wheat			Barley			Comparison of yields		
	SI	R	Diff	SI	R	Diff	SI	R	Diff	SI	R	Diff
North	42.7	24.3	18.4	46.3	30.5	15.8	27.8	15.0	12.8	42.4	24.9	17.5
Beja	44.0	25.6	18.4	51.0	30.7	20.3	—	—	—	45.4	26.6	18.8
Jendouba	40.8	22.1	18.7	41.6	30.0	11.6	27.8	15.0	12.8	39.0	22.8	16.2
Central	21.6	5.8	15.8	20.3	8.4	11.9	19.8	5.3	14.5	20.6	5.7	14.9
Kairouan	25.0	7.5	17.5	24.6	9.7	14.9	24.6	5.3	19.3	24.7	6.9	17.8
Kasserine	19.7	4.9	14.8	17.0	10.5	6.5	20.0	6.2	13.8	19.7	5.9	13.8
Sidi Bouzid	19.9	4.7	15.2	10.3	5.2	5.1	14.0	3.8	10.2	16.7	4.4	12.3

In the central regions, the average increases in yield were from 5.7 q/ha in rainfed farming to 20.6 q/ha with supplemental irrigation, i.e. a difference of 14.9 q/ha or an increase of 261%. As for durum wheat, the improvement of average yields amounted to 75% in the north and 272% in the center of the country.

Improvement of yields with supplemental irrigation is also important in favorable rainfall years (70% improvement for cereals and 75% for durum wheat in the north). It is even more important when water shortages are more acute (261% improvement for all kinds of cereals in the central region). In addition to this evaluation of yields provided by the Director of Planning and Statistics and Economic Analysis, yields have been estimated by collecting samples from farms growing durum wheat under rainfed farming and farms with supplemental irrigation of wheat. These results are presented in Table 27.7.

Table 27.7. Comparison of yields of durum wheat with supplemental irrigation (SI) and under rainfed (R) conditions, 1986-87.

Location Governorate	Farmer sample	Rainfall Sept-May (mm)	Irrigation applied (mm)	Ave. yield SI (q/ha)	R	Diff	% diff	Water prod. (kg/m³)
Bizerte	3	729	53	46.3	28.5	17.8	62	3.3
Ariana*	1	450	100	62.6	52.3	10.3	20	1.0
Zaghouan	8	426	48	33.2	13.8	19.4	140	4.0
Beja	7	511	100	53.0	28.5	24.5	86	2.4
Jendouba	16	513	61	57.3	43.0	14.3	33	2.3
Kef	8	360	63	36.4	12.0	24.4	203	3.9
Siliana	8	416	134	49.8	30.7	19.1	62	1.4
Average yields				48.4	29.4	18.6	86	
Kairouan	19	175	135	27.6	4.0	23.6	590	1.7
Sidi Bouzid	11	175	200	23.3	3.0	20.3	676	1.0
Average yields				25.4	3.5	21.9	633	

* Experimental plot.

Results show that yield improvement on farms with supplemental irrigation varies in the north between 33% and 203% increase in production despite fairly good climatic conditions. In the center of the country where rainfall is low, irrigation resulted in increases in yields of 7 times the rainfed production yields at Kairoum and 8 times the rainfed production yields in Sidi Bouzidi.

Yield improvement in the north is 86% and in the central areas, 33%. Water productivity with supplemental irrigation, generally, has been superior to 1 kg of grain/m³ water. The productivity has amounted to 1.7 kg of grain/m³ at Kasserine and 1.0 kg of grain/m³ at Sidi Bouzidi. In the north, this value has ranged from 1.4 kg of grain/m³ where supplemental irrigations of 1,340 m³ of water has resulted in an additional yield of 19.1 q/ha to 4 kg of grain/m³ where an additional yield was 19.4 q/ha with only 480 m³ of water.

These results, recorded in farmers' fields, prove that supplemental irrigation helps considerably towards improving yields in the center of the country and, in favorable years, even in the north.

Conclusions

Cereal cultivation holds a very important place in the agriculture and national economy of Tunisia. Despite the progress achieved during the last two decades, cereal production has remained in deficit and is very unstable from one year to the next. This is a consequence of rainfed cultivation of cereals which is highly dependent upon the climatic conditions which result in variable rainfall.

Results recorded at experimental plots as well as on farmers' fields emphasize the necessity of implementing supplemental irrigation for cereal production in the irrigable zones in the center and north of Tunisia with the goal of improvement of yields. The north does not need huge quantities of water, only 100-150 mm, and the central area needs moderate quantities of 200-250 mm. The productivity of water with irrigation is in general superior to 1 kg of grain/m^3 water with the most common values ranging between 1.1-2.4 kg of grain/m^3.

It must be stated that improvement in production does not only depend on water but on the cultural conditions of cropping patterns and on the socio-economic environment.

References

Bouzaidi, A., 1982. On cereal production improvement; summary of studies on rainfed and irrigated wheat.

Bouzaidi, A. and Z. Chaabouni, 1985. Soil-water relation for wheat cropping; (in French) Forum International Sur Classification des Sols, 2-14 Sept. 1985.

C.R.G.R./CATID, 1972. Irrigation of annual cropping of wheat; (in French) Doc. no. 6.

El Amami, S., A. Bouzaidi, and E. Bouaziz, 1985. Supplemental irrigation of wheat; summary of a decade of experiments. (in French) C.R.G.R. Records no. 17.

FAO, 1987. Yields response to water; FAO Irrigation and Drainage Bulletin No. 32. pp 201-209.

Koop, E., 1976. The potential of production in the semi-arid area of Mejerdah High Valley in Tunisia under sprinkler irrigation; Vol. 38. German Office for Technical Cooperation.

Lobert, M. A., 1964. Wheat and water yields in the semi-arid zone; Note H.A.D. - 15 CREGR No. 83, May.

Minister of Agriculture, Cereals Office, 1986. La cerealiculture - Le Bilan (1964-1985); January (in French).

Minister of Agriculture, Ministry of Agric. Prod. and Agro- Alimentation, FAO, Investment Center, 1986. Development program of the cereal sector; Report No. 142/86 TATUN 54 - TCP/TUN/4505 - 27 November (in French).

Ministry of Agriculture, 1987. VIIth econ. and social development plan, 1987-91; Agriculture and Fishing, March; (in French).

Ministry of Agriculture, 1987. Supplemental irrigation of cereals; Project Presentation. Progress of the first part; Presentation of the second part, May; (in French).

Ministry of Agriculture, PSAE Directorate, 1987. Note on the results of the enquiry on irrigated cereals; August; (in French).

Montenay, B. A., 1970. Water and energy balance of a wheat cropping in the semi-arid area; problems of irrigation and productivity. INRAT Annals, Vol. 43; Fasc; (in French). 1036 pp.

Souissi, A., 1986. Maps of valid soils for cereal cropping in Northern Tunisia. -E -S 232; (in French); Direction des Sols, Min. of Agric.

Vernet, A., P. Mousset, and P. Saglio, 1969. Nitrogen nutrition and water consumption of irrigated wheat under Tunisian climate. Bulletin Ecole Nationale Superieure Agric., Tunisia (19 pp; in French). pp 77-102.

Chapter 28

Supplemental Irrigation in Turkey

NECATI GULBAHAR

Geographical Characteristics

Turkey has a land area of 779,452 sq km and a population of 50,664,000. The country has influential geo-political status because its location serves as a natural bridge between Asia and Europe. The country is bounded on 3 sides by water – the Black Sea, the Aegean Sea, and the Mediterranean Sea – and is divided into 2 parts – Anatolia with 755,688 sq km and Thrace with 23,786 sq km.

Turkey is a mountainous country. Two parallel mountain ranges descend from east to west forging a closed basin in the Central Anatolia Plateau. These mountains isolate the central region from the influence of the seas and a typical continental climate is observed.

Climate

Turkey is in a warm climatic belt. Geographical location and geological formation cause climatic conditions suitable for the production of many kinds of crops in the 9 agricultural regions of the Anatolian Plain, Aegean Sea, Black Sea, Mediterranean Sea, and Marmara (Figure 28.1).

In the interior, the climate is arid and semi-arid with sub-tropical Mediterranean-type climates dominating the coastal plains. Three major climatic conditions occur from the influences of the seas and the inland region.

Black Sea: The Black Sea climate occurs along the northern coast. Average annual temperature is 14-15°C; summers are cool and winters are mild. Average annual rainfall is more than 1,000 mm with an even distribution throughout the season.

Mediterranean Sea: This climate is manifested in the coastal regions adjacent to the Mediterranean, Aegean, and Marmara seas. Average annual temperature is 18-20°C in areas of the Mediterranean, 14-16°C around the Aegean, and

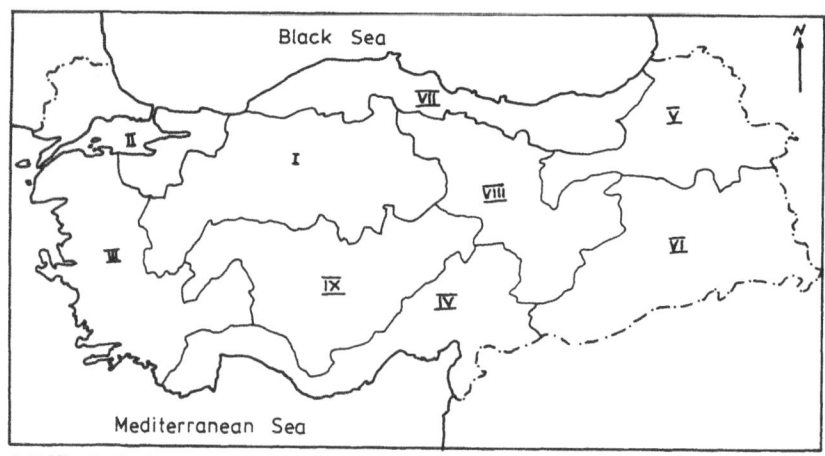

I Middle North, II Marmara, III Aegean, IV Mediterranean, V North East, VI South East, VII Black Sea, VIII Middle East, IX Middle South.

Figure 28.1. Map of Turkey showing 9 agricultural regions.

12-14° in the areas of the Marmara. Average annual rainfall is 650-1,000 mm which occurs in 3 seasons: autumn, winter, and spring.

Inland region: In the interior lands removed from effects of the sea; winters are cold and average annual temperature is 11°C with average annual rainfall of 360 mm. Summers are hot and dry and evaporation is high.

Land Resources

The total land area of Turkey is 77,945,200 ha, separated into 27,699,003 ha of arable land; 21,745,690 ha of pastures and common grazing lands; 1,102,396 ha of water surface areas; 23,468,463 ha of shrubs and forests; 3,360,248 ha of unused land; and, 569,400 ha of residential areas.

Of the arable land, 25,305,444 ha is irrigable (Figure 28.2). This arable land can be further divided as shown in Figure 28.3. With current available water resources, it would be economically feasible to irrigate an estimated 8,500,000 ha with major and minor irrigation works. A further 16,805,444 ha could be irrigated using advanced technology.

Both groundwater and surface flow are used to irrigate 3,176,330 ha. Using surface water, 4 different interests presently operate irrigation systems (Figure 28.4):
1. the public, 1,000,000 ha;
2. the State, 1,841,435 ha;
3. the General Directorate of State Hydraulic Works (DSI), 1,102,625 ha; and,
4. the General Directorate of the Rural Services (Koy Hizmetleri), 738,810 ha.

Supplemental Irrigation in Turkey

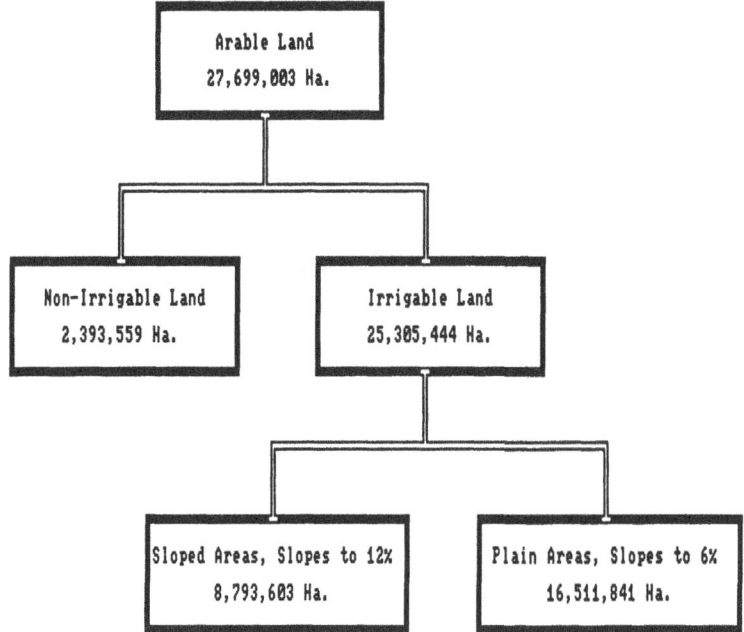

Figure 28.2. Arable land for irrigation.

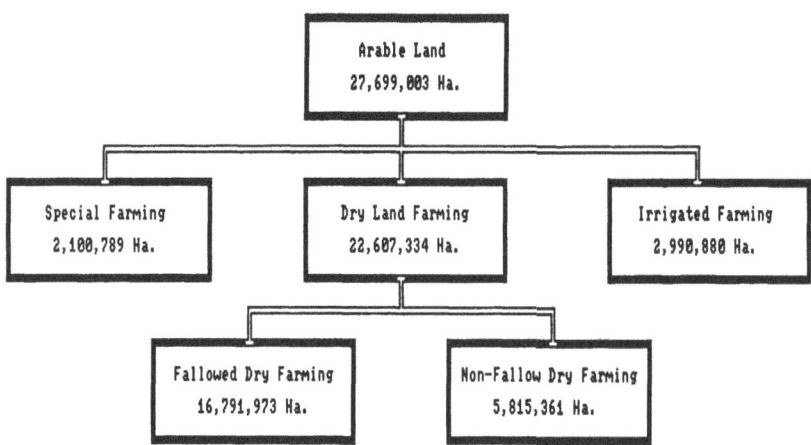

Figure 28.3. Arable land use.

Groundwater is an important source for irrigation (Figure 28.5) Public wells irrigate 55,000 ha and government wells irrigate 279,895 ha. An irrigation network constructed by DSI serves 56,580 ha.

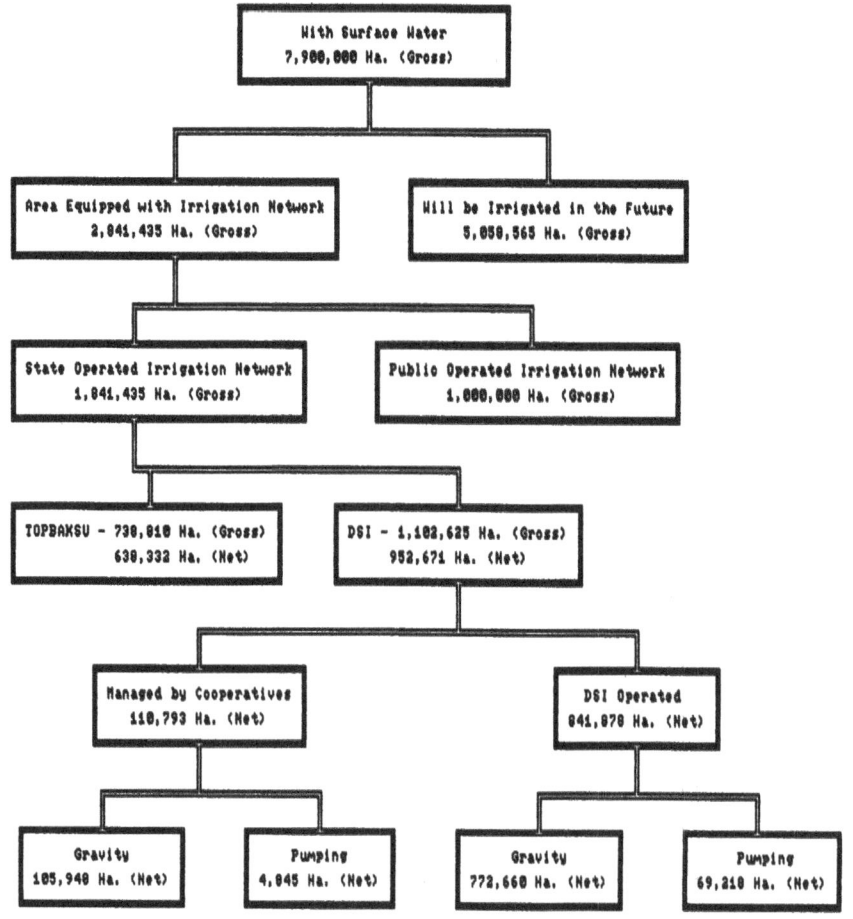

Figure 28.4. Surface water use.

Water Resources

Surface water

Twenty six basins were classified with respect to drainage patterns after problems were determined for the development of water resources in these basins. Data indicates the intensity of study given to this subject (Addendum 28A).

The average annual precipitation of 653 mm corresponds to a water potential of 509 B m³/yr. Runoff amounts to 206 mm or 185 B m³, an average rate of 36%, and the remaining 64% is lost to evapotranspiration. Runoff flows into neighboring countries: the Coruh, Arpacay, and Aras rivers into Russia; Sarisu

Supplemental Irrigation in Turkey

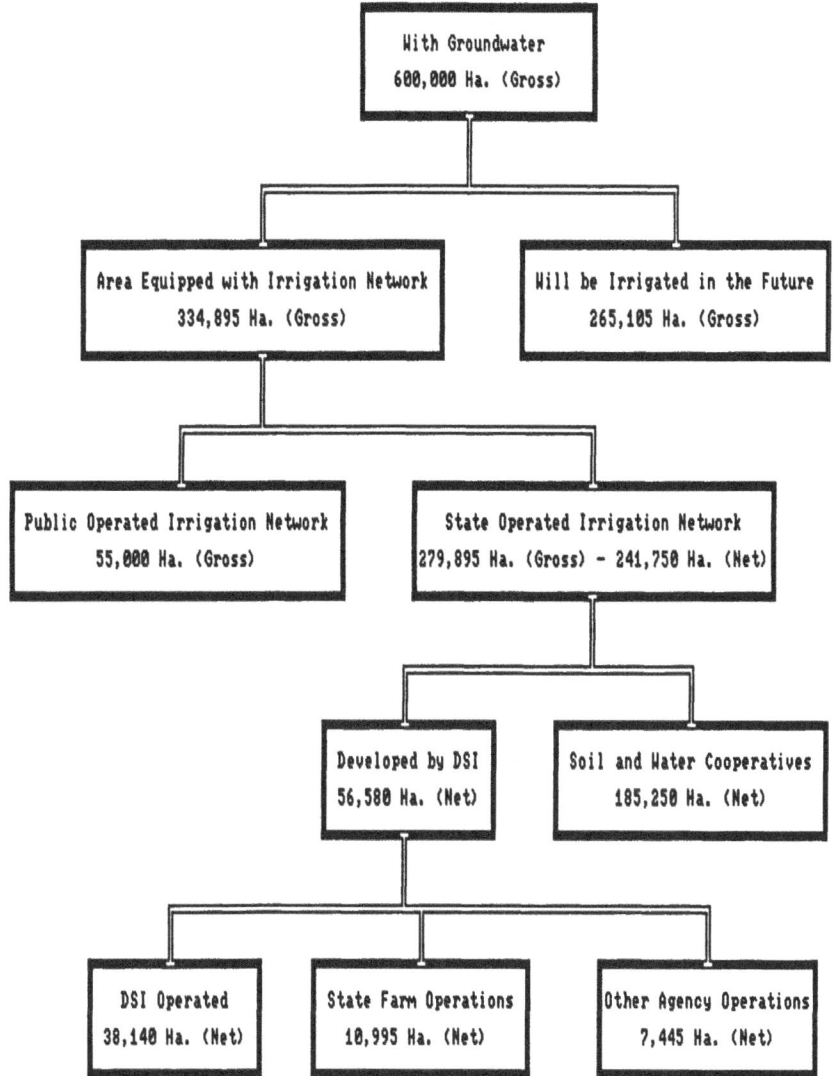

Figure 28.5. Groundwater use.

River into Iran; Dicle (Tigris) into Iraq; and the Firat (Euphrates) into Syria and Iraq.

An estimated 185 B m³ of runoff, annually, is not fully utilized. After consideration of adequate runoff to meet the water rights and requirements of Turkey's neighbors, 95 B m³/year of runoff still remains available for national consumption purposes.

Groundwater

By the end of 1982, research was completed for estimating the volume of groundwater reserves. An estimated 9.5 B m^3/yr are available for exploitation. The potential of total available water resources from surface flow and groundwater would amount to 104.5 B m^3/yr.

Reconaissance surveys of basin-wide supplies of the rivers will be conducted in the near future to determine the possible scope of water and land resource development. A summary of this research is presented in Appendix 28A. The data demonstrates that to regulate the water regime of Turkey, consideration should be given to the construction of 498 dams. The water supplies from these dams would be regulated to achieve the following:
1. irrigation of 5,925,032 ha;
2. drainage of 165,367 ha;
3. flood control of 512,320 ha;
4. conveyence of 2,520.1 m^3 of water to urban areas; and,
5. generation of 109,684 M Kwh of electric power.

Hydroelectric power plants would be constructed with a total capacity of 30,911.48 MW of generated electricity.

Overview of Supplemental Irrigation Practices

Since the concept of supplemental irrigation is new in Turkey, no particular document on this concept is available for review or evaluation. Farmers in the central areas have been observed to use a method of irrigation which can be seen to apply supplemental irrigation to winter crops. Farmers in other regions apply water 1-2 times each growing season with the number of applications based on the annual rainfall, the principal factor used for deciding when to irrigate. Farmers in other regions do not add additional water because the rainfall is considered adequate.

Among the farmers with supplemental irrigation systems and those without, differences materialize in income, standard of living, mechanization, farming technology, and agronomic inputs. Some explanation can be offered for these differences: for example, farmers who cultivate sugar beets are well organized in a cooperative for sugar beet growers and, as members, are furnished with many agricultural inputs such as seed, pumps, pipe, fertilizer, herbicide, and, pesticide, as well as being provided with credit for agricultural purposes. This support to irrigated production contributes to the other observed differences.

Groundwater is the major source for supplemental irrigation; the farmers use a well and a pumping system to extract water. After pumping water out from the well, if topographic conditions are suitable, field application is done by gravity. This method is commonly practiced in the Central Anatolia Region where Ankara, Eskisehir, and Konya Provinces are located.

Supplemental irrigation is used by farmers in the inland region without

Supplemental Irrigation in Turkey 535

sufficient knowledge of irrigation technology and the requisite farming practices essential for efficient and effective agricultural production.

National Goals and Objectives

Supplemental irrigation could be applied both in arid and humid areas when rainfall is a problem. In arid regions, the scarcity of rainfall is a major problem for the farmers and supplemental irrigation could prevent the negative effects of drought. In this case, the crop water requirement would be met during critical periods by applications of supplemental irrigation; thus, stabilizing crop production in the area.

When the evaporation rates are high during plant growth, keeping soils at field capacity can be difficult. In regions such as Southeast Anatolia, supplemental irrigation applied at a critical point in growth would be an effective means for significantly increasing yields.

In the Central Anatolia Region, an area possessing the greatest capacity for cereal production, supplemental irrigation could increase total yields and contribute substantially to the irrigation economy. In this region, spring (April & May) and fall (October) are critical seasons of water use. If supplemental irrigation could be implemented, this would lead to tremendous increases in agricultural productivity.

In the humid areas along the coasts, rainfall distribution during the year is a major problem for agricultural activities, especially during the vegetative growth stage for crops. Application of additional water is necessary for a higher yield. For example, in the Black Sea Region which has more than 1,000 mm average annual rainfall, scheduling supplemental irrigation at critical stages of plant growth would significantly increase yields as well as stabilize production.

Evaluation of Existing Farming Systems

The existing farming systems within each agro-ecological zone are described. After an overall national preview of Turkey, district levels are assessed. All surveyed districts, Ankara-Cubuk, Eskisehir-Central, and Konya-Karapinar, are located in the Central Anatolia Region. Overall, no significant differences in farming practices or levels of mechanization exist among the 3 districts.

Weather Station Data

In most districts surveyed at least 1 or 2 weather stations are established and data are available for designing supplemental irrigation projects. A 10-year data set of precipitation from different climatic regions is given in Table 28.1. During periods of crop growth, the inadequacy and uneven distribution of rainfall is recognized. For instance, in the Central Anatolia Region, annual rainfall is divided between winter (34%) and spring (32%) but maximum precipitation is in October. These percentages are relative and vary with location in the region. The map showing the distribution of annual rainfall gives a good assessment equivalent to geographical regions of Turkey.

Table 28.1. Average precipitation for geographical regions of Turkey, 10-yr data set.

Geographical regions	Jan	Feb	Mar	Apr	May	Jun	Jul	Aug	Sep	Oct	Nov	Dec
							(mm)					
Marmara												
Max.	102.7	72.5	55.4	60.0	43.4	61.5	89.1	73.0	91.2	81.3	81.4	100.1
Ave.	54.0	50.5	50.0	50.4	41.9	33.4	40.3	35.0	42.5	59.9	78.1	51.5
Black Sea												
Max.	86.8	52.4	54.0	59.8	47.8	48.2	113.4	103.4	75.7	147.9	72.9	78.0
Ave.	46.6	39.6	40.0	31.3	45.8	40.2	47.4	45.9	43.7	82.3	58.5	58.1
Aegean												
Max.	134.6	91.0	83.2	59.8	53.2	45.8	23.6	42.2	47.7	134.1	110.0	137.4
Ave.	105.4	89.0	75.4	53.9	37.5	21.2	7.6	4.0	11.3	54.3	102.9	95.7
Central Anatolian												
Max.	65.9	33.0	42.0	73.5	71.7	32.9	23.7	35.6	29.6	88.7	56.9	54.9
Ave.	50.6	31.4	39.8	53.3	52.3	23.6	10.1	8.0	11.2	31.8	31.5	39.2
Mediterranean												
Max.	180.6	125.0	139.2	66.5	78.8	106.2	51.0	39.9	38.3	121.3	220.2	170.2
Ave.	160.0	99.4	71.7	62.9	31.6	17.3	4.2	3.0	9.1	59.1	100.8	131.8
South East Anatolian												
Max.	140.1	145.5	85.4	73.7	50.5	30.0	6.7	2.3	19.2	81.9	83.0	104.0
Ave.	100.9	90.6	71.6	67.5	34.7	5.0	0.6	0.0	1.8	26.1	56.2	94.0
East Anatolian												
Max.	56.2	41.8	73.9	71.8	71.4	36.9	58.2	30.9	23.5	43.6	70.0	48.3
Ave.	44.1	39.8	55.3	56.4	38.3	36.1	19.2	11.4	10.4	43.0	42.4	38.1

Table 28.2 presents other climatic data of temperature, hours of sunshine, and average annual percentage relative humidity for Turkey and its 9 geographical regions. The minimum temperature for East Anatolia is −45.6°C with the maximum temperature of 48°C in the Southeast Anatolian Region. Turkey has normal values of wind speeds; for example, in the Central Anatolia Region, the dominant winds occur in the north-east with an average speed of 1.6-4.9 m/sec. Maximum wind speeds of 39.5 m/sec have been recorded in the south and south-west.

Table 28.2. Climatic data.

Geographical regions	Temperature (°C)							Sunshine (hr.min/day)	Relative humidity %
	Ave.	Min	Site	Date	Max.	Site	Date		
Mediterranean	15.6	−29.0	Goksun	1-1968	45.6	Adana	8-1958	8.06	66
East Anatolian	9.9	−45.6	Agri	1-1972	43.9	Kulp	8-1966	7.15	63
Aegean	13.9	−28.1	Kutahya	7-1948	47.0	Marmaris	8-1958	8.08	65
South-East Anatolian	16.9	−24.2	Diyarbakir	1-1933	48.0	Cizre	7-1978	8.03	53
Central Anatolian	11.0	−34.4	Sivas	2-1950	41.8	Cankiri	8-1954	7.30	63
Black Sea	11.9	−34.0	Bolu	2-1929	44.2	Gokhoyuk	7-1962	4.22	77
Marmara	13.5	−29.2	Bursa	1-1972	43.7	Balikesir	8-1958	6.32	75
Turkey	13.2	−45.6	Agri	1-1972	48.0	Cizre	7-1978	7.02	66

Current Farming and Cropping Practices

With the exception of the cultivated lands along the coasts, two thirds of the cultivated land is semi-arid and arid as defined in the Martonne Aridity Index. The most serious constraint to agricultural activities is rainfall with only one region, the East Black Sea, where rainfall occurs during each season. Farming practices in all other regions are widely subjected to conditions of climatic variation.

Reviewing the values for 1982, rainfed farming comprises 81.7% of the total 27.7 M ha of cultivable land, irrigated agriculture, 10.8%, horticulture, 3.8%, and, other crops, 3.7%. Even though irrigated land under production has been increasing, rainfed production in the arable lands remains the dominant farming system.

In a large proportion of the country, the fallow period is of vital importance to crop production. The objective of establishing fallow periods is to accumulate adequate water and plant nutrients in the soil for crop production. For this reason, the wheat – fallow rotation system is widely practiced for rainfed farming. In Ankara, Eskisehir, and Konya Provinces in the Central Anatolia Region, the rate of fallow is 44% in contrast to a prevailing rate of 33% for fallow in the total rainfed land areas. The total land cultivated and in fallow is correlated with the average annual rainfall in Table 28.3.

Whereas some benefits may accrue from fallow – crop rotations, an economic loss occurs in total productivity. For this reason, the government initiated a fallow reduction project in 14 provinces (including Ankara, Eskisehir, and Konya) when average annual rainfall is more 400 mm. Production of pulses and fodder crops is encouraged, especially lentils, chickpeas, and vetch (seinfoin), with the goal of further reducing fallow

Table 28.3. Cultivated and fallow land with average annual rainfall in each region.

Geographical regions	Land area		Applied rates of fallow		Average annual rainfall (mm)
	Cultivated	Fallow	Cult.	Fall.	
	(1,000 ha)		(%)		
Central Anatolia	5,607	4,367	44	53	420
Southeast Anatolia	1,668	1,048	39	13	744
Inter zones	2,327	1,207	34	15	584
East Anatolia	1,489	937	39	11	605
Coastal Regions	5,391	701	12	8	828
Turkey	16,354	8,200	33	100	652

rotation. Since implementation of this project, the amount of land in fallow has been reduced to an average of 3 M ha in the last 5 years.

Other important agricultural activities and projects which have been implemented or supported by the Ministry of Agriculture, Forestry and Rural Affairs are discussed.
1. Developments in agricultural support include credit, fertilizer, and animal breeding.
2. Three major inputs, irrigation, seeds, and fertilizer, have had a significant effect on crop production. More than 50% of investments in agriculture allocated by the government are devoted to irrigation projects. In spite of this high proportion of investment, a large part of land that is irrigable remains under rainfed cultivation and crop production still relies upon fluctuating weather conditions. In the government's program, the importance of the seed industry has been stressed as a means for advancing crop production. Although fertilizer consumption has steadily increased in recent years, use has not reached the desired scale.
3. Five prominent programs for agriculture have been implemented: second cropping, fallow reduction, forestry, livestock development, and fisheries.

Central Anatolia region: This region is considered an arid area and includes Ankara, Eskisehir, and Konya Provinces with 420 mm average annual rainfall and an average temperature of 11°C. Not only a deficiency in average annual rainfall but also the distribution within the year has influenced farming practices. The rainy seasons are fall, winter, and spring; therefore, rainfed farming is viewed as the compulsory system for crop production.

Table 28.4 shows the total land sown with the crop produced, actual area harvested, total production, and the crop yields. Data shows that the greatest proportion of land is sown to wheat with the highest production in the region. Other major crops are barley, oats, pulses, industrial crops, tuber crops, and fruit trees.

Generally, tillage operations in this region are done in spring and summer; but in other regions, tillage is done in autumn. The moldboard plow, duckfoot

Table 28.4. Major agricultural products: area sown, area harvested, and yields, 1984.

		Province/District				
		Ankara		Eskisehir	Konya	
Crop	*	Bala	Cubuk	Central	Karapinar	Central
Cereals						
Wheat	A	99,477	34,434	46,868	71,738	149,215
	B	98,989	34,434	46,869	71,738	149,215
	C	232,026	70,006	112,610	119,328	289,568
	D	2,344	2,033	2,403	1,663	1,941
Barley	A	29,353	9,295	30,821	63,599	47,944
	B	29,353	9,295	30,821	63,599	47,944
	C	82,203	24,296	86,313	118,738	100,699
	D	2,815	2,615	2,800	1,867	2,100
Oats	A	----	----	1,291	----	5,960
	B	----	----	1,291	----	5,960
	C	----	----	1,584	----	10,660
	D	----	----	1,227	----	1,789
Maize (No data)						
Pulses						
Chickpea	A	399	1,098	299	200	1,996
	B	399	1,098	299	200	1,996
	C	402	1,106	271	131	3,318
	D	1,008	1,007	906	655	1,662
Lentil	A	3,984	896	199	299	4,980
	B	3,984	896	199	299	4,980
	C	3,203	901	120	150	7,508
	D	804	1,006	603	502	1,508
Dry bean	A	196	279	275	16	944
	B	196	279	263	16	944
	C	251	356	263	20	1,807
	D	1,281	1,276	956	1,250	1,914
Industrial crops						
Cotton (lint), (No data)						
Sugar Beet	A	----	----	9,560	2,478	5,364
	B	----	----	9,546	2,478	5,364
	C	----	----	374,543	68,358	1,316
	D	----	----	39,236	27,586	245
Tuber crops						
Potato	A	----	----	494	49	1,186
	B	----	----	494	49	1,186
	C	----	----	7,675	384	19,186
	D	----	----	15,536	7,837	16,177
Dry onion	A	----	----	196	29	343
	B	----	----	196	29	343
	C	----	----	1,538	433	5,385
	D	----	----	7,847	14,931	15,700

Table 28.4. (Continued.)

Crop	*	Province/District				
		Ankara		Eskisehir	Konya	
		Bala	Cubuk	Central	Karapinar	Central
Oil seeds						
Cotton (seed), (No data)						
Soybean (No data)						
Sunflower	A	501	----	----	----	----
	B	501	----	----	----	----
	C	706	----	----	----	----
	D	1,409	----	----	----	----
Fruits						
Oranges (No data)						
Mandarins (No data)						
Grapes	A	488	14	137	239	1,610
	B	----	82	98	698	13,428
	C	2,441	----	----	----	----
Apples	B	492	----	----	----	----
	C	----	6,222	590	----	----
	D	27,768	----	----	98	7,365
	E	----	114,049	39,768	11,137	198,346
Pears	B	344	----	----	----	----
	C	----	5,324	----	----	----
	D	14,385	----	103	78	7,365
	E	----	119,542	13,789	4,167	99,205

* A = Area sown (ha)   B = Area harvested (ha)
  C = Production (tons)   D = Yield (kg/ha)   E = No. of bearing trees

harrow, and hoe are the standard equipment to achieve a depth of 18-20 cm. Some farmers use a disk-type plow for this purpose even though extension workers have warned against the practice. This equipment mixes the soil too much, causes a single grainy structure at too shallow a depth, and besides, requires a more powerful tractor.

Plant varieties which are well adapted to local agro-ecological conditions and resistant to periods of drought and winter cold should be grown to obtain maximum yields. This factor is regarded as vital to improving agricultural production. The government initiated the Seed Production and Distribution Project to fortify vigorous varieties of seeds.

Two different fertilizers, Di-Ammonium Phosphate and Ammonium Nitrate, are applied: the first is applied at sowing and the second during the spring. For protection of the crop, pesticide and herbicide are widely applied in the region.

Although most farmers sow certified seeds, some continue to use the same seeds for more than 5 years. Durum wheat is the major variety in the region but the Club variety is second in popularity. Usually, sowing is done in October using a seed drill, although some small farmers still sow manually. The farmers who sow manually may obtain lower yields than those who use the seed drill. Cereals are harvested in July with a combine harvester.

Irrigation applied before sowing ensures the germination and emergence of the plant; for this reason, the spring irrigation is the most important for plant growth. The amount of water applied for germination is 30-40 mm in a single application but never more than 50 mm. Moreover, additional irrigations to meet the plant water demand when rainfall is deficient reduce the need for fallow while increasing yields significantly. Plant roots grow healthier with this irrigation and the tillering stage begins early giving sufficient plant growth to tolerate the cold winter and the droughty spring. This supplemental irrigation leads to good seedling emergence, retards weed growth, and produces an even and vigorous crop stand.

Although an irrigation in the fall ensures yields, crop growth depends on the spring rainfall when the growth stages of flowering and grain filling occur. Higher yields can be obtained if an irrigation is applied in spring or when annual rainfall is adequate in fall and spring. In summary, two critical periods of plant growth need uninterrupted water availability to obtain the maximum yield in the Central Anatolian Region. Scheduling of supplemental irrigations to coincide with these two periods requires:
1. an irrigation in October to enable germination and seedling emergence; and,
2. an irrigation scheduled from 15 May to 15 June to support flowering and the filling of grain heads.

Conversely, since scheduling of soil tillage operations affect yields, farmers who till early and farmers who till late attain different yields. Operations in this region must be timed when soil has good tilth and before clods form; extremes, early or late tillages, seriously affect the water storage capacity of the soil. When the soil is tilled early, clods form; when it is tilled late, soil moisture is reduced; but, farmers who till the soil when tilth is adequate obtain higher yields. Some farmers even do tillage operations in fall, an unnecessary practice which is harmful here because the soils are erodible.

Coastal region: Ceyhan is located in the Adana Province in the Mediterranean Region. The area is conducive for growing various kinds of crops and has a favorable agro-ecological and climatic environment supportive of polycultural agricultural activities typical of the Mediterranean Region.

The major agricultural crops grown according to data for 1984 are given in Table 28.5. Industrial crops, oil seeds, and citrus predominate to provide greater profits to the farmer than other crops. The agricultural activities in the region reflect the cultivation of these crops.

Cereals are irrigated when they are in a rotation system; otherwise, they receive no irrigation. Supplemental irrigation systems are not common but conventional irrigation is practiced. A rotation system of wheat – cotton or wheat – soybean is widely practiced.

Table 28.5. Major agricultural products, Ceyhan district: area sown, area harvested, yield, and production.

Crop	*	
Cereals		
Wheat	A	59,495
	B	59,495
	C	164,937
	D	2,772
Barley	A	1,088
	B	1,088
	C	2,843
	D	2,613
Maize	A	440
	B	400
	C	1,880
	D	4,273
Industrial crops		
Cotton	A	49,595
	B	49,595
	C	42,813
	D	863
Oil seeds		
Cotton (seed)	A	49,595
	B	49,595
	C	68,501
	D	1,381
Soybean	A	7,644
	B	7,644
	C	19,207
	D	2,513
Fruit		
Orange	A	————
	B	————
	C	1,649
	D	————
	E	10,834
Mandarin	A	————
	B	————
	C	80
	D	————
	E	1,763

* A = Area sown (ha)
 B = Area harvested (ha)
 C = Production (t)
 D = Yield (kg/ha)
 E = Number of bearing trees

Supplemental Irrigation in Turkey 543

Mechanization

Mechanization, defined as a way of utilizing mechanical equipment in agricultural activities, has made significant progress in recent years in Turkey. Various types of agricultural machines and compatible equipment have been manufactured applying new technologies and are exported to the Middle East as well as meeting domestic demands. Since their introduction in 1950, an ever increasing number of tractors are in operation as shown in Table 28.6. The country can be placed in a high-use category throughout all regions. Table 28.7 presents the distribution of tractor parks showing a 64% rate of increase from 1978-84 in the Middle North Region and a minimum rate of 30% in the Mediterranean Region.

Table 28.6. Tractor parks, 1985.

Year	No. of tractor parks
1950	16,585
1955	40,282
1963	50,844
1967	74,982
1973	156,139
1977	320,578
1980	436,369
1982	491,001
1984	556,784

Table 28.7. Distribution of tractor parks, 1978-84.

Agricultural regions	No. of parks		Rate of increase (%)
	1978	1984	
Middle North	59,873	98,045	63.8
Aegean	74,484	112,876	51.5
Marmara	62,176	95,187	52.8
Mediterranean	52,727	68,706	30.3
North East	10,431	15,763	51.1
South East	14,927	21,849	46.4
Black Sea	24,425	37,054	51.7
Middle East	27,027	41,720	54.4
Middle South	44,089	65,580	48.7
Total	370,259	556,784	50.4

Although some regions have improved enough to enable a comparison to developed countries, other regions are well below this caliber of operations as can be observed in Table 28.8. Mechanization in Turkey is greater than a score of 0.5-0.8 as determined by the UN Commission on Mechanization.

Table 28.8. Indicators of mechanization, 1984.

Number	Agricultural region	No. of tractor, 1,000 ha	Horsepower, ha
I	Middle North	27.64	1.21
II	Marmara	58.88	2.45
III	Aegean	54.65	2.22
IV	Mediterranean	33.25	1.44
V	North East	18.03	0.84
VI	South East	9.43	0.43
VII	Black Sea	28.23	1.24
VIII	Middle East	29.12	1.27
IX	Middle South	19.70	0.88
Average for Turkey		23.36	1.00

Equipment for tractors has paralleled the incidence of tractors. Progress towards equipping farmers at a pace equivalent to the number of tractors is compared for the years of 1961-84 in Table 28.9. A large increase in cultivators occurred from 1961-84 with values reaching an optimal number comparable to existing tractors. Threshers are second to tractors; however, other equipment has yet to reach desired numbers.

Table 28.9. Development of tractor parks, 1961-84.

Kind of equipment	No. of tractor parks 1961	1984	Rate of increase (%)
Tractor	42,183	556,784	1,320
Plough	43,990	662,041	1,414
Disc harrow	18,173	99,348	547
Land roller	3,652	39,014	1,068
Cultivator	5,208	210,176	4,036
Trailer	25,719	508,471	1,977
Tractor drawn grain drill	8,734	110,560	1,266
Combine drill	12,092	40,355	334
Thresher	3,803	118,955	3,128
Binder	3,539	10,778	305
Centrifugal pump	11,869	64,730	545

Efficient productivity relies on a significant relationship between tractors and available equipment. To utilize tractors for agricultural activities leading to effective crop production is impossible without proper equipment in optimal supply. When farmers are forced to use tractors for non-agricultural purposes, this situation affects mechanization in a negative way.

Central Anatolia Region: When annual rainfall is more than 400 mm, the tendency is to reduce the amount of land in fallow. Central Anatolia is the region where the practice of fallowing has been widespread. Tillage operations are examined with reference to mechanization and its effect on the farmers.

Although the moldboard is the most effective plow for tilling the soil in the area, some farmers do use a swivel type tractor plow. Soil moisture is lessened and tilth is diminished when other types of plows are used. Also many farmers in the irrigated lands tend to over-till the soil; it is not economical given the cost of each operation and tends to compact the soil.

Another practice is the burning of stubble in the field which destroys potential organic matter and removes a protective covering from the soil surface. It is technically harmful to agriculture, reducing the quality of production as well as being environmentally unsound, causing desertification of arable land.

Utilization of suitable equipment for the local soils is critical because yields are reduced when equipment does not match soil characteristics. For improvement of current situations to achieve better yields, certain measures should be exercised:
1. plows are definitely to be used to upturn the soil;
2. bottom-side tillage equipment should be used for rainfed farming practices;
3. reliable equipment should reduce crop stubble to particle sizes; and,
4. use the most suitable tillage equipment available to match local soil characteristics.

Coastal region: The Ceyhan District is located in the Mediterranean Region, an area where adequate rainfall usually occurs. Sometimes farmers may not have enough time between rains to till the soil and complete sowing activities on time. To hurry the seedbed preparation for putting the seeds into the soil, they till in haste. As a result, the farmers' objective of a better seedbed for higher yields with the least possible operations suffers and may well affect economical and technical aspects of production.

Under existing practices, farmers, after harvesting cereals, grow a second crop (soy beans or maize) which requires 105-135 days to complete vegetative growth. Within this period, farmers have to accomplish several serious farming activities, from tillage to harvest. Time is limited; seedbed preparation for a single crop can require 6-7 tillage operations; and these activities take a comparatively long time, shortening the time left for producing a second crop.

Several measures can be taken to improve this situation but the following can be foreseen:
1. tillage operations should be the least possible number;
2. farmers should choose suitable combinations of equipment for achieving activities in the shortest amount of time; and,
3. farmers need more powerful tractors for tillage.

Fertilizer Use

Although there are scant discernible differences in fertilizer use when comparing practices in all regions, nationally applications have not yet reached desired levels. Nonetheless, some impressive achievements have been realized and now fertilizer is applied to 30% of the land area. Total fertilizer consumption for Turkey is given in Table 28.10.

Table 28.10. Total fertilizer consumption, 1981-86.

Chemical fertilizer	Consumption (t/yr)					
	1981	1982	1983	1984	1985	1986
Nitrogen (N)	776,408	847,241	990,806	998,384	920,568	953,563
Phosphorous (P_2O_5)	495,308	569,624	617,975	574,728	476,013	519,885
Potassium (K_2O)	37,626	33,325	24,571	31,902	33,902	47,330
Total	1,309,342	1,450,190	1,633,351	1,604,554	1,430,483	1,520,778

Total fertilizer consumption in Turkey was 1.31×10^6 t in 1981 and increased to 1.52×10^6 t in 1986. The Planning Period was initiated in 1963 and, for a better evaluation, this year can serve as a baseline year. In 1963, fertilizer consumption was 86,892 t. By 1986, this figure had increased to 1,520,778 t showing that development of fertilizer use happened swiftly.

Corresponding to fertilizer consumption, a substantial increase was manifested in agricultural production: in 1970, the average yield of wheat was 1,000 kg/ha and, by 1985, this average doubled to 2,000 kg/ha. While these increases were sustained, fertilizer consumption increased dramatically. Farmers in the districts have used fertilizers without careful attention to soil analyses or timing of application to crop requirements even though extension services have tried to deliver the necessary training.

Government regulations have liberalized import and distribution of chemical fertilizer to broaden competition in this sector of agronomic inputs. Systems of distribution are expected to improve with this new policy. Since July, 1986, producers and private companies can distribute their own products and imported fertilizers in addition to the Turkish Agriculture Supply Organization (TZDK) and the Agricultural Credit Cooperatives, the previously designated agencies for importation and distribution of chemical fertilizer.

Weed and Pest Control

Numerous factors cause many different diseases in the regions; therefore examination of the activities of plant protection is useful.

Central Anatolia region: Since cereals are the major crop, plant protection activities focus on wheat control programs and some herbicides are used. The insect called *Aelia Spp.* causes a lot of damage to wheat crops and control of these infestations is accomplished with technical aid from the government in furnishing equipment as well as public labor assistance.

Other plant protection activities are control of *Tilletia spp*, *Ascochyta spp.*, *Leptinotarsa spp.*, *Microtus spp.*, all pests and diseases which infest fruit and vegetables in Turkey. All plant protection activities are performed in accordance with the current regulations. Three methods relate to protection activities:
1. state control which could not be executed by the growers, such as the control of *eurygaster Integriceps* and *schistocerca sp.*;
2. state aid furnishes technical aid, pesticides and insecticides, and equipment for application but the labor is supplied by the farmers; and,
3. guidelines and technical back-up are developed by the State and farmers provide other requisite assistance.

Some mistakes have been made by farmers practicing plant protection in the Central Anatolia Region: untimely control activities; inaccurate selection of pesticide or insecticide; excess or insufficient applications for control; inadequate adaptation of the techniques for control; wrong calibration of equipment; inattention to the mixing of pesticides; and mistakes made in choice of equipment.

Coastal region: Major crops in Ceyhan District are the industrial crops of cotton, citrus fruits, soya bean, groundnut, and sesame. Cotton is the most important because of its high profitability and wheat is a rotation crop with cotton or soya bean. So, conventional irrigation is more commonly used for cereals than is supplemental irrigation.

Most errors in farming practices that are made in Central Anatolia Region apply here as well but many special pest control activities are specific to this Region:
1. in cotton *Xanthomonas spp.*, *Rhizoctoni spp.*, *Agrotis spp.*, and *Pestinophora spp.* are problems;
2. in Soybean *Heliotis spp.* and *Nazara spp.* are controlled;
3. in Groundnut *Aspergillus spp.* is vital to control for plant protection;
4. in Sesame *Antigastra spp.* is prominent; and,
5. in Sunflower *Plasmophora spp.* and *Agrotis spp.* are treated with controls.

Yield Relationships

The 1984 values (see Table 28.4) are the official yield data for Turkey's agricultural production. A comparison of these values gives a standard against which to compare yield data collected from farmers using rainfed practices and from farmers with supplemental irrigation. Table 28.11 presents data to assess, in detail, yields obtained in the winter season for the major cereals of wheat and barley employing supplemental irrigation.

Table 28.11. Wheat and barley grain yields.

Cereal	Yield (kg/ha)	Province/District							
		Konya/ Karapinar		Eskisehir/ Central		Ankara/ Cubuk		Adana/ Ceyhan	
		*R	SI	R	SI	R	SI	R	SI
Wheat	Min	900	3,000	1,100	2,500	1,250	3,000	1,500	2,500
	Ave	1,700	3,750	2,000	4,500	1,500	4,000	3,000	4,000
	Max	2,500	4,500	3,200	6,250	2,000	5,000	4,000	6,000
Barley	Min	1,100	3,200	1,000	3,500	1,250	3,500	2,000	3,000
	Ave	1,700	4,000	1,750	4,550	1,500	4,250	2,500	3,500
	Max	2,700	5,000	2,750	5,600	2,000	5,500	3,000	4,000

* R = Rainfed farming system
SI = Supplemental irrigation farming system

The relationships between yields of rainfed production and those grown with supplemental irrigation are dramatic. For instance, Eskisehir Central District has a maximum yield of 6,250 kg/ha for wheat and 5,600 kg/ha for barley under supplemental irrigation. In the Central District, even for rainfed farming, the yields are larger than the other 3 regions with maximum yields of 3,200 kg/ha for wheat and 2,750 kg/ha for barley. The exception, of course, is in the Ceyhan District of Adana Province where average annual precipitation is 750.6 mm. Here, the annual rainfall would be adequate for wheat production (487 mm) if the distribution were less variable during the growing season.

From the point of view of plant physiology wheat production requires a low temperature at the beginning stage of growth; if this occurs, crop development proceeds successfully. Temperature differences might be the crucial factor to consider for Eskisehir Province, Central District with the ramification of greater yields with supplemental irrigation.

The data indicate that priority should be given to Eskisehir Province but serious consideration should be given to the other districts for the implementation of supplemental irrigation. The Karapinar District in Konya Province and Cubak District in Ankara Province display similar trends in yield data for wheat and barley, offering a potential for development.

Livestock and Forage

Development of animal husbandry has not reached the desired levels of production even though natural climatic conditions of Turkey are suitable. When the demand for animal products is contemplated, increased development efforts should be directed to livestock production as a sector of agricultural enterprise.

To develop and sustain livestock production, many components need attention: refined breeding practices and enhancement of breeding conditions, increased production of animal feed, grassland improvement, advanced practices of herd maintenance, determination of nutritional feeding allowances, and disease control. Concurrently, government policy needs to support satisfactory prices and offer available credit to breeders, as well as establishing adequate marketing networks and related distribution services. Furthermore, all needs must be considered together.

During the period of 1972-86, 5 Livestock Development Projects were implemented with foreign financial contributions. The goals of these projects were to upgrade the native breeds of cattle and sheep with the objective of increasing the outputs of meat and milk products. All projects have been completed except the poultry sub-project, a part of the 5th project, but this is planned for completion by the end of the year (1987). Table 28.12 using 1984 data presents total population of livestock and poultry in Turkey. Reviewing this information, a judgment can be reached of the status of livestock production in different agro-climatic zones.

Table 28.12. Population of livestock and poultry, 1984.

Kind of livestock	Population (1,000 head)
Sheep	40,391
Goat	11,391
Angora goat	1,973
Cattle	12,410
Buffalo	544
Horse	623
Donkey	1,226
Mule	213
Poultry	56,616

Central Anatolia region: This region is of great importance because of wide-ranging pastures, cereal stubble, and the natural vegetation and shrubs on the hillsides. The vegetation growing on the rangeland is suitable for raising sheep and goats and the structure of the population for supporting agricultural enterprises is traditionally family oriented. Tables 28.13 and 28.14 present the status of animal husbandry and livestock production for the area. The maximum number of sheep and goats are located in Konya Province where the agro-ecological conditions are acceptable for these animals. Ankara possesses the most cows because of the market demand for the by-products as well as adequate agro-ecological conditions for breeding cattle.

Table 28.13. Livestock population in Provinces of the Central Anatolia region, 1984.

Type of animal	Livestock population (head)		
	Ankara	Eskisehir	Konya
Cows	228,320	68,580	199,102
Sheep	1,328.630	763,600	2,889,910
Goat			
Ordinary	33,550	22,230	456,820
Angora	564,020	298,300	210,600
Buffalo (cow-male)	12,579	1,043	4,117
Poultry (hen-cock)	2,588,359	639,426	2,552,915
Apiculture (beehives)	88,020	15,360	78,764
Sericulture (no. silkworm boxes opened)	1,496	3,875	74

Table 28.14. Production of livestock by-products in the Central Anatolia region, 1984.

Type of by-product	By-product production (t)		
	Ankara	Eskisehir	Konya
Milk			
Cow	39,955	25,720	98,755
Sheep	17,215	14,815	59,775
Goat	2,040	6,245	16,210
Meat			
Lamb/Mutton	9,525	1,830	4,325
Beef/Veal	10,785	1,515	3,255
Goat	475	125	910
Buffalo	935	65	180
Hides			
Sheep	749,010	181,420	426,620
Goat	37,890	11,000	54,110
Cattle	209,430	55,520	71,790
Buffalo	8,990	1,460	1,770

Coastal region: Livestock production in the Mediterranean Region, Ceyhan District, will be discussed from the perspective of Adana Province because this area has no rangeland or pasture for breeding animals. The land is valuable for cultivation and agronomic production having a poly-culture farming system and practicing second cropping agriculture. All these agricultural activities are currently progressing in the Ceyhan District.

Table 28.15 is presented to clarify the situation of livestock production in the region. Athough there are many cows in Adana Province, the number is not comparable to that of say Ankara, Konya Province. Regarding the number of sheep, Ceyhan has the fewer number with the exception of Eskisehir which has a variety of livestock but in smaller numbers than the other provinces. In contrast, the number of goats is large, reaching almost that of Ankara and Konya which have the largest populations. The number of beehives is large.

Table 28.15. Livestock population and production of animal by-products, Mediterranean Region, Adana Province, 1984.

Type of animal	Livestock population (no. of head)	Animal by-products (t)		
		Milk	Meat	Hide
Cows	131,943	58,060	2,490	35,330
Sheep	314,240	14,990	9,645	874
Goat		23,840	980	79,000
Ordinary	438,810			
Angora	130			
Buffalo (cow-male)	1,248		35	700
Poultry (hen-cock)	1,443,324			
Apiculture (beehives)	74,193			
Sericulture (no. silkworm boxes opened)	393			

Cost of Production

When evaluating farming expenses and calculating costs of production, major components of the farming system have to be considered. These components include soil tillage operations, sowing practices, crop growth patterns, threshing and harvesting activities, agronomic inputs including irrigation water, as well as transportation and other general expenses of product marketing.

Since the Central Anatolia Region has an important status in wheat production in Turkey, this region will be used to calculate costs of wheat both for rainfed farming and supplemental irrigation systems. These costs of wheat production with supplemental irrigation will then be generalized from the calculations of the Central Anatolia Region as an indication of the costs in the Ceyhan District even though wheat is grown under conventional irrigation.

Costs for wheat production for the Konya Province can be calculated to show total costs of production per hectare (970 TL = 1%US). These values are representative of Turkey's cereal production both in rainfed farming and with supplemental irrigation. Tables 28.16 (rainfed) and 28.17 (supplemental irrigation) present these calculations.

Table 28.16. Costs for wheat production in the Konya Province for rainfed farming.

1. Seedbed preparation:	
1st tillage, plow	$ 1.44/ha
2nd tillage, duckfoot/cultivator	4.12/ha
3rd tillage, duckfoot/cultivator	4.12/ha
2. Sowing:	5.46/ha
3. Input used:	
Seed, 20 kg/ha × 101 TL/kg	2.08/ha
Fertilizer,	
P_2O_5, 55 kg/ha × 107 TL/kg	
N, 50 kg/ha × 44 TL/kg	
4. Plant protection:	
Herbicide, appl. rate 2,000 g/ha	4.12/ha
Pesticide,	
Seed dusting and fumigation,	
appl. rate 2,000 g/t seeds	.52/ha
Control of *microtus spp.*,	
appl. rate 2,000 g/100 kg/seed	.52/ha
5. Harvesting and threshing:	9.10/ha
6. Fertilizer application costs:	
Tractor 75 HP, 4,150 TL/hr / 4 ha	
Fertilizer spinners, 100 TL/ha	1.17/ha
7. Transportation:	
2 TL/kg × yield (1700 kg/ha)	3.51/ha
8. Field rent:	32.98/ha
PRODUCTION COSTS	$ 82.21/ha
General Expenses 3%	37.00/ha
Interest rate 45%	37.00/ha
TOTAL PRODUCTION COSTS	$121.68/ha
By-products	$ 10.31/ha
Cost/ha	111.37/ha
Cost/kg	0.035/ha

Characterization of Existing Supplemental Irrigation Systems

Integration of supplemental irrigation techniques into rainfed farming systems is an innovative method in irrigation which might cause winter crop production to rise. When the activities related to this system at the district level were surveyed and evaluated, data showed that 15% of the farmers who are members of agricultural cooperatives have adopted this system in Turkey. Application of this technique could have a pivotal role in sustaining cultivation of winter crops especially in the Central Anatolia Region. Supplemental irrigation has been used in rotation cropping in the coastal areas where industrial crops are the major crops. Data presented in tables and figures of earlier sections should be recalled for clarification of the following discussion.

Table 28.17. Costs for wheat production in Konya Province with supplemental irrigation.

1. Seedbed preparation:	
1st tillage, plow	$ 18.56/ha
2nd tillage,	4.64/ha
3rd tillage,	4.64/ha
4th tillage,	4.64/ha
planker,	2.94/ha
2. Sowing:	
(using mechanical equipment)	11.34/ha
3. Input used:	
Seed, 23 kg/ha × 101 TL/kg	2.39/ha
Fertilizer,	7.30/ha
P_2O_5, 60 kg/ha × 107 TL/kg	
N, 55 kg/ha × 44 TL/kg	
Irrigation water,	10.30/ha
4. Plant protection:	
(including costs of application)	5.16/ha
Herbicide, appl. rate 2,000 g/ha (4,000 TL/ha)	
Pesticide,	
Seed dusting and fumigation,	
appl. rate 2,000 g/t seeds (500 TL/ha)	
Control of *microtus spp.*,	
appl. rate 2,000 g/100 kg/seed (500 TL/ha)	
5. Harvesting and threshing:	
(calculating increase in yields)	18.20/ha
6. Fertilizer application costs:	
Tractor 75 HP, 4,150 TL/hr / 4 ha	
Fertilizer spinners, 100 TL/ha	1.17/ha
7. Irrigation application:	25.77/ha
7. Transportation:	
2 TL/kg × yield (4,500 kg/ha)	9.28/ha
8. Field rent:	56.70/ha
PRODUCTION COSTS	$177.98/ha
General Expenses 3%	5.34/ha
Interest rate 45%	80.09/ha
TOTAL PRODUCTION COSTS	$263.41/ha
By-products	$ 10.31/ha
Cost/ha	253.10/ha
Cost/kg	0.041/ha

Karapinar District, Konya Province

The population of Karapinar District is 62,411 with most people depending upon agriculture as the main economic activity. Farmers are heads of 10,378 households. Illiteracy among farmers is low, 2.2%, and this is gradually decreasing even more. The roads are adequate and railway lines provide transporation during the entire year. Each village is connected to the National Electrical Network System. The majority of farmers with families have land to cultivate; therefore, migration does not occur. Usually, marketing of produce is no problem.

Total land area is 431,345 ha of which 128,810 ha are non-arable, leaving 302,535 ha (84%) of cultivable land. Of this cultivable land, rainfed farming occupies 160,487 ha; fallow land, 97,920 ha; and, 44,128 ha are farmed with supplemental irrigation.

Major soil characteristics can be summarized: color is brown; slope and infiltration are moderate; soil is stone-free and of good tilth but with saline conditions. Groundwater is the major source for domestic water and for supplemental irrigation. Surface water is of poor quality, but groundwater is suitable for all purposes. Where farmers (15%) use supplemental irrigation, surface irrigation is the most commonly chosen method of application.

Agricultural activities focus around cereal production; wheat, barley, and rye are cultivated as winter crops. Generally, a 2-course rotation system of wheat – fallow is practiced. Sheep, goats, and poultry are more appropriate than other farm animals. The vast rangelands and pastures in the hilly terrain contain various plants and bushes which, in addition to cereal stubble, furnish ready sources for livestock grazing.

Central District, Eskisehir Province

Eskisehir Province, Central District has a population of 63,905 distributed in and around 146 villages. The area is of great importance to the agriculture of the Central Anatolia Region. Most of the 13,404 households are associated with agricultural activities. The illiteracy among farmers is 19% but this rate is decreasing.

The total land area is 382,300 ha: 182,914 ha are non-arable with the remaining 199,286 ha, cultivated. The rainfed land in fallow is 123,651 ha. In addition, 51,554 ha are farmed using conventional irrigation, with 12% (24,181 ha) of the cultivable land under supplemental irrigation (24,181 ha).

The existing roads, railway, air transport, and communications systems are well developed when a comparison is made with international standards. No problems are related to the infrastructure from the standpoint of the farmer. Electrical power and water supplies are available and deemed adequate for the demand.

Although soils vary from one zone to another, generally, they are considered to be deep with a good tilth, moderate infiltration, low salinity, and a moderate slope. The source for supplemental irrigation is surface water. Farmers have exploited groundwater resources by means of wells. The major method of irrigation uses gravity flow but pumping can also be needed to deliver the water.

Supplemental irrigation is used on 24,181 ha; 9,672 ha are irrigated by gravity flow and the remaining by sprinklers. Supplemental irrigation is not commonly applied in the district. Land leveling is extensively done in order to ensure efficient application of water during irrigation.

The major crops are cereals, pulses, industrial crops, tuber crops, oil seeds, and fruits. Cereals and industrial crops are cultivated more often than the other crops. Industrial crops are cultivated with conventional irrigation. Wheat – sugar beet and wheat – fallow are the current rotation courses. No significant differences in practices of animal husbandry exist between Konya Province, Karapinar District, and Eskisehir Province, Central District.

Eskisehir Province, Central District has important status in agricultural production and is the most developed sector in Turkey. No migration and no unemployment has been faced as yet.

Ceyhan District, Adana Province

Ceyhan has 83 villages with a population of 147,497 distributed in 27,709 households. Farmers widely apply a poly-cultural farming system. The percentage of illiteracy among farmers is 5%; most educated farmers who tend to learn new agricultural technology live in this district. Communication and transportation networks are good.

Total land area is 142,700 ha of which 31,070 ha are non-arable. Rainfed farming is practiced on 25,000 ha of the 111,630 ha of cultivated land.

The red soils are fertile and deep without a drainage problem. Infiltration is moderate; there are no severe problems of slope or salinity. Groundwater is of poor quality. The major source for supplemental irrigation is surface water which has a good quality and is suitable for all purposes. Dams and diversion dams have been constructed to convey water for delivery to the farmer.

Supplemental irrigation systems and rainfed farming are both considered secondary activities. The excellent environmental conditions are suitable for a poly-cultural farming system which supports diversification of agricultural production.

Cereals are grown in a rotation system applying supplemental irrigation; however, the major agricultural actitities are the cultivation of industrial crops under conventional irrigation. These crops include cotton, citrus, sesame, ground nuts, and soya bean. Poultry, sheep, goats, and cattle are bred in the district. Crop residues are the major feed supply for the livestock. There is no rangeland or pasture in the district so the flocks move to neighboring areas for their forage requirements.

Cubuk District, Ankara Province

Food and industrial processes are concentrated here, with a population of 57,916 and 7,010 households. The illiteracy rate among farmers in the district is high (15%) even though most farmers have a primary certificate.

Total land area is 150,200 ha, half of which is non-arable (70,100 ha). The remaining 80,100 ha is cultivated for agricultural production. The soil is stony but of fair tilth. Rainfed farming is applied to the hillsides where shallow soils are predominant. Supplemental or conventional irrigation systems have been applied where soils are deeper. The soil is black with no salinity.

Groundwater is the principal source for irrigation water. Surface flow and sprinklers are commonly used as methods of irrigation. Domestic water requirements are provided by the municipalities.

Infrastucture poses no problem for development of the area. All roads and communications have been provided including storage facilities for agricultural products.

The total land under supplemental irrigation is 10,000 ha. Fruit tree orchards are dominant and cereals are considered of secondary importance for production. Surface irrigation and sprinklers are used at the same rate for winter crops. There is no rangeland or pasture in this region but there is some poultry production and cattle breeding.

Farmers participate in the various categories of agriculture including the agro-industrial sector. Activity abounds in both agriculture and industry and a heavy demand occurs the year around for labor; therefore, migration does not occur.

Potential Areas for Supplemental Irrigation Development

In all cases, some form of irrigation is needed for meeting crop water requirements. A restrictive factor for the implementation of supplemental irrigation is that winter cereals, wheat, barley, and rye, be included as crops for irrigation. An essential criterion for choosing to irrigate is careful selection of economically important crops. From this viewpoint, supplemental irrigation does offer the alternative of irrigating more crops because less water is applied for production.

Based on this argument, the Central Anatolia Region would be the most profitable to implement supplemental irrigation. This is an area recognized for cereal production. Some of the inter zone areas, between the coast and central region, should be counted as possible sites for development of supplemental irrigation techonology.

Conclusions

Water resources are limited; water is a constraint to agricultural productivity in comparison with the extent of existing irrigable land resources. If all water resources were regulated, 5,925,032 ha of land could be irrigated in addition to those already in production. This total would mean that 9,101,362 ha would be in irrigated crop production: 5,925,032 ha of new land plus 3,176,330 ha already under cultivation with irrigated crops.

In actuality, there are 25,305,444 ha of irrigable land in Turkey. Implementation of supplemental irrigation would seem to be economically feasible for agricultural production throughout this land. Under these circumstances, the integration of supplemental irrigation technology into existing farming systems of Turkey will be placed on the agenda for all regions in the near future especially with an eye to profitability.

References

Devlet Istatistik Enstitusu, 1984. "Tarimsal Yapi ve Uretim".
Devlet Istatistik Enstitusu, 1986. "Turkiye Istatistik Cep Yilligi".
General Directorate of State Hydraulic Works, 1983. "Statistical Bulletin with Maps".
Sarigedik, U., 1987. "Agriculture in Turkey", Republic of Turkey Ministry of Agriculture, Forestry, and Rural Affairs.
Tarim Orman ve Koy Isleri Bakanligi, 1986. "Developments in Turkish Agriculture 1984-1986".
Tarim Orman ve Koy Isleri Bakanligi, 1981 "Turkiye II. Tarim Kongresi".

Addendum 28A: Water and Land Resource Potential

Table 28A.1. Distribution of water resources showing the population and area and average amount of water, including average annual precipitation and annual runoff and number of dams with total storage potential of each basin.

				Average annual of water storage capacity			
					Runoff		
Basin	Population (1980)	Area (km^2)	Precip. (mm)	(l/sec/ km^2)	Total (Bm^3)	No. of dams	Volume (Mm^3)
1. Meniq Epoene	859,191	14,560	640	2.72	1.25	11	616.55
2. Marmara	5,914.407	34,100	766	10.03	7.62	32	1,592.50
3. Susurlok	1,797,982	23,763	730	7.14	5.35	21	3,629.22
4. North Aegean	498,490	9,032	730	7.72	2.20	13	508.94
5. Oediz	1,598,541	17,110	639	3.35	1.81	8	2,723.40
6. Kucuk Xenderes	1,258,521	7,165	740	4.96	1.12	2	228.10
7. Buiuk Xenderes	1,451,101	24,903	656	3.76	2.95	11	2,489.90
8. West Mediterranean	678,819	22,615	865	10.88	7.76	11	1,886.90
9. Antalya	882,162	14,518	910	24.55	11.24	9	1,419.30
10. Burbour Lakes Basin	181,647	8,764	436	1.12	0.31	7	696.05
11. Akarcat	400,077	8,377	472	1.70	0.45	5	205.56
12. Sakaria	4,118,628	56,504	534	3.38	6.03	31	5,517.49
13. West Black Sea	1,880,155	29,682	803	10.73	10.04	25	1,417.57
14. Yesilirmak	2,515,194	36,129	556	4.86	5.54	56	7,969.64
15. Mizilirmak	3,255,076	78,646	459	2.53	6.28	54	14,569.07
16. Konya Closed Basin	1,873,876	56,554	437	1.88	3.36	17	671.10
17. East Mediterranean	1,179,093	22,484	669	17.30	12.27	3	5,986.50
18. Seyhak	1,000,180	20,731	629	10.86	7.06	13	5,639.40
19. Asi	1,037,295	10,885	837	3.50	1.20	2	255.00
20. Ceyhan	917,932	21,222	758	10.77	7.21	13	6,105.20
21. Firat	4,932,905	120,917	582	8.77	33.48	60	114,198.72
22. East Black Sea	2,291,806	24,072	1,291	18.48	14.00	35	922.07
23. Coruh	620,782	19,894	540	10.30	6.46	10	3,064.00
24. Aras	1,253,995	27,548	462	6.38	5.54	14	3,336.05
25. Lake Van Closed Basin	541,820	15,254	507	5.38	2.59	12	1,081.99
26. Dycle	1,797,272	51,489	814	13.43	21.81	23	7,448.27
Total	44,736,957	766,870[a]	653[b]	206.42	184.93	498	194,178.99

[a] Parts of the River Basins outside boundaries of Turkey are not included
[b] Average value

Table 28A.2. Distribution of land resources and necessary investment for realizing development for agricultural purposes of the land within the basins, calculated using 1980 and 1993 values.

Basin	Land resources plains area (ha)			Investment needed for development (1,000 TL)	
	Total	Non-irrigable	Irrigable	1980	1993
1. Meniq Epoene	1,381,229	157,966	1,223,263	4,128,308	67,524,219
2. Marmara	319,020	81,830	237,190	15,292,105	457,548,609
3. Susurlok	529,455	133,382	396,073	9,268,762	162,765,452
4. North Aegean	904,117	266,753	637,364	8,233,816	210,567,846
5. Oediz	521,472	135,459	386,013	2,251,780	67,496,627
6. Kucuk Xenderes	202,415	27,649	174,766	1,124,400	19,597,172
7. Buiuk Xenderes	812,000	222,300	589,700	12,607,097	188,239,085
8. West Mediterranean	322,000	110,500	211,500	10,463,468	225,474,915
9. Antalya	444,260	93,951	350,309	22,256,867	195,308,924
10. Burbour Lakes Basin	220,025	38,217	181,808	945,630	18,552,305
11. Akarcat	323,900	51,800	272,100	2,734,813	44,979,899
12. Sakaria	2,075,100	403,500	1,671,600	62,693,954	797,795,692
13. West Black Sea	392,400	164,500	227,900	9,601,156	174,730,676
14. Yesilirmak	1,326,046	431,661	594,385	13,947,375	338,015,542
15. Mizilirmak	3,528,500	1,134,985	2,393,515	32,692,110	646,702,371
16. Konya Closed Basin	2,702,383	630,543	2,071,840	13,118,272	289,045,437
17. East Mediterranean	212,256	73,763	138,493	70,109,702	256,050,280
18. Seyhak	485,466	170,841	314,625	77,448,091	1,925,124,385
19. Asi	442,260	277,270	165,010	7,029,400	104,629,080
20. Ceyhan	734,472	74,600	659,872	35,115,062	1,078,021,636
21. Firat	4,947,640	3,171,065	1,776,575	620,175,620	3,088,101,353
22. East Black Sea	736,998	653,029	83,969	158,733,450	913,943,118
23. Coruh	157,600	75,200	82,400	1,408,200	83,337,276
24. Aras	810,900	327,600	483,300	11,126,249	203,313,453
25. Lake Van Closed Basin	229,000	33,600	195,400	7,525,268	57,429,549
26. Dycle	1,950,898	1,548,046	402,852	360,143,800	1,211,952,136
Total	26,712,132	10,490,010	16,222,122	1,570,174,755	12,826,347,237

Table 28A.3. Benefits to be obtained after development of water and land resources in the 26 basins.

Basin	Irrigation (ha)	Flood protection (ha)	Classification and drainage (ha)	Installed capacity (MW)	Ave. energy (GHW)	Domestic water (Mm³)
1. Meniq Epoene	63,164	47,248	9,018	— —	— —	60.0
2. Marmara	50,023	8,190	— —	— —	— —	9.5
3. Susurlok	170,474	5,377	— —	548.44	1.330	147.2
4. North Aegean	83,959	2,600	— —	19.60	.052	18.0
5. Oediz	130,830	22,300	5,475	219.00	.328	0.5
6. Kucuk Xenderes	3,150	— —	— —	— —	— —	128.0
7. Buiuk Xenderes	232,665	737	— —	205.93	.580	— —
8. West Mediterranean	117,392	7,963	3,290	525,20	1.606	— —
9. Antalya	192,879	14,267	17,910	1,109.65	3.942	— —
10. Burbour Lakes Basin	31,216	2,000	— —	— —	— —	— —
11. Akarcat	74,830	12,000	— —	— —	— —	— —
12. Sakaria	266,413	100,337	15,811	1,026.44	2.341	822.0
13. West Black Sea	95,168	8,100	308	450.60	1.973	18.6
14. Yesilirmak	326,118	73,050	50,000	1,215.30	5.007	127.0
15. Mizilirmak	497,320	61,094	30,000	2,257.48	6.715	96.0
16. Konya Closed Basin	243,924	— —	3,100	30.90	.080	123.9
17. East Mediterranean	71,213	4,570	5,265	918.25	3.226	66.6
18. Seyhak	342.633	22.144	— —	1,964.50	6.641	— —
19. Asi	116,576	8,480	20,000	8.00	.015	— —
20. Ceyhan	361,548	78,360	2,200	949.60	3.530	— —
21. Firat	1,506,867	1,220	— —	8,752.36	35.119	82.5
22. East Black Sea	924	1,283	— —	3,755.69	12.035	— —
23. Coruh	36,417	— —	— —	2,876.86	10.109	— —
24. Aras	286,352	28,000	— —	642.00	1.874	— —
25. Lake Van Closed Basin	67,641	3,000	— —	54.80	.223	— —
26. Dycle	525,336	— —	— —	3,405.68	12.644	— —
Total	5,925,032	512,320	165,367	30,911.48	109.684	2,520.1

Chapter 29

Supplemental Irrigation in Yemen Arab Republic (YAR)

ABDULRAHMAN M. BAMATRAF

Three quarters of Yemen's (Yemen Arab Republic, YAR) cultivated land depends totally on rainfall to replenish crop available soil moisture. Rainfall patterns are erratic and dependent on physiography, orientation and distance from the Red Sea and Gulf of Aden. Environmental conditions dictate a strict regime of cropping patterns and yield. In many instances, crop production is not possible under prevalent rainfall and other means such as supplemental irrigation are crucial for a successful cropping season. Supplementing rainfall and increasing its effectiveness has been the main concern of the Yemeni farmer since the dawn of history. The centuries-old mountainous bench terraces are living evidence of a prosperous agricultural civilization.

Traditional measures of water harvesting and spate irrigation to supply water for rainfed farming are of limited benefit in meeting increased demand for food and as techniques to diversify cropping patterns. An in-depth review of supplemental irrigation practices and review of the present status of agriculture are important for identifying potential areas to develop the state-of-the-art.

Geographical Characteristics

North Yemen is located in the south-western corner of the Arabian Peninsula, between latitudes 12°40' and 17°25' north, and longitudes 42°30' and 46°31' east. The country is bordered on the north and north-west by Saudi Arabia, on the south and south-east by the South Yemen (PDR), and on the west by the Red Sea. There is a total land area of 200,000 sq km which can be characterized by extremely diverse physiography, climate, soils, and agricultural activities (Senykoff, 1983). Elevation alone varies from sea level to nearly 4,000 m, where 3 major climatic zones can be encountered in less than 100 km along a traverse from east to west. Also, other environmental, demographic, and cultural features vary significantly from place to place.

E.R. Perrier and A.B. Salkini (eds), Supplemental Irrigation in the Near East and North Africa, 561-598.
© 1991 ICARDA.

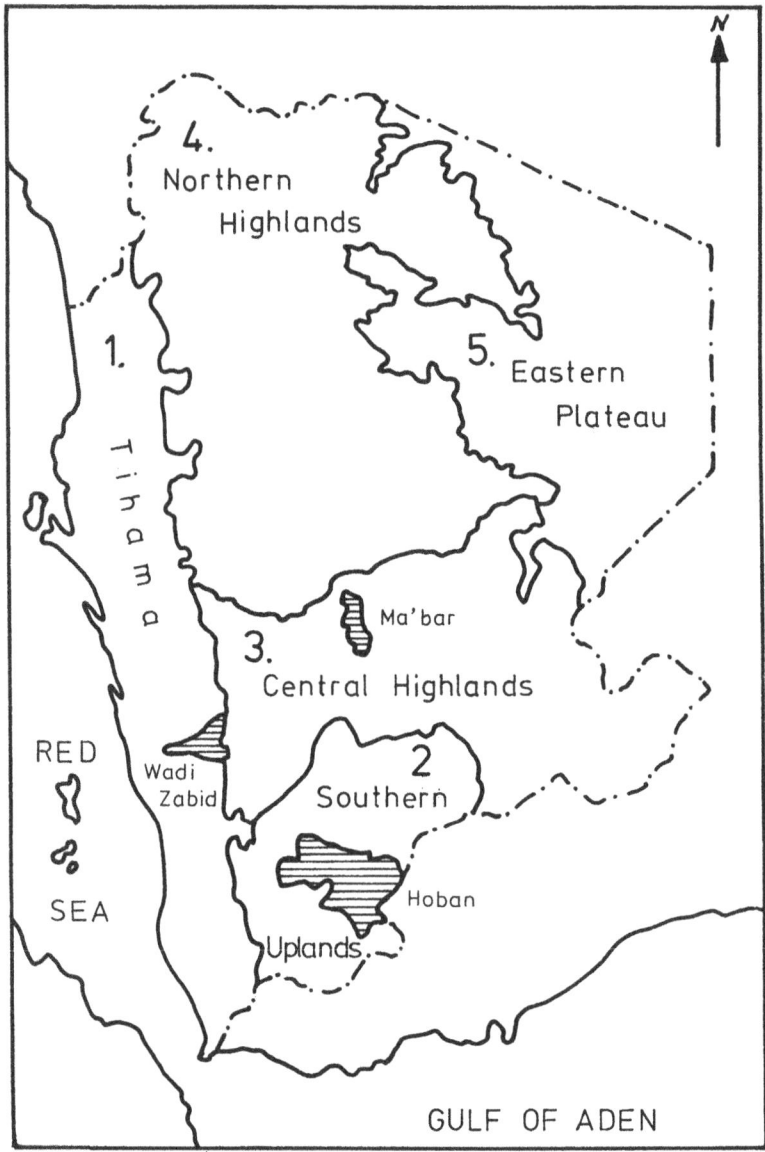

Figure 29.1. General map of Yemen Arab Republic showing agro-ecological zones and locations of sample districts.

Agro-Ecological Zones

Topography is of significance to agricultural land use. The orographic variations, the differences between rain-exposed and rain-shaded slopes and the distance from the Red Sea and the Gulf of Aden have an influence on local micro-climates (Swiss, 1978). Hydrologic divides partition the land into three main escarpments (the Western, the Southern, and the Eastern) incorporating the main watersheds (Grolier, *et al*, 1981).

Various authors have given different grouping to the agro-ecological zones (Swiss, 1978; Grolier *et al*, 1981; King, *et al*, 1983). For purposes of agricultural research, the country has been divided into 5 physiographic provinces (Bamatraf and Al-Sakkaf, 1987). Although conditions are not uniform within these provinces, this apportionment provides a means of allowing for major climatic variations and supports regional research management. Figure 29.1 shows these divisions, depicting the 5 agro-ecological zones distinctly specified to utilize relevant natural resources.

Zone 1. Tihama Coastal Plain is the tropical lowland bordering the Red Sea ranging in elevation from 0-500 m. This region has an arid tropical climate extending in the western escarpment over an area of 20,000 km^2 with a cultivable area of 235,000 ha. The marginally cultivated area, that land cultivated once in every 3-5 yr depending upon rainfall, is much larger.

Zone 2. Southern Upland is a subtropical region covering all the southern escarpment and the southern part of the western escarpment. Elevation generally ranges from 500-1,500 m but some land in the southern escarpment rises above 1,500 m. The zone extends to an area of 20,000 km^2 with cultivable land of 550,000 ha, most of which is on mountain terraces.

Zone 3. Central Highlands is a temperate but rather dry region including most of the intermontane plains. The elevation of these plains is more than 1,500 m. The highest mountain crests are located in this zone rising 3,000 m above sea level. This zone includes most of Dhamar and Al-beida Provinces.

Zone 4. Northern Highlands arbitrarily include areas within the western and eastern escarpments varying in elevation from 500 m to above 1,500 m. The area is 50,000 km^2 including the Provinces of Sana'a, Hajja, Sa'ada, and Mahweet. The delineation is administrative rather than geographic. Thus, the zone has differing land use potential because the rugged mountains have little land suitable for farming. Until recently, access was limited which further contributed to the region's isolation from main urban centers.

Zone 5. Eastern Plateau encompasses the Marib and Al-Jowf on the arid eastern escarpment. The region slopes gently from 1,500 m into the fringes of the empty quarter (Al Rub-al-Khali) which is a vast expansive sand desert.

Climate

The country experiences an arid to semi-arid tropical climate. Situated between the Red Sea and Rub-al-Khali Desert, agro-ecological zones are subject to a variety of climatic conditions contingent on elevation. Mean monthly climatic

Table 29.1. Mean monthly climatic data.

District (elev.)	Air temp. (°C)	Precipitation (mm)	ET* (mm)	Sunshine (hr)
Ma'bar (2,583 m)				
Jan	11.5	1.7	37.4	9.3
Feb	12.8	6.7	40.1	8.5
Mar	15.2	55.0	59.1	8.4
Apr	16.7	61.0	68.3	6.8
May	17.6	38.7	77.6	8.5
Jun	18.2	4.0	80.2	7.8
Jul	18.8	40.0	85.9	7.4
Aug	18.2	97.0	81.0	6.1
Sep	16.8	6.3	68.3	8.6
Oct	14.7	38.7	55.6	10.3
Nov	12.1	12.3	39.6	10.6
Dec	11.3	0.0	36.0	9.5
Taiz (1,500 m)				
Jan	19.2	3.2	53.5	9.3
Feb	22.3	12.1	75.0	8.0
Mar	23.7	19.8	101.3	7.2
Apr	24.8	70.0	115.5	7.9
May	25.9	91.5	137.3	8.5
Jun	26.9	96.2	143.1	8.7
Jul	26.3	60.9	143.5	7.0
Aug	26.1	71.3	139.0	6.7
Sep	24.7	95.8	113.0	8.4
Oct	22.8	92.6	89.7	9.1
Nov	20.9	16.2	67.0	8.9
Dec	19.8	8.2	59.9	9.2
Zabid (240 m)				
Jan	25.5	5.8	98.8	7.2
Feb	26.3	1.7	106.9	7.0
Mar	28.3	6.2	139.7	6.6
Apr	30.7	19.8	163.1	8.0
May	32.3	29.2	165.4	7.8
Jun	33.5	6.2	169.4	6.6
Jul	33.1	35.2	176.8	4.6
Aug	32.0	82.6	142.7	5.7
Sep	31.8	112.0	128.8	6.8
Oct	29.4	46.1	120.1	8.2
Nov	27.1	2.4	106.9	8.9
Dec	25.6	1.2	99.5	7.8

* ET is calculated using Thornthwaite for Ma'bar and Taiz and Pan Evaporation × 0.6 for Zabid.

Supplemental Irrigation in Yemen Arab Republic (YAR)

data from the station nearest to the 3 sample districts are presented in Table 29.1. Salient features are compiled from different sources (Bamatraf and Aldomi, 1987).

Average annual rainfall varies from less than 50 mm in the coastal area of Tihama to more than 1,500 mm rainfall around Ibb in the southern upland (Figure 29.2). Annual rainfall in the Tihama increases with distance from the

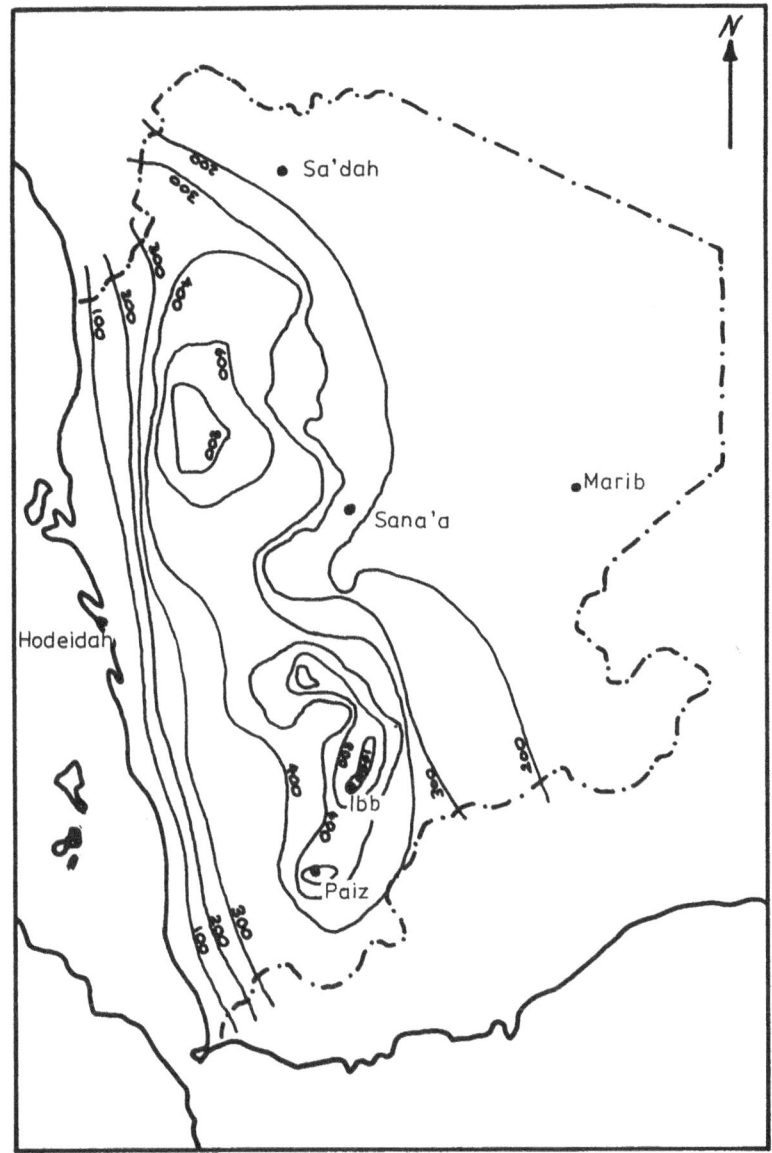

Figure 29.2. Generalized rainfall map (isohyets in mm).

Red Sea; near the foothills, rainfall increases to 300-400 mm. In the southern and central highlands, a consistant variation with altitude is not evident. Total rainfall increases with distance from the Gulf of Aden, reaches the highest average around the Ibb area, then decreases towards Sana'a where average annual rainfall is 200 mm. In the northern highlands, rainfall increases gradually from Sana'a towards Hajja and Sa'ada. The Eastern Plateau exhibits a pattern of decreasing rainfall downslope of the eastern escarpment towards the Rub-al-Khali Desert. There are two rainy seasons: one in spring which starts late in March and continues until the beginning of June; and one, with heavier rainfall, in summer during July through September.

Mean annual air temperatures at weather stations are less than 15°C in the central highland (Dhamar) to more than 30°C in the Tihama area (Hodeidah). Temperatures over 40°C have been recorded in summer at the Tihama. Temperature can fall below freezing in the highlands imposing a frost hazard to crops in the winter months. Mean relative humidity ranges from 30% in the Eastern Plateau (Merib) to 80% in the coastal areas of the Tihama (Hodeidah): the months from January to April are slightly more humid.

Diurnal patterns of wind speed vary: often, nights can be windless. Wind starts in the morning hours and reaches a maximum speed during afternoon. Maximum wind speed occurs during the summer between July and September when monsoon winds prevail. In the afternoons, sandstorms in the lowlands and duststorms in the highlands are common events. These foster soil erosion and damage sensitive crops and are the repercussion of dry, unprotected soil surfaces.

Duration of sunshine is recorded only to a limited extent, which means generalization can be confusing; however, possible duration of sunshine at latitude 15° N (which splits the country in half) is estimated at 4,420 hr. The ratio of actual and possible sunshine hours is yet to be determined.

Geology

There are 4 major geologic groups: the Precambrian shield; sedimentary rocks of Paleozoic and Mesozoic Ages; Tertiary and Quaternary volcanics; and, Quaternary alluvial deposits (Grolier and Overstreet, 1978).

Soil Resources

Detailed soil surveys have been carried out in scattered areas and are of limited use for wide interpretation. A 1:500 000 soil map of two thirds of the land area was completed 4 years ago (King, *et al.*, 1983). One fourth of these mapped areas are classified as predominantly rocky outcrops. The map identifies the predominant soils and presents them at the sub-group level according to the USDA Soil Taxonomy.

Due to the rugged nature of the country, most soils are subject to alternating processes of erosion and deposition with the result that 60% of the mapped area is dominated by soils of the Entisol order with sub-orders of orthents, fluvents, and psamments. The incidence of the sub-order, *orthents*, can be associated with steep slopes in arid mountainous regions; *fluevents*, with plains and wadis; and *psamments*, with sandstone mountains or sand dune plains. Greater groups of these sub-orders are classified as ustic or torric type.

Other orders include Aridisols (8.7%), Mollisols (2.6%), Inceptols (4.3%), and for a limited extent, Vertisols and Alfisols. Aridisols are either calcic or gypsic groups of the orthid sub-order. Mollisols are present in the most cool and moist areas of the country. In the mountain plains, most Mollisols are buried; whereas, on the mountain slopes, they have shallow-lithic contact. At this level of the map's scale, generalizations concerning the remaining orders may cause discrepancies and more detailed soil surveys and classification work is imperative.

Moisture regimes encountered are Udic, Ustic, and Aridic: (Soil moisture regimes are defined in Soil Taxonomy, USDA-SCS, 1975). The soil moisture regime (SMR) of the various soil resources could be a limiting factor for irrigation management. Supplemental irrigation is occasionally needed when the soil has an ustic moisture regime. The soil survey report (King, *et al.*, 1983) utilized a calculations model for soil climate estimation from available meteorological data on which a soil moisture regime map was constructed (Figure 29.3). The authors discussed the limitations of the model and concluded that ustic areas along the Tihama mountain fronts might be drier and wadi areas near the front might be more moist than predicted by the model. They contributed this to significant losses and gains of rainfall due to runoff and inflow. In mountainous areas, however, prediction of SMR for soils on terraced slopes would probably be more accurate due to higher rainfall effectiveness. In both, the lowlands and the highlands, water harvesting areas might be more moist than predicted.

Water Resources

Rainfall is the major source of all water in Yemen. That portion of precipitation which is not used in locations where it falls flows as surface runoff and some percolates downward to replenish the aquifers. Except for the newly constructed reservoir behind Marib Dam, there are no surface bodies of water in the country. Unfortunately, long term records are lacking for rainfall and other forms of atmospheric water, stream discharge, and groundwater reserves.

Precipitation

The mountains have a critical effect on rainfall which is, therefore, seasonal

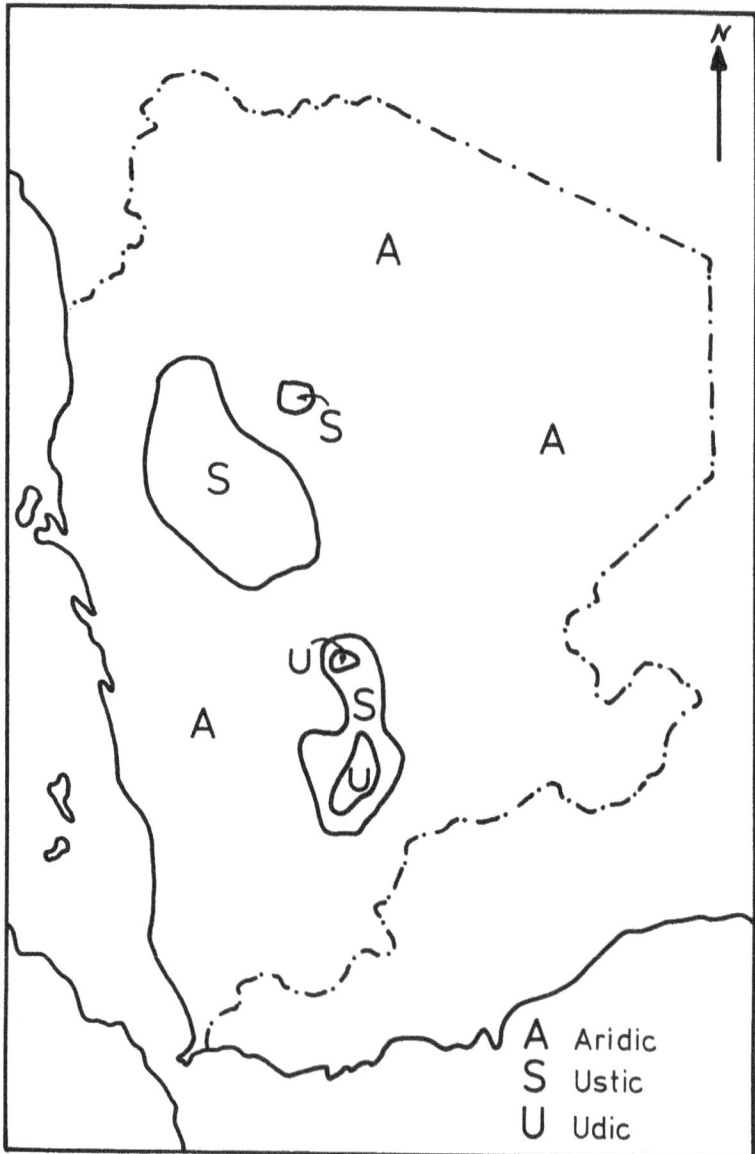

Figure 29.3. Soil moisture regimes.

and variable. Distribution patterns and amounts vary with altitude, latitude, and distance from the Red Sea and the Gulf of Aden (Grolier, *et al.*, 1981). For purposes of supplemental irrigation, not only is the total volume of rainfall important but also the nature of the rainfall event and number of rainy days per season. High intensity storms of short duration are not uncommon and are destructive to soils and crops alike especially on the steeper slopes.

Supplemental Irrigation in Yemen Arab Republic (YAR)

Surface water

Surface water originates as rainfall runoff from mountain slopes of the highlands. There are 3 main drainage basins (Grolier, *et al.*, 1981) which have been demarcated on aerial photographs (Swiss, 1978), (Figure 29.4):

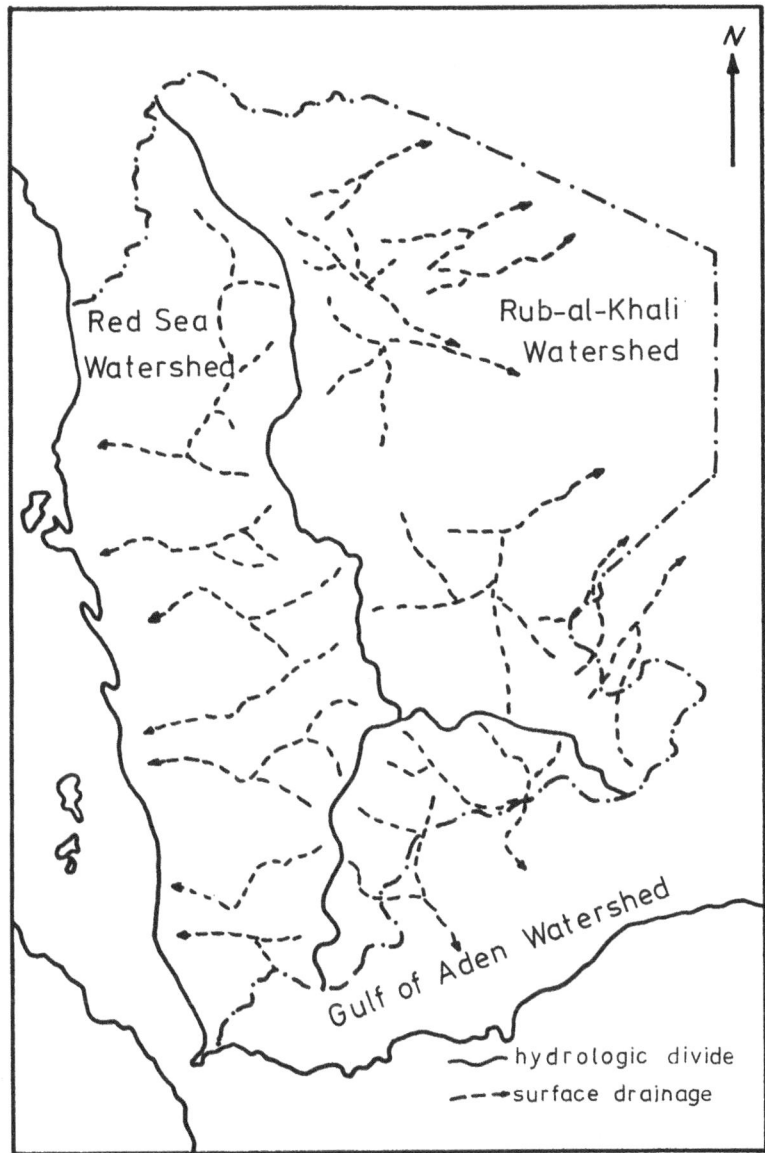

Figure 29.4. Drainage basins.

1. the Red Sea watershed, 23,000 km^2, on the western escarpment from which the westward wadis flow onto the Tihama;
2. Rub-al-Khali watershed, 31,000 km^2, on the eastern escarpment; and,
3. the Gulf of Aden watershed, 9,500 km^2, on the southern escarpment from which the southward wadis flow across the border into South Yemen.

Table 29.2 presents the land areas of the major wadis. Streamflow is greatest in the summer months and early fall. Drainage basins of Wadi Mawr, Wadi Zabid, Wadi Tuban, and Wadi Bana probably have the largest surface and groundwater resources in the country (Al-Eryani, 1979; Grolier, *et al.*, 1981).

Table 29.2. Watershed areas of major wadi courses.

Escarpment	Wadi		Area (km^2)
Red Sea (total land area = 23,150 km^2)	Wadi	Mawr	7,500
		Surdud	2,450
		Siham	3,200
		Rimaa'	2,450
		Zabid	4,500
		Rasyan	1,750
		Mawza'	1,300
Gulf of Aden (total land area = 9,450 km^2)	Wadi	Tuban	3,550
		Bana	5,900
Rub-al-Khah (total land area = 30,650 km^2)	Wadi	Amlah	1,000
		Khabb	1,400
		Awban	800
		Madhab	2,700
		Hirran	3,000
		Kharid	7,000
		Saba	11,300
		Harib	1,000
		Bayhan	2,450

Source: Swiss Team, 1978

Groundwater

Groundwater is an important water source with a reliable year-round supply of clean water for domestic, agricultural, and industrial use. However, exploitation of this resource is confronted with problems of overpumping, deterioration of quality, and contamination. Monitoring these problems is essential for protection of the supply and development of this resource. Unfortunately, studies on the aquifers in the YAR are not extensive. Recharge of groundwater in the valley and coastal plains occurs from streamflow in the lower reaches of the many streams crossing this land. The ratio of this contribution to water loss through evaporation remains to be measured.

Water rights

Distribution and use of diverted, harvested, or pumped shares of groundwater are still governed by custom. Several authors have presented and discussed the customary and Sharia'h (Islamic) water laws in Yemen (Caponers, 1973; Aldomi, 1983). Laws, law enforcement, and arbitration systems are different in the rainy season (Sayl) than they are in the dry season (Ghayl), and slightly variable from one region to another (Bamatraf and Aldomi, 1978). In the rainy season, water is first available to lands nearest the source; then it is released for use further downstream. This principle is called the highest first. On the other hand, dry season water (permanent springs or Ghayl) is divided between lands located in the upper reaches of the wadi. No comprehensive regulations exist for a system of extracting groundwater for agriculture and there is no rational allocation of water. Water for crop irrigation is sold to users who do not possess wells. A charge is paid to meet pumping costs rather than water used and payment (usually a share of the crop yield produced) is arranged between the well owner and the buyer.

Assessment of Present Agricultural Activities

An intricate terrain accentuates the diversity of agricultural activities. Cropping land varies from large, intensively cultivated fields to narrow terraces of a few hundred square meters built on steep mountainsides. Statistical data on agricultural activities range from detailed analysis of limited areas to generalized conflicting results which rely on estimation and projection. In this section, the major components of agriculture, compiled and assessed from available information sources, are reviewed as they relate to existing farming systems within various agro-ecological zones. This review relies heavily upon data from the Report of the Agricultural Census (Ministry of Agriculture and Fisheries, 1983).

Farming Systems

Since early history, farmers have realized that agricultural production is only possible if plant available soil-water is replenished from scarce water resources that are difficult to control. Ruins of dams and reservoirs, water conservation and conveyance structures, as well as, intricate patterns of spectacular mountain terraces reflect efforts which were initiated thousands of years ago to manage flooding and rainfall runoff (Bamatraf and Aldomi, 1987). This evidence accentuates the contrast between rainfed and irrigated farming systems.

Land areas cultivated with different farming systems presented in Table 29.3 illustrate that, of the total land area, only 1.35 M ha is cultivable with a cropping intensity of about 75%. Rainfed farming totals 790,000 ha or 58.5%

of the cropped land. Irrigated land accounts for 17% and fallow land, 24.5%. Water for irrigation is from groundwater or harvested from surface runoff and these supplies are applied equally to the irrigated land (MAF Agricultural Census, 1983). The values demonstrate that in contrast to the eastern region, the southern upland zone has the highest percentage of rainfed farming (82.6%) and the lowest irrigated (7.9%) with fallow land of 9.5%.

Table 29.3. Distribution of cultivated land.

Agro-ecological zones *	Rainfed areas (ha;%)	Irrigated areas (ha;%)	Fallow areas (ha;%)	Total area (ha;%)
Tihama Plain	151,055 (45.3%)	80,022 (24.0%)	102,530 (30.7%)	333,607 25.7%
Southern Upland	185,933 (82.6)	17,883 (7.9)	21.173 (9.5)	255,089 20.0%
Central Highland	82,381 (65.8)	23,689 (19.2)	18,544 (15.0)	123,623 9.1%
Northern Highland	369,157 (63.6)	59,707 (10.3)	151,598 (26.1)	580,462 38.7%
Eastern Plateau	2,757 (3.1)	48,275 (54.9)	36,962 (42.0)	87,995 6.5%
Total lands	790,283	229,585	330,907	1,350,775
Total %	58.5%	17.0%	24.5%	100%

Source: "Summary of the final results of the argricultural census in eleven provinces," Ministry of Agriculture and Fisheries, YAR. April, 1983. (Table 11-B).

* Provinces were included in agro-ecological zones based on agricultural administration in each zone. Tihama plain included Hodeidah Province; Southern upland included Taiz and Ibb; Central highland included Dhamar and Albeida; Northern highlands included Sana'a, Sa'adah, and Al Mahweet; and Eastern plateau included Marib and Al Jawf.

Supplemental irrigation is practiced on less than 13% of the total agricultural holdings (Table 29.4) with more than 75% of these holdings using supplemental irrigation to fulfill less than 50% of the crop water requirement. This information might lead to confusion about the actual need for supplemental irrigation; however, estimated data in Table 29.5 illustrates that 50% of the rainfall is less than 450 mm, an amount thought to be inadequate for prevailing climatic conditions. Sub-dividing the farming systems points out the relative importance of each system and allows identification of research needs for developing supplemental irrigation within the rainfed farming systems. However, the data in Table 29.5 fails to locate sub-systems or identify into which province or agro-ecological zone these sub-systems belong. As such, it becomes difficult to differentiate between rainfed, water harvesting, or supplemental irrigation farming systems at the province or district level.

Table 29.4. Farm type number and percentage of holdings.

Agro-ecological zone	Rainfed 100%	Suppl. irrig. < 50%	Suppl. irrig. > 50%	Irrig. 100%	Total
Tihama Plain	36,438 (62.2)	3,603 (6.2)	1,685 (2.9)	16,805 (28.7)	58,531 10.0
Southern Upland	197,086 (88.7)	9,772 (4.4)	2,628 (1.2)	12,690 (5.7)	222,176 38.0
Central Highland	54,258 (67.7)	16,818 (21.0)	3,964 (5.0)	5,037 (6.3)	80,077 13.0
Northern Highland	162,050 (76.5)	27,889 (13.2)	8,250 (3.9)	13,658 (6.4)	211,847 36.0
Eastern Plateau	795 (4.3)	231 (1.3)	65 (0.4)	17,349 (94.1)	18,440 3.0
Total	450,627 (76.3)	58,313 (9.9)	16,592 (2.8)	65,539 (11.0)	591,071 (100%)

Source: Compiled from Table 11.A of the Agricultural Census

Table 29.5. Estimated cropping areas under different farming systems and sub-systems.

Farming system	Cropped area (ha)	(%)
Rainfed (total)	790	77.5
Low rainfall (< 450 mm)	380	37.3
Med. rainfall (450-600 mm)	220	21.6
High rainfall (> 600 mm)	110	18.6
Irrigated (total)	230	22.5
Groundwater	119	11.6
Surface water (spate)	111	10.6
Total farming systems	1,020	100.0

Source: Compiled and modified from Table 3, World Bank, 1985

Cultural Practices

Most farmers typically use traditional cultural practices at a subsistence level of farming. Common constraints to agriculture are a lack of water, varieties of low productivity, and a minimal use of fertilizer and pesticides because of cost and availability. The principal crops are cereals which occupy 84% of the cropped land with sorghum being the main crop. Average cropping patterns for all farming systems in different agro-ecological zones are shown in Table 29.6. The areas of land which are left to fallow comprise 25% of the cultivable land;

however, only one third of this fallowed land was included in a crop–fallow rotation. The remainder of the fallow land has been abandoned because of lack of capital and labor for production. Data is not available on the significance of fallowing practices to productivity. However, these uncultivated lands even excluding the abandoned lands, offer valuable potential for expanding cultivated land.

Sorghum is planted under all farming systems as a popular grain for human consumption with other parts of the plant used for livestock feed. Millet as a food grain is sown to a land area equal to that sown to wheat and barley combined, and is usually grown under low rainfall conditions and in areas using water harvesting of flood waters, especially for spate irrigation in Tihama. Some farmers in Tihama produce millet with supplemental irrigation. Although this practice is not recommended (Kambal, 1986), it can be justified considering the high prices paid for millet grain.

Table 29.6. Cropping patterns and associated yields.

Crop	Land area (ha)	(%)	Yield* (kg/ha)
Cereals	864,561	64.0	———
Sorghum	622,901	46.1	———
Millet	87,989	6.5	———
Barley	47,188	3.5	0.6
Wheat	43,833	3.2	0.6
Maize	32,064	2.4	0.8
Mixed cereals	30,586	2.3	———
Legumes	22,556	1.7	0.6
Industrial crops	18,257	1.3	———
Vegetables	13,219	1.0	———
Alfalfa	11,226	0.8	12.3
Coffee	16,214	1.2	0.4
Date palm	9,887	0.7	———
Grapes	7,588	0.6	5.2
Other tree crops	56,350	4.2	———
Fallow	330,907	24.5	———
Total	1,350,775	100.0	———

Source: Agriculture Census, 1983

* Yield figures are averages of 1982-86 seasons; for some crops or groups of crops no yield could be given.

Maize is an all-region crop and usually is grown under irrigation or supplemental irrigation except in areas with more than 800 mm of rainfall. The demand for maize is increasing with expansion of the poultry industry and each

Supplemental Irrigation in Yemen Arab Republic (YAR)

year imports are required to augment domestic production. Wheat and barley are grown in the highlands and the Eastern Plateau. In the central and northern highlands, these two crops are grown under all farming systems with barley replacing wheat as moisture availability is reduced. High consumption of wheat, coupled with limited production of low yields, results in wheat grain being the principal imported cereal. Except for some legumes, other crops listed in Table 29.6 are either fully or partially irrigated. Average yields shown are estimates, the compiled results encompassing a wide range of values.

Most crops are sown manually with the help of draft animals. The usual method for cereal is closely spaced corrugations of less than 15 cm height. Crops sown mechanically are grown in furrows or rows in basins. Intercropping with legumes (particularly cowpea) is a common practice in sorghum fields under all farming systems. Estimated land of intercropped sorghum was more than 43,000 ha in 1983.

A representative crop calendar for different regions is shown in Figure 29.5. Typically, sorghum and wheat have more than one growing season in a single region. The environment and, therefore the existing growing seasons, permit cropping diversity with crops harvested at one site while sown in another if water supply is adequate.

Cultural practices which rely on rainfall are vulnerable to the vagaries of the

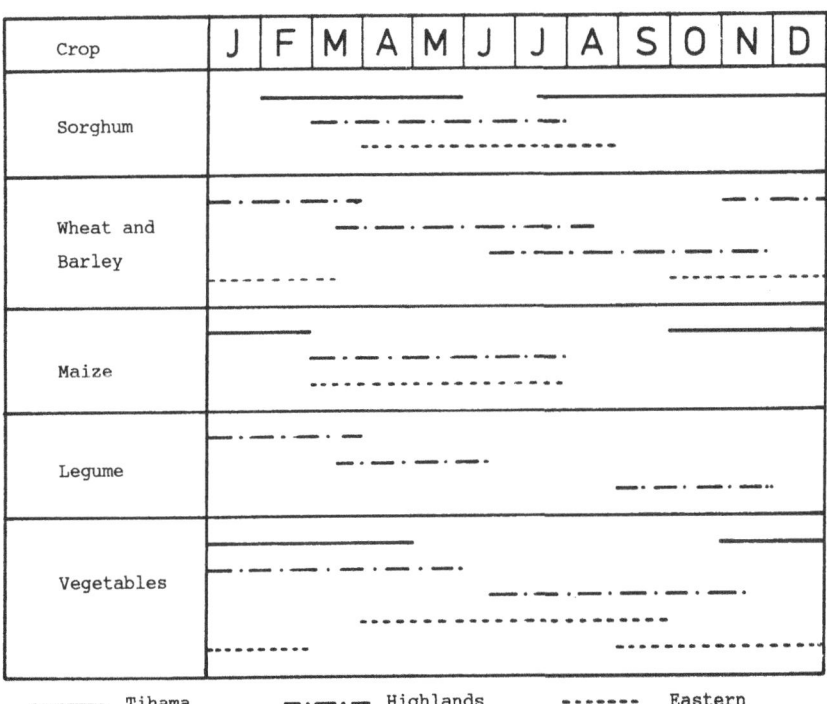

Figure 29.5. Cropping calendar for major crops.

weather. In combination with the farmer's inability to forecast if and when water will be available for plant growth, this results in more land sown to crops than is actually harvested. Undoubtedly, this contributes significantly to lower yields unless some provision is considered to supplement rainfall. Harvesting is manually performed with substantial yield losses as well as higher labor requirements.

Mechanization

Farm mechanization is relatively recent (beginning within the last 25 yr) with an ever increasing demand for mechanical power. Central Bank Reports reveal that in 1974 there were 700 tractor units in operation; by 1984 this number had increased ten-fold. By 1983, 40% of the agricultural holdings used tractors for cultivation (Table 29.7); however, draft animals still provided farm power for 84.8% of all farms. Use of tractors was highest (70.2%) in the Eastern Plateau (Marib and Al-Jawf) and was lowest in Tihama (9.8%). On the other hand, differences in use of draft animals in all zones are not appreciable. Manual tools are used in 15.2% of all farms. An interesting relationship pointed to by the 1983 Agricultural Census is that, of those farmers using tractors, only 7.5% owned them, with the remainder using rented equipment.

Table 29.7. Source of farm power for cultivation.

Zone	Source of power (%)		
	Manual	Draft animal	Mechanical*
Tihama	16.4	83.6	9.8
Southern Upland	19.1	80.9	41.5
Central Highland	8.4	91.6	40.8
Northern Highland	13.3	86.7	41.6
Eastern Region	11.5	87.7	70.2
YAR	15.2	84.9	39.2

Source: Compiled from Agriculture Census, 1983

* Exclusively tractors; most holdings that used tractors had also used draft animals; for that reason figures sum up to more than 100%.

Tractors are used mainly for land preparation, predominately plowing. Mechanical power is used for transport and for pumping of irrigation water. Regrettably, the Agricultural Census (1983) reported data on the number of holdings that used pumps and motor vehicles in only 4 provinces. As such, these data are only an indication of country-wide participation.

Drawbacks to farm mechanization are the lack of skills in the use and maintenance of equipment, land fragmentation and size of holdings, and the topography. These drawbacks should diminish with development of mechanization otherwise much capital investment could be wasted.

Fertilizer Use

A generalized status of macronutrient and organic matter in the soils is shown in Table 29.8. In all agro-ecological zones, a shortage of soil nitrogen and phosphorus is a major factor contributing to low crop yields. Soil organic matter is noticeably low which may be due to environmental factors; but, more likely is due to improper management of crops and soils. Management practices of single crop cultivation, removal or burning of crop residues, and inadequate applications of manure are aggravating the problem. Micronutrient status in soils has not been well studied and investigators at the Agricultural Research Authority (ARA) were probably the first to report micronutrient deficiency in some fruit trees (ARA, 1980). Their report attracted attention to the need for further investigations. Al-Fouli (1986) identified micronutrients that are most likely to be deficient, i.e. iron, manganese, zinc, and copper. He recommended a detailed research program, part of which has been adapted by ARA specialists. Considerable efforts are imperative for improving soil fertility through recommended rates of fertilizer application based on research findings.

Recommended application rates of fertilizers for major crops under different farming systems in the ARA areas of work are given in Table 29.9. These research results show that with the exception of sorghum to phosphorus (no response with spate irrigated sorghum) cereal crops need additions of nitrogen and phosphorus fertilizers.

Data supporting actual quantities of different fertilizers for application on various crops are not yet available. Table 29.10 shows the percentages of cultivable land areas that received fertilizers: nationally, 18% of the land area has received inorganic and 42% has received organic fertilizers. These values also reflect the number of holdings which have received some fertilizer: 42% received inorganic fertilizer; whereas, 77% received organic fertilizer. The data

Table 29.8. Generalized soil fertility status.

Zone	Range of soil texture	pH	Organic matter	Nutrient level		
				N	P	K
Tihama	sand to silt loam	8-8.5	vl	l	vl	h/vh
S. Upland	loam to clay loam	7.5-8	l	l/m	l	vh
Highlands	silt loam to silt	7-8	l	l/m	l	vh
E. Plateau	sand to loam	8+	vl	l	vl	vh

l = low, vl = very low, l/m = low to medium, h = high, and vh = very high.

Source: Compiled partially from ARA Soil and Water Lab Records, and from Sallam and Noaman, 1981.

Table 29.9. Recommended rates of N-P fertilizers for sorghum, maize, and wheat.

Farming system	Crop		
	Sorghum	Maize	Wheat
Rainfed			
450-600 mm	60-40	60-40	90-60
600-1,200 mm	80-40	80-40	100-80
Irrigated			
Spate	60-0	60-40	—
Well	80-40	120-60	80-80

Source: Saleh, 1986

show that Tihama soils received the least fertilizer treatment; whereas, the Southern Upland was treated with the most. This could be attributed to the common belief held by farmers that crops with spate irrigation do not need fertilizer; and, also that half of the cultivable area in Tihama (Hodeidah Province) has low rainfalls of less than 400 mm. The seeds sown are local varieties of summer sorghum and millet which have low responses to fertilizer.

Table 29.10. Distribution of areas treated and percentage of holdings using fertilizers and pesticides.

	% Cultivable area			% Holding		
	Fertilizers*		Pesticides	Fertilizers		Pesticides
Zone	I	O		I	O	
Tihama	0.1	0.0	3.5	4.0	2.0	8.4
S. Upland	46.5	86.3	3.7	66.9	95.8	5.6
C. Highland	9.4	56.1	0.6	20.3	80.2	0.4
N. Highland	22.3	47.1	6.1	35.7	77.5	4.0
E. Plateau	6.7	34.0	0.7	21.2	56.9	4.1
Total YAR	17.9	42.0	4.2	41.7	76.7	4.6

* Inorganic (I), Organic (O)
Source: Compiled from Agriculture Census, 1983

Weed and Pest Control

Weeds are strong competitors with crops for soil, water, and nutrients because their vigorous growth habits and root systems are more intensive than those of cultivated crops. Traditional cultural practices of repeated hand weeding, cultivation, and livestock grazing are used to control weeds. Often hand weeding is not efficient since most weeds are fed to animals; the removal practice is selective and some dangerous weeds are left uncontrolled. Hand weeding for livestock consumption is practiced in the areas of Tihama and the Southern Upland; however, in the highlands, the weeds are sometimes burned in the field after harvesting to reduce the labor requirement. A list of common weeds, insects, and diseases for the various crops is presented in Table 29.11.

Table 29.11. Common weeds, insects, and diseases.

Crops	Major pests (zone)*		
	Weeds	Insects	Diseases
Cereals	*Solanum dubium* (T, SU) *Cynadon dactylon* *Cyperus notendus* *Flaveria trinervia*	Armyworm (SU) Sorghum midge (T) Aphids (H); Termites (T)	Rust
Legumes	*Cynadon dactylon*	Aphids	Rust
Vegetables	[same as cereals]	Potato tuber moth (H) US bollworm (T, SU) Onion thrips (SU) Flea beetle Cabbage aphids	Virus Lf Blight P. Mildew Rust
Fruit trees	[same as cereals]	Thrips/banana Lesser date moth Coffee berry borer Scale insects	Canker (T) Greening (SU) Virus

* Agro-ecological zones are given in brackets: Tihama (T), Southern Upland (SU), Highlands (H); when no brackets are given, all zones are affected.

Source: ARA Plant Protection Section (Personal communication), 1987.

Comprehensive illustrated studies on important insects and their control in YAR was prepared by Al-Humiari (1982); similarly, a study on plant diseases was completed by Kamal and Al-Aghbari (1985). When practiced, pest control is done with pesticides; however, the level of chemical use is low (Table 29.10). Less than 5% of all farms use pesticides. As in many developing countries, misuse of pesticides is common despite efforts by government agencies.

Livestock

A recent review of the status of livestock deals with different aspects of the subject (Hasnain, 1985). In 1983, livestock totaled 5 M chickens, 3 M sheep, 2 M goats, one M donkeys, and 0.1 M camels (Table 29.12). The data show also the distribution of farm animals in different zones.

Table 29.12. Distribution of farm animals.

Zone	Type of livestock (1,000 head)					
	Cattle	Sheep	Goats	Camels	Donkeys	Chickens
1. Tihama	153	300	280	22	92	489
2. S. Upland	345	386	373	13	90	2,231
3. C. Highland	140	692	290	14	70	493
4. N. Highland	318	1,287	722	18	189	1,503
5. E. Plateau	19	377	419	23	18	136
Total livestock	976	3,042	2,084	90	459	4,852

Source: Compiled by YAR Agriculture Census, 1982

Most cattle are in Taiz and Ibb Provinces in the Southern Upland; the highest sheep and goat populations are found in the Provinces of Hajjah, Mahweet, Sa'adah, and Sana'a in the Northern Highlands. Fifty percent of the chickens are in the Southern Upland. A strong correlation exists between actual cropped land and number of farm animals (see Table 29.3). This might be attributed to availability of more crop residues and livestock feed. The relation is quite different when correlating sheep, goat, and cattle to the human population. Hasnain (1985) pointed out that more sheep and goats were available per capita in the Eastern Plateau (Marib and Al-Jawf) than other zones and as much as 5 times the national average. Differences between provinces seem to be insignificant for cattle. There is a trend towards mixed farming and away from pure crop production. On a national basis, about 90% of the agricultural holdings practice mixed farming on 1.3 M ha.

Major problems of livestock production are related to an inadequate data base, lower performance of livestock, improper management due to shortage of qualified labor, inadequate infrastructure, poor management of ranges and pastures as indicated by overgrazing, and little research backup (Hasnain, 1985). However, there are indications that livestock production as an industry promises higher productivity and profitability. These indications are supported by consumer preference for domestic products, vast rangelands, and enhancement of free enterprise.

Socio-Economic Aspects

Small farm size is a characteristic of all farming systems. According to the 1983 Agricultural Census, 71.6% of all holdings are less than 2 ha, spread over one fifth of the cultivated land. Farms of more than 10 ha constitute only 4.1% of the total; these represent 33.9% of the cultivated land. Wide variations in size of holding occur throughout the 5 agro-ecological zones. For instance, in Tihama (Hodeidah), only 31% of the farms are less than 2 ha; whereas, in the

Southern Upland (Ibb), 90.6% are less than 2 ha. Large farms are rare: 0.15% of all farms are more than 50 ha and these are mostly in the Provinces of Albeida, Hajja, and Hodeidah. The proportion of holdings and cultivated land for various sizes of holding is presented in Table 29.13. Average farm size is 2.3 ha varying from 0.9 ha in Ibb to 5.7 ha in Hodeidah (Table 29.14).

Table 29.13. Percentage of holdings and percentage cultivated area by size of holdings.

Size of holding	% Holding	% Cultivated land area
< 2	71.6	20.1
2-5	16.9	23.5
5-10	7.4	22.5
> 10	4.1	33.9

Source: Agricultural Census, 1983

Table 29.14. Average size of holding and number of parcels per holding.

Province	Average size of holding (ha)	Average number of parcels per holding
Hodeidah	5.7	2.8
Ibb	0.9	3.7
Taiz	1.2	4.0
Al-Beida	2.2	4.7
Dhamar	1.3	7.1
Hajjah	2.8	4.1
Mahweet	1.1	3.2
Sa'ada	2.3	4.6
Sana'a	3.1	5.8
Al-Jawf	3.4	7.1
Marib	5.3	6.2
YAR	2.3	4.6

Source: Agricultural Census, 1983.

Land fragmentation is another feature of the farm structure. Of the total number of holdings (591,071) only 16.3% are not fragmented holdings. Sixty percent of the remaining holdings have 2-5 parcels, and 8% have more than 10 parcels. The national average is 4.6 parcels per holding (see Table 29.14). Hodeidah's farms have the least number of fragments (2.8).The farms in Dhamar and Al-Jowf have the highest (7.1). The small size of holdings and fragmented parcels are negative influences on agricultural development.

Land tenure data show that 77.4% of the total cultivated land is farmer owned, 3.5% is share-cropped, 0.3% is rented from "wakf" land, and 18.8% is under two or more types of land tenure systems. Land tenure systems vary from 100% total ownership in Marib to 47.3% in Dhamar. Sharecropping is most common in Mahweet (16.8%). Interestingly, mixed land tenure systems are common in the highlands.

Cost of production in many areas is high because of intensive labor requirements, particularly during planting and harvesting. The cost of mechanization for land preparation is also high; but, this could be attributed to the small size of holdings and irregular shape of fields as well as land fragmentation which requires a longer time for execution of work. The cost of irrigation for fuel consumption and labor is considered too high when compared with traditional flood irrigation that wastes both water and time but appears to be cheaper to the farmer.

Evaluation of Selected Supplemental Irrigation Systems

To gain an in-depth understanding of the variability of existing supplemental irrigation systems in different agro-ecological zones, a precoded questionnaire (developed by ICARDA) was used to collect field information. The questionnaire includes all aspects affecting farming at the district level. District, for the purpose of this assessment, is defined as the smallest administrative organization for farm extension services. Farming systems of interest are rainfed, where crop production is managed with only rainfall; supplemental irrigation, when more water is added to rainfall to stabilize and improve yield; and, water harvesting, where modification or treatment of the land surface is done to either maximize or minimize runoff from natural rainfall. These 3 farming systems are difficult to distinguish from one another because, in most rainfed farming, some form of water harvesting supplements natural rainfall.

On a basis of data availability, 3 districts, located in different agro-ecological zones, were selected for evaluation (Figure 29.1):
1. Wadi Zabid district in the tropical lowland, Tihama;
2. Hoban district in the sub-tropical upland, Southern Upland; and
3. Ma'bar district in the temperate highland, Central Highlands.

Wadi Zabid, part of the District of Zabid, is an Extension Supervisory within the southern region of Tihama run by Tihama Development Authority (TDA). Hoban, serving the Districts of Taizzia, Mawya, and Khadir, is an Extension Block Center in the Taiz Province. And Ma'bar, also the District of Ma'bar, is served by one extension center in Qa'Jahran in the Province of Dhamar. Within the boundary of each district there is a Regional Research Station, a branch of ARA.

Tropical Lowland

Total land is estimated at 25,000 ha upon which 150,000 inhabitants live in 210 villages. Households in the district total 30,000 with an average of 5 members per household. Tihama is not an exemplary site for supplemental irrigation but has a potential for investigation since 45% of existing cultivated land is farmed under rainfed conditions.

Rural support services

Support services are provided by government, semi-government, and private agencies. Most of the educational facilities, including illiteracy control, are financed by the Ministry of Education. Medical care is supported by the Ministry of Health and Local Councils for Cooperative Development (LCCD). Agricultural services, extension and research, are administered by the Ministry of Agriculture and Fisheries (MAF).

There is 1 seed multiplication center, 1 credit bank, 1 veterinary care clinic, 10 extension centers, and 1 regional research station in the district. Water supply and electrical power projects are managed by either cooperative or private sectors and most rural road networks were constructed with government funds (Tihama Development Authority). Zabid and Al-Jerrahi towns are the main markets for the district.

Land use

Seventy two percent of the land, or 18,000 ha is arable, while the remaining surface is desert shrub, range pasture, and sand. Seven hundred hectares are rainfed, 500 ha have water harvesting (spate irrigation), and 9,300 ha are with supplemental irrigation. The remaining, 7,500 ha, is not in crop production. Estimated number of farms is 4,300, with an average size of 4.2 ha. Major problems with management of the farm units are land fragmentation and multiple ownership. Rainfed farming commonly suffers from water shortages. Agricultural land is 82% totally owned by individuals or in joint ownership, with only 18% rented. Rented land is usually sharecropped, agreements depending on the source of water.

Soils

Soils are predominantly deep, of alluvial parent material deposited in the flood plain of the Wadi Zabid. Except in the foothills where most rainfed farming occurs, average soil depth is more than 1 m; but, soils of the alluvial fan can be as shallow as 50 cm. Towards the western areas, soils are light brownish, level to moderately sloping, and most are stone-free. Soils under rainfed conditions or spate irrigation are of low salinity unless improperly irrigated by groundwater at which time salinity problems can occur. Infiltration rate is moderate and soils are non-crust forming or only moderately crust forming with a good to fair tilth.

Water resources

Average annual rainfall in the wadis is 0-350 mm with increasing rainfall near the foothills. The major source of water for supplemental irrigation as well as for conventional irrigation is the tubewell. Water quality of the tubewells is 0.9-1.6 mS/cm but in some instances can be higher than 1.6 mS/cm. Land cultivated using tubewells is estimated to be 9,500 ha.

There are 983 productive wells with an average depth to water table of 30 m, with a pumping depth of 45 m. Pump flow is estimated at 10 l/sec, producing a general decline in water table; and consequently, in pumping capacity. Lack of funds for capital investment is the major constraint; a new pumping unit can cost 150,000 rials for drilling plus 200,000 rials for pump and engine. Well water is often shared for a payment of half the crop yield.

Supplemental irrigation

Well water is the principal water supply for supplemental irrigation. Supplemental irrigation farming systems are difficult to separate from those land areas completely irrigated because all data records include, as a single category, those areas irrigated from wells regardless of rainfall amount. Therefore, survey responses from participants are drawn from their experience of irrigated farming. No actual data is available on supplemental irrigation with spate water.

Irrigation is applied in basins and furrows on a fixed time schedule. When sorghum is grown with supplemental irrigation from well water, 2-4 irrigations are applied compared with only 1-2 irrigations when using spate water. Neither surface nor sub-surface drainage is provided for irrigated lands despite the fact that farmers usually favor over-irrigating. Runoff is a common problem in irrigated fields as well as salt build-up with well water and silt accumulation with spate water.

Water harvesting

Spate irrigation is the main method of water harvesting which utilizes flood water from the seasonal surface flows in wadis diverted onto terraced fields surrounded by high earth bunds of higher than 70 cm: a frequent practice in Tihama. Fields are filled with flood water; farmers are prone to over-irrigation, initiating loss of water and causing soil erosion. Suspended silts accumulate on the soil surface and surface crusting occurs but the silts do furnish varying amounts of plant nutrients. Traditionally, large field terraces were built and levelled with the help of draft animals; but now, heavy machinery is gradually being substituted for the animal traction.

Crop production

Sorghum and millet grain crops and fodder sorghum are reduced in summer with rainfall. With irrigation, crops like cotton and intercropped cowpea are usually cultivated. No specific crop rotation is followed for reasons related to tradition and habit. Cropping patterns of the 3 farming systems depend upon several factors besides the traditional cropping season. Major factors that determine size of area cropped include:
1. total rainfall up to sowing time for rainfed farming;
2. cost and availability of water used in conjunction with rainfall for supplemental irrigation; and,
3. availability of flood water and depth of water in the soil profile in water harvesting farming.

Farmers tend to sow seeds in moist soil after the first rain or flood but, with irrigation, they are more likely to put seeds in dry soil. Sowing dates for sorghum and cowpea are February, March, and July; whereas, for millet and cotton, sowing is in July and August. Interestingly, as water becomes available, fodder sorghum is sown anytime except December and as many as 3 ratoons are possible. Seeding rates are 35 kg/ha for sorghum, 20 kg/ha for millet, 40 kg/ha for cotton, and 15 kg/ha for cowpea. Sowing is ordinarily done with oxen drawn tillers or mechanically if seeders are available.

Chemical fertilizer for cereal production is limited particularly for rainfed and water harvesting farming of the Tihama; there are recommended rates of fertilizers (see Table 29.9).

Hand weeding is the major method for control and the weeds are often collected for livestock feed. Cereal rusts and smut are the major diseases and the midge is the main insect on sorghum. Termites are by far the most serious pest on crops during all seasons in Tihama; but, local varieties of sorghum and millet are reasonably resistant. However, cotton cultivation is not possible without chemical control of these destructive insects. Cotton seeds provided to farmers by the Cotton Company are treated with insecticide against termites.

Harvest dates are June, July, and October for sorghum, October for millet, and, December and January for cotton. Harvesting is done manually creating a labor intensive operation, and a critical time for conflicts in labor supply. Agronomic factors that should be considered when intending to increase cereal production include water availability as well as crop seed variety, fertilizer use, sowing dates, and seedbed preparation.

Livestock

Most farmers own livestock and the estimated population of livestock is 80,000 sheep, 50,000 cattle, and 40,000 goats. Available sources of animal feed are sorghum straw, natural pasture, and crop residues, most of which are low in nutritional value. Flocks are usually moved in winter from the rainfed areas to relatively wetter (irrigated) areas where ample supplies of crop residues and pasture are available.

Socio-economic factors

Marketing procedures for cereal crops, sorghum and millet, are very simple. Products go from farmer to local retailers who store the grain then sell the stored grain throughout the year. Alternatively, products go from farmers to wholesalers who undertake marketing of the crop in locations outside the district. For cotton, all produce should go to the Cotton Company. An average return on summer crops is given in Table 29.15.

Table 29.15. Average return to the farmer (rials/ha).

Farming system	Sorghum	Millet	Cotton
Rainfed	2,140	1,835	—
Water harvesting	2,625	2,360	1,950
Well irrigation	3,310	—	2,000

Data clearly show that irrigated crops give a higher net return than do those crops cultivated under other farming systems. Also, sorghum grown in summer is the best cash crop from all farming systems.

A large proportion of farmers rely on farm production for total income when farming with well irrigation (90%) or water harvesting (80%); but, only 20% of the farmers rely on production revenues from rainfed farming for their total income.

Labor requirements are provided by family labor and draft animals. The major problem with hired labor is supply, particularly at harvest time. Labor supply is a function of both temporary and permanent migration which does occur. Most of the permanent migrants still have farms which are operated by sharecroppers, tenants, or family members.

Sub-Tropical Upland

The second district is Hoban Block Center (Hoban District) in Taiz Province. A block center is a term used by the Southern Upland Rural Development

Project (SURDP) to describe an agricultural administration unit for a group of extension centers. Hoban District as a block center serves the Districts of Taizzia, Mawiya, and Khadir and has 400 villages populated by 210,000 inhabitants representing 35,000 households. The district is a typical supplemental irrigation area with farmers relying on rainfall with supplemental irrigation using both well water and water harvesting. Unlike the Zabid district, water harvesting in Hoban is not well recognized and cannot be separated from the rainfed farming system. All field terraces are constructed manually to increase effectiveness of rainfall and intercept runoff.

Rural support services

Like Zabid district, support services are provided by government, semi-government, and private agencies. There is 1seed multiplication center, 1 credit bank, 1 veterinary care unit, 1 regional research station, and 12 extension centers. In the district, there are 14 medical care facilities excluding those located in Taiz city. Each of the districts of Hoban has a LCCD which is a semi-government cooperative group responsible for different aspects of non-agricultural development.

Many villages have potable water most of which is provided by SURDP. Electrical power is furnished by the villagers themselves. Rural road networks are present in most parts of the district and the secondary roads with gravel and dirt surfaces are in good to fair condition suitable for all year transportation. The city of Taiz is the major market but other towns have good markets.

Land use

Total land is 180,000 ha with only 22% arable. Most of the cultivated land, 33,000 ha, is rainfed with the remaining 7,000 ha using supplemental irrigation: all land is distributed to 35,000 farms. A higher percentage of the rainfed farms are fragmented with an average of 4-5 parcels; fragmented farms with supplemental irrigation average only 2-3 parcels. The major problems with management of farms are size and fragmentation of holdings as well as inadequate water supplies.

Soils

Shallow soils are not predominant in the district: average soil depth of rainfed farms is 51-75 cm with this increasing on holdings with supplemental irrigation. Soils are of colluvial and alluvial origin, most having a brown color, nearly level, slightly stony to stone-free, originally non-saline, with moderate infiltration, thin crusting, and a fair tilth.

Water resources

Average annual rainfall is 400-600 mm with an average maximum intensity of 30 mm/hr lasting for 1 hr or less. The main source of water for supplemental irrigation is tubewells although harvested water is probably used in addition to rainfall. Groundwater is of medium to high salinity which can reduce crop yields. The average depth to water table is 25 m but pumping depth is at 70 m with an average flow of 5-6 l/sec. The total number of productive wells dug in the district is 157, 13 of which have since gone out of production.

Supplemental irrigation

Pumping is needed to convey water to the cultivated fields. Land levelling is done by mechanical and draft animal power. Supplemental irrigation is applied 2 times to maize and 1 time to sorghum and millet in below normal rainfall seasons. Pre-sowing irrigation is not common but an irrigation is applied immediately after sowing. Most farmers use flood irrigation (75%) with the remaining farmers using furrow irrigation. Available water supply is the main criterion used to schedule irrigation but a neighbor's scheduling, weather factors, and condition of the plants are used as well. Problems of deep percolation and fertilizer leaching occur from over-irrigation. Internal drainage was not reported as a problem and few waterlogged wastelands (Khazaja) can be found in the district.

Water harvesting

Although the Southern Upland is famous for diverse water conservation techniques including bench terraces, data records on water harvesting farming are not kept separately from rainfed farming. A database for water harvesting should be established in this zone.

Crop production

Three major summer crops, sorghum, maize, and millet, are grown. Farmers use different criteria to determine the cropping pattern for any particular year. Farmers in the rainfed areas either plant the entire cultivated area while monitoring the total rainfall to sowing time or determine the number of intensive storms before sowing in order to decide their cropping pattern for the season. On the other hand, farmers with supplemental irrigation use criteria such as availability of water for irrigation, cost of water, and total rainfall up to sowing time to determine their cropping pattern for the season. The estimated area allocation for summer crops is given in Table 29.16.

Table 29.16. Land area allocation for summer crops in the Hoban District.

Crop	Rainfed (ha)	Suppl. irrig. (ha)
Sorghum	28,790	4,912
Maize	3,570	1,462
Millet	370	—

Crops are usually sown after the first rain. Average seed rates are 20-25 kg/ha for sorghum, 35-40 kg/ha for maize, and 10-12 kg/ha for millet but somewhat higher seed rates are used when crops are cultivated with rainfall only. Sowing dates for these crops are May for sorghum, mid-June for maize, and July for millet.

More farmers use fertilizer for crops grown with supplemental irrigation (80%) than for rainfed conditions (50%); the trend is increasing towards applying fertilizer. The major method of application is broadcasting using average rates for rainfed conditions of 60 kg/ha for nitrogen, 40 kg/ha of P_2O_5, and 5 t/ha of organic fertilizer. With supplemental irrigation, 80 kg/ha nitrogen, 60 kg/ha P_2O_5, and 10 t/ha of organic fertilizer is applied. No specific crop rotation is followed under rainfed farming and, only 30-40% of the farmers follow a 2-course rotation of sorghum – maize with supplemental irrigation. Reasons for using rotation are possibly related to tradition and habit.

Weed control is done manually as research recommendations on herbicide use are not yet available. Control of diseases and insect pests with pesticide is limited to 20% of the farmers in rainfed farming and 40% of the farmers using supplemental irrigation. The major plant diseases are cereal rust, smut, and leaf blight; the major insects are the midge on sorghum and the stem borer on other crops. Army worm outbreaks are not uncommon but a national control campaign has been organized, financed, and implemented by the MAF.

Normally, manual harvesting of sorghum begins in October and November, in September or October for maize, and in September for millet. The expected average grain yield of rainfed sorghum is 2.8 t/ha and 2.5 t/ha for maize. Higher grain yields, 3.5 t/ha are expected for maize cultivated with supplemental irrigation.

Increasing cereal production could possibly be achieved by considering factors such as sowing date, seedbed preparation, seed variety, and fertilization for rainfed farming, and fertilization, seed variety, and irrigation after sowing for the supplemental irrigation farming system. Conflicts occur between cereals and other crops of supply and scheduling of supplemental irrigation because of labor needs, harvest operations, and land preparation.

Livestock

Most farmers own livestock with an estimated population of 70,000 goats, 54,000 sheep, and 32,000 cattle. Farmers grow feed for livestock; but, variability in supply still occurs. Hay, straw, and cereal grains are the sources of livestock feed. Cereal straw contributes about 85-90% of the total feed requirements and crop residues are grazed. No payment is given for grazing but a charge is collected for taking the animals to graze.

Socio-economic factors

Marketing procedures are not different from those of Zabid district. Net returns from sorghum and maize for the farming systems are presented in Table 29.17. It is evident that rainfed sorghum gives more return per hectare than does maize. This could be attributed to higher yields obtained from local varieties of sorghum which are better adapted to rainfed conditions in contrast to improved maize varieties that are susceptible to shortages of water and fertilizer. Maize production with supplemental irrigation added to rainfall has doubled its net return.

Table 29.17. Average net returns for sorghum and maize.

Farming system	Sorghum (rials/ha)	Maize (rials/ha)
Rainfed	5,200	2,300
Suppl. irrig.	NA	5,400

Most farmers (70-80%) in both farming systems rely on farm production for their total income. Agronomic constraints to the farmer's income are related to harvesting and fertility. Often, irrigation technology introduces problems to those who supplement rainfall with irrigation.

Labor requirements are met either by sharecroppers or hired labor. In the case of sharecropping, arrangements are 50-50 shares for land and labor under rainfed conditions, and 25-37.5 shares for irrigation, land, and labor with supplemental irrigation. Hired labor is mostly from within the district (80%). Most labor is provided by the family members (50-60%), but some is supplied from hired labor (10-20%), draft animals (20%), and mechanical power (10%). Problems with labor relate to higher costs (rainfed farming) and timing conflicts (supplemental irrigation).

Migration occurs, both permanent (rainfed) and temporary (supplemental irrigation), for reasons related to work opportunity, family committments, and seeking of education and training; however, the level of migration is decreasing. For both farming systems, migrating farmers usually travel outside the country for temporary migration and may migrate permanently to urban centers. They still retain farms which are operated by sharecroppers (rainfed) or family (supplemental irrigation). Temporary migration happens in winter (rainfed) and in fall (supplemental irrigation) but the time away does not last more than 3 months for the supplemental irrigation systems and lasts from 3-6 months in the rainfed farming systems.

Mechanized farming is not common, as only 2-10% of the farmers own tractors. Tractors have been used for 10-15 yr only for land preparation; pesticide sprayers have been used for 7 yr. Rainfed farms were the last to use tractors.

The major constraints to implementing supplemental irrigation in this district, farmed under rainfed conditions for centuries, are the lack of experience in irrigation farming practices and lack of training in irrigation techniques.

Temperate Highlands

The third district is representative of the arid central highlands. The District of Ma'bar, also called Qa' (plain) Jahran, is located in the Dhamar Province. The district's center is the town of Ma'bar, 25 km to the north of Dhamar city. The plain lies at an altitude of 2,300-2,500 m. The total area is 37,000 ha populated by 20,000 persons distributed in 4,000 households. The district was chosen for the relative availability of data as it houses the Dhamar area extension supervisory of the Central Highlands Rural Development Project (CHRDP).

Rural support services

Support services are provided by public and private organizations all in coordination with the LCCD in the town of Ma'bar. Agricultural services are maintained through one extension center established by CHRDP which is planning a block center in Ma'bar. Seed Potato and Seed Multiplication Projects have their agricultural field work here and the Regional Research Station for the Central Highlands is located on the southern limits. Soon the ARA will have its headquarters and housing complex here.

The district is bisected by the main highway between Dhamar and Sana'a. Secondary roads are of gravel and dirt surfaces but in fair condition. Education and health facilities are adequate. Almost all villages have electrical power provided by private generators and soon the area will receive power from the All-Yemen central power station. Potable water is provided by private and LCCD projects. The main market is Ma'bar town located in the middle of the district.

Land use

Of the total cropped area, 16,410 ha, 60% are estimated to be under rainfed farming and the remaining under irrigation. The major problems with management of the farm unit are insufficient water, fragmentation of holdings, and inadequate farm size. The holdings are either owned or rented. Most land is under individual and family ownership with no land rented and is farmed under rainfed conditions.

Soils

Soils are of a lacustrine sediment mixed with alluvium, colluvium of volcanic rock outcrop and loess cap. Moderately shallow soils are predominant in rainfed and water harvesting areas with an average depth of 51-75 cm. Soil depth increases in areas with irrigation. The major soils of the plain have a yellowish brown color, level slopes, slightly stony, of low salinity with calcareous concretions, slow infiltration, moderate crusting (silt cap) and fair tilth.

Water resources

Average annual rainfall is 300-350 mm with an average maximum intensity of 20 mm/hr lasting for less than an hour. Rain falls in two seasons, the first in March/April and the second in July/August. Frost is commonplace, with the first frost coming as early as October and the last in March. Because of low rainfall and limited water harvesting, wells for irrigation are widespread. Groundwater quality is suitable for all uses. The average depth to water table is 90-120 m with an average pumping depth of 100-140 m with a flow rate of 5-10 l/sec; but, seasonal declines and a continuous decline in pumping capacity are experienced. Often the well owners sell water to neighbors for supplemental irrigation. The average cost of digging a well is 150,000 rials and the cost of a pumping unit (pump and engine) is 200,000 rials. All engines use diesel fuel. Major constraints when deciding to buy a new pump are initial capital investment, credit availability, and availability of parts and repair service.

Supplemental irrigation

Farmers irrigate their fields using unlined irrigation channels and crops are grown in basins which are flood irrigated. Land levelling of these basins is done using attachments and tractors, animal traction equipment, and hand tools.

Three to four supplemental irrigations are given to wheat in an average rainy season with more irrigations needed in below average seasons. Presowing irrigation is a newly introduced practice that 40% of the farmers have adapted. The average presowing irrigation is 80-100 mm. No surface or sub-surface drainage is provided although farmers tend to over-irrigate and runoff, waterlogging, and salt build-up are potential problems.

Water harvesting

Traditionally, farmers tend to use water harvesting techniques because rainfall is not sufficient. Tied ridges or micro-catchments, conservation bench terraces, and bunding are major methods used for water harvesting. Collected runoff is used to irrigate field crops, even though other crops could benefit from such a practice. In good water harvesting systems, the actual volume collected is several times that of rainfall but actual contribution to production remains unestimated. Records on water harvesting areas are not definitive; therefore, the data presented are not exclusively for this type of farming system. Data are usually collected for rainfed conditions regardless if water harvesting is used or not.

Observed problems of water harvesting facilities are erosion, maintenance, and reliable collection of rainfall. The fields that receive harvested water are prone to erosion and waterlogging.

Crop production

Rainfed crops are barley, sorghum, lentils, and a local variety of wheat. Irrigated crops are wheat, maize, and potatoes. Unfortunately, there are no records on cropping patterns. In determining the size of area sown it is assumed that the entire cultivated area is planted in rainfed farming and that the area sown depends upon availability of water supply for supplemental irrigation agriculture.

Farmers sow cereal crops after the first rain. The average seed rates are: 110 kg/ha for wheat, 90 kg/ha for barley, 15 kg/ha for sorghum, and 35 kg/ha for maize. The average sowing dates for these crops are July for wheat and barley and March for sorghum and maize. No specific crop rotation is followed; and yet, farmers have been observed growing winter legumes and alfalfa when using supplemental irrigation production.

Farmers use fertilizers with supplemental irrigation: these are mostly organic and nitrogen fertilizers; but, there are reports on response to phosphate fertilizers in the Central Highland zone. Market availability and price of fertilizers are the main difficulties for farmers who want to apply fertilizers; nonetheless, fertilizer use is increasing.

The common weed is *Cynadon dactylon*. Rusts of wheat and barley are the major diseases and aphids on cereal crops are the major insect pest.

Harvesting is ordinarily done by hand and, therefore, is a labor intensive process. Harvesting dates are October and November for wheat and barley, mid-October for sorghum, and mid-September for maize. Average grain yields obtained under rainfed conditions for wheat are 1.2 t/ha and for barley are 1.1 t/ha but these increase to 2.3 t/ha for wheat and to 4.0 t/ha for barley with supplemental irrigation.

Major agronomic factors for increasing cereal production are seed variety, sowing date, seedbed preparation, and fertilizer use. The major conflicts between cereals and other crop production are harvest operations and labor requirements.

Livestock

Most farmers have livestock specifically cattle, sheep, and goats. Sheep and goats are herded together and goats account for one tenth of the flock. The average flock size is large and, in general, flocks are herded by family labor. Animals are taken for grazing on hill slopes and uncultivated land but, after harvesting, they graze crop residues, for which, there is no charge. Farmers grow some animal feed such as alfalfa but cereal straw contributes a large proportion of the feed requirement.

Socio-economic factors

Farmers market their crops locally at Ma'bar and Dhamar. Vegetables are taken to markets at further distances, and occasionally are bought by traders at the farm gate which means the farmer's share of the final price is much lower. Average net returns (rials/ha) obtained by farmers from growing summer cereal crops are presented in Table 29.18. The data show that crops produced with supplemental irrigation give higher net returns than do crops under rainfed conditions. Barley with supplemental irrigation doubled the net return.

Table 29.18. Average net returns (rials/ha) obtained from cultivating summer cereal crops in Ma'bar District.

Farming system	Wheat	Maize	Barley	Sorghum
Rainfed	—	—	2,938	4,487
Suppl. irrig.	8,150	7,193	5,496	—

Most farm income derives from cereal and livestock production in rainfed and water harvesting farming systems. However, vegetable crops contribute more to total income of farmers who use supplemental irrigation.

Most farm labor is provided by family members. Problems of using hired labor are cost and supply, especially at harvest time. All hired labor comes from within the district. Both temporary and permanent migration occur but is declining. Migrants still have farms that are operated by family or sharecroppers.

Major constraints for management of supplemental irrigation are the lack of training in irrigation technology and the lack of experience with farming practices with irrigation.

Potential for Supplemental Irrigation

In the past two decades in YAR, exploitation of groundwater for irrigation in rainfed areas has demonstrated a potential for changing and improving patterns of agricultural production. The survey of farming systems in selected districts emphasizes this but results of the questionnaire identify that water scarcity and reliable supplies are severe limitations to increasing production.

Available data on the agricultural activities indicate that fallowed arable land constitutes one fourth of the land area; and, two thirds of this is abandoned for reasons other than inadequate water resources (see Table 29.3). Areas depending on low and medium rainfall for their water requirements are half the cropped land (see Table 29.5). However, the results of the Agricultural Census (MAF, 1983) disclose that only 13% of all farms supplement natural rainfall to increase crop yields (see Table 29.4). Bringing the abandoned arable lands back into production and improving and changing water resource management practices on areas receiving less than 600 mm of rainfall are the main prospects for developing supplemental irrigation in these areas.

Normally, for supplemental irrigation of crop production, groundwater is the first source to exploit. This action seems logical only if sufficient groundwater is available (not true in most areas of YAR); and therefore, comprehensive regulations governing groundwater exploitation are required to protect this resource from over-exploitation.

Therefore, components for development would entail:
1. a database of land and water resources;
2. adaptive research to convert data into experience to identify alternative technologies that could be adopted in the rainfed areas of YAR; and,
3. dissemination of experience and technology to potential users of supplemental irrigation.

So far, little information is available on the nature of rainfall, amounts and occurrence of runoff from watershed areas or distribution of different soil water regimes. Those sources which are available are of limited value for water management at the farmer level. Water harvesting systems have not been intensively studied. Likewise, groundwater resources need to be studied further to establish discharge/recharge relations and explore new aquifers.

Research data on supplemental irrigation management are rare. Data for crop water requirements in different agro-ecological zones would be of great value to farmers who must schedule supplemental irrigation for crop production. Improving irrigation by adapting more efficient methods than those the farmers are using (flooding) would help both in economics and conservation of water use and contribute to increasing cultivated land with supplemental irrigation. Other elements of irrigation technology integrated into a farming system research study would be worthwhile. Testing the feasibility of water harvesting systems is also a possible field of neglected research.

An active and practical field extension service is complementary to research activities. Utilizing the existing structure of agricultural extension in disseminating research findings would be a step towards increasing crop production.

Varieties which have been introduced and consistently out-yield local varieties should be a part of alternative production practices. Drought tolerance of such varieties should become a criterion of the screening process for released varieties and for plant breeding. Investigating cultural practices with minimal cultivation and simple mechanized harvesting operations are alternatives that should be considered to evaluate implementation of supplemental irrigation.

Summary and Conclusions

This study on the status of farming systems in the rainfed areas of the YAR was prepared primarily as a basis for future development of supplemental irrigation. The main objectives of the study were to describe the farmers' environment and farming practices of rainfed and supplemental irrigation agriculture; to assess the current farming practices for acceptance of innovations and improvements using water management techniques; to evaluate the effect of supplemental irrigation on rainfed agriculture in selected areas; and, to identify prospects and major constraints for development of supplemental irrigation in potential areas.

The physical environment and its influence on agricultural activities, was briefly described with regard to the agro-ecological divisions, climate, geology, and soil and water resources. Existing agricultural activities were assessed which established that most agriculture is dependent on erratic and limited rainfall. Most arable land annually receives less than 600 mm and evapotranspiration exceeds double the volume of rainfall (1,200 mm). Traditional cultural practices are still predominant; but, with a trend which promises change and improvement. Major constraints to rainfed farming and supplemental irrigation agriculture have been discussed. Case studies in 3 districts, representing different agro-ecological zones and characterized by major differences in physiography, climate, and soils were used to evaluate these constraints and the impact of supplemental irrigation on rainfed farming. The chapter ends by summarizing prospects for development of supplemental irrigation by identifying potential areas, evaluating available water resources, and proposing development component needs and technological alternatives.

As a result of the survey, the following concluding remarks can be stated.

1. Differences in physiography support major ecological changes that are reflected in diverse agricultural systems. This diversity depends upon unique practices for conservation of soil and water resources when water is the most restrictive resource. For agriculture, most land is terraced.
2. Of the total 1.35 M ha of arable land, only 75% is cultivated annually, depending on rainfall. Seventy five percent of the rainfed farming receives less than 600 mm average annual rainfall. In most years, these areas experience water shortages.
3. Limited experience, shortage of manpower, small size and irregular shape of fields, and land fragmentation are typical features which constrain improved farm management. Increased costs of production, limited use of fertilizers, lack of herbicides and pesticides, and, cultivation of low yielding varieties are all contributing factors to decreasing agricultural productivity. Poor management of livestock is detrimental to the industry; however, potential for improvement is evident.
4. The case studies demonstrate that farming practices in the sample districts are not too different from other parts of the country. Cropping patterns and crop calendars are different between districts as a consequence of significant differences in climate. Commonly, farmers do not keep records on cultural practices because tradition and habit are the major criteria for decision making. Although the trend is towards change and improvement of agro-techniques, traditional cultural practices are still followed.
5. Documentation on farming systems as related to water source for agriculture is limited and conflicting. For instance, in the Tihama Plain (Zabid District), records are kept on groundwater irrigation and water harvesting (spate water) with few or no records on rainfed farming even though half the plain (Hodeidah Province) practices this type of agriculture. In comparison, records available on Hoban and Ma'bar Districts failed to mention information on water harvesting farming, notwithstanding that crop production, in several locations is possible only because of harvested rainfall.
6. The major constraints to implementing supplemental irrigation in the selected districts are related to farmers' lack of experience and training in irrigation.
7. Vast areas of the country have potential for development of supplemental irrigation provided adequate data is obtained on water resources with special reference to water harvesting.
8. Planned development could not be achieved without a well defined design with objectives from prioritized constraints and input from the extension service to execute technology transfer.
9. Besides improving on-farm water management, alternative technology for investigation and transfer could include agronomic improvements and introduction of crop varieties that are drought resistant.

Acknowledgements

This chapter would not have been possible without the benefit of work previously done by various scientists, agencies, and organizations, some of which are cited in the list of references. The administrative and logistics support for this study was provided by the Agricultural Research Authority (ARA) of YAR and financial support was given by ICARDA. The assistance provided by Dr. E. Perrier and Mr. A. B. Salkini (both of ICARDA) at the preparation stage of this paper is gratefully acknowledged.

I was most fortunate to have the interest, cooperation and assistance from the extension agencies, especially Mr. Jafar H. Alawi, Deputy Director of Agriculture for Tihama Development Authority (TDA) in Hodeidah. Also at the TDA, a special thanks to Mr. Abdulla M. Saif, Director, and Mr. Najib M. Ali, Head, Extension Service in the Southern Region (Tihama). The author is grateful to the management and staff of Southern Upland and Central Highland Rural Development Projects (SURDP and CHRDP), especially Mr. Mohamed Izzo, Extension Expert (SURDP) and Mr. Mohamed Al-Marwani, Extension Agronomist, and Mr. Ali Alswaidi, Farming Systems Agronomist (CHRDP).

The cooperation and assistance of the Local Councils for Cooperative Development (LCCD) in the districts sampled is duly acknowledged, especially that of Mr. Abdulla Ateya of Zabid LCCD. The author would sincerely like to show appreciation and duly acknowledge the major contribution of Mr. Ghazi Alsakkaf, ARA Agricultural Economist, to the field work portion of this study. Thanks are extended also to Dr. Abdulwali Agbari, Plant Scientist and Assistant Professor, Faculty of Agriculture, University of Sana's; for his critical review of this manuscript. Thanks to Ms Najat Mohammed Qassim for preparing the manuscript on the word processor.

References

Agricultural Research Authority (ARA), 1980. Annual report. ARA, Ministry of Agric. and Fisheries, Taiz, YAR. (formerly Central Agric. Res. Sta.)

Al-domi, I. A., 1983. Study on the historical stages of water techniques in Yemen Arab Republic. Presented at First Symp. on Water in YAR, 10-15 Sept.; Sana'a.

Al-eryani, M. L., 1979. Hydrology and groundwater potential of the Tihama, Yemen Arab Republic. Master's thesis, Univ. of Ariz., Tuscon, Arizona.

Al-fouli, M. M., 1986. Report on studying the problems of micro-nutrients in Yemen Arab Republic. Egyptian Nat'l. Res. Center and GTZ; Cairo, Egypt.

Al-humiari, A. A., 1982. A first guide to the agric. insect pest of the Yemen Arab Republic and their management. Master's thesis; Univ. of Ariz., Tuscon, Arizona.

Bamatraf, A. M., and J. A. Al-domi, 1987. Water conservation techniques in Yemen Arab Republic. Paper presented at the Second Int'l Soil Correlation Mtg., 6-15 Jan., 1987. Sana'a YAR (in press).

Caponera, D. A., 1973. Water laws in Moslem countries. FAO Irrig. and Drain. Paper 20. FAO, Rome.

Chapter 30

Conclusions

Traditional agriculture is an extractable process where all resources – human, water, and land – are taken and applied to immediate use. Modern agriculture uses planned technology and emphasizes management practices of conservation and renewability of resources. Modernization forces the growth of an infrastructure concomitant with rural development, urbanization, and industry. Overall economic development depends on the effective use of population and conservation of water and land as vital resources of the environment. President John F. Kennedy's inaugural address on January 20, 1961 pertains to the situations found within the region, "....ask not what your country can do for you; ask what you can do for your country."

Technology does exist for managing water and land as renewable resources. It is possible to develop more efficient utilization of rainfall and groundwater for land-use diversification: the challenge is to create an economic water supply of an optimal quality and quantity with a capacity for delivery on demand. It is important, for example, to improve national averages on rainfed cereal production by increasing yields from 1-2 t/ha by fertilization, varietal improvement, and water use efficiency techniques. The potential for a forward leap in production of 4-6 t/ha is attainable using modern irrigation technology. Throughout the region the various options of such technology can be used to expand the potential for stabilizing and increasing agricultural production. At specific locations, there are excellent opportunities to utilize and maintain surface and groundwater supplies, install large isolated irrigation systems using center-pivot and side-roller equipment, and develop long-term water harvesting systems.

The concepts of probability, water balance, and rainfall-runoff techniques integrated with socio-economic research methodologies can be used to evaluate the potential for the integration of these techniques into the modernization process. The steppe and pasture region with its traditional farming system of rainfed barley-livestock suffers from continuous barley cropping and overgrazing of natural pastures which causes water loss and soil degradation to progress at an alarming rate. New strategies for managing existing water and land resources, in harmony with traditional practices, can be formulated to intensify the effectiveness of human resources using capital investment in equipment to implement varying scales of irrigation systems and water harvesting farming.

The task will not be easy. There is need for flexibility in the system; if water restricts the farmer, then other farm management operations are restricted. Too often policy is a mere reaction to specific problems and does not grasp the total picture or perceive system needs. Planners and scientists are trained for a cautious, precise, controlled experimental research environment rather than the reality and art of farm production which introduces flexibility and change with diminished control.

Farmers in arid regions cannot afford the prolonged risk involved with changing their traditional farming practices without external support. The integration of multiple levels of interest for the successful exploitation of agricultural development in the Near East and North Africa and the conservation of water resources can be achieved by continuous assessment and evaluation. Influencing the technical and social constraints to the modernization of agriculture requires a systems approach emphasizing stepwise technology transfer for sustainable food security.

Recommendations

In addition to the recommendations given by the author of each country, the 7 recommendations listed below generalize the trends needed for changes in the political, administrative, scientific, and farming system.

1. Need to determine governmental policies that affect water resource development and the scope of the enacted programs for support of agriculture.
2. Design and implement alternative types of irrigation technology such as supplemental irrigation and water harvesting to meet specific needs for increasing agricultural production.
3. Implement and sustain training programs for extension personnel and farmers to encourage the technical development of efficient and effective irrigation technology as well as agronomic alternatives within the region.
4. Evaluate current trends in land-use to quantify the amount and extent of misuse or degradation in the environment.
5. Design and implement a livestock management program with alternative animal husbandry practices to interface with water harvesting technology.
6. Complete a hydrogeologic and topographic mapping of the region to encourage rational exploitative development of groundwater, fossil water, and recharge of aquifers.
7. Develop an interactive climatological and demographic database to provide mathematical modeling of the Near East and North Africa for planning and design of regional solutions to short- and medium-term problems.

Conclusions

Food for Thought

Riddle: When it comes to reading the report of an empirical scientific study, what is the difference between a layman, a researcher, and a methodologist?

Answer: The layman reads the text and skips the tables; the researcher reads the tables and skips the text; and, the methodologist does not care very much about either the tables or the text, as long as they agree with each other.

Appendix

Source Materials for Chapters 1-16

Allen, R. G., 1988, "Short course notes on scheduling irrigation water with the aid of computers", Irr. Systems Mgmt. Res. Project, University of Idaho, Moscow, Idaho.
Arar, A., "Report on consultancy to ICARDA on supplementary irrigation", January, 1984, Water Res., Dev. and Mgmt. Service, Land & Water Div., FAO, Rome, Italy.
Basawan, Sinha, and Ramesh Bhatia, 1982. "Economic appraisal of irrigation projects in India"; Agricole Publ. Academy, Institute of Econ. Growth, Delhi. 487 pp.
Baver, L. D., W. H. Gardner, and W. R. Gardner, 1972, "Soil physics", John Wiley & Sons, New York.
Beltran, J. M., 1978, "Drainage and Reclamation of Salt-Affected Soils", Int. Inst. Land Rec. Impr., Wageningen, Netherlands.
Blalock, H. M., 1972, "Social Statistics", McGraw-Hill Book Co., New York.
Bolton, F. E., 1981, "Optimizing the Use of Water and Nitrogen through Soil and Crop Management", pp. 231-248, In Soil Water and Nitrogen in Mediterranean-type Environments, Eds. J. Monteith and C. Webb, Martinus Nijhoff/Dr. W. Junk Publs., The Hague.
Boulding, Kenneth, 1956. "General systems theory – the skeleton of science, Mgmt. Sci., April:197-208.
Buckman, H. O. and N. C. Brady, 1960, "The nature and properties of soils", The MacMillan Co., New York.
Burt, C. M., R. E. Walker, and S. W. Styles, 1985, "Irrigation system evaluation manual", Dept. of Agri. Engr., Cal Poly State Univ., San Louis Obispo, California.
Campbell, S., 1988, "The home water supply", Storey Communications, Inc., Pownal, Vermont, USA.
CBS (Central Bureau of Statistics), "Statistical abstracts of Aleppo Province", 1984, Department of Statistics, Aleppo, Syria.
Churchman, C. W., "The systems approach", 1968, Dell Publs. Co., New York.
Churchman, C. W., "The design of inquiring systems: basic concepts of systems and organization", 1971, Basic Books, Inc., New York.
CIMMYT, 1980. "Planning Technologies Appropriate to Farmers", Economics Program, Londres 40, Apartado postal 6-641, Mexico 6, D.F., Mexico.
Collett, J., and J. Boyd, "Eight simple surveying levels", No. 42, Intermediate Technology Publications, Ltd., 9 King Street, London, England.
Cooper, P.J.M., 1983, "Crop Management in Rainfed Agriculture with Special Reference to Water Use Efficiency", Proc. 17th Coll. Int. Potash Inst., Bern, pp. 63-79.
Davidson, M., 1983. "Uncommon sense: the life and thought of Ludwig von Bertalanffy (1902-1972), father of general systems theory." J. P. Tarcher, Inc. Los Angeles.
Davis, J. C., 1973. "Statistics and Data Analysis in Geology", John Wiley & Sons, New York, USA.
Dixon, W. J., Ed., 1985, "BMDP Statistical Software", University of California Press, Berkeley, CA.

Doneen, L. D., and D. W. Westcot, 1984, "Irrigation Practice and Water Management," FAO Irrigation and Drainage Paper 1, Rev. 1, FAO, Rome.

Doorenbos, J., and Pruitt, W. O., 1984. "Guidelines for predicting crop water requirements." FAO Irrigation and Drainage Paper No. 24, FAO/UNDP, Rome.

Downing, T. E. and M. Gibson, "Irrigation's impact on society", Anthropological papers of the Univ. of Ariz., 1974, No. 25, Univ. of Ariz. Press, Tucson.

Dregne, H.E., 1976, "Soils of Arid Regions", Elsevier Scientific Publ. Co., New York.

Edelman, J. H., 1972, "Groundwater hydraulics of extensive aquifers", ILRI, Wageningen, The Netherlands. 216 pps.

FAO, "Food balance sheets 1979-81 average", 1984, FAO, Rome, Italy.

Finkel, H. J.(Ed.), 1982, "CRC Handbook of Irrigation Technology: Vol. 1", CRC Press, Inc., Boca Raton, Florida 33431.

Ford, W. B. III., and E. R. Perrier, 1980, "Markov models of hydrologic parameters for water quality loadings", USAE, Waterways Expt. Sta., Proceedings Nat'l Symposium on Hydrologic transport modeling, Amer. Soc. Agri. Engrs., St. Joseph, Mich.

Freeze, R. A., and J. A. Cherry, 1979, "Groundwater", Prentice-Hall, Inc., Englewood Cliffs, New Jersey, USA.

Gilbert, E.G., D.W. Norman, and F.W. Winch, "Farming systems research: A critical appraisal", 1980, MSU Rural Development Paper, No. 6, Michigan, p 17.

Gischler, C. E., 1979, "Water resources in the Arab middle east and north africa", Middle East & North African Studies press LTD., Gallipoli House, The Cottons, Outwell, Wisbech, Cambridge PE14 8TN, England. 132 pp.

Gittinger, J. Price, 1983. Economic analysis of Agricultural Projects, John Hopkins University Press, Baltimore.

Hagin, J. and B. Tucker, 1982, "Fertilization of Dryland and Irrigated Soils", Springer-Verlag, Berlin, Germany.

Hanks, R. J., and R. W. Hill, 1980, "Modeling Crop Responses To Irrigation: In Relation To Soils, Climate, and Salinity", Intl. Irr. Inf. Center, Pergamon Press, Ltd. England.

Harris, J., and E. R. Perrier, 1978, "A comparison of runoff and reservoir water quality functions to predict annual loadings from an agricultural watershed", USAE, Waterways Expt. Sta., EEL, Technical Report No. DS-78-42, Presented Amer. Soc. Agron. Meetings, Chicago, Illinois.

Hillel, D., 1971, "Soil and water: Physical principles and processes", Academic Press, New York, USA.

Hirschi, T., and H. C. Selvin, 1973, "Principles of Survey Analysis", The Free Press, New York.

Hjelmfelt, A. T., Jr., and J. J. Cassidy, 1975, "Hydrology for engineers and planners", Iowa State Univ. Press, Ames, Iowa.

Israelsen, O. W., and V. E. Hansen, 1962, "Irrigation principles and practices", John Wiley & Sons, New York.

Jensen, M. E. (Editor), 1981, "Design and operation of farm irrigation systems", ASAE Monograph, (3), American Society of Agricultural Engineers, St. Joseph, MI 49085.

Johl, S. S. (ed.), 1980. "Irrigation and Agricultural Devlopment"; Publ. for UN by Pergamon Press. 369 pp

Kay, M., 1986, "Surface irrigation, systems and practice", Cranfield Press, Cranfield Bedford MK43 0AL England, U.K.

Kirkham, M. B., and E. T. Kanemasu, 1983, "Wheat", pp 481-520, In Crop-Water Relation, Eds. I. D. Teare and M. M. Peet, John Wiley & Sons, New York.

Kohnke, H., 1968, "Soil Physics," TATA McGraw-Hill Publ. Co., LTD., New Delhi, India.

Kruseman, G. P., and N. A. De Ridder, 1979, "Analysis and evaluation of pumping test data", ILRI, P.O. Box 45, 6700 AA, Wageningen, The Netherlands. 200 pps.

LeMert, R. D., and E. R. Perrier, 1977, "Water Balance in an Imperial Valley cotton field", Field Station Report, Southwestern Irrigation Field Station, USDA/ARS, Technical Report No. 3-76, Brawley, California.

Li, J. C. R., 1966, "Statistical Inference: I", Edward Bros., Inc., Ann Arbor, Michigan.

Appendix

Little, T. M., and Hills, F. J., 1978. "Agricultural Experimentation: Design and Analysis", John Wiley & Sons, New York, USA.

Madsen, A., and M. Haider, 1983, "Farm enterprise budgeting," *In* "Diagnostic analysis of irrigation systems, Vol. 2," *Eds.* C. A. Podmore and D. G. Eynon, Water Mgmt. Syn. Proj., Colorado State University, Fort Collins, CO.

Marttin, D., and J. van Brocklin, 1985, "The Risk and Return with Deficit Irrigation," ASAE No. 85-2594, 28 pp.

Mellor, J. W., 1966. "The economics of agricultural development," Cornell University Press, Ithaca, NY.

Menz, K. M. and H. C. Knipscheer, 1981. "The location specificity problem in Farming Systems Research"; Agricultural Systems Vol. 7:93-103.

Milligan, S., 1963, "The little pot boiler," The Anchor Press Ltd, Tiptree, Essex, Great Britain.

Nakayama, F. S. and D. A. Bucks, 1986, "Trickle irrigation for crop production,"Elsevier, Amsterdam, The Netherlands, 383 pps.

Nelson, Vithange, 1982. "Sensitivity analysis and risk analysis"; pp 83-95, In "Evaluation of Irrigation Projects: A Case Study of the Mahavillachchiya Scheme, Sri Lanka"; Jour. of Agrarian Studies, Vol. 3(2).

NOAA, 1972, "National weather service: observing handbook No. 2", Data Acquisition Division, Office of Meteorological Operations, Silver Spring, Md., USA. 77 pps.

Norman, Colin, 1981. "The God that limps: science and technology in the eighties". A Worldwatch Institute Book, W. W. Norton & Co., New York. 224 pp.

Norusis, M. J., 1985, "SPSSx: Advanced Statistics Guide," McGraw-Hill Book Co., New York.

Pan, A. Yolopoulos, 1980. "The strategy of irrigated agricultural development"; pp 31-40, *In* "Irrigation and Agricultural Development"; S. S. Johl (ed.); Publ. by Pergamon Press.

Perrier, E. R., and K. R. Stockinger, 1961, "Diffusion of thermal neutrons in a soil-water system using the Fluggie mathematical model", Field Station Report No. 5-14-61, Southwestern Irrigation Field Station, ARS, Brawley, Calif, Presented at the Amer. Soc. Agron. Meetings, Univ. of Ohio, Columbus, Ohio.

Perrier, E. R., and D. D. Evans, 1961, "Soil moisture evaluation by tensiometers", Soil Sci. Soc. Amer. Proc. 25:173-175.

Perrier, E. R., and W. R. Johnston, 1962, "Distribution of thermal neutrons in a soil-water system", Soil Science 93:104-112.

Perrier, E. R., and K. R. Stockinger, 1964, "Using neutron and γ ray probes to measure soil moisture and density", Field station report No. 5-04-64, Southwestern Irrigation Field Station, USDA/ARS, Brawley, Calif., Presented at the Western soil Sci. Soc. meetings, Washington State Univ., Pullman, Washington.

Perrier, E. R., K. R. Stockinger, and R. V. Swain, 1966, "Scintillation Counter for measuring thermal neutrons in a soil-water system", Soil Science 101:125-199.

Perrier, E. R., and A. J. MacKenzie, 1973, "Sprinkler irrigation of cotton with saline water in the Imperial Valley, California", Field Station Report No. 5-23-73, Southwestern Irrigation Field Station, USDA/ARS, Presented Amer. Soc. Agron. Meetings, Las Vegas, Nevada.

Perrier, E. R., 1976, "Stochastic processes of water flow in a hydrologic system", USAE, Waterways Expt. Sta., EEL, Technical Paper No. EMB-2-76.

Perrier, E. R., J. Harris, and W. B. Ford III, 1977, "An evaluation of deterministic mathematical watershed models", USAE, Waterways Expr. Sta., EEL, Technical Paper No. DCDB-21-77, Vicksburg, Miss..

Perrier, E. R., 1980, "Artificial Groundwater Recharge Report", USAE, Waterways Expt. Sta., EEL, Division Report No. 7, Vicksburg, MS.

Perrier, E.R., 1986, "Small Scale Water Harvesting Techniques", 222-241, *In*: "Rainfed Agriculture in the Near East Region," *Eds.* Whitman, C.E., J. R. Parr, R. I. Papendick, and R. E. Meyer, Proceedings of the Workshop at Amman, Jordan, USAID/USDA, Washington, DC, October, 1989, pp. 343.

Perrier, E. R., and A. B. Salkini, 1986, "Evaluation Research for Management Planning in

Supplemental Irrigation," for Presentation at ASA-CSSA-SSSA Annual Meetings, 30/11 – 5/12, New Orleans, LA, USA.

Perrier, E. R., 1986, "Adaptation of water management practices to rainfed agriculture on alfisols in the sahel", *In* "Food Grain Production in Semi-Arid Africa", *Eds.* J.M. Menyonga, T. Bezuneh, and A. Youdeowei, The Coordination Office, OAU/STRC-SAFGRAD, Ouagadougou, Burkina Faso. 683 pps.

Perrier, E. R., 1987, "Catchment area runoff evaluation, CARE". FRMP, ICARDA, Aleppo, Syria.

Perrier, E. R., and A. B. Salkini, 1987, "Supplemental Irrigation as an Integral Component of Farming Systems", Inter. Congress of Irrigation/Drainage, Rabat, Morocco.

Perrier, E. R., 1988, "Probability analysis with water balance calculation for water harvesting farming", FRMP, ICARDA, Aleppo, Syria.

Perrier, E. R., 1988, "Opportunities for the productive use of rainfall normally lost to cropping for temporal or spatial reasons," *In* "ICRISAT Drought research priorities for the dryland tropics", *Eds.* Bidinger, F. R., and Johansen, C., Patancheru, A. P. 502 324, India.

Perrier, E. R., 1989, "Lecture Series: Supplemental Irrigation in the Near East and North Africa; for use by National Scientists", Prepared for Consultancy with the Ministry of Agriculture, Tunis, Tunisia, January.

Perrier, E. R., and A. B. Salkini, 1989, "Scheduling of supplemental irrigation on spring wheat using water balance methods", *In:* "Irrigation, Theory and Practice", *eds.*, J. R. Rydzewski and C. F. Ward, Pentech Press, London, England. 919 pps.

Perrier, J. R., 1975, "Evaluational research for management planning in criminal justice", *In:* "Criminal Justice Research", *ed.*, E. Viano, Lexington Books, Massachusetts, USA.

Podmore, C.A., and D. G. Eynon, 1983, "Diagnostic analysis of irrigation systems. Volume 2: Evaluation Techniques". Water Management Synthesis Project, University Services Center, Colorado State University, Fort Collins, Colorado.

Ritchie, I. J., J. B. Dent, and M. J. Blackie, 1978. "Irrigation management: an information system approach", Agricultural Systems, Vol.3:67-74. Applied Sci. Publ. Ltd., England.

Ritchie, J. T., 1972, "A model for predicting evaporation from a row crop with incomplete cover". Water Resources Research 8(5):1204-1213.

SCS, 1972, "National Engineering Handbook", National Technical Information Service Springfield, Virginia, USA.

Shalhevet, J., A. Mantell, H. Bielorai, and D. Shimshi, 1976, "Irrigation of Field and Orchard Crops Under Semi-Arid Conditions", International Irrigation Information Center, POB 8500, Ottawa K1G 3H9 Canada.

Shaner, W. W., P. F. Philipp, and W. R. Schmehl, 1982, "Farming systems research and development: Guidelines for developing countries", West View Press, Bolder, Colorado, USA.

Shipley, J.L., 1977, "Scheduling Irrigations with Limited Water", 5th Annual High Plains Grain Conference, Bushland, Texas.

Soil Science Division, 1980, "Soil Map of the Arab Countries: Vol. 1 Syria and Lebanon", ACSAD/SS/P15/1980, ACSAD, Damascus, Syria.

Stapper, M., 1984, "Simulations assessing the productivity of wheat maturity types in a Mediterranean climate", PhD thesis, University of New England, Armidale, N.S.W., Australia.

Suchman, E., "Evaluative research", 1967, New York: Russell Sage Foundation.

Tanner, C. B., 1963, "Basic instrumentation and measurements for plant environment and micrometeorology", Soils Bulletin No. 6, Department of soil science, University of Wisconsin, Madison, WI.

Tiffen, Mary, 1989. "Designing for sustainability or for a high internal rate of return"; pp 71-82, *In* "Irrigation Theory and Practice"; J. R. Rydzewski and C. F. Ward (*ed.*); Pentech Press, London.

Trout, T. J., I. G. Garcia-Castillas, and W. E. Hart, 1982, "Soil-water engineering: field and laboratory manual", Water Management synthesis Project, Dept. of Agric. & Chem. Engr, Colorado State University, Fort Collins, Colorado, USA.

Viera-Gallo, J. A., 1986, "The right to food, a contribution to peace", an essay, FAO Paper I/S2001

Appendix

prepared for world food day, 1986.

Wagner, W. (Ed.), 1982, "Geology and hydrogeology, 1978-1982: parts I-IV", Tech. Coop. Project No. 77.2176.4,Fed. Rep. of Germany & ACSAD, Damascus, Syria.

Watt, S. B., and W. E. Wood, 1977, "Hand dug wells and their construction", Intermediate Technology Publ. Ltd, 9 King St., London WC2E 8HN, U.K. 253 pps.

Wetherby, J. C., 1984. "Systems analysis and design: traditional, structural, and advanced concepts and techniques". West Publ. Co., New York.

Wilson, Kathleen K., Perry F. Philipp, and W. W. Shaner, 1986. "Socio-cultural effects on the farming systems research and development approach", Agricultural Systems Vol. 19:pp 83-110.

Withers, B., and S. Vipond, 1974, "Irrigation: design and practice", B. T. Batsford LTD., 4 Fitzhardinge St., London W1H 0AH, England.

Zandstra, H. G., Price, E. C., Litsinger, J. A., and Morris, R. A., 1981. "A Methodology for On-Farm Cropping Systems Research", The International Rice Research Institute, IRRI, Los Banos, Laguna, Philippines, P. O. Box 933, Manila, Philippines.

Zeller, R. A., and E. G. Carmines, 1980, "Measurement in the Social Sciences: The Link Between Theory and Data", Cambridge University Press, Chelsea, Michigan.

Index

active root depth 50, 299, 301–302
aquastats 159
aquicludes 159
aquifer recharge 157, 158
aquifers 157, 158
aquifers, confined 159, 161
aquifers, leaky 162
aquifers, perched 158
aquifers, semi-confined 162
aquifers, unconfined 161
available water 50–52, 120

benefit-cost analysis 204
benefit-cost ratio 214
border strip irrigation 5–6, 264–266
bulldozers 191

capillarity 113
capillary fringe 156
cash flow budget 221–222
center-pivot systems 7, 452
cereal production, Algeria 319, 323–324
cereal production, Cyprus 317, 348–349
cereal production, Syrian Arab Republic 505
cereal production, Tunisia 513–515
CERES 82
cone of depression 166
conventional irrigation 4
conveyance losses 259
crop coefficient 49
crop coefficient curves 298–299
crop simulation model 82
crop water requirement 45

Darcy's Law 109, 118
decision field 282
discount rate 209, 213
distribution uniformity 183, 259
drainage, methods 10

drawdown 166
drawdown, measurement 256–257
drip irrigation 8, 187–189, 273–278

effective rainfall 41
effective rainfall estimates 41–42
emitter uniformity 187–188
emitters 9
emitters 187–188, 274–276
emitters, clogging 187, 278
evaporation, measurement 42–44
evaporation pan 42–43
evaporometers, Piche 44, 49
evapotranspiration 48
evapotranspiration, estimation 44, 67–70
experimental design 20

farm budget 207, 218–221
farmer acceptance 12, 16
fertilizers 347, 405–406, 473, 474, 545–546, 577–578
field capacity 118–120
field size, measurement 198
flood water 322, 451, 522
flow rates, measurement 181
frost protection 182
fuel requirements 252
furrow irrigation 262–264
furrows 5

grain yield 302–305, 357–358, 383–384, 453, 519, 523–525
grain yields 308, 348–349
Green and Ampt Equation 142–144
groundwater 155
groundwater, age 162–163
groundwater, locating 155–156
groundwater, pollution 171–172
groundwater, quality 180
growth stages 86–87

growth stages, cereals 88–95
growth stages, spring wheat 294, 297–299
gun sprinklers 7
gypsum blocks 129

hydraulic conductivity 109, 117–118
hydraulic conductivity, measurement 110–112, 113

infiltration 133
infiltration, factors affecting 138
infiltration rate 134, 138–142
infiltration rate, measurement 145
infiltration, units 144
infiltrometer, double ring 146–150
infiltrometer, furrow 151–153
infiltrometer, single ring 145–146
internal rate of return 210
interviewer 234
irrigation application 53
irrigation efficiency 4, 53, 181, 182–186
irrigation of cereals, Algeria 315–325
irrigation scheduling 39, 45, 46, 79–86, 293
irrigation water, quality 177
irrigations, number 45

Kostiakov-Lewis Equation 153

labor, costs 212
land grading 5, 191
land leveling 191
land ownership 393
land plane 192
land tenure 333–335, 477–480, 490–492, 582
land value 212
laser-leveler 192
leaching 187
leveling devices 193–198
leveling devices, A-frame 195
leveling devices, garden hose 196
leveling devices, water manometer 193
linear-move systems 7
livestock 336, 350–352, 477, 490, 548–551, 579–580, 586, 590, 594
livestock production 320, 339
logic of inquiry 18

Marriotte siphon 146
matric potential 107
mechanization 347, 475–476, 543–545, 576
micro-irrigation 187–189

net present value 209–210
neutron probe 126

opportunity cost 209, 212, 213
osmotic potential 109
over-irrigation 185

pan coefficient 48
partial budgeting 207, 222–224, 310
Penman's Equation 44, 67
plant height 300–302
ponding basins 151
precipitation 40
present value 225
present values 208
prices, farm-gate 208
prices, market 208
prices, shadow 211
probability analysis 58–67, 74–77, 83
psychrometers 130
pumping 252
pumping, costs 253
pumping tests 157–158, 255
pumps 173, 484

questionnaire 13, 234, 236–250, 440

rainfall measurement 40–41
raingages 40–41
record keeping 234, 250
reliability 21, 23–24
research 341, 370, 371, 386, 452–453, 506–507
research, at ICARDA 29–37, 294–312
research plot, design 29–30, 31–33, 36–37, 294–295, 507
research results 452–454, 518–521
research results, Iran 384–385
research results, Iraq 390–391
road graders 191
root zone moisture 50–52
runoff 135, 259, 322

salinity control 5, 135
sample selection 22–23
soil bulk density 100–105
soil improvement 135–138
soil moisture characteristic 114
soil porosity 100, 104
soil structure 101
soil texture 97
soil water 97
soil water, measurement 123
soil water potential 107
soil water potential, units 115

Index

soils 333, 353, 432, 444, 485–486, 566–567, 583, 587, 592
soils, data collection 259
sprinkler irrigation 6, 178, 266–273, 394
sprinkler irrigation, measurement 179
sprinkler irrigation, uniformity 179
sprinkler systems 6–8, 395
sprinkler systems, micro-sprinklers 187–189
subsidied 342, 419
subsidies 466
supplemental irrigation 4
supplemental irrigation, costs 205–206, 225, 509–510
supplemental irrigation, economics of 5, 203–227, 309–311, 509–510
supplemental irrigation, Pakistan 481–485
supplemental irrigation, Tunesia 522–523
supplemental irrigation systems, Cyprus 327, 335, 345, 356
supplemental irrigation systems, Syrian Arab Republic (SAR) 504–506
supplemental irrigation, timing 299
surface irrigation 5, 394
surveying 191
systems approach 15–18
systems theory 15–18

tensiometer 124–126
trickle irrigation 8, 273–278

under-irrigation 186
unsaturated flow 113

validity 21, 23–24

water applied 257–258

water balance 46
water balance, applications 79
water balance, calculation 46–58, 70–74, 83–85
water, cost 258
water harvesting 4, 356, 369, 370–371, 409, 413, 414, 437, 445, 584, 593
water quality 172, 257
water requirements 86, 87
water resources, 435–436
water resources, Algeria 317–318, 321
water resources, Cyprus 335, 354, 361–362
water resources, Indus Plain 465, 467, 470–471, 484–485, 488, 492
water resources, Iran 369
water resources, Jordan 400, 408, 410, 412, 415–417
water resources, Libya 425–428, 430–432, 441, 442–445
water resources, Morocco 450
water resources, Syrian Arab Republic 499
water resources, Turkey 532–534, 558
water resources, Yemen 567–570, 584, 588, 592
water rights 571
water storage 86
water table 156
water use efficiency 305
weed control 348, 407, 546–547, 578–579
well drilling 169–171
wells 164–165
wells, artesian 159
wells, cost 225
wells, design life 171
wells, methods of construction 167–171
wilting point 120

The manufacturer's authorised representative in the EU is Springer Nature Customer Service Centre GmbH, Europaplatz 3, 69115 Heidelberg, Germany. If you have any concerns regarding our products, please contact ProductSafety@springernature.com

Printed and bound by CPI Group (UK) Ltd, Croydon, CR0 4YY

24/04/2026

02096308-0014